QUANTUM EFFECTS IN BIOLOGY

Quantum mechanics provides the most accurate microscopic description of the world around us, yet the interface between quantum mechanics and biology is only now being explored. This book uses a combination of experiment and theory to examine areas of biology believed to be strongly influenced by manifestly quantum phenomena.

Covering subjects ranging from coherent energy transfer in photosynthetic light-harvesting, to spin coherence in the avian compass and the problem of molecular recognition in olfaction, the book is ideal for advanced undergraduate and graduate students in physics, biology, and chemistry seeking to understand the applications of quantum mechanics to biology.

MASOUD MOHSENI is a Senior Research Scientist at Google, where he develops novel machine learning algorithms that fundamentally rely on quantum dynamics. He has made significant contributions to quantum transport, quantum measurement, and open quantum system approaches to biology.

YASSER OMAR is Assistant Professor at the University of Lisbon and leads the Physics of Information Group at Instituto de Telecomunicações. His research is centered on quantum information theory, quantum transport, and their interface with biology.

GREGORY S. ENGEL is an Associate Professor at the Institute for Biophysical Dynamics and the James Franck Institute, The University of Chicago. His group is working on creating synthetic mimics of biological systems.

MARTIN B. PLENIO is Alexander von Humboldt Professor and Director of the Institute of Theoretical Physics, Ulm University. His main contributions are in the fields of quantum information, quantum optics, and the development of the theory of quantum effects in biology.

QUANTUM EFFECTS IN BIOLOGY

Edited by

M. MOHSENI
Google

Y. OMAR
Instituto de Telecomunicações and ISEG, University of Lisbon

G. ENGEL
University of Chicago

M. B. PLENIO
Ulm University

CAMBRIDGE
UNIVERSITY PRESS

CAMBRIDGE
UNIVERSITY PRESS

University Printing House, Cambridge CB2 8BS, United Kingdom

One Liberty Plaza, 20th Floor, New York, NY 10006, USA

477 Williamstown Road, Port Melbourne, VIC 3207, Australia

4843/24, 2nd Floor, Ansari Road, Daryaganj, Delhi - 110002, India

79 Anson Road, #06-04/06, Singapore 079906

Cambridge University Press is part of the University of Cambridge.

It furthers the University's mission by disseminating knowledge in the pursuit of education, learning and research at the highest international levels of excellence.

www.cambridge.org
Information on this title: www.cambridge.org/9781107010802

First published 2014
Reprinted 2015

A catalogue record for this publication is available from the British Library

ISBN 978-1-107-01080-2 Hardback

Contents

Foreword page x
Contributors xiii
Preface xvii

Part I Introduction

1 Quantum biology: introduction 3
GRAHAM R. FLEMING AND GREGORY D. SCHOLES
1.1 Introduction 3
1.2 Excited states in biology 4
1.3 Light particles and tunnelling 7
1.4 Radical pairs 8
1.5 Questions for the present 9
1.6 Some wide-reaching questions 12

2 Open quantum system approaches to biological systems 14
ALIREZA SHABANI, MASOUD MOHSENI, SEOGJOO JANG, AKIHITO ISHIZAKI, MARTIN PLENIO, PATRICK REBENTROST, ALAN ASPURU-GUZIK, JIANSHU CAO, SETH LLOYD AND ROBERT SILBEY
2.1 Quantum mechanics concepts and notations 15
2.2 Open quantum systems: dynamical map approach 17
2.3 Open quantum systems: master equation approach 21
2.4 Formally exact QME 22
2.5 QME in the weak system–bath coupling limit 24
2.6 QME for weak coupling to a Markovian bath 26
2.7 QMEs beyond weak and Markovian limits 28
2.8 Second-order cumulant time-non-local equation and its hierarchical representation 33
2.9 A post-perturbative time convolution QME 35
2.10 QME in the polaron picture 38

2.11 Path integral techniques 44
2.12 DMRG based approaches 47

3 Generalized Förster resonance energy transfer 53
SEOGJOO JANG, HODA HOSSEIN-NEJAD AND GREGORY D. SCHOLES
3.1 Introduction 53
3.2 Förster's rate expression: a complete derivation 56
3.3 Transition density cube method 62
3.4 Generalized Förster theories 65
3.5 Important computational issues in an actual application 72
3.6 Applications of MC-FRET 77
3.7 Summary 81

4 Principles of multi-dimensional electronic spectroscopy 82
TOMÁŠ MANČAL
4.1 Photo-induced dynamics of molecular systems 82
4.2 Non-linear response of multi-state systems 89
4.3 Cumulant expansion of a non-linear response 99
4.4 Selected non-linear spectroscopic methods 106
4.5 Conclusions 119

Part II Quantum effects in bacterial photosynthetic energy transfer

5 Structure, function, and quantum dynamics of pigment–protein complexes 123
IOAN KOSZTIN AND KLAUS SCHULTEN
5.1 Introduction 123
5.2 Light-harvesting complexes from purple bacteria: structure, function and quantum dynamics 125
5.3 Optical transitions in pigment–protein complexes 131
5.4 Electron transfer in pigment–protein complexes 140

6 Direct observation of quantum coherence 144
GREGORY S. ENGEL
6.1 Detecting quantum coherence 144
6.2 Observation of quantum coherence using 2D electronic spectroscopy 147
6.3 Identifying and characterizing quantum coherence signals 151
6.4 Quantum coherence in reaction centres using two colour electronic coherence photon echo spectroscopy 155
6.5 Observing quantum coherences at physiological temperatures 156
6.6 Outlook for future measurements of coherence 157

7 Environment-assisted quantum transport 159
MASOUD MOHSENI, ALÁN ASPURU-GUZIK, PATRICK REBENTROST, ALIREZA SHABANI, SETH LLOYD, SUSANA F. HUELGA AND MARTIN B. PLENIO
7.1 Introduction 159
7.2 Master equations for quantum transport 161
7.3 Quantum transport in a two-chromophore system 162
7.4 The principles of noise-assisted quantum transport 164
7.5 Quantum transport in the Fenna–Matthews–Olson protein complex 167
7.6 Optimality and robustness of quantum transport 170
7.7 Conclusion 175

Part III Quantum effects in higher organisms and applications

8 Excitation energy transfer and energy conversion in photosynthesis 179
ELISABET ROMERO, VLADIMIR I. NOVODEREZHKIN AND RIENK VAN GRONDELLE
8.1 Photosynthesis 179
8.2 Photosynthetic energy conversion: charge separation 180
8.3 Light-harvesting 190

9 Electron transfer in proteins 198
SPIROS S. SKOURTIS
9.1 Introduction 198
9.2 The rate for a single-step electron transfer reaction mediated by elastic through-bridge tunnelling 201
9.3 Dependence of tunnelling on protein structure: tunnelling pathways and their interferences 205
9.4 Tunnelling matrix element fluctuations in deep-tunnelling ET reactions 208
9.5 Vibrational quantum effects and inelastic tunnelling 211
9.6 Biological ET chains with tunnelling and hopping steps through the protein medium 214
9.7 Conclusions 216
9.8 Acknowledgements 217

10 A chemical compass for bird navigation 218
ILIA A. SOLOV'YOV, THORSTEN RITZ, KLAUS SCHULTEN AND PETER J. HORE
10.1 Introduction 218
10.2 Theoretical basis for a chemical compass 221

10.3 In vitro magnetic field effects on radical pair reactions 227
10.4 Evidence for a radical pair mechanism in birds 232
10.5 Conclusion 236

11 Quantum biology of retinal 237
KLAUS SCHULTEN AND SHIGEHIKO HAYASHI
11.1 Introduction 237
11.2 Retinal in rhodopsin and bacteriorhodopsin 238
11.3 Quantum physics of excited state dynamics 242
11.4 Regulation of photochemical processes for biological function 245
11.5 Potential energy crossing and conical intersection 246
11.6 Electronic structure of protonated Schiff base retinal 252
11.7 Mechanism of spectral tuning in rhodopsins 255
11.8 Photoisomerization of retinal in rhodopsins 257
11.9 Summary and outlook 262
11.10 Acknowledgement 263

12 Quantum vibrational effects on sense of smell 264
A.M. STONEHAM, L. TURIN, J.C. BROOKES AND A.P. HORSFIELD
12.1 Phonon assisted tunnelling in olfaction 264
12.2 Important processes and timescales 267
12.3 Quantum rate equations 271
12.4 Putting in numbers 272
12.5 Can we make predictions? 275
12.6 Extensions of the theory for enantiomers 275

13 A perspective on possible manifestations of entanglement in biological systems 277
HANS J. BRIEGEL AND SANDU POPESCU
13.1 Introduction 277
13.2 Entanglement 278
13.3 Non-local correlations 281
13.4 Entanglement in biology 289
13.5 Open driven systems and entanglement 292
13.6 Conclusions 310

14 Design and applications of bio-inspired quantum materials 311
MOHAN SAROVAR, DÖRTHE M. EISELE AND K. BIRGITTA WHALEY
14.1 Potential applications of bio-inspired quantum materials 311
14.2 Progress in designing biomimetic quantum materials 315

15 Coherent excitons in carbon nanotubes 335
LEONAS VALKUNAS AND DARIUS ABRAMAVICIUS
15.1 Structure 335
15.2 Electronic properties in 1D systems 335
15.3 Exciton–exciton interactions 339
15.4 Non-linear optical response of excitons 343
15.5 Simulations of intensity-dependent 3PEPS 345
15.6 Discussion and conclusions 347
15.7 Acknowledgement 349

References 350
Index 396

Foreword

When the revolutionary conceptual structure for the description of the physical world which we know as quantum mechanics was first formulated nearly 90 years ago, and its predictions tested in the laboratory, most of the experiments in question were on systems which were both very well characterized and reasonably well isolated from their environments, such as single electrons and atoms, small molecules and near-perfect crystalline solids.

While from the very start most physicists have taken it for granted that the formalism of quantum mechanics, when combined with appropriate system-specific information, can "in principle" account for all phenomena occurring in the physical world, including those usually regarded as the subject-matter of biology, until quite recently the overwhelmingly majority opinion has been that in a biological context the role of quantum theory is confined to elucidating the equilibrium structures of the relevant molecules and their reaction processes, and that subtle phenomena such as superposition and entanglement, of which we can now routinely exhibit spectacular effects at the level of a few well-isolated photons or atoms, play at most a very indirect role in any phenomena of biological interest. A major reason conventionally given for this view has been that biological systems, at least working ones, are by their very nature "warm and wet" – a phrase which in the physicist's lexicon translates to "prone to massive decoherence"; it looks as though any interesting superposition, say of different energy eigenstates of one's system, would be rapidly decohered by the ever-present, and usually microscopically very complex, environment.

Over the last decade or so, however, it has become clear that this conclusion may be (from the point of view of the quantum theorist!) unduly pessimistic. There is by now rather convincing evidence that non-trivial quantum superpositions occur in the first stages of photosynthesis, and a plausible argument that they play a role also in the intriguing phenomenon of bird navigation; as argued in some of the chapters of this book, there are other biological contexts in which a significant role for superposition and possibly even entanglement are not excluded by the traditional arguments.

This situation poses a major challenge to our understanding of the more general and long-studied subject of "open quantum systems," that is, systems which may show subtle quantum effects even while in intimate contact with a complex and noisy environment, which may not even be reproducible from shot to shot of the experiment. Over the last few years many have risen to this challenge on the experimental, theoretical and computational fronts, with intriguing results; one particularly striking observation has been that under certain circumstances a noisy environment may actually enhance the characteristically quantum-mechanical features of the behavior of the system. This book should provide the reader with a broad-based introduction to this fascinating and rapidly developing field of research.

Tony Leggett,
Department of Physics, University of Illinois at Urbana-Champaign

Contributors

Darius Abramavicius
Department of Theoretical Physics, Faculty of Physics, Vilnius University, Vilnius, Lithuania.

Alán Aspuru-Guzik
Department of Chemistry and Chemical Biology, Harvard University, Cambridge, MA, USA.

Hans J. Briegel
Institute for Theoretical Physics, University of Innsbruck, Innsbruck, Austria, and Institute for Quantum Optics and Quantum Information of the Austrian Academy of Sciences, Innsbruck, Austria.

Jennifer C. Brookes
Center for Biomedical Engineering, Massachussetts Institute of Technology, Cambridge, MA, USA.

Jianshu Cao
77 Massachusetts Avenue, Department of Chemistry, Massachusetts Institute of Technology, Cambridge, MA, USA.

Dörthe M. Eisele
Center for Excitonics, Research Laboratory of Electronics, Massachusetts Institute of Technology, Cambridge, MA, USA.

Gregory S. Engel
Department of Chemistry and The James Franck Institute, The University of Chicago, Chicago, IL, USA.

Graham R. Fleming
Department of Chemistry, University of California Berkeley, and Physical Biosciences Division, Lawrence Berkeley National Laboratory, Berkeley, CA, USA.

Rienk van Grondelle
Department of Biophysics, Faculty of Sciences, VU University, Amsterdam, The Netherlands.

Shigehiko Hayashi
Department of Chemistry, Graduate School of Sciences, Kyoto University, Kyoto, Japan.

Peter J. Hore
Department of Chemistry, University of Oxford, Physical and Theoretical Chemistry Laboratory, Oxford, UK.

Andrew P. Horsfield
Department of Materials, Imperial College London, London, UK.

Hoda Hossein-Nejad
University College London, Department of Physics & Astronomy, London, UK.

Susana F. Huelga
Institut für Theoretische Physik, Universität Ulm, Ulm, Germany.

Akihito Ishizaki
B77 Hildebrand Hall, Department of Chemistry, University of California Berkeley, Berkeley, CA, USA.

Seogjoo Jang
Department of Chemistry and Biochemistry, Queens College of the City University of New York, NY, USA.

Ioan Kosztin
Department of Physics and Astronomy, University of Missouri, Columbia, MS, USA.

Seth Lloyd
Department of Mechanical Engineering, Massachusetts, Institute of Technology, Cambridge, MA, USA.

Tomáš Mančal
Institute of Physics, Faculty of Mathematics and Physics, Charles University, Prague, Czech Republic.

Masoud Mohseni
Google Research, Venice, CA, USA.

Vladimir I. Novoderezhkin
A. N. Belozersky Institute of Physico-Chemical Biology, Moscow State University, Leninskie Gory, Moscow, Russia.

Yasser Omar
Physics of Information Group, Instituto de Telecomunicações and CEMAPRE, ISEG, University of Lisbon, Portugal.

Martin B. Plenio
Institut für Theoretische Physik, Universität Ulm, Ulm, Germany and Quantum Optics and Laser Science group, Blackett Laboratory, Imperial College London, London, UK.

Sandu Popescu
H.H. Wills Physics Laboratory, University of Bristol, Bristol, UK.

Patrick Rebentrost
Department of Chemistry and Chemical Biology, Harvard University, Cambridge, MA, USA.

Thorsten Ritz
Department of Physics and Astronomy, Irvine, CA, USA.

Elisabet Romero
Department of Biophysics, Faculty of Sciences, VU University, Amsterdam, The Netherlands.

Mohan Sarovar
Sandia National Laboratories, Livermore, CA, USA.

Gregory D. Scholes
Department of Chemistry, Institute for Optical Sciences, and Centre for Quantum Information and Quantum Control, University of Toronto, Toronto, Ontario, Canada.

Klaus Schulten
Department of Physics, University of Illinois at Urbana-Champaign, and Beckman Institute for Advanced Science and Technology, Urbana-Champaign, Illinois, USA.

Alireza Shabani
Department of Chemistry, University of California, Berkeley, CA, USA.

Robert Silbey
Department of Chemistry, Massachusetts Institute of Technology, Cambridge, MA, USA. Chapter 2 is being offered in honor of Professor Robert Silbey who sadly passed away during its writing.

Spiros S. Skourtis
Department of Physics, University of Cyprus, Nicosia, Cyprus.

Ilia A. Solov'yov
Beckman Institute for Advanced Science and Technology, Urbana-Champaign, IL, USA, and Department of Physics, Chemistry and Pharmacy, University of Southern Denmark (SDU), Odense M, Denmark.

A. Marshall Stoneham
Formerly of University College London. Chapter 12 is being offered in honor of Professor Marshall Stoneham, who died unexpectedly during its writing.

Luca Turin
BSRC Alexander Fleming, Vari, Greece.

Leonas Valkunas
Department of Theoretical Physics, Faculty of Physics of Vilnius University, Vilnius, Lithuania, and Center for Physical Sciences and Technology, Institute of Physics, Vilnius, Lithuania.

K. Birgitta Whaley
Department of Chemistry, University of California, Berkeley, CA, USA.

Preface

Recent progress in science and technology has led to the revival of an old question concerning the relevance of quantum effects in biological systems. Indeed Pascual Jordan's 1943 book, *Die Physik und das Geheimnis des Lebens* had already posed the question "Sind die Gesetze der Atomphysik und Quantenphysik für die Lebensvorgänge von wesentlicher Bedeutung?" (Are the laws of atomic and quantum physics of essential importance for life?) and coined the term Quanten-Biologie (quantum biology). At the time this question was essentially of a theoretical nature as the technology did not yet exist to pursue it in experiment.

Indeed quantum biology has been benefiting considerably from the refinement in experimental tools which is beginning to provide direct access to the observation of quantum dynamics in biological systems. Indeed, we are increasingly gaining sensitivity towards quantum phenomena at short lengths and timescales. In recent years, these newly found technological capabilities have helped to elevate the study of quantum biology from a mainly theoretical endeavour to a field in which theoretical questions, concepts and hypotheses may be tested experimentally and thus verified or disproved. We should stress here that experiments are essential to verify theoretical models because biological systems already have a complexity and structural variety that prevents us from knowing and controlling all of the aspects. Results obtained using these refined experimental techniques lead to new theoretical challenges and thus stimulate the development of novel theoretical approaches. It is this mutually beneficial interplay between experiment and theory that promises accelerated developments within the field.

Biological systems tend to be warm, wet and noisy (the latter because they are exposed to environmental fluctuations), conditions which are normally expected to result in rapid decoherence and thus suppression of quantum features. Therefore quantum phenomena may at first sight seem to be unlikely to play a significant role in biology. Note, however, that at the level of molecular complexes and proteins, important biological processes can be very fast (taking place within picoseconds) and well localised (extending across a few nanometres, the size of proteins) and may thus exhibit quantum phenomena before the environment has had an opportunity to destroy them. Hence the possible existence of significant quantum dynamics is a question of length and timescale; indeed quantum phenomena such as electron tunneling have been observed in biological systems and there is some evidence for proton tunneling in enzymes. As such, tunneling phenomena are not intimately related to biology and the question therefore remains whether on the one hand biological systems will exhibit more complex quantum-dynamical phenomena that may either involve several interacting particles or multiple interacting components of a network, or on the other hand whether the specifics of the biological systems and their environments will play a crucial role in allowing or supporting certain quantum-dynamical phenomena in biology. Only then could we call these 'non-trivial' quantum effects in biological systems. Indeed, it appears that there are biological processes such as energy and electron transfer in photosynthesis, magneto-reception in birds or the olfactory sense that rely at a fundamental level on such 'non-trivial' quantum-dynamical processes. Thus quantum effects in biology may well be possible and more importantly relevant towards function.

This book reports on quantum biology, its theoretical foundations, experimental findings and future possibilities as they have emerged over the past few years. Needless to say not all subjects can be covered and we have had to make a sub-selection that has been driven by several objectives. Firstly, given that the basis of the field is the fruitful interplay between experiment and theory, we have endeavoured to choose subjects that are either already under experimental investigation or for which it could be expected that technology will give access to these theoretical predictions in the foreseeable future. This has led us to exclude subjects such as quantum conciousness or the speculations concerning the origin of life. Secondly, it is our aim to provide a reasonably coherent set of chapters, starting from experimental and theoretical foundations and leading on to specific topics of interest. Finally, of course, personal preferences and tastes do also play a role.

The original plan for this book was hatched during the first conference on Quantum Effects in Biological Systems (QuEBS 2009), held from 7–10 July, 2009 in Lisbon, Portugal, which has become the first of the annual QuEBS conferences. Subsequent QuEBS meetings were held at Harvard University in 2010, Ulm University in 2011, Berkeley in 2012 and Vienna in 2013, and their ever growing attendance attests to a growing interest in the field.

This steady development has convinced us that the time is right for an introductory book on quantum effects in biology and we do hope that the present text will help scientists, especially young and adventurous scientists, during or shortly after their PhD, to gain a first insight into the field of quantum biology. It is our hope that in this way we can assist the further development of the field by converting an increasing number of scientists into becoming quantum biologists.

We would like to acknowledge financial support from the Natural Sciences and Engineering Research Council of Canada, the DARPA QuBE Program, the MIT-ENI Energy Initiative, the Center for Excitonics at RLE, and the Quantum Artificial Intelligence Lab at Google, to Masoud Mohseni; the Fundação para a Ciência e a Tecnologia (Portugal), namely through the programme POCTI/POCI/PTDC (PEst-OE/EGE/UI0491/2013, PEst-OE/EEI/LA0008/2013, PTDC/EEA-TEL/103402/2008), partially funded by FEDER (EU), and from the EU FP7 (LANDAUER, PAPETS), to Yasser Omar; The Searle Foundation and DARPA QuBE Program to Greg Engel; the Alexander von Humboldt Professorship, the SFB TR/21, the European Research Council and the European Union, to Martin Plenio. There have also been a number of friends, students and colleagues who have supported us in the completion of this book. Here, we would especially like to thank Robert Rosenbach for his considerable help in compiling the final edition of the book.

A website hosted by the editors is available at www.cambridge.org/9781107010802.

Part I

Introduction

1

Quantum biology: introduction

GRAHAM R. FLEMING AND GREGORY D. SCHOLES

1.1 Introduction

Key features of quantum mechanics are the uncertainty principle, wave–particle duality, quantization of energies and the modification of classical probability laws. Biology is concerned with how natural systems function – from understanding how genetically coded information is replicated, to attaining a mechanistic model for complex multistep reactions. Recently researchers have been asking whether quantum mechanics, normally the domain of physics, is also needed to understand some biological processes. This field includes fascinating developments in theory and experiment, as well as multidisciplinary discussion, and the state-of-the-art is documented in this book. Erwin Schrödinger, in his famous book *What is Life*? (Schrödinger, 1944), noted that quantum mechanics accounts for the stability of living things and their cellular processes because of our understanding, via quantum mechanics, of the stability and structure of molecules. The fact that quantum effects create, sometimes large, energy gaps between different states of a chemical system is also important. Such energy gaps, between electronic energy levels, enable living organisms to capture and store the energy carried from the sun by photons, and to visualize the world around them via optically induced chemical reactions. Davydov's view in *Biology and Quantum Mechanics* (Davydov, 1982) was that quantum mechanics is most relevant for isolated systems in pure states and therefore is of little importance for biological systems that are in statistical states at thermal equilibrium.

If we set aside the fact that quantum mechanics is required to explain the properties of molecules and their reactions – obviously important in biochemical processes ranging from the action of enzymes to genetic expression of phenotypes and the very construction of a living organism – then *quantum biology* identifies

Quantum Effects in Biology, ed. Masoud Mohseni, Yasser Omar, Gregory S. Engel and Martin B. Plenio. Published by Cambridge University Press.

biological phenomena that make explicit use of quantum mechanics to attain functionality or to carry out a process. Of course the application of quantum mechanics to the study of biological problems is still important, but we do not include that in our definition of quantum biology. Early work pursuing such directions was the main focus of the, now discontinued, *International Journal of Quantum Chemistry: Quantum Biology Symposium* series. However, there remains the question of precisely where we draw the line separating unsurprising manifestations of quantum mechanics from those of interest for quantum biology. For example, intermolecular forces, including hydrogen bonding and van der Waals forces, are prevalent in defining protein structures, the helical motif of DNA; they account for enzyme–substrate association, aid protein synthesis, and cause membrane formation, to name just a few examples. In general, such non-bonding interactions, while described by quantum-mechanical models, do not appear to introduce special properties or function because of quantum mechanics, so we would exclude them from quantum-biological examples. Instead the field is mostly concerned with excited electronic states and their dynamics, long-range tunnelling through barriers, interference effects and other abilities conferred by quantum mechanics.

A molecular basis for biology, including atomic-scale descriptions of complex processes, has transformed biology over the past decades. One motivation for the emerging field of quantum biology is to elucidate new insights into biological function that may emerge from a quantum perspective. Hence the discovery and objective assessment of examples of quantum biology is important. Key criteria for identifying quantum-biological processes include their amenability to experimental verification and predictions or theoretical modelling using appropriate and rigorous frameworks. It is worthwhile distinguishing between phenomena where the detailed dynamics of the process are susceptible to experimental and theoretical investigation, and those in which only overall rates are observable and the challenge is to explain the magnitudes and trends found from experiments. The chapters that follow in this book will explore these examples in greater depth.

1.2 Excited states in biology

In 1962, Longuet-Higgins wrote in his paper entitled 'Quantum mechanics and biology' that quantum mechanics only helps us to understand a few biological processes that involve radiation (Longuet-Higgins, 1962). Research since then has revealed that this rather conservative view requires modification. Indeed, quantum phenomena in biological systems that require explicit reference to quantum theory abound. For example, the 'energy wealth' of a molecule defines the energy stored in quantum states; usually chemical bonds that can be transformed in reactions, thereby storing or releasing a quantum of energy. There are many examples that are of importance for biology, including nucleoside phosphates (ATP), thioesters, imidazoles and others (Pullman and Pullman, 1963). The special properties of molecules such as quinones that are important in electron transport chains can also be attributed to their quantum-mechanical energy gaps. Research at present, as documented in this book is, however, largely restricted to light-induced processes.

1.2.1 Photosynthetic light-harvesting

Light-harvesting in photosynthesis has recently become the paradigmatic model for quantum effects in biology. Photosynthetic pigment–protein complexes collect sunlight and transfer energy in the form of electronic excitation to the reaction centre, where charge separation initiates a web of biochemical processes. Light-harvesting in photosynthetic organisms occurs with remarkable quantum efficiency; usually a quantum efficiency of >90% for initiation of charge separation per absorbed photon is found under conditions of low irradiance. Therefore it is of great interest to investigate the design principles of this extremely efficient process (Scholes *et al.*, 2011).

The success of natural light-harvesting depends on ultra fast excited state dynamics including energy transfer and charge separation, where quantum superposition and coherence dynamics turn out to play roles (Ishizaki and Fleming, 2012). Thus, models based on quantum theory are crucial for the understanding of the primary process of photosynthesis. Moreover, the recent development of two-dimensional electronic spectroscopy and its utility for examining these systems has placed photosynthetic light-harvesting in a unique position for investigating quantum-dynamical phenomena in biological systems.

Many quantum phenomena are often regarded as exceedingly delicate and not likely to survive over relevant timescales in 'warm, wet and noisy' living things. Thus, the experimental observation by Engel *et al.* of long-lived quantum-electronic coherence in a photosynthetic protein (the FMO complex) (Engel *et al.*, 2007) produced widespread interest. The initial experiments were carried out at 77 K, but subsequent work by Scholes and co-workers on a light-harvesting protein from marine algae (Collini *et al.*, 2010) confirmed the persistence of quantum coherence at physiological (room) temperature. Engel and co-workers then demonstrated that quantum coherence in the FMO protein survives up to room temperature (Panitchayangkoon *et al.*, 2010). These studies used femtosecond duration laser pulses. The relevance of the coherent phenomena observed in such experiments to the behaviour of systems illuminated by sunlight requires careful clarification (Jiang and Brumer, 1991). It might be worth saying that, while the experiments are carried out with coherent excitation, the underlying Hamiltonian probed by these experiments is the same Hamiltonian that governs the dynamics under sunlight irradiance. A key point is that the simulations inspired by these experiments do not need to assume coherent excitation, and it is really these simulations that ultimately give us an insight into how coherences modify the dynamics.

1.2.2 Other excited state processes: from vision to circadian clocks

One way to answer the question of biological relevance would be to argue that the process would simply not work without quantum mechanics. Vision is perhaps one such example, because the quantum-mechanical arrangement of electronic states and their symmetries is responsible for light-activated isomerization. Light is absorbed by a chromophore in the rhodopsin protein, which initiates a photochemical isomerization. This is the quantum-mechanical process. After that, a series of enzymes are activated, culminating in hydrolysis of cyclic guanosine 3′–5′ monophosphate which causes Na^+ ion channels to close. The resulting hyperpolarization induces an electrical impulse that is transferred to a nerve cell. This really is a remarkable sequence of events – initiated by a comparatively simple quantum-mechanical process.

Photochemical reactions like the photo-induced isomerization of retinal in rhodopsin protein are a subset of chemical reactions in general, and the mechanism of many ground state reactions can also only be understood from a quantum-mechanical viewpoint. For example, the Woodward–Hoffman rules for orbital symmetry explain how large barriers to chemical transformations in the electronic ground state can result from the way orbital symmetry must change in a discontinuous fashion during the transformation from reactant to product (Woodward and Hoffmann, 1969). In photochemical reactions (Michl, 1990), these kinds of barriers are greatly reduced by using photoexcitation to change the symmetry of the initial state and provide internal energy to surmount the thermal barrier. Similarly, the catalytic activity of enzymes can sometimes best be understood on a quantum-mechanical basis. It is therefore difficult to decide which chemical reactions, whether light-induced or not, constitute important examples of quantum-mechanical optimization in biology.

Lastly, it is worth documenting that vision is not the sole photochemical process important in biology. Other examples include deactivation of excited states in DNA, thus avoiding photodamage to the genetic code, vitamin D (calciferol) biosynthesis, photoinduced electron transfer in photosynthesis, cell photoprotection using melanin, bioluminescence and sophisticated sunscreens used by coral. An important class of proteins that operate by light activation are the phytochromes. They play a role in a multitude of processes including phototropism and phototaxis (growth/movement determined by response to a light source), photoperiodism, seed germination and circadian clocks.

1.3 Light particles and tunnelling

1.3.1 Electron tunnelling

Electron flow between distant (greater than 15 Å) redox-active cofactors is central to the operation of aerobic respiration and photosynthesis. Proteins are not electrical conductors, yet charge needs to move in such biological processes. The quantum-mechanical process of tunnelling allows classical energy barriers to be circumvented to some extent, basically by virtue of the time–energy uncertainty principle. Electrons can thereby be directed through space by protein backbones. Such long-range electron transfer in the biologically relevant timescales of milli- to microseconds plays a key role in the energy transduction pathways of life.

Tunneling in biological systems was first reported for electron-transfer reactions in proteins (Marcus and Sutin, 1985). Investigations on ruthenium-modified proteins carried out by Gray and co-workers in the past two decades have yielded a remarkably detailed description of the distance- and driving-force dependences of long-range electron tunnelling rates in proteins (Gray and Winkler, 2003). Experimental observations of weak temperature dependence in rates and exponential decay of the transfer rate as distance increases indicate that long-range electron transfer in proteins occurs by single-step electron tunnelling across a long distance. This mechanism is in contrast to a multistep hopping mechanism. The protein medium provides electronic states associated with pathways along its backbone that help donor and acceptor wavefunctions delocalize towards each other. The result is a 'superexchange' mechanism that speeds up electron tunnelling rates, usually by >10 orders of magnitude compared with similar distances through a vacuum. Theoretical analysis has even predicted the dominant pathways the electron transfer will take through proteins, and these involve not only covalent bonds, but also hydrogen bonds or even van der Waals contacts. In spite of the prediction of tunnelling pathways, Dutton and co-workers have shown that an empirical model based on average protein density, effectively treating the protein as a structureless random medium, also explains the experimental data (Page *et al.*, 2003). Whether or not proteins have evolved efficient and specialized pathways for electron transduction is still an open question.

1.3.2 Proton tunnelling

In many enzymatic catalytic reactions, the rate-determining step involves the transfer of a proton, hydride or hydrogen atom (Allemann and Scrutton, 2009). The simultaneous transfer of a proton and an electron from different sites (so-called proton-coupled electron transfer) also plays an important role in a wide range of biological functions. Moreover, in some enzymes, quantum effects may contribute to catalytic rates because of the energy shift due to the zero-point energy that gives a quantum correction to the classical activation free energy and the H-tunnelling effects (Nagel and Klinman, 2006). Such nuclear quantum effects represent another class of quantum phenomena in biological systems. It is the short de Broglie wavelength of the proton wavefunction that makes H-tunnelling extremely sensitive to distance fluctuations, leading to the possibility of strong coupling between protein motions and the H-tunnelling kinetics. A hotly debated issue in enzymatic catalysis is whether or not the couplings to slow protein dynamics serve to control the quantum tunnelling and contribute to an increase in the catalytic reaction rate. Such 'gating dynamics' have been proposed to explain anomalous temperature dependences found in several enzymes.

1.3.3 Olfaction

A related area of interest is olfactory reception (Turin, 1996; Brookes *et al.*, 2007; Brookes, 2010). Olfaction is the sense of smell, or the activation of related sensory cells in invertebrates (for instance to detect pheromones). How these systems work and particularly how such a range of odorants can be discriminated by a restricted number of distinct olfactory receptors has not been fully elucidated. It has been suggested that odorants are not discriminated solely by their shape and therefore how they bind to olfactory receptors. Among the various theories, it has been suggested that a second ingredient acts together with the lock and key model; that is, the vibrational spectrum of the odorant is important. It has been suggested that the mechanism in play here is a phonon assisted tunnelling of an electron between two receptor sites via the odorant. Here is an example where a more detailed understanding of the protein, as well as its interactions with a bound analyte, are critical for testing this hypothesis for the operation of olfactory receptors and the possible role of quantum-mechanical tunnelling. A recent detailed study of the Drosophila odorant receptor mechanism could not find good evidence for the vibrational theory (Guo and Kim, 2010), yet other (Franco *et al.*, 2011) clinical studies show fascinating evidence in its favour.

1.4 Radical pairs

1.4.1 Magnetoreception

There is compelling evidence that numerous organisms, including magnetotactic bacteria, insects, amphibians, birds, fish, sharks and rays, and some animals orient themselves using the Earth's magnetic field (Wiltschko and Wiltschko, 1995a; Kirschvink *et al.*, 2001; Rodgers and Hore, 2009; Wajnberg *et al.*, 2010). It has been shown, for example, that homing pigeons can be trained to recognize a weak magnetic anomaly. Their response after training can be upset by attaching a magnet to a part of their beak known to contain a biogenic magnetite body, suggesting that this structure – also found in the organisms mentioned above – is involved in sensing the magnetic field. Other studies have established how the magnetic field lines are sensed in order to derive direction. There is now no doubt that many creatures can navigate using a magnetic sense.

One hypothesis for the mechanism underlying magnetoreception is that magnetite bodies (arranged as an oriented string) are coupled to special receptors so that mechanical torque in response to magnetic field variations activates an ion channel to initiate signalling. This hypothesis explains the presence and use of the chains of magnetic bodies that have been clearly identified. Nevertheless, precisely how the signalling happens is unknown. In some studies it has been found that magnetoreception, at least in the case of newts and birds, is light dependent. This work has strengthened the case for a second hypothesis for the mechanism of magnetoreception, whereby a light-initiated chemical reaction, possibly occurring in a cryptochrome photoreceptor, is tuned by changes in magnetic field. Changes in rate of a reaction involving radical pairs, caused by changes in magnetic field orientation, are suggested to provide magnetic field transduction. A substantial amount of experimental evidence shows that the light dependence includes wavelength specificities and is evidently complex.

The magnetic sense has obvious biological relevance because it aids navigation, orientation and long-range migration, but is the underlying mechanism quantum mechanical? The answer depends on which mechanism is ultimately found to underpin magnetoreception. The first mechanism described above is based on classical electromagnetism and can therefore be anticipated without resort to quantum mechanics. The second mechanism builds on the idea that ladders of electronic states are prevalent in biological examples of quantum mechanics, but in this case those ladders, specifically the relative energies of singlet and triplet states, are used to sense an external stimulus. The explicit dependence of this proposed mechanism on electron spin means that quantum mechanics lies at the heart of the explanation. The possibility that biological systems are performing a kind of magnetic resonance experiment to guide their seasonal migration patterns or other navigation is fascinating. Crucial advances, however, are needed to obtain compelling experimental connections between the behaviour of organisms and the molecular level mechanisms underlying the traits.

1.5 Questions for the present

1.5.1 Do quantum effects introduce new functions in biology?

This is a critical question and one that is being actively pursued by researchers at present. Finding new ways that quantum mechanics helps biological function will drive the field forward. Examples might include sensors, catalysis, photoprotection

Table 1.1 *Examples of biological phenomena and putative underlying quantum-mechanical processes*

Biological phenomenon	Quantum process
Vision, bioluminescence, light-sensing/ response (e.g. phytochromes), vitamin D biosynthesis	Large energy gaps because of electronic excited states
Respiration, photosynthesis	Electron tunnelling
Magnetoreception	Radical pairs and electron spin
Light-harvesting in photosynthesis	Electronic energy transfer involving quantum coherence

and the use or production of light. Thinking in terms of realistic high-level biological function, this may be one of the hardest questions to answer.

Viewed as a 'control knob' or design tool in the construction of biological devices, one important question is: do quantum effects enable the introduction of qualitatively new functions in biological systems? In other words, are some biological phenomena operating in a regime where classical approximations break down (see Table 1.1)? Answering this question in general terms is a goal of the present volume. More specifically, taking photosynthetic light-harvesting as an example, the answer is clearly yes. A natural pigment–protein system at finite temperature will of necessity have variations in local energies, producing a rough energy landscape. Quantum coherence allows temperature insensitivity and robustness to trap states. It also enables construction of energy flow rectifiers (Ishizaki and Fleming, 2009a), thereby influencing the relative rates of forward and backward energy flow.

Coherent energy transfer, in principle, allows constructive or destructive interference between multiple pathways. We are not aware of specific examples of this phenomenon in natural systems, but it clearly provides a new type of control feature. For maximum efficiency of transport processes, the key point is to correctly balance coherent behaviour and dephasing timescales (Rebentrost *et al.*, 2009a). Exact models of quantum transport as a function of coupling strength to the environment (proportional to dephasing rate) show a maximum in the rate for a given reorganization energy (Ishizaki and Fleming, 2009b). Our present understanding is that natural photosynthetic light-harvesting systems operate in a parameter range at or around this maximum value.

1.5.2 Are our experimental methods adequate?

Single-molecule studies of enzyme action have produced remarkable new insights into enzyme function (Min *et al.*, 2005). These phenomena were previously hidden in the ensemble average common to, and necessary for, all earlier measurements. Studies of quantum-dynamical phenomena at the single molecule level are very challenging because of the timescales involved. However, this is encouraging progress in this direction. Brinks *et al.* (2010) recently reported the observation of vibrational coherence in individual molecules at ambient temperatures by means of a phase-locked spontaneous fluorescence technique. Application of this technique to detection of electronic coherence in photosynthetic complexes may provide important new insights into quantum energy flow in biology and motivate the development of new theoretical studies.

1.5.3 Are our theoretical methods adequate?

Again, taking photosynthesis as our example, we note that standard theoretical approaches are based on the quantum master equation which describes the time evolution of reduced density matrices. The ensemble (statistical) average inherent in such an approach may wash out details of the quantum behaviour (Dawlaty *et al.*, 2012; Scholes *et al.*, 2012). Recent work using a mixed quantum/classical approach strongly suggests this is so (Ishizaki and Fleming, 2011; Wong *et al.*, 2012). The simulations show that quantum-coherent motion is robust in the individual realizations of the environment-induced fluctuations, contrary to intuition obtained from the reduced density matrices. Indeed our results imply that experimentally detected delocalized states (excitons) in the ensemble averaged behaviour indicate the existence of wave-like energy flow in individual complexes. Taking such recent results as a first approach, it would seem that investigating quantum phenomena at the individual protein level for a variety of biological processes should prove very rewarding.

There has been significant interest from the quantum information community, and those researchers concerned with exploration of non-classical correlations from more formal perspectives than are usual in the chemical community. For example, Wilde *et al.* state, “Agreement between quantum theoretical models does not irrevocably demonstrate the presence of quantum effects” (Wilde *et al.*, 2010), while Bradler *et al.* comment, “Standard measures of quantum behaviour are more convincing than, say, a claim that wave-like motion in population elements in a density matrix is a signature of quantumness” (Bradler *et al.*, 2010). However, it remains to be seen whether the methods and insights of quantum information theory bring new understandings and predictions to quantum biological phenomena.

1.5.4 What opportunities do an understanding of quantum-biological phenomena bring?

Biology shows the existence of capabilities, efficiencies and emergent phenomena that would perhaps be impossible to imagine otherwise. Biological energy, communication and self-repair, for example, are all emergent functionalities. Many, if not all of these, rely on phenomena that are important only at the nanoscale and on ultra-precise construction achieved through self-assembly. The methods of synthetic biology may well offer the best opportunities for making progress in implementing nature's design principles in synthetic devices based on quantum phenomena. Potential areas for quantum devices are in sensing, in energy harvesting and storage, biosynthesis and in quantum simulation, but likely there are many more possibilities remaining to be discovered.

1.6 Some wide-reaching questions

1.6.1 What is life and consciousness?

Perhaps one of the biggest questions in all of science is that concerning the essence of life. How does the brain work, what is consciousness, free will and the sense of self? How does the central nervous system work and how do certain drugs influence it?

As a consequence of the importance of such questions, and the confluence of interest from researchers with a range of specializations, one of the most prominently recognized debates on a possible role played by quantum mechanics in biology is that concerning brain function. The ethereal notion of consciousness is particularly perplexing and there have been numerous speculations about the special roles played by quantum mechanics. Most famously, for example, Hameroff and Penrose have hypothesized that quantum-mechanical superposition states of microtubules can rationalize brain activities leading to thoughts, feelings, sense of self and transitions of consciousness (Hameroff, 1998). Tantalizing questions have been examined in that work, such as the abrupt transition to unconsciousness caused by anaesthetic drugs (Hameroff, 2006). Nevertheless, the rather qualitative arguments presented so far are not easily assessed by experimental approaches (Davies, 2004; Schuster, 2009). Most criticism focuses on decoherence (Tegmark, 2000). Reimers and co-workers recently examined the Penrose–Hameroff orchestrated objective-reduction proposal for human consciousness in more detail, and concluded that it is not feasible (McKemmish *et al.*, 2009). To be fair, Hameroff has noted objections to a number of interpretations of that paper and it is certainly not easy to dismiss outright the role of microtubules in brain function (Hameroff *et al.*, 2010), although the proposed quantum aspects have yet to be proven or justified (Litt *et al.*, 2006).

1.6.2 Did quantum mechanics play a role in the origin of life?

The confluence of events that led to initiation of life on earth are unknown, but it is clear that the revolution was the production of a molecular template that can be replicated. In the very simple manifestations of life at its outset, was quantum mechanics more prevalent than now? If photochemistry played a role in producing the molecules upon which life is based, is this an instance of quantum biology? These questions are all very speculative, but could feasibly unveil interesting hypotheses and might one day be tested by experiments mimicking the dawn of life.

1.6.3 Is quantum-mechanical mechanism intrinsic or was it selected through evolution?

What approaches can be used to assess the biological relevance or necessity of quantum-mechanical phenomena? Biological significance is not easy to quantify directly. For example, quantum-coherent light-harvesting does not necessarily translate into increased photosynthetic activity because other processes (e.g. CO_2 fixation) can be limiting under various environmental conditions. An approach is to ask whether or not biological systems recognize quantum-mechanical mechanisms as a trait that confers fitness. In other words, it would be compelling to discover whether quantum-mechanical aspects of processes like light-harvesting have been fine tuned during evolution. The traits of living organisms, their relationships and their evolution from one species to another are described by phylogenetic reconstruction. Phylogenetic analysis has played an important role in establishing relationships among species of organisms and between phyla. Perhaps similar analyses can be used to examine biophysical traits and to establish how photophysical and molecular-level mechanisms have evolved or have contributed to species diversification.

Finally, sometimes it is said that if quantum mechanics provides the optimum means of performing a function, then biology will have discovered and harnessed that ability during billions of years of evolution. That is a misconception. Nature is not considered to be perfectly optimized. Evolution tends to accumulate new functions or optimizations into existing motifs, rather than designing from scratch for a particular task. For example, the respiratory enzymes in mitochondria evolved from the extant machinery of photosynthetic reaction centres. Therefore, quantum mechanics will not *necessarily* be discovered by biology, even if it confers important new abilities, if the essential precursor infrastructure does not already exist and serve a function.

2

Open quantum system approaches to biological systems

ALIREZA SHABANI, MASOUD MOHSENI, SEOGJOO JANG, AKIHITO ISHIZAKI, MARTIN PLENIO, PATRICK REBENTROST, ALAN ASPURU-GUZIK, JIANSHU CAO, SETH LLOYD AND ROBERT SILBEY

Quantum biology, as introduced in the previous chapter, mainly studies the dynamical influence of quantum effects in biological systems. In processes such as exciton transport in photosynthetic complexes, radical pair spin dynamics in magnetoreception, and photo-induced retinal isomerization in the rhodopsin protein, a quantum description is a necessity rather than an option. The quantum modelling of biological processes is not limited to solving the Schrödinger equation for an isolated molecular structure. Natural systems are open to the exchange of particles, energy or information with their surrounding environments that often have complex structures. Therefore the theory of open quantum systems plays a key role in dynamical modelling of quantum-biological systems. Research in quantum biology and open quantum system theory have found a bilateral relationship. Quantum biology employs open quantum system methods to a great extent while serving as a new paradigm for development of advanced formalisms for non-equilibrium biological processes.

In this chapter, we overview the basic concepts of quantum mechanics and approaches to open quantum system (or *decoherence*) dynamics. Here, we do not intend to discuss all aspects of about a century-old theory of open quantum systems that dates back to the original work of Paul Dirac on atomic radiative emission and absorption (Dirac, 1927). Instead, we mainly focus on the integro-differential equations that are commonly used for modelling quantum-biological systems. Interested readers can learn more about open quantum systems in various books and review articles in both physics and chemistry literature, including the references (Kraus, 1983; Breuer and Petruccione, 2002; Kubo *et al.*, 2003; Weiss, 2008; May and Kühn, 2011).

Quantum Effects in Biology, ed. Masoud Mohseni, Yasser Omar, Gregory S. Engel and Martin B. Plenio. Published by Cambridge University Press.

2.1 Quantum mechanics concepts and notations

Quantum mechanics provides a mathematical framework that allows us to envision and describe the microscopic world of molecules, atoms, and sub-atomic particles. Arguably, it is the most successful physical theory known. Here, we briefly state the basis of quantum mechanics composed of four postulates on (i) system state representation, (ii) time evolution, (iii) measurement by an observer living in a classical world,[1] and (iv) systems with multiple degrees of freedom. In the following statement of the postulates, we avoid rigorous mathematical definitions to make it accessible to readers with some linear algebra background.

Postulate 1 (State Space) Associated with any physical system is a linear space of complex vectors called Hilbert space $\mathcal{H}$. The state of a system is described by a *density operator* ρ that is a trace one positive operator defined on the Hilbert space. A density operator is often called a *density matrix* for a finite dimension Hilbert space, e.g. an electron two-state spin space. Specifying the Hilbert space sets the stage for a mathematical description of a physical system in a quantum regime.

A quantum system is in a pure state if the density matrix is rank-one $\rho = |\psi\rangle\langle\psi|$ (here we use the bra-ket notation for a vector $|\psi\rangle$ and its Hermitian conjugate $\langle\psi|$). A pure state can be simply represented by the vector $|\psi\rangle$, often referred to as a *wavefunction*, with normalization property $\langle\psi|\psi\rangle = 1$. A quantum state that is not pure is called a *mixed* state. A simple test for the purity of a density matrix ρ is the equality $\mathrm{Tr}(\rho^2) = \mathrm{Tr}(\rho)$.

Postulate 2 (Dynamical Evolution) The evolution of an isolated or closed quantum system is governed by the *Liouville–von Neumann* equation, a first-order linear differential equation,

$$\frac{d\rho(t)}{dt} = -\frac{i}{\hbar}[H(t), \rho(t)]. \tag{2.1}$$

The Hermitian operator H is the system *Hamiltonian* containing all information about the interactions and forces driving the system dynamics. The coefficient $\hbar \approx 6.63 \times 10^{-34}$ m^{2}kg/s is Planck's constant which relates the dynamical timescales to the system energy.

For the special case that a system is in a pure state, $\rho = |\psi\rangle\langle\psi|$, Equation (2.1) reduces to the Schrödinger equation,

$$\frac{d|\psi(t)\rangle}{dt} = -\frac{i}{\hbar}H(t)|\psi(t)\rangle. \tag{2.2}$$

[1] Here, we use the orthodox language of quantum mechanics that distinguishes a quantum world from a classical one. In the context of this book, we generally follow the standard 'shut up and calculate' approach to quantum mechanics, as expressed by David Mermin.

We can also, equivalently, express the postulate on a closed-system dynamic, based on the solution of the Liouville–von Neumann equation. That is, Equation (2.1) guarantees a unitary evolution,

$$\rho(t) = U(t)\rho(0)U^{\dagger}(t), \tag{2.3}$$

with the unitary operator $U(t) = T_{+} \exp[-i \int_0^t H(t')dt']$. The symbol T_{+} denotes a time-ordering operator.[2]

Postulate 3 (Observable and Measurement) Associated with any physically measurable property q is a Hermitian operator Q defined on the Hilbert space. Consider the spectral decomposition $Q = \sum_m q_m P_m$ with projective operators P_m satisfying $P_m = P_m^2$. A measurement apparatus randomly outcomes a value q_m for the property q. For a system in a pre-measurement state ρ, such a random outcome occurs with probability $\mathrm{Tr}(\rho \mathrm{P_m})$. As a result the expectation value of the physical property q is given by $\langle Q \rangle = \sum_m q_m \mathrm{Tr}(\rho \mathrm{P_m}) = \mathrm{Tr}(\rho \mathrm{Q})$.

Any quantum measurement inevitably changes the state of a quantum system that is not an eigenstate of the measured observable. A quantum state ρ *collapses* into the state $\rho_m = P_m \rho P_m / \mathrm{Tr}(\rho \mathrm{P_m})$ upon recording outcome q_m by the measurement apparatus. The quantum state collapse process is not instantaneous in general and happens in a timescale determined by the measurement protocol.[3]

In a real experiment, the average value $\langle Q \rangle$ is obtained either by a statistically large number of measurement repetitions on a single system every time prepared in the same state ρ, or by simultaneous measurement on a large ensemble of systems all in the same state ρ. A quantum biology related example for these two different approaches is electronic spectroscopy with a single molecule versus an ensemble of molecules.

Postulate 4 (Composite Systems) Suppose we have two quantum systems A and B with Hilbert spaces $\mathcal{H}_A$ and $\mathcal{H}_B$. The Hilbert space of the combined system $A + B$ is constructed as $\mathcal{H}_{A+B} = \mathcal{H}_A \otimes \mathcal{H}_B$, where $\otimes$ denotes a tensor product. In another scenario, the combined space of two different degrees of freedom (DOF) of a system can be formed in a similar way. For instance, a proper Hilbert space to describe an electron in an electromagnetic field is $\mathcal{H}_{spin} \otimes \mathcal{H}_{\vec{r}}$, where $\mathcal{H}_{spin}$ is the two-dimensional spin space and $\mathcal{H}_{\vec{r}}$ is the space of a spatial wavefunction.

After this short review on quantum mechanics postulates, we delve into the theory of open quantum systems. We try to adhere to the mathematical notations used above.

[2] The operator T_{+} is defined as $T_{+}A(t)B(\tau)\{= B(\tau)A(t) \text{ if } \tau > t, = A(t)B(\tau) \text{ if } \tau \leq t\}$.

[3] A more general measurement scheme is to indirectly measure the system by coupling it to a probe, also a quantum system, and to perform projective measurement on the probe. See (Breuer and Petruccione, 2002) for such a *generalized quantum measurement*.

2.2 Open quantum systems: dynamical map approach

We have learned that the time evolution of an isolated quantum system is governed by the Liouville equation (2.1). However, no ideally isolated system exists in the real world except the universe as a whole. Then the question arises of how to formulate the evolution of an open quantum system. The general approaches to answering this question can be divided into two categories: *Quantum Dynamical Maps* and *Quantum Master Equations.* In the former approach, the system evolution is described as a mapping between the system states at two different times, ignoring the intermediate moments. In the latter approach, an integro-differential equation captures all infinitesimal variations of the system state over time. The counterparts of these two approaches for closed quantum systems are the unitary map (2.3) and the Schrödinger equation (2.2), respectively. This analogy may imply that quantum maps are simply the solutions of master equations. However, in general, this is not the case for open quantum systems. The quantum map and master equation formalisms have been developed mostly independently. A quantum map was known as the right description for sudden changes in the system state aftermath of a measurement event (Kraus, 1983; Alicki and Lendi, 1987), while quantum master equations were first developed in quantum optics to describe photon emission processes (Scully and Zubairy, 1997).[4] An example of where the two approaches meet is the famous Gorini–Kossakowski–Sudarshan–Lindblad master equation (Section 2.6), which is derived from the complete positivity property of quantum maps. In this chapter, we briefly review quantum-dynamical maps and their classification as completely positive and non-completely positive maps.

In a full quantum-mechanical treatment of an open quantum system, the environment (or commonly called the *bath*) is also modelled as a second quantum system coupled to the primary system of interest. The ideal model of the bath as the world minus the system is mathematically intractable, therefore bath is usually modelled as an effective set of external DOF, for instance, a set of harmonic oscillators (Leggett *et al.*, 1987) or a set of spins (Prokof'ev and Stamp, 2000).

Consider a quantum system S and a bath B, with respective Hilbert spaces $\mathcal{H}_S$ and $\mathcal{H}_B$, such that together they form one isolated system, described by the joint initial state (density matrix) $\rho_{SB}(0)$. Applying Equation (2.3), the system–bath state at time t is given by

$$\rho_{SB}(t) = U_{SB}(t)\rho_{SB}(0)U_{SB}^{\dagger}(t). \tag{2.4}$$

The propagator $U_{SB}(t)$ is a unitary operator, the solution to the Schrödinger equation $\dot{U}_{SB} = -(i/\hbar)H_{SB}U_{SB}$, where H_{SB} is the total system–bath Hamiltonian.

[4] We exclude the Pauli rate equation which describes population transfer only.

The state of the system of interest S is found by the standard quantum-mechanical prescription

$$\rho_S(t) = \mathrm{Tr}_B[\rho_{SB}(t)] = \mathrm{Tr}_B[U_{SB}(t)\rho_{SB}(0)U_{SB}^{\dagger}(t)], \tag{2.5}$$

where Tr_B represents partial trace operation, corresponding to an averaging over the bath DOF. The quantum-dynamical process (QDP) (2.5) is a transformation from $\rho_{SB}(0)$ to $\rho_S(t)$. However, since we are not interested in the state of the bath, it is natural to ask: under which conditions is the QDP a map from $\rho_S(0)$ to $\rho_S(t)$? A simple answer to this question is an initial uncorrelated system–bath state: $\rho_{SB}(0) = \rho_S(0) \otimes \rho_B(0)$. This condition always holds if the system is in a pure state $\rho_S = |\psi\rangle\langle\psi|$ (Shabani, 2009), otherwise the physics of the system determines the validity of this assumption. The corresponding quantum-dynamical map can be derived as follows. Consider the diagonal form of the bath initial state $\rho_B(0) = \sum_k \lambda_k |\phi_k\rangle\langle\phi_k|$ which is usually assumed to be a thermal equilibrium state. In this case Equation (2.5) reduces to

$$\rho_S(t) = \sum_{l,k} \lambda_k \langle\psi_l|U_{SB}(t)|\phi_k\rangle \rho_S(0) \langle\phi_k|U_{SB}^{\dagger}(t)|\psi_l\rangle, \tag{2.6}$$

where $\{|\psi_l\rangle\}$ is a complete basis for the bath Hilbert space. In the case of a closed system, this equation simply reduces to a unitary transformation $\rho_S(t) = U_S(t)\rho_S(0)U_S^{\dagger}(t)$. A compact representation of the above equation, known as Kraus representation, is $\rho_S(t) = \sum_\alpha E_\alpha(t)\rho_S(0)E_\alpha^{\dagger}(t)$ (Kraus, 1983). The property $\sum_\alpha E_\alpha^{\dagger}(t)E_\alpha(t) = I_S$ guarantees the unity trace of the final state $\rho_S(t)$. The above map (2.6) was independently studied by Choi on the classification of completely positive (CP) maps (Choi, 1975). A linear map Φ acting on the space of complex matrices, $\Phi : \mathbb{C}^{n\times n} \to \mathbb{C}^{m\times m}$, is called positive semi-definite if $\Phi(A) \geq 0$ for all positive matrices $A \in \mathbb{C}^{n\times n}$. The map Φ is called CP if the extended map $I_k \otimes \Phi$ acting on the extended Hilbert space $\mathbb{C}^{k\times k} \otimes \mathbb{C}^{n\times n}$ is positive semi-definite for all k (I_k denotes the identity operator acting on $\mathbb{C}^{k\times k}$). Choi showed that a map is CP if and only if there exist operators E_α such that $\Phi(.) = \sum_\alpha E_\alpha . E_\alpha^{\dagger}$. The CP property has an interesting interpretation in the context of quantum dynamics. A real system is not isolated and evolves in the presence of other systems. Using a CP map to represent the dynamics of the system, not only preserves the positivity of the system density matrix but also the density matrix of all present systems. This property was so appealing that it led some quantum physicists to assume it as a necessary condition for any dynamical map, either obtained directly from the Liouvillian equation (2.5), or indirectly as a solution to a master equation. Next, we challenge this idea and go beyond Kraus/CP maps and discuss an approach for obtaining non-CP quantum-dynamical maps.

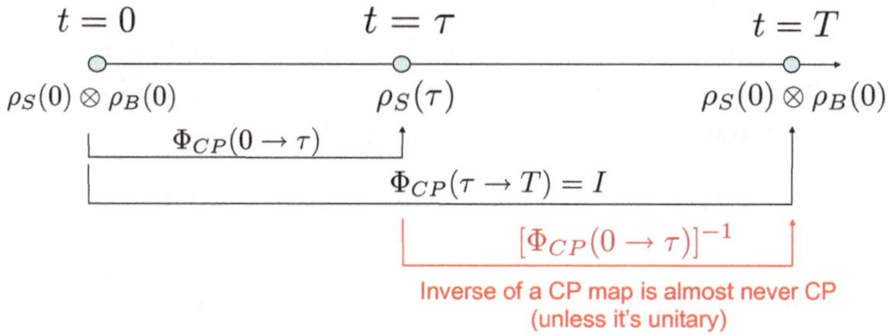

Figure 2.1 A periodic dynamics of coupled two-level systems with time period T. The dynamics of a TLS from time origin to some time τ, $\Phi_{CP}(0 \to \tau)$, can be described by a CP map while the evolution between times τ and the end of a period, $\Phi(\tau \to T)$, must be a non-CP mapping.

2.2.1 Non-completely positive quantum maps

As stated in the preceding section, the old school of open quantum system theory assumes a CP map as the only acceptable form of a quantum dynamical map. This belief has been seriously challenged both theoretically and experimentally. For a historical background, we refer the interested reader to Shabani (2009).

Here we exemplify the violation of CP for a simple quantum system. Consider two coupled two-level systems (TLS) interacting via the Hamiltonian $H = \hbar\gamma\sigma_z \otimes \sigma_z$ for some coupling strength γ.[5] We treat one of the TLSs as a system and the other as a bath. The dynamics of these TLSs is periodic in time with a period of γ^{-1}, which means that if they start from a product state $\rho_{SB}(0) = \rho_S \otimes \rho_B$, after time $T = \gamma^{-1}$ system and bath return to state $\rho_{SB}(T) = \rho_S \otimes \rho_B$. This scenario is depicted in Figure 2.1. The evolution of the system state between the time origin and some intermediate time τ can be represented by a CP map $\Phi_{CP}(0 \to \tau)$, as explained before. On the other hand, the total evolution from the time origin to the end of a period T is the trivial identity map, $\Phi_{CP}(0 \to T) = I_S$. Now we ask what is the dynamical map between the states $\rho_S(\tau)$ and $\rho_S(T)$? A simple answer is $[\Phi_{CP}(0 \to \tau)]^{-1}$. Note that the inverse of a CP map can be a CP map if and only if the map is a unitary transformation. Hence in this example, the dynamical map $\Phi_{CP}(\tau \to T)$ exists and is non-CP.

Now let us go back to the question we originally asked. Is it always possible to describe an arbitrary evolution of a quantum system Equation (2.5) as a map from $\rho_S(0)$ to $\rho_S(t)$? A constructive affirmative answer to this question was given in Shabani and Lidar (2009a, b), which we outline here. For a comprehensive answer to this question see Dominy *et al.*, 2013.

We first construct a quantum map for an initial system–bath state of the form,

$$\rho_{SB}(0) = \sum_{ij} \alpha_{ij} |i\rangle\langle j| \otimes \varphi_{ij}, \tag{2.7}$$

[5] The Pauli operator σ_z is defined as $\sigma = \begin{pmatrix} 1 & 0 \\ 0 & -1 \end{pmatrix}$ in a basis representing the levels of a TLS.

with bath operator φ_{ij} satisfying $\mathrm{Tr}[\varphi_{ij}] \neq 0$ or $\varphi_{ij} = 0$ ($\{|i\rangle\}$ is an orthonormal basis for $\mathcal{H}_S$). A state of this form is called special linear (SL). What is interesting about a SL state is that any state not in this form can be approximated arbitrarily close by a SL state. The dynamical equation (2.5) for a SL state (2.7) is

$$\rho_S(t) = \sum_{ij} \alpha_{ij} \mathrm{Tr}_B[U_{SB}(t)|i\rangle\langle j| \otimes \varphi_{ij} U_{SB}^\dagger(t)]. \tag{2.8}$$

This relation can be equivalently written as a map Φ_{SL} between $\rho_S(0) = \sum_{ij} \alpha_{ij}|i\rangle\langle j|$ and $\rho_S(t)$:

$$\rho_S(t) = \Phi_{\mathrm{SL}}(t,0)[\rho_S(0)] \equiv \sum_{i,j,k,\alpha} \lambda_\alpha^{ij} V_{kij}^\alpha P_i \rho_S(0) P_j \left(W_{kij}^\alpha\right)^\dagger, \tag{2.9}$$

where λ_α^{ij} are the singular values in the singular value decomposition $\phi_{ij} = \sum_\alpha \lambda_\alpha^{ij} |x_{ij}^\alpha\rangle\langle y_{ij}^\alpha|$. Also, $V_{kij}^\alpha \equiv \langle\psi_k|U_{SB}|x_{ij}^\alpha\rangle$ and $W_{kij}^\alpha \equiv \langle\psi_k|U_{SB}|y_{ij}^\alpha\rangle$, with $\{|\psi_k\rangle\}$ being an orthonormal basis for the bath Hilbert space $\mathcal{H}_B$. The quantum map Φ_{SL} is constructed as a function of the bath operators $\{\varphi_{ij}\}$ only and does not depend on the system state $\rho_S(0)$ specified by the coefficients α_{ij}. In this sense, for fixed bath operators $\{\varphi_{ij}\}$, the map Φ_{SL} is a linear map acting on density matrices $\rho_S = \mathrm{Tr}_B[\rho_{SB}]$ corresponding to the system–bath states $\rho_{SB} = \sum_{ij} \alpha_{ij}|i\rangle\langle j| \otimes \varphi_{ij}$. This is somehow similar to the linearity of the Kraus map representation (2.6), as fixing the bath state $\rho_B(0)$ ensures the linearity of the map.

Next we describe how to derive a map representation for an arbitrary initial system–bath state. The density matrix $\rho_{SB}(0)$ can be split into two parts: one is a summation of the SL product terms $\alpha_{ij}|i\rangle\langle j| \otimes \varphi_{ij}$ with bath operator φ_{ij} satisfying $\mathrm{Tr}[\varphi_{ij}] \neq 0$ or $\varphi_{ij} = 0$, the other includes product terms $\beta_{ij}|i\rangle\langle j| \otimes \psi_{ij}$, of which $\psi_{ij} \neq 0$ and $\mathrm{Tr}[\psi_{ij}] = 0$:

$$\rho_{SB}(0) = \sum_{ij(\mathrm{SL})} \alpha_{ij}|i\rangle\langle j| \otimes \varphi_{ij} + \sum_{ij(\mathrm{non-SL})} \beta_{ij}|i\rangle\langle j| \otimes \psi_{ij}. \tag{2.10}$$

The set of traceless matrices $\{\psi_{ij}\}$ is a zero measure in the space of all bath operators. This explain why any non-SL state can be well approximated by a SL state.

The dynamics for a non-SL state is also described by a map. By inserting the decomposition (2.10) into Equation (2.5), we find,

$$\begin{aligned}\rho_S(t) = &\sum_{ij(\mathrm{SL})} \alpha_{ij} \mathrm{Tr}_B[U_{SB}(t)|i\rangle\langle j| \otimes \varphi_{ij} U_{SB}^\dagger(t)] \\ &+ \sum_{ij(\mathrm{non-SL})} \beta_{ij} \mathrm{Tr}_B[U_{SB}(t)|i\rangle\langle j| \otimes \psi_{ij} U_{SB}^\dagger(t)]. \end{aligned} \tag{2.11}$$

Notice that the second term in decomposition (2.11) has no contribution to the system, only state $\rho_S = \mathrm{Tr}_B[\rho_{SB}]$, thus can be treated as a constant $K_{\mathrm{non-SL}}$. Therefore

Equation (2.11) can be expressed as an affine quantum map,

$$\Phi_Q(t,0)[\rho_S(0)] = \Phi_{SL}(t,0)[\rho_S(0)] + K_{\text{non-SL}}, \tag{2.12}$$

where the linear map Φ_{SL} is given in Equation (2.9). It is argued in Shabani and Lidar (2009a) that the above affine map is actually linear, considering the map acting only on the space of the density matrices. Furthermore, this map is Hermitian, meaning that it preserves the Hermiticity of its input state. The generalized Kraus representation for a non-CP Hermitian map is

$$\rho_S(t) = \sum_\alpha E_\alpha(t)\rho_S(0)E_\alpha^\dagger(t) - \sum_\beta E_\beta(t)\rho_S(0)E_\beta^\dagger(t), \tag{2.13}$$

with the trace preserving condition $\sum_\alpha E_\alpha^\dagger(t)E_\alpha(t) - \sum_\beta E_\beta^\dagger(t)E_\beta(t) = I_S$. This completes an explicit expression for a non-CP map based on the system–bath parameterization in Equation (2.10). However, other derivations of a quantum map are possible upon different representations for the state $\rho_{SB}(0)$ that can reflect experimental constraints (Dominy *et al.*, 2013).

A CP or non-CP dynamical process corresponds to different physical conditions. A biology related example is the case of the light absorption process by electronic state of a molecule, where the applicability of the Franck–Condon condition can distinguish a CP from a non-CP dynamical process of electronic states. The Franck–Condon principle states that the internal dynamics of electronic states can happen over much faster timescales than the dynamics of nuclear states, such that nuclear DOF can be considered to be dynamically frozen during the light absorption process (Atkins and Friedman, 1999). Under these circumstances, there would be no initial correlation between the electronic and nuclear DOF. Therefore the dissipative dynamics of electronic states after initial excitement can be described by a CP quantum map. Conversely, violation of the CP property can be evidence for the violation of the Franck–Condon approximation.

2.3 Open quantum systems: master equation approach

Although quantum-dynamical maps provide a full description for the dynamics of an open system, this is not an effective formalism to represent continuous-time dynamics. To be specific, in order to model the dynamics for a time period $[0, T]$, we need to assign a quantum map for every time instance $t \in [0, T]$. In the context of closed systems, that means specifying the set of unitary transformations $U_{t\in[0,T]}$, which is clearly redundant as all dynamical information can be expressed by a single Liouville (2.1) or Schrödinger (2.2) equation. A proper approach to formulating continuous open quantum system dynamics is to use quantum-dynamical equations, generally known as quantum master equations (QME). In the following sections, we review different QMEs employed in modelling quantum-biological systems. We exemplify the application of QMEs by modelling energy transfer in photosynthetic complexes. Therefore, we focus on bosonic environments representing vibrational modes of a protein structure or an electromagnetic reservoir. The other class of environmental models is spin-bath (Prokof'ev and Stamp, 2000), which can be relevant to some biological processes such as magnetoreception. This chapter does not address this latter class, although some of the presented master equations, in principle, can be used to describe decoherence dynamics induced by a spin-bath.

The rest of this chapter is organized as follows. We first present a formally exact QME based on the projection operator technique. Then we consider its perturbative limits in weak system–bath coupling and Markovian regimes. After this we focus on the energy transfer process in pigment–protein complexes and discuss a number of QMEs that can be applied in the regimes of strong decoherence with non-Markovian characters.

2.4 Formally exact QME

The total Hamiltonian governing the dynamics of system and bath can be written as follows:

$$H = H_S + H_{SB} + H_B. \tag{2.14}$$

The total density operator of system and bath, denoted as $\rho_{SB}(t)$, is governed by the following quantum Liouville equation:

$$\frac{d}{dt}\rho_{SB}(t) = -i\mathcal{L}\rho_{SB}(t) = -\frac{i}{\hbar}[H, \rho_{SB}(t)]. \tag{2.15}$$

Let us introduce the zeroth order Hamiltonian as follows:

$$H_0 = H_S + H_B. \tag{2.16}$$

Then, in the interaction picture of H_0, Equation (2.15) becomes

$$\frac{d}{dt}\rho_I(t) = -i\mathcal{L}_{SB,I}(t)\rho_I(t) = -\frac{i}{\hbar}[H_{SB,I}(t), \rho_I(t)], \tag{2.17}$$

where $\rho_I(t) = e^{iH_0 t/\hbar}\rho_{SB}(t)e^{-iH_0 t/\hbar}$ and $H_{SB,I}(t) = e^{iH_0 t/\hbar}H_{SB}e^{-iH_0 t/\hbar}$. In general, exact numerical solution of Equation (2.17) is impossible because the bath consists of virtually infinite degrees of freedom. On the other hand, the dimension of the system is relatively small. Thus, it is more plausible and economical to solve for the system part, which is the primary interest, and to make approximations for the bath part if necessary. For this, let us define the following reduced system density operator:

$$\rho_{S,I}(t) = \mathrm{Tr}_B[\rho_I(t)]. \tag{2.18}$$

Let us define a projection operator and its complement as follows:

$$\mathcal{P}(\cdot) = \mathrm{Tr}_B[\cdot] \otimes \rho_B(0), \tag{2.19}$$

$$\mathcal{Q} = 1 - \mathcal{P}. \tag{2.20}$$

Here, $(\cdot)$ and $[.]$ represent the same arbitrary operator. For a bath at temperature T, we assume $\rho_B(0) = e^{-\beta H_B}/\mathrm{Tr}[e^{-\beta H_B}]$ with $\beta = 1/k_B T$.[6] Applying $\mathcal{P}$ and $\mathcal{Q}$ to Equation (2.17) and inserting the identity, $\mathcal{P} + \mathcal{Q} = 1$, between $\mathcal{L}_{SB,I}(t)$ and $\rho_I(t)$, we obtain the following time evolution equations for the projected and unprojected portions of the total density operator:

$$\frac{d}{dt}\mathcal{P}\rho_I(t) = -i\mathcal{P}\mathcal{L}_{SB,I}(t)(\mathcal{P} + \mathcal{Q})\rho_I(t), \tag{2.21}$$

$$\frac{d}{dt}\mathcal{Q}\rho_I(t) = -i\mathcal{Q}\mathcal{L}_{SB,I}(t)(\mathcal{P} + \mathcal{Q})\rho_I(t). \tag{2.22}$$

First, consider the equation for the unprojected portion, Equation (2.22), which can be written as

$$\frac{d}{dt}\mathcal{Q}\rho_I(t) + i\mathcal{Q}\mathcal{L}_{SB,I}(t)\mathcal{Q}\rho_I(t) = -i\mathcal{Q}\mathcal{L}_{SB,I}(t)\mathcal{P}\rho_I(t). \tag{2.23}$$

A formal solution of this equation can be shown to be

$$\begin{aligned}\mathcal{Q}\rho_I(t) &= T_+ \exp\left[-i\int_0^t d\tau\, \mathcal{Q}\mathcal{L}_{SB,I}(\tau)\right]\mathcal{Q}\rho_I(0) \\ &\quad - i\int_0^t d\tau\, T_+ \exp\left[-i\int_\tau^t d\tau' \mathcal{Q}\mathcal{L}_{SB,I}(\tau')\right]\mathcal{Q}\mathcal{L}_{SB,I}(\tau)\mathcal{P}\rho_I(\tau).\end{aligned} \tag{2.24}$$

Inserting this into Equation (2.21), we obtain the following time evolution equation for the projected portion of the total density operator:

$$\begin{aligned}\frac{d}{dt}\mathcal{P}\rho_I(t) &= -i\mathcal{P}\mathcal{L}_{SB,I}(t)\mathcal{P}\rho_I(t) \\ &\quad - i\mathcal{P}\mathcal{L}_{SB,I}(t)T_+ \exp\left[-i\int_0^t d\tau\, \mathcal{Q}\mathcal{L}_{SB,I}(\tau)\right]\mathcal{Q}\rho_I(0) \\ &\quad - \int_0^t d\tau\, \mathcal{P}\mathcal{L}_{SB,I}(t)T_+ \exp\left[-i\int_\tau^t d\tau' \mathcal{Q}\mathcal{L}_{SB,I}(\tau')\right]\mathcal{Q}\mathcal{L}_{SB,I}(\tau)\mathcal{P}\rho_I(\tau).\end{aligned} \tag{2.25}$$

[6] We should explain a subtle point in modelling an open quantum system. We assume an open system is coupled to a second quantum system, bath B. However, this is not the whole story as under actual physical conditions, the bath is also coupled to the rest of the world not accounted for in defining B. Such a secondary environmental influence manifests itself in the choice of bath initial state or bath fluctuation characteristics.

Taking a trace of this over the bath and using the fact that $\mathcal{P}\rho_I(t) = \rho_{S,I}(t) \otimes \rho_B(0)$, we find that

$$\begin{aligned}
\frac{d}{dt}\rho_{S,I}(t) \\
&= -i\mathrm{Tr}_B[\mathcal{L}_{SB,I}(t)\rho_B(0)]\rho_{S,I}(t) \\
&\quad - i\mathrm{Tr}_B[\mathcal{L}_{SB,I}(t)T_+ \exp\left[-i\int_0^t d\tau\,\mathcal{Q}\mathcal{L}_{SB,I}(\tau)\right]\mathcal{Q}\rho_I(0)] \\
&\quad - \int_0^t d\tau \mathrm{Tr}_B[\mathcal{L}_{SB,I}(t)T_+ \exp\left[-i\int_\tau^t d\tau'\mathcal{Q}\mathcal{L}_{SB,I}(\tau')\right]\mathcal{Q}\mathcal{L}_{SB,I}(\tau)\rho_B(0)]\rho_{S,I}(\tau).
\end{aligned} \tag{2.26}$$

This formally exact QME, as shown above, is in a closed form but it is not yet amenable to computation because $T_+ \exp[-i\int_\tau^t d\tau'\mathcal{Q}\mathcal{L}_{SB,I}(\tau')]$ cannot be determined without full information on the system and bath degrees of freedom. Rather, it serves as the starting point to make systematic approximations leading to computable equations.

2.5 QME in the weak system–bath coupling limit

For weak system–bath interaction, an approximation can be made such that only terms exact up to the second order of $\mathcal{L}_{SB,I}(t)$ are kept in Equation (2.26). The resulting equation is the well-known *second-order perturbative time-convolution* (TC2) QME shown below:

$$\begin{aligned}
\frac{d}{dt}\rho_{S,I}(t) = &-i\mathrm{Tr}_B[\mathcal{L}_{SB,I}(t)\rho_B(0)]\rho_{S,I}(t) \\
&- i\mathrm{Tr}_B[\mathcal{L}_{SB,I}(t)\mathcal{Q}\rho_I(0)] - \int_0^t d\tau \mathrm{Tr}_B[\mathcal{L}_{SB,I}(t)\mathcal{Q}\mathcal{L}_{SB,I}(\tau)\mathcal{Q}\rho_I(0)] \\
&- \int_0^t d\tau \mathrm{Tr}_B[\mathcal{L}_{SB,I}(t)\mathcal{Q}\mathcal{L}_{SB,I}(\tau)\rho_B(0)]\rho_{S,I}(\tau).
\end{aligned} \tag{2.27}$$

Alternatively, we can replace $\rho_{S,I}(\tau)$ with $\rho_{S,I}(t)$ in Equation (2.27), which does not affect the accuracy up to the second order of $H_{SB}(t)$. Thus, the following *second-order perturbative time-convolutionless (TCL2)* QME can be obtained:

$$\begin{aligned}
\frac{d}{dt}\rho_{S,I}(t) = &-i\mathrm{Tr}_B[\mathcal{L}_{SB,I}(t)\rho_B(0)]\rho_{S,I}(t) \\
&- i\mathrm{Tr}_B[\mathcal{L}_{SB,I}(t)\mathcal{Q}\rho_I(0)] - \int_0^t d\tau \mathrm{Tr}_B[\mathcal{L}_{SB,I}(t)\mathcal{Q}\mathcal{L}_{SB,I}(\tau)\mathcal{Q}\rho_I(0)] \\
&- \int_0^t d\tau \mathrm{Tr}_B[\mathcal{L}_{SB,I}(t)\mathcal{Q}\mathcal{L}_{SB,I}(\tau)\rho_B(0)]\rho_{S,I}(t).
\end{aligned} \tag{2.28}$$

Further simplification of Equations (2.27) and (2.28) is possible if the bath can be modelled as a set of harmonic oscillators (bosonic bath) and the system–bath interaction is linear in the displacements of the oscillators as follows:

$$H_B = \sum_{\xi} \hbar\omega_{\xi} \left(b_{\xi}^{\dagger} b_{\xi} + \frac{1}{2} \right), \tag{2.29}$$

$$H_{SB} = \sum_{l,\xi} \hbar\omega_{\xi} g_{\xi,l} (b_{\xi} + b_{\xi}^{\dagger}) S_l, \tag{2.30}$$

with some system operators S_l. For these, $\mathrm{Tr}_B[\mathcal{L}_{SB,I}(t)\rho_B(0)] = 0$. In addition, if we assume that the initial density operator is given by

$$\rho_I(0) = \rho_S(0) \otimes \rho_B(0), \tag{2.31}$$

all the inhomogeneous terms in Equation (2.27) or (2.28) disappear because $\mathcal{Q}\rho_I(0) = 0$. As a result, the time non-local equation, Equation (2.27), reduces to

$$\begin{aligned} \frac{d}{dt}\rho_{S,I}(t) &= -\int_0^t d\tau \mathrm{Tr}_B[\mathcal{L}_{SB,I}(t)\mathcal{L}_{SB,I}(\tau)\rho_B(0)]\rho_{S,I}(\tau) \\ &= -\sum_{l}\sum_{l'}\int_0^t d\tau \left\{ C_{ll'}(t-\tau)[S_l(t), S_{l'}(\tau)\rho_{S,I}(\tau)] \right. \\ &\qquad \left. - C_{l'l}(\tau - t)[S_l(t), \rho_{S,I}(\tau) S_{l'}(\tau)] \right\}, \end{aligned} \tag{2.32}$$

where $S_l(t) = e^{iHt/\hbar} S_l e^{-iHt/\hbar}$ and

$$\begin{aligned} C_{ll'}(t) &= \sum_{\xi} \omega_{\xi}^2 g_{\xi,l} g_{\xi,l'} \mathrm{Tr}_B\left\{ \left(b_{\xi} e^{-i\omega_{\xi} t} + b_{\xi}^{\dagger} e^{i\omega_{\xi} t}\right)\left(b_{\xi} + b_{\xi}^{\dagger}\right)\rho_B(0) \right\} \\ &= \int_0^{\infty} d\omega \mathcal{J}_{ll'}(\omega) \left(\coth\left(\frac{\beta\hbar\omega}{2}\right)\cos(\omega t) - i\sin(\omega t) \right) \end{aligned} \tag{2.33}$$

are bath correlation functions. In the above equation, the following *spectral density* has been defined,

$$\mathcal{J}_{ll'}(\omega) = \sum_{\xi} g_{\xi,l} g_{\xi,l'} \delta(\omega - \omega_{\xi}) \omega^2, \tag{2.34}$$

which carries information about the bath physical structure. The spectral density can be specified either via direct calculations (Weiss, 2008) or spectroscopy experiments (Mukamel, 1995). We will learn more about bath spectral density in Section 2.7.1 in the context of electronic energy transfer. Alternatively, starting from Equation (2.28), we obtain the following time local equation:

$$\frac{d}{dt}\rho_{S,I}(t) = \sum_{l,l'} \mathcal{R}_{ll'}(t)\rho_{S,I}(t), \tag{2.35}$$

where $\mathcal{R}_{ll'}(t)$ is a super-operator defined as

$$\mathcal{R}_{ll'}(t)(\cdot) = -\int_0^t d\tau\,\{C_{ll'}(t-\tau)[S_l(t), S_{l'}(\tau)(\cdot)] - C_{l'l}(\tau - t)[S_l(t), (\cdot)S_{l'}(\tau)]\}. \tag{2.36}$$

Equations (2.32) and (2.36) are the commonly used forms of TC2 and TCL2. Note that these equations can be obtained for any bath model as long as the conditions $\mathrm{Tr}_B[\mathcal{L}_{SB,I}(t)\rho_B(0)] = 0$ and $\rho_I(0) = \rho_S(0) \otimes \rho_B(0)$ are satisfied (Breuer and Petruccione, 2002). It is often believed that the TCL2 equation is less accurate than the TC2 equation because the former appears to involve an additional approximation. However, in general, it is not true because the additional step needed to get the time local equation does not affect the accuracy up to the second order of $H_{SB,I}$ (at the level of time evolution equation) and sometimes may include correct fourth-order terms (Jang *et al.*, 2002a). Thus care should be taken in assessing which one is appropriate for the nature of the problem.

2.6 QME for weak coupling to a Markovian bath

The presented exact master equation (2.26) and its perturbative reductions (2.32) and (2.36) describe the influence of a bath with memory. The notion of memory refers to the fact that the instantaneous bath action $\dot{\rho}_{S,I}$ either explicitly depends on the history of the system states $\rho(\tau < t)$ as in TC2, or the dynamical generator is time dependent (in the Schrödinger picture) as $\sum_{l,l'} \mathcal{R}_{ll'}(t)$ in TCL2. The latter can be interpreted as the bath remembers when the dynamics has started. Such a dynamical property is known as *non-Markovianity*.[7] In contrast, a quantum bath can be so rigid that the system back-action on the bath is negligible. In this so-called Markovian regime, the QME is represented by some time-independent generator $\mathcal{R}$,

$$\frac{d}{dt}\rho_S(t) = \mathcal{R}[\rho_S(t)]. \tag{2.37}$$

At a microscopic level such a condition is realized when the bath is so large that it quickly returns to its equilibrium state upon any system induced disturbance. The correlation functions $C_{ll'}(t)$ represent the corresponding bath non-equilibrium

[7] The notions of Markovian or non-Markovian are concepts borrowed from classical random process literature, however, the definitions in the quantum realm do not necessarily follow their classical counterparts.

behaviour. A Markovian or memoryless decoherence dynamics is effective if τ^{rxn}, the decay timescale of correlation function $C_{ll'}(t)$, is much shorter than decoherence timescale τ_D.

The Markovian limit of the TC2 and TCL2 equations is the famous *Bloch–Redfield* equation.[8] Consider the system Hamiltonian $H_S = \sum \hbar\Omega_m|m\rangle\langle m|$ and the density matrix ρ_S with elements $\rho_{nm} = \langle n|\rho_S|m\rangle$ in the energy eigen-basis. The Bloch–Redfield equation for a single coupling operator $S_l = S$ is

$$\frac{d\rho_{nm}(t)}{dt} = -i\hbar\Omega_{nm}\rho_{nm} + \sum_{n'm'} \Gamma_{nmn'm'}\rho_{n'm'} \tag{2.38}$$

for $\Omega_{nm} = \Omega_n - \Omega_m$ and the Redfield tensor,

$$\Gamma_{nmn'm'} = \Lambda_{m'mnn'} + \bar{\Lambda}_{nn'm'm} - \sum_k \Lambda_{nkkn'}\delta_{mm'} + \bar{\Lambda}_{kmm'k}\delta_{nn'}, \tag{2.39}$$

with $\Lambda_{m'mnn'} = \frac{1}{2}S_{nm}S_{n'm'}\tilde{C}(\Omega_{m'} - \Omega_{n'})$, $\bar{\Lambda}_{m'mnn'} = \frac{1}{2}S_{nm}S_{n'm'}C^*(\Omega_{n'} - \Omega_{m'})$, where $\tilde{C}(\omega) = \int_0^\infty dt e^{i\omega t}C(t)$.

Equation (2.38) can be recast into the compact form (2.37) describing a Markovian dynamics. The Bloch–Redfield equation has been widely used in different areas of physics and chemistry to model decoherence and dissipation processes, however, it suffers from one technical problem that Equation (2.38) does not necessarily yield, a valid density matrix $\rho_S(t)$. We can resolve this problem by applying secular approximation which ignores some high-frequency terms in (2.38), which is valid if $|\Omega_m - \Omega_n|^{-1} \ll \tau_D, \forall n, m$.[9] In this regime we find the following QME in a Lindblad form:

$$\begin{aligned}\frac{d\rho_S(t)}{dt} &= -i\left[H_S + \sum_\Delta \text{Im}[\tilde{C}(\Delta)]S^2(\Delta), \rho_S(t)\right] \\ &\quad + \frac{1}{2}\sum_\Delta \text{Re}[\tilde{C}(\Delta)]\left(S(\Delta)\rho_S(t)S(\Delta) - \frac{1}{2}\{S^2(\Delta), \rho_S(t)\}\right) \triangleq \mathcal{R}_L\rho_S(t)\end{aligned} \tag{2.40}$$

for the system operator $S(\Delta) = \sum_{(\Omega_\xi - \Omega_m = \Delta)} S_{nm}|n\rangle\langle m|$. The quantum map $\Phi(t) = e^{\mathcal{R}_L t}$, representing the solution of Equation (2.40), is CP for any time t. This property guarantees a valid density matrix $\rho_S(t)$. From an abstract mathematical approach, Gorini–Kossakowski–Sudarshan (Gorini *et al.*, 1976) and Lindblad (Lindblad, 1976) derived the following general form for the superoperator $\mathcal{R}$ in

[8] The TCL2 master equation is often called the time-dependent Redfield equation.
[9] See Breuer and Petruccione (2002) for other parameter regimes resulting in a Lindblad equation.

Equation (2.37) that generates an always CP dynamical map $e^{\mathcal{R}t}$:

$$\mathcal{R}(.) = -i[H, .] + \sum_{\alpha} A_\alpha . A_\alpha^\dagger - \frac{1}{2}\{A_\alpha^\dagger A_\alpha, .\}, \tag{2.41}$$

where H is Hermitian. The operators A_α and H can be specified not only from a microscopic picture but also in a phenomenological way. QMEs of the Lindblad type have been extensively used for simulating energy transfer in light-harvesting systems, either derived from a microscopic model (Mohseni *et al.*, 2008; Rebentrost *et al.*, 2009c) or applied phenomenologically (Plenio and Huelga, 2008; Cao and Silbey, 2009; Caruso *et al.*, 2009; Rebentrost *et al.*, 2009b). We review some of these studies in Chapter 7 of this book.

2.7 QMEs beyond weak and Markovian limits

In the rest of this chapter, we review various mathematical frameworks for dynamical modelling of quantum systems that interact non-perturbatively with their surrounding environment. As mentioned before, we focus on electronic energy transfer (EET) in photosynthetic complexes as a biological paradigm to study QMEs. To this end, we begin by discussing the physical modelling of pigment–protein complexes in photosynthetic systems.

2.7.1 Quantum modelling a pigment–protein complex

Photosynthesis provides the energy source for essentially all living things on Earth, and its functionality has been one of the most fascinating mysteries of life. Photosynthetic conversion of the energy of sunlight into its chemical form suitable for cellular processes involves a variety of physicochemical mechanisms. The conversion starts with the absorption of a photon of sunlight by one of the light-harvesting pigments, followed by transfer of electronic excitation energy (EET) to the reaction centre, where charge separation is initiated.

To describe EET, we consider a pigment–protein complex (PPC) consisting of N pigments. We restrict the electronic spectra of the mth pigment in a PPC to the ground state $|\phi_{mg}\rangle$ and the first excited state $|\phi_{me}\rangle$. When the mth and nth pigments are situated in close proximity and the mth pigment is excited, the excitation energy may be transferred to the nth pigment. We assume that there is no orbital overlap between the two molecules so that electrons can be assigned unambiguously to one molecule or the other. In this situation the PPC Hamiltonian for describing photosynthetic EET can be expressed as (Renger *et al.*, 2001; May and Kühn,

2011),

$$H_{\text{PPC}} = \sum_{m=1}^{N} \sum_{a=g,e} H_{ma}(\mathbf{x})|\phi_{ma}\rangle\langle\phi_{ma}| + \sum_{m,n} \hbar J_{mn}|\phi_{me}\rangle\langle\phi_{mg}| \otimes |\phi_{ng}\rangle\langle\phi_{ne}|. \quad (2.42)$$

Here, $H_{ma}(\mathbf{x}) = \epsilon_{ma}(\mathbf{x}) +$ (nuclear kinetic energy) ($a = g, e$) is the Hamiltonian describing the nuclear dynamics associated with an electronic state $|\phi_{ma}\rangle$, where $\epsilon_{ma}(\mathbf{x})$ is the potential energy as a function of the set of the relevant nuclear coordinates including protein DOF, $\mathbf{x}$. The electronic coupling J_{mn} is also influenced by the nuclear motions and local polarizations of the protein environment (Scholes, 2003; May and Kühn, 2011). However, we assume that nuclear dependence of J_{mn} is vanishingly small as usual. The Hamiltonian of the single-exciton manifold composed of $\{|m\rangle \equiv |\phi_{me}\rangle \prod_{k(\neq m)}|\phi_{kg}\rangle\}_{m=1,\ldots,N}$ takes the form,

$$H_{\text{PPC}} = \sum_{m=1}^{N} \Big[H_{me}(\mathbf{x}) + \textstyle\sum_{k\neq m} H_{kg}(\mathbf{x})\Big] |m\rangle\langle m| + \sum_{m,n} \hbar J_{mn}|m\rangle\langle n|. \quad (2.43)$$

The normal mode treatment is usually assumed for PPC nuclear dynamics, because anharmonic motion with large amplitudes and long timescales produces static disorder on timescales irrelevant to photosynthetic EET (Mukamel, 1995; Renger *et al.*, 2001). Furthermore, it may be assumed that nuclear configurations for the electronic excited states of pigments are not greatly different from those for the ground state, owing to the absence of large permanent dipoles on the pigments. Thus, $H_{mg}(\mathbf{x})$ and $H_{me}(\mathbf{x})$ can be modelled as a set of displaced harmonic oscillators:

$$H_{mg}(\mathbf{x}) = \epsilon_{mg}\big(\mathbf{x}_{mg}^{0}\big) + \sum_{\xi} \frac{\hbar\omega_{m\xi}}{2}\big(p_{m\xi}^{2} + q_{m\xi}^{2}\big), \quad (2.44)$$

$$H_{me}(\mathbf{x}) = H_{mg}(\mathbf{x}) + \hbar\Omega_m - \sum_{\xi} \hbar\omega_{m\xi} d_{m\xi} q_{m\xi}, \quad (2.45)$$

where $\mathbf{x}_{mg}^{0}$ is the equilibrium configuration of the nuclear coordinates associated with the electronic ground state of the mth pigment, and $q_{m\xi}$ is the dimensionless normal mode coordinate with accompanying frequency $\omega_{m\xi}$ and momentum $p_{m\xi}$; $d_{m\xi}$ is the dimensionless displacement. For later convenience, we set $\epsilon_{mg}(\mathbf{x}_{mg}^{0}) = 0$ without loss of generality. The Franck–Condon transition energy, $\hbar\Omega_m = \epsilon_{me}(\mathbf{x}_{mg}^{0}) - \epsilon_{mg}(\mathbf{x}_{mg}^{0})$, is also termed the site energy in the literature. After electronic excitation in accordance with the vertical Franck–Condon transition, reorganization takes place from the nuclear configuration $\mathbf{x}_{mg}^{0}$ to the actual equilibrium configuration in the excited state $\mathbf{x}_{me}^{0}$, with dissipating reorganization energy defined as $\hbar\lambda_m = \epsilon_{me}(\mathbf{x}_{mg}^{0}) - \epsilon_{me}(\mathbf{x}_{me}^{0})$. This reorganization proceeds on a finite timescale, τ_m^{rxn}. Therefore, we can write the PPC Hamiltonian (2.43) as a system (electronic DOF) coupled to a bosonic bath (nuclear DOF):

$$H_{\text{PPC}} = H_{el} + H_{el-nuc} + H_{nuc}. \tag{2.46}$$

The first term on the right-hand side is the electronic excitation Hamiltonian with respect to the equilibrium nuclear configuration of the electronic ground state, $\{\mathbf{x}^0_{mg}\}$,

$$H_{el} = \sum_m \hbar\Omega_m |m\rangle\langle m| + \sum_{m,n} \hbar J_{mn} |m\rangle\langle n|, \tag{2.47}$$

where we have set $\sum_m \epsilon_{mg}(\mathbf{x}^0_{mg}) = 0$. The second part describes the coupling of nuclear motion to the electronic excitations,

$$H_{el-nuc} = \sum_m S_m V_m, \tag{2.48}$$

with $S_m \equiv |m\rangle\langle m|$ and $V_m = -\sum_\xi \hbar\omega_\xi d_{m\xi} q_\xi$. The last term in Equation (2.46) is the ensemble of the normal mode Hamiltonians, i.e. the phonon Hamiltonian expressed as $H_{nuc} = \sum_\xi \hbar\omega_\xi (p_\xi^2 + q_\xi^2)/2$. It should be noted that the explicit expressions of H_{el} and H_{el-nuc} depend on the choice of a reference nuclear configuration. To reduce the total PPC density operator, we suppose that the total system at the initial time $t = 0$ is in the factorized product state of the form, $\rho_{\text{PPC}}(0) = \rho_{el}(0) \otimes \rho_{nuc}(0)$, where $\rho_{nuc}(0) = e^{-\beta H_{nuc}}/\text{Tr}_{nuc}[e^{-\beta H_{nuc}}]$. This factorized initial condition corresponds to the electronic ground state or an electronic excited state generated in accordance with the vertical Franck–Condon transition.

Because $\{q_\xi\}$ are normal mode coordinates or phonon modes, Wick's theorem (Rammer, 2007) yields the following *Gaussian property* (Kubo, 1962; Kubo *et al.*, 2003) of u_m:

$$\langle \mathcal{T} \tilde{V}_m(t_{2n}) \tilde{V}_m(t_{2n-1}) \cdots \tilde{V}_m(t_2) \tilde{V}_m(t_1) \rangle_{mg} = \sum_{\text{a.p.p}} \prod_{k,\ell} \langle \mathcal{T} \tilde{V}_m(t_k) \tilde{V}_m(t_\ell) \rangle_{mg}, \tag{2.49}$$

where $\tilde{V}_m(t) \equiv e^{iH_{mg}t/\hbar} V_m e^{-iH_{mg}t/\hbar}$, and $\langle \ldots \rangle_{mg}$ denotes averaging over $\rho_{mg} \equiv e^{-\beta H_{mg}}/\text{Tr}[e^{-\beta H_{mg}}]$. The sum is over all possible ways of picking pairs (a.p.p) among $2n$ operators. $\mathcal{T}$ denotes an ordering operator which orders products by some rule, e.g. time ordering $\mathcal{T} = T_+$. Therefore, all the phonon-induced relaxation processes can be quantified by two-point correlation functions of $\tilde{V}_m(t)$. We assume the fluctuation–dissipation processes in one pigment are not correlated with those in others. Fluctuations in the electronic energy of the mth pigment are described by the *symmetrized correlation function* of $\tilde{V}_m(t)$ as

$$\mathcal{S}_m(t) \equiv \frac{1}{2} \langle \{ \tilde{V}_m(t), \tilde{V}_m(0) \} \rangle_{mg}. \tag{2.50}$$

Information on this function can be obtained by means of three-pulse photon echo peak shift measurement (Fleming and Cho, 1996; Joo *et al.*, 1996). In addition, environmental reorganization involving the dissipation of reorganization energy can be understood as the response to a sudden change in electronic state via the vertical Franck–Condon transition, and thus can be characterized by the *response function*,

$$\chi_m(t) \equiv \frac{i}{\hbar}\langle[\tilde{V}_m(t), \tilde{V}_m(0)]\rangle_{mg}. \tag{2.51}$$

The reorganization dynamics can, in principle, be measured by the time-dependent fluorescence Stokes shift experiment (Fleming and Cho, 1996; Joo *et al.*, 1996), where the direct observable quantity is the *relaxation function* defined by

$$\Gamma_m(t) \equiv \int_t^\infty d\tau\, \chi_m(\tau), \tag{2.52}$$

with $\Gamma_m(0) = 2\hbar\lambda_m$ being the Stokes shift magnitude. The quantum fluctuation–dissipation relation (Kubo, 1966; Kubo *et al.*, 2003) allows us to express the symmetrized correlation function and the response function as

$$\mathcal{S}_m(t) = \frac{\hbar}{\pi}\int_0^\infty d\omega\, \mathcal{J}_m[\omega] \coth\frac{\beta\hbar\omega}{2}\cos\omega t, \tag{2.53}$$

$$\chi_m(t) = \frac{2}{\pi}\int_0^\infty d\omega\, \mathcal{J}_m[\omega]\sin\omega t. \tag{2.54}$$

Here, $\mathcal{J}_m[\omega]$ is the imaginary part of the Fourier–Laplace transform of the response function, $\mathcal{J}_m[\omega] \equiv \mathrm{Im}\int_0^\infty dt e^{i\omega t}\chi_m(t)$, the spectral density (Kubo *et al.*, 2003).[10] If the environmental phonons can be described classically, the symmetrized correlation, response and relaxation functions satisfy the classical fluctuation–dissipation relation (Zwanzig, 2001),[11]

$$\mathcal{S}_m(t) = \frac{1}{\beta}\Gamma_m(t) \quad \chi_m(t) = -\frac{d}{dt}\Gamma_m(t). \tag{2.55}$$

Owing to the fluctuation–dissipation relation, the symmetrized correlation function, the response function, the relaxation function and the spectral density contain

[10] The term 'spectral density' may mislead unless we draw attention to its definition. It does not in itself give the distribution of phonon modes.

[11] Generally, fluctuation tends to drive any system to an 'alive' state, while dissipation tends to relax the system to a 'dead' state, as described in Zwanzig (2001). The balance between fluctuation and dissipation is required to guarantee a thermal equilibrium state at long times. This is the physical significance of the fluctuation–dissipation theorem expressed in Equations (2.53)–(2.55). In reality, stochastic models without any dissipative effects correspond to unphysical pictures where the fluctuation continues to activate the system toward an infinite temperature. The Haken–Strobl model (Haken and Strobl, 1973) is in this category.

the same information on the phonon dynamics, whose characteristic timescale is given by (Kubo, 1963)

$$\tau_m^{\mathrm{rxn}} \equiv \frac{1}{\Gamma_m(0)} \int_0^\infty dt\, \Gamma_m(t), \tag{2.56}$$

although the relaxation function and the associated spectral density may have complicated forms involving various components (Joo *et al.*, 1996). In fact, an arbitrary relaxation function can be numerically decomposed to the sum of exponential functions, as discussed by Meier and Tannor (1999).

2.7.2 Electronic energy transfer in photosynthetic complexes: a case of failure for perturbative QMEs

The electronic coupling $\hbar J$ between pigments and the electron–nuclear coupling characterized by the reorganization energy $\hbar\lambda$ are two fundamental interaction mechanisms determining the nature of EET in photosynthetic complexes. The transfer processes are usually described in one of two perturbative limits. When the electronic coupling $\hbar J$ is small in comparison with the electron–nuclear coupling $\hbar\lambda$, the original localized electronic state is an appropriate representation and the inter-pigment electronic coupling can be treated perturbatively. This treatment yields the Förster theory (Förster, 1948) (see the next chapter for a review on the Förster theory). In the opposite limit, when the electron–nuclear coupling is small, it is possible to treat it perturbatively to obtain a quantum master equation. The most commonly used approach for this limit in the literature of photosynthetic EET is the Redfield equation (2.38) (Redfield, 1957, 1965).

Ordinarily, photosynthetic EET is discussed only in terms of the mutual relation between magnitudes of the two couplings, as just described. However, we should not overlook that the nature of EET is also dominated by the mutual relation between the two timescales, the characteristic timescale of the nuclear reorganization, τ^{rxn}, and the inverse of the electronic coupling, J^{-1}, that is, the time the excitation needs to move from one pigment to another neglecting any additional perturbations. In the case of $\tau^{\mathrm{rxn}} \ll J^{-1}$, it is impossible to construct a wavefunction straddling multiple pigments. The nuclear reorganization introduces fast dephasing, and hence EET occurs *after* the nuclear equilibration associated with the excited pigment. In this situation, EET is described as a diffusive motion similar to the classical random walk; it follows classical rate laws where the transition rate is given by the Förster theory. In the contrary case of $J^{-1} \ll \tau^{\mathrm{rxn}}$, the excitation can travel almost freely from one pigment to others according to the Schrödinger equation, until the nuclear configurations are quenched by the reorganization. The excitation travels through photosynthetic complexes as a quantum-mechanical wave packet, keeping its phase coherence. Thus, this process is termed *coherent transfer*. It is worth noting that the timescale of energy transport does not exceed that of J^{-1} whenever $\tau^{\text{rxn}} \ll J^{-1}$ or $J^{-1} \ll \tau^{\text{rxn}}$. Obviously, there exist regimes of EET where the two coupling magnitudes and/or the two timescales compete against one another, i.e. $\lambda \sim J$ and/or $\tau^{\text{rxn}} \sim J^{-1}$. These intermediate regimes are typical situations for photosynthetic EET (van Amerongen *et al.*, 2000; Cho *et al.*, 2005), and therefore they are of considerable interest. As a result, perturbative or Markovian master equations fail to provide a proper description of EET in natural systems. We refer readers to review articles (Ishizaki *et al.*, 2010; Ishizaki and Fleming, 2012) for detailed numerical analyses of EET using TC2 and TCL2 equations.

2.8 Second-order cumulant time-non-local equation and its hierarchical representation

As discussed above, it is crucial to consider and describe the dynamics of environmental phonons in a more appropriate fashion in order to elucidate quantum coherence and its interplay with the protein environment in EET. In particular, areas to be commented on include (Ishizaki *et al.*, 2010):

(A) site-dependent reorganization dynamics of the environment
(B) an appropriate description of Gaussian fluctuations in electronic energies of pigments to produce optical lineshapes.

To treat both issues (A) and (B), we cast a spotlight on the fact that the cumulant expansion up to second order is rigorous for phonon operators owing to the Gaussian property in Equation (2.49). Thus, the formally exact expression for the reduced density operator can be derived as

$$\tilde{\rho}_{\text{el}}(t) = \text{T}_+ \prod_{m=1}^{N} \exp\left[\int_0^t d\tau_1 \int_0^{\tau_1} d\tau_2\, \tilde{\mathcal{K}}_m^{(2)}(\tau_1, \tau_2)\right] \tilde{\rho}_{\text{el}}(0), \tag{2.57}$$

with the non-Markovian relaxation kernel,

$$\tilde{\mathcal{K}}_m^{(2)}(t, \tau) = -\frac{1}{\hbar^2} \tilde{S}_m(t)^\times \left[S_m(t-\tau) \tilde{S}_m(\tau)^\times - i\frac{\hbar}{2} \chi_m(t-\tau) \tilde{S}_m(\tau)^\circ \right]. \tag{2.58}$$

Here, the tilde refers to the interaction picture and we denote $O^\times f \equiv Of - fO$ and $O^\circ f \equiv Of + fO$ for any operators O and f. The time evolution corresponding with Equation (2.57) is described by the following equation of motion (Ishizaki and Fleming, 2009b):

$$\frac{d}{dt}\tilde{\rho}_{\text{el}}(t) = \text{T}_+ \sum_{m=1}^{N} \int_0^t d\tau\, \tilde{\mathcal{K}}_m^{(2)}(t, \tau) \tilde{\rho}_{\text{el}}(t). \tag{2.59}$$

Although Equation (2.59) is derived in a non-perturbative fashion, its expression shows close resemblance to those of the second-order perturbative quantum master equations such as the TC2 and TCL2 equations. Indeed, Equation (2.59) as well as the TC2 and TCL2 equations reduce to the Redfield equation in the operator form if the Markov approximation is employed:

$$\frac{d}{dt}\tilde{\rho}_{\mathrm{el}}(t) = \sum_{m=1}^{N} \int_0^{\infty} ds\, \tilde{\mathcal{K}}_m^{(2)}(s, 0)\, \tilde{\rho}_{\mathrm{el}}(t); \tag{2.60}$$

which can be recast into the traditional form of the Redfield equation when the energy–eigenstate representation is employed. This fact implies that it is significant to discuss timescales of the environment or non-Markovian effects in an appropriate manner in addition to system–bath coupling strength.

The first point to notice here is that Equation (2.59) is a time-non-local equation, unlike the TCL2 equation, because the chronological time ordering operator T_+ resequences and mixes the hyper-operators $\tilde{S}_m(t)^{\times}$ and $\tilde{S}_m(t)^{\circ}$ comprised in $\tilde{\mathcal{K}}_m^{(2)}(t, s)$ and $\tilde{\rho}_{\mathrm{el}}(t)$. Secondly, when the excitonic coupling $\hbar J_{mn}$ is vanishingly small, Equation (2.57) leads to an analytically exact expression of absorptive lineshape of a monomer, unlike the TC2 equation and the Redfield equation. These two features are significant for the above features described in (A) and (B) in elucidating the quantum aspects of EET processes in protein environments (Ishizaki *et al.*, 2010).

To perform practical numerical calculations beyond the formally exact structure, we model the relaxation function by an exponential decay form, $\Gamma_m(t) = 2\hbar\lambda_m e^{-\gamma_m t}$, in order to focus on the timescale of fluctuation–dissipation processes induced by environmental phonons. For this modelling, the timescale of the fluctuation–dissipation processes is simply $\tau_m^{\mathrm{rxn}} = \gamma_m^{-1}$, and the spectral density is expressed as an Ohmic form with a Lorentz–Drude regularization (Breuer and Petruccione, 2002), i.e. $\mathcal{J}_m[\omega] = 2\hbar\lambda_m\gamma_m\omega/(\omega^2 + \gamma_m^2)$. Although this spectral density has been successfully employed for the analysis of experimental results (Zhang *et al.*, 1998; Novoderezhkin *et al.*, 2005; Zigmantas *et al.*, 2006; Read *et al.*, 2008), it may produce qualitatively different vibrational sidebands from the experimental results in the zero temperature limit. For this choice of spectral density and classical fluctuation–dissipation relation in Equation (2.55), the non-Markovian relaxation kernel, Equation (2.58), leads to

$$\tilde{\mathcal{K}}_m^{(2)}(t, \tau) = \tilde{\Phi}_m(t) e^{-\gamma_m (t-\tau)} \tilde{\Theta}_m(\tau), \tag{2.61}$$

where we have defined the relaxation operators as

$$\Phi_m \equiv i S_m^{\times}, \quad \Theta_m \equiv i \left(\frac{2\lambda_m}{\beta\hbar} S_m^{\times} - i\lambda_m\gamma_m S_m^{\circ} \right).$$

Owing to the exponential functions in Equation (2.61), the formal expression in Equation (2.59) can be represented as

$$\frac{d}{dt}\sigma(\mathbf{n}, t) = -\left(i\mathcal{L}_{el} + \sum_{m=1}^{N} n_m\gamma_m\right)\sigma(\mathbf{n}, t) + \sum_{m=1}^{N}\left[\Phi_m\sigma(\mathbf{n}_{m+}, t) + n_m\Theta_m\sigma(\mathbf{n}_{m-}, t)\right] \tag{2.62}$$

for sets of non-negative integers, $\mathbf{n} \equiv (n_1, n_2, \ldots, n_N)$. $\mathbf{n}_{m\pm}$ differs from $\mathbf{n}$ only by changing the specified n_m to $n_m \pm 1$, i.e. $\mathbf{n}_{m\pm} \equiv (n_1, \ldots, n_m \pm 1, \ldots, n_N)$. In Equation (2.62), only the element $\sigma(\mathbf{0}, t)$ is identical to the reduced density operator $\rho_{\mathrm{el}}(t)$, while the others $\sigma(\mathbf{n} \neq \mathbf{0}, t)$ are auxiliary operators defined as (Takagahara *et al.*, 1977; Tanimura and Kubo, 1989)

$$\tilde{\sigma}(\mathbf{n}, t) \equiv \mathrm{T}_{+}\prod_{m=1}^{N}\left[\int_0^t d\tau\, e^{-\gamma_m(t-\tau)}\tilde{\Theta}_m(\tau)\right]^{n_m}\tilde{\rho}_{\mathrm{el}}(t). \tag{2.63}$$

Equation (2.62) is a multidimensional extension of the hierarchical equation of motion derived in Tanimura and Kubo (1989) and Tanimura (2006), with the use of the path integral influence functional formalism (Feynman and Vernon, 1963; Caldeira and Leggett, 1983). The hierarchically coupled equations, Equation (2.62), continue to infinity. However, the numerical calculations can converge at a finite depth of hierarchy for a finite timescale of phonon dynamics. Note that quantum correction terms need to be included into Equation (2.62) if the quantum fluctuation–dissipation theorem, Equation (2.53), is to be applied (Ishizaki and Tanimura, 2005; Tanimura, 2006; Ishizaki and Fleming, 2009b). For more detailed numerical analysis for photosynthetic EET processes, we refer readers to Ishizaki *et al.* (2010) and Ishizaki and Fleming (2012).

2.9 A post-perturbative time convolution QME

In Section 2.5, we learned that the TC2 non-Markovian master equation can be obtained in the regime of a weak system–bath coupling. Here we discuss how to avoid the weak coupling assumption and to derive the TC2 Equation (2.32) in a near Markovian regime. We refer readers to Shabani *et al.* (2012) for a detailed derivation.

Consider a pigment–protein system model with Hamiltonian (2.46). The exact system dynamics in interaction picture is given by the Liouville equation,

$$\frac{\partial\tilde{\rho}_{\mathrm{el}}(t)}{\partial t} = -i\langle[\tilde{H}_{\mathrm{el-nuc}}, \tilde{\rho}_{\mathrm{el-nuc}}(t)]\rangle_{\mathrm{nuc}} = -i\langle\tilde{\mathcal{L}}_{\mathrm{el-nuc}}\tilde{\rho}_{\mathrm{el-nuc}}(t)\rangle_{\mathrm{nuc}}, \tag{2.64}$$

where $\rho_{\text{el-nuc}}$ denotes the system–bath state. The time evolved density operator of the system can be expressed by the propagator,

$$\tilde{\rho}_{\text{el}}(t) = \left\langle T_+ \exp\left[-i\int_0^t \tilde{\mathcal{L}}_{\text{el-nuc}}(\tau)d\tau\right]\right\rangle_{\text{nuc}} \rho_{\text{el}}(0). \tag{2.65}$$

Expanding the above time-ordered exponential function, we arrive at the Dyson–Cumulant expansion for time evolution of the density operator (Breuer and Petruccione, 2002) (from here on the subscripts el–nuc and nuc are dropped for convenience),

$$\frac{\partial}{\partial t}\tilde{\rho}_{\text{el}}(t) = \Big[-\int_0^t dt_1 \langle\tilde{\mathcal{L}}(t)\tilde{\mathcal{L}}(t_1)\rangle + \int_0^t dt_1 \int_0^{t_1} dt_2 \int_0^{t_2} dt_3 \langle\tilde{\mathcal{L}}(t)\tilde{\mathcal{L}}(t_1)\tilde{\mathcal{L}}(t_2)\tilde{\mathcal{L}}(t_3)\rangle + \cdots\Big]\rho_{\text{el}}(0). \tag{2.66}$$

The n-time correlation superoperator has the following form for $t_1 > t_2 \cdots > t_n$:

$$\langle\tilde{\mathcal{L}}(t_1)\ldots\tilde{\mathcal{L}}(t_n)\rangle\rho(0)_{\text{el}} = \sum_{j_1..j_n}\sum_{i_1..i_n}(-1)^{n-k}\langle\tilde{V}_{j_{k+1}}(t_{i_{k+1}})\cdots\tilde{V}_{j_n}(t_{i_n})\tilde{V}_{j_1}(t_{i_1})\cdots\tilde{V}_{j_k}(t_{i_k})\rangle \times \tilde{S}_{j_1}(t_{i_1})\cdots\tilde{S}_{j_k}(t_{i_k})\rho(0)_{\text{el}}\tilde{S}_{j_{k+1}}(t_{i_{k+1}})\cdots\tilde{S}_{j_n}(t_{i_n}), \tag{2.67}$$

with site indices $\{j_1, \ldots, j_n\}$ and the second summation is over all indices $\{i_1, \ldots, i_n\} \in \{1, \ldots, n\}$ such that $t_{i_1} > \cdots > t_{i_k}$ and $t_{i_{k+1}} < \cdots < t_{i_n}$.

Next consider $\mathcal{T} = I_+$ in the Gaussian relation (2.49), where I_+ is an operator ordering j indices:

$$\langle\tilde{V}_{j_1}(t_{i_1})\cdots\tilde{V}_{j_{2n}}(t_{i_{2n}})\rangle = \sum_{\substack{all\\pairs}}\prod_{l,k}\langle I_+\tilde{V}_{j_l}(t_{i_l})\tilde{V}_{j_k}(t_{i_k})\rangle. \tag{2.68}$$

Relation (2.68) is a generalized Wick's theorem in the form of Wightman functions (van Vliet, 2008).

In the limit of fast decaying correlations, we keep the leading term $\langle\tilde{V}_j(t_1)\tilde{V}_{j'}(t_2)\rangle$ in the above expansion (2.68):

$$\langle\tilde{V}_{j_1}(t_{i_1})\ldots\tilde{V}_{j_{2n}}(t_{i_{2n}})\rangle \approx \langle\tilde{V}_j(t_1)\tilde{V}_{j'}(t_2)\rangle\langle I_+ \text{ rest of operators}\rangle, \tag{2.69}$$

where index $j(j')$ corresponds to time $t_1(t_2)$. Now we can factor out $\int_0^t dt_1$ $\langle\tilde{\mathcal{L}}(t)\tilde{\mathcal{L}}(t_1)\rangle$ from Equation (2.66) and obtain a QME in TC2 form:

$$\begin{aligned}\frac{\partial}{\partial t}\tilde{\rho}_{\text{el}}(t) &= -\int_0^t dt_1\langle\tilde{\mathcal{L}}(t)\tilde{\mathcal{L}}(t_1)\rangle\left[I - \int_0^{t_1} dt_2\int_0^{t_2} dt_3\langle\tilde{\mathcal{L}}(t_2)\tilde{\mathcal{L}}(t_3)\rangle + \cdots\right]\rho_{\text{el}}(0)\\ &= \sum_m \int_0^t d\tau\tilde{\mathcal{K}}_m^{(2)}(t,\tau)\tilde{\rho}_{\text{el}}(\tau).\end{aligned} \tag{2.70}$$

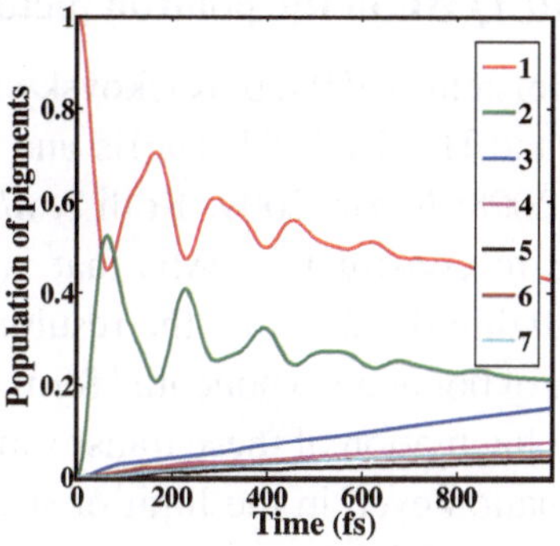

Figure 2.2 Room temperature dynamics of the FMO complex for initial excitation at pigment 1. In agreement with the experimental results (Panitchayangkoon *et al.*, 2010) simulation of the TC2 equation predicts oscillation in populations lasting for a few hundred femtoseconds.

The above derivation provides new insights into the conditions for a TC2 equation to be applicable beyond the conventional weak system–bath coupling limit. In addition, the inherent time convolution form can be more appropriate in capturing certain types of non-Markovian effects. Cao (1997) and Shabani *et al.* (2012) provide more detailed and in-depth analysis of this issue. Thus, while the genuine capability and limitations of the TC2 equation remain to be understood, recent computational studies by Fenna–Matthews–Olson (FMO) (Wu *et al.*, 2010; Shabani *et al.*, 2012; Mohseni *et al.*, 2014) suggest that the TC2 equation can serve as a highly efficient and qualitatively reliable method for addressing key issues in quantum biology. Indeed, a heuristic error analysis employed in these works demonstrates that a TC2 equation may even be quantitatively reliable in the intermediate regime depending on the quantities being calculated (Shabani *et al.*, 2012). An explicit example addressing these points is provided below.

A FMO unit consists of seven pigments on a protein scaffold that acts as a wire to transfer energy from antenna to reaction centre in some bacteria. To model the protein environment of the FMO, we choose overdamped Brownian oscillator model $\mathcal{J}[\omega] = 2\hbar\lambda\gamma\omega/(\omega^2 + \gamma^2)$ with parameter values $\lambda = 50\ \mathrm{cm}^{-1}$, $\gamma = 35\ \mathrm{cm}^{-1}$ at temperature $T = 298$ K. We use the FMO Hamiltonian given in Cho *et al.* (2005). Figure 2.2 shows the time evolution of electronic state populations for the system initially populated in pigment 1. The simulation illustrated in Figure 2.2 predicts some oscillations lasting for a few hundred femtoseconds, in agreement with the experimental results at room temperature (Panitchayangkoon *et al.*, 2010). See Chapter 7 for the application of TC2 to exploring environment-assisted quantum transport phenomena.

2.10 QME in the polaron picture

Polaron transformation (Holstein, 1959a, b; Rackovsky and Silbey, 1973; Jackson and Silbey, 1983; Silbey and Harris, 1984; Harris and Silbey, 1985; Jang *et al.*, 2008; Cheng and Silbey, 2008; Nazir, 2009; Kolli *et al.*, 2011; McCutcheon and Nazir, 2011) can be used to construct a QME that is applicable beyond weak system–bath coupling, utilizing the fact that the resulting system–bath couplings after the transformation are of bounded exponential form. The second-order approximation with respect to the fluctuation of these transformed system–bath couplings serves as a good approximation even in the limit of strong system–bath coupling of the original Hamiltonian.

Consider the system Hamiltonian

$$H_S = \sum_{m=1}^{N} \Omega_m |m\rangle\langle m| + \sum_{m\neq m'}^{N} J_{mm'} |m\rangle\langle m'|, \tag{2.71}$$

where in the context of energy transfer in photosynthetic systems, $|m\rangle$ represents the state where only the mth chromophore is excited and Ω_m is its excitation energy. $J_{mm'}$ is the electronic coupling between states $|m\rangle$ and $|m'\rangle$. For convenience we divide it into population and coherence terms, $H_S = H_S^p + H_S^c$, where $H_S^p = \sum_{m=1}^{N} \Omega_m |m\rangle\langle m|$ and $H_S^c = \sum_{m\neq m'}^{N} J_{mm'}|m\rangle\langle m'|$. Suppose chromophore m is coupled to the nth bath oscillator with strength $g_{\xi,m}$, then $S_m = |m\rangle\langle m|$ in Equation (2.30). The generating operator of the polaron transformation is the following site-dependent momentum operator (except for a factor of i):

$$G = \sum_{m=1}^{N} \sum_{\xi} g_{\xi,m} (b_\xi^\dagger - b_\xi) |m\rangle\langle m|. \tag{2.72}$$

Then, application of the unitary operator e^{-G} and its Hermitian conjugate to H leads to the following polaron transformed Hamiltonian:

$$\tilde{H} = e^G H e^{-G} = \tilde{H}_S^p + \tilde{H}_S^c + H_B. \tag{2.73}$$

In the above expression, $H_B = \sum_\xi \hbar\omega_\xi (b_\xi^\dagger b_\xi + \frac{1}{2})$, $\tilde{H}_S^p = \sum_{m=1}^{N} \tilde{\Omega}_m |m\rangle\langle m|$ with $\tilde{\Omega}_m = \Omega_m - \sum_\xi \hbar\omega_\xi g_{\xi,m}^2$, and $\tilde{H}_S^c = \sum_{m\neq m'}^{N} J_{mm'} \theta_m^\dagger \theta_{m'} |m\rangle\langle m'|$ with $\theta_m = \exp(-\sum_\xi g_{\xi,m}(b_\xi^\dagger - b_\xi))$. Accordingly, the total density operator in the polaron picture becomes $\tilde{\rho}_{SB}(t) = e^G \rho_{SB}(t) e^{-G}$. The time evolution of this is governed by the following transformed Liouville equation:

$$\frac{d}{dt}\tilde{\rho}_{SB}(t) = -i\tilde{\mathcal{L}}\tilde{\rho}_{SB}(t) = -i\big(\tilde{\mathcal{L}}_S^p + \tilde{\mathcal{L}}_S^c + \mathcal{L}_B\big)\tilde{\rho}_{SB}(t), \tag{2.74}$$

where $\tilde{\mathcal{L}}_s^p$, $\tilde{\mathcal{L}}_s^c$ and $\mathcal{L}_B$ are quantum Liouville operators corresponding to $\tilde{H}_S^p$, $\tilde{H}_S^c$, and H_B.

The transformed Hamiltonian $\tilde{H}$ of Equation (2.73) can be divided into the zeroth and the first-order terms as follows: $\tilde{H} = \tilde{H}_0 + \tilde{H}_1$, where

$$\tilde{H}_0 = \tilde{H}_S^p + \langle \tilde{H}_S^c \rangle_B + H_B \equiv \tilde{H}_{0,s} + H_B, \tag{2.75}$$

$$\tilde{H}_1 = \tilde{H}_S^c - \langle \tilde{H}_S^c \rangle_B, \tag{2.76}$$

with $\langle \tilde{H}_s^c \rangle_B = \mathrm{Tr}_B[\tilde{H}_s^c \rho_B(0)]$. This term represents the average effect of the system–bath interaction and can be calculated explicitly. In Equation (2.75), $\tilde{H}_{0,s} = \tilde{H}_s^p + \langle \tilde{H}_s^c \rangle_B$ represents the system part of the zeroth order Hamiltonian. An explicit expression for this can be found as follows.

$$\tilde{H}_{0,s} = \sum_{m=1}^{N} \tilde{\Omega}_m |m\rangle\langle m| + \sum_{m\neq m'}^{N} J_{mm'} w_{mm'} |m\rangle\langle m'|, \tag{2.77}$$

where

$$w_{mm'} = \langle \theta_m^\dagger \theta_{m'} \rangle = \langle \theta_{m'}^\dagger \theta_m \rangle = e^{-\sum_\xi \coth(\frac{\beta\hbar\omega_\xi}{2})\delta g_{\xi,mm'}^2/2}, \tag{2.78}$$

with $\delta g_{\xi,mm'} = g_{\xi,m} - g_{\xi,m'}$. The first-order term $\tilde{H}_1$, defined by Equation (2.76), can also be calculated and has the following expression,

$$\tilde{H}_1 = \sum_{l\neq m'}^{N} J_{mm'} \tilde{B}_{mm'} |m\rangle\langle m'|, \tag{2.79}$$

where

$$\tilde{B}_{mm'} = \theta_m^\dagger \theta_{m'} - w_{mm'} = e^{\sum_\xi \delta g_{\xi,mm'}(b_\xi^\dagger - b_\xi)} - w_{mm'}. \tag{2.80}$$

Now let us consider the dynamics in the interaction picture of $\tilde{H}_0$ (not H_0). First, the interaction picture representation of $\tilde{H}_1$ becomes

$$\tilde{H}_{1,I}(t) \equiv e^{i\tilde{H}_0 t/\hbar} \tilde{H}_1 e^{-i\tilde{H}_0 t/\hbar} = \sum_{m\neq m'}^{N} J_{mm'} \tilde{B}_{mm'}(t) \mathcal{T}_{mm'}(t), \tag{2.81}$$

where

$$\tilde{B}_{mm'}(t) = e^{iH_B t/\hbar} \tilde{B}_{mm'} e^{-iH_B t/\hbar}$$
$$\mathcal{T}_{mm'}(t) = e^{i\tilde{H}_{0,s} t/\hbar} |m\rangle\langle m'| e^{-i\tilde{H}_{0,s} t/\hbar}. \tag{2.82}$$

Then, the interaction picture representation of the polaron transformed total density operator, $\tilde{\rho}_I(t) = e^{i\tilde{\mathcal{L}}_0 t} \tilde{\rho}_{SB}(t)$, evolves according to the following quantum Liouville equation:

$$\frac{d}{dt}\tilde{\rho}_I(t) = -\tilde{\mathcal{L}}_{1,I}(t)\tilde{\rho}_I(t) = -\frac{i}{\hbar}[\tilde{H}_{1,I}(t), \tilde{\rho}_I(t)], \tag{2.83}$$

where the second equality serves as the definition of $\tilde{\mathcal{L}}_{1,I}(t)$.

As in the case of Equation (2.18), we can define the reduced system density operator as follows:

$$\tilde{\rho}_{S,I}(t) \equiv \mathrm{Tr}_B[\tilde{\rho}_I(t)]. \tag{2.84}$$

Then, employing the same projection operator $\mathcal{P}$ and $\mathcal{Q}$, as defined by Equations (2.19) and (2.20), we can obtain the following second-order time-non-local equation:

$$\frac{d}{dt}\tilde{\rho}_{S,I}(t) = -i\mathrm{Tr}_B\big[\tilde{\mathcal{L}}_{1,I}(t)\mathcal{Q}\tilde{\rho}_{SB}(0)\big] - \int_0^t d\tau \mathrm{Tr}_B\big[\tilde{\mathcal{L}}_{1,I}(t)\tilde{\mathcal{L}}_{1,I}(\tau)\mathcal{Q}\tilde{\rho}_{SB}(0)\big]$$
$$- \int_0^t d\tau \mathrm{Tr}_B\big[\tilde{\mathcal{L}}_{1,I}(t)\tilde{\mathcal{L}}_{1,I}(\tau)\rho_B(0)\big]\tilde{\rho}_{S,I}(\tau), \tag{2.85}$$

where we use $\mathrm{Tr}_B[\tilde{\mathcal{L}}_{1,I}(t)\rho_B(0)] = 0$ and $\tilde{\rho}_I(0) = \tilde{\rho}_{SB}(0)$.

In Equation (2.85), we can replace $\tilde{\rho}_{S,I}(\tau)$ with $\tilde{\rho}_{S,I}(t)$ without affecting the accuracy up to the second order of $\tilde{\mathcal{L}}_{1,I}(t)$. The resulting QME has the following form (Jang, 2009, 2011):

$$\frac{d}{dt}\tilde{\rho}_{S,I}(t) = -\tilde{\mathcal{R}}(t)\tilde{\rho}_{S,I}(t) + \mathcal{I}^{(1)}(t) + \mathcal{I}^{(2)}(t), \tag{2.86}$$

where

$$\tilde{\mathcal{R}}(t) = \int_0^t d\tau \mathrm{Tr}_B\big[\tilde{\mathcal{L}}_{1,I}(t)\tilde{\mathcal{L}}_{1,I}(\tau)\rho_B(0)\big], \tag{2.87}$$

$$\mathcal{I}^{(1)}(t) = -i\mathrm{Tr}_B\big[\tilde{\mathcal{L}}_{1,I}(t)\mathcal{Q}\tilde{\rho}_{SB}(0)\big], \tag{2.88}$$

$$\mathcal{I}^{(2)}(t) = -\int_0^t d\tau \mathrm{Tr}_B\big[\tilde{\mathcal{L}}_{1,I}(t)\tilde{\mathcal{L}}_{1,I}(\tau)\mathcal{Q}\tilde{\rho}_{SB}(0)\big]. \tag{2.89}$$

In the above expressions, $\mathcal{I}^{(1)}(t)$ and $\mathcal{I}^{(2)}(t)$ are the first-order and the second-order inhomogeneous terms, respectively. Each term involves $\mathcal{Q}\tilde{\rho}_{SB}(0)$, which is the unprojected portion of the initial transformed density operator given by

$$\mathcal{Q}\tilde{\rho}_{SB}(0) = e^G\rho_{SB}(0)e^{-G} - \mathrm{Tr}_B[e^G\rho_{SB}(0)e^{-G}] \otimes \rho_B(0). \tag{2.90}$$

The expression for $\tilde{\mathcal{R}}(t)\tilde{\rho}_{S,I}(t)$ in Equation (2.86) can be found by inserting Equations (2.81)–(2.82) into the Liouville operator in Equation (2.83) and making a further calculation. It can be expressed as

$$\tilde{\mathcal{R}}(t)\tilde{\rho}_{S,I}(t)$$
$$= \frac{1}{\hbar^2}\sum_{l\neq l'}^{N}\sum_{m\neq m'}^{N} J_{ll'}J_{mm'}\int_0^t d\tau \langle \tilde{B}_{ll'}(t)\tilde{B}_{mm'}(\tau)\rangle \left[\mathcal{T}_{ll'}(t), \mathcal{T}_{mm'}(\tau)\tilde{\rho}_{S,I}(t)\right] + \text{H.C.}, \tag{2.91}$$

where $\langle \tilde{B}_{ll'}(t)\tilde{B}_{mm'}(\tau)\rangle$ is the time correlation function of the fluctuation of the polaronic bath response $\tilde{B}_{mm'}(t)$, which is defined by Equations (2.80) and (2.82). Employing these definitions, the correlation of $B_{mm'}(t)$ can be calculated and has the following expression:

$$\langle \tilde{B}_{ll'}(t)\tilde{B}_{mm'}(\tau)\rangle = w_{ll'}w_{mm'}\left(e^{-\mathcal{K}_{ll',mm'}(t-\tau)} - 1\right), \tag{2.92}$$

where $w_{ll'}$ or $w_{mm'}$ have been defined by Equation (2.78), and

$$\mathcal{K}_{ll',mm'}(t) = \sum_{\xi} \delta g_{\xi,ll'}\delta g_{\xi,mm'}\left(\coth\left(\frac{\beta\hbar\omega_\xi}{2}\right)\cos(\omega_\xi t) - i\,\sin(\omega_\xi t)\right). \tag{2.93}$$

From the above definition, it is clear that $\mathcal{K}_{ll',mm'}(t) = -\mathcal{K}_{ll',m'm}(t) = -\mathcal{K}_{l'l,mm'}(t) = \mathcal{K}_{l'l,m'm}(t)$.

For the calculation of inhomogeneous terms in Equations (2.88) and (2.89), the initial condition of the exciton state needs to be specified first. The assumed general initial exciton state is $|I\rangle = \sum_{p=1}^{N} I_p|p\rangle$, where the I_ps are complex numbers that can be arbitrary except for the normalization condition $\sum_p |I_p|^2 = 1$. On the other hand, the bath is assumed to be in canonical equilibrium for the ground electronic states of chromophores. Thus, the total initial density operator is given by

$$\rho_{SB}(0) = |I\rangle\langle I| \otimes \rho_B(0) = \sum_{p,p'=1}^{N} I_p I_{p'}^{*}|p\rangle\langle p'| \otimes \rho_B(0). \tag{2.94}$$

Physically, the above initial condition corresponds to sudden creation of a single exciton in a system originally prepared in the ground electronic state in thermal equilibrium with the environments. After some algebra we find,

$$\mathcal{I}^{(1)}(t) = -\frac{i}{\hbar}\sum_{m\neq m'}^{N}\sum_{p,p'=1}^{N} J_{mm'}w_{mm'}\left[\mathcal{T}_{mm'}(t), \tilde{\sigma}^{pp'}(0)\right]\left\{e^{-\mathcal{K}_{mm',pp'}(t)}f_{p',mm'}(t) - 1\right\}, \tag{2.95}$$

where $\mathcal{K}_{mm',pp'}(t)$ has been defined by Equation (2.93) and

$$f_{p',mm'}(t) = e^{2i\sum_\xi g_{\xi,p'}\delta g_{\xi,mm'}\sin(\omega_\xi t)}. \tag{2.96}$$

The second-order inhomogeneous term defined by Equation (2.89) can be expressed as

$$\mathcal{I}^{(2)}(t) = -\sum_{p,p'=1}^{N}\sum_{l\neq l'}^{N}\sum_{m\neq m'}^{N}\frac{J_{ll'}J_{mm'}}{\hbar^2}\int_0^t d\tau \mathcal{F}_{ll',mm'}^{pp'}(t,\tau)\big[\mathcal{T}_{ll'}(t), \mathcal{T}_{mm'}(\tau)\tilde{\sigma}^{pp'}(0)\big], + \text{H.C.}, \tag{2.97}$$

where

$$\mathcal{F}_{ll',mm'}^{pp'}(t,\tau) = w_{ll'}w_{mm'}\big\{e^{-\mathcal{K}_{ll',mm'}(t-\tau)}\big(e^{-\mathcal{K}_{ll',pp'}(t)-\mathcal{K}_{mm',pp'}(\tau)}f_{p',ll'}(t)f_{p',mm'}(\tau)-1\big) - e^{-\mathcal{K}_{ll',pp'}(t)}f_{p',ll'}(t) - e^{-\mathcal{K}_{mm',pp'}(\tau)}f_{p',mm'}(\tau)+2\big\}. \tag{2.98}$$

This completes the second-order time local QME in the polaron picture, which is applicable in the regime that the Hamiltonian $\tilde{H}_{1,I}(t)$ (2.81) is perturbative.

As an illustration for the application of the polaron QME, results of the calculation are presented for a model of a donor–bridge–acceptor chromophore system, denoted as 1, 2 and 3. The population and coherence terms of the system Hamiltonian and the system–bath interaction Hamiltonians are as follows:

$$H_s^p = E_1|1\rangle\langle 1| + E_2|2\rangle\langle 2| + E_3|3\rangle\langle 3|, \tag{2.99}$$

$$H_s^c = J_{12}(|1\rangle\langle 2| + |2\rangle\langle 1|) + J_{23}(|2\rangle\langle 3| + |3\rangle\langle 2|), \tag{2.100}$$

$$H_{sb} = \sum_{l=1}^{3}\sum_n \hbar\omega_n\big(b_n + b_n^\dagger\big)g_{n,l}|l\rangle\langle l|. \tag{2.101}$$

Although it is not necessary, for simplicity, it is assumed that each chromophore is coupled to an independent set of bath modes and there are no common modes. It is also assumed that all the spectral densities are of super-Ohmic form:

$$\mathcal{J}_{ll}(\omega) \equiv \sum_n g_{n,l}^2\delta(\omega-\omega_n)\omega_n^2 = \frac{\eta}{3!}\frac{\omega^3}{\gamma_l^2}e^{-\omega/\gamma_l}, \quad l = 1, 2, 3. \tag{2.102}$$

The following sets of parameters were used in all of the calculations: $T = 300$ K, $J_{12} = J_{23} = 100\ \text{cm}^{-1}$, $J_{23} = 50\ \text{cm}^{-1}$, $E_2 - E_1 = E_1 - E_3 = 100\ \text{cm}^{-1}$, $\gamma_l/2\pi = 200\ \text{cm}^{-1}$. Three different choices of $\eta = 0.2$, 1 and 5 were considered and the results are shown in Figure 2.3.

For each choice of parameter, Equation (2.85) was solved numerically in the exciton basis (Jang, 2009), including all inhomogeneous terms. In order to assess the effect of quantum coherence, exciton populations based on the following Pauli master equation were also calculated:

$$\frac{dP_l(t)}{dt} = \sum_{m\neq l}\left\{k_{m\to l}^F(t)P_m(t) - k_{l\to m}^F(t)P_l(t)\right\}, \tag{2.103}$$

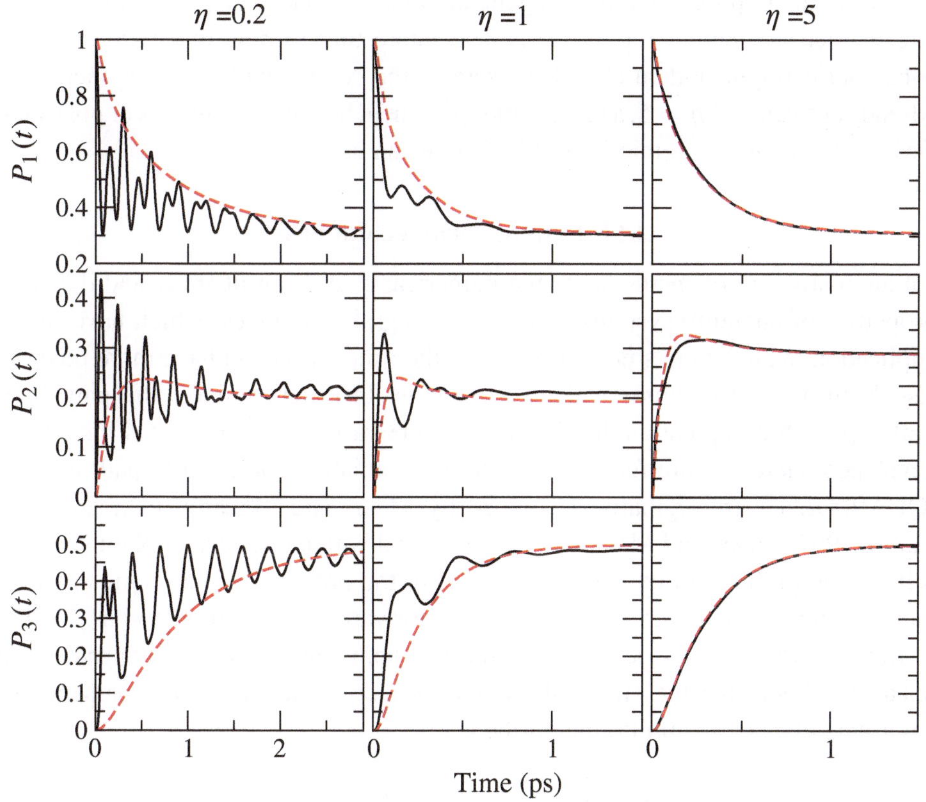

Figure 2.3 Time-dependent populations of excitons for different values of system–bath coupling. The sold black lines are the results for the polaronic QME including inhomogeneous terms and the grey dashed lines are the results for FRET, Equation (2.104).

with the rate kernel given by the following non-Markovian version of Förster resonance energy transfer theory (FRET) (Jang *et al.*, 2002b):

$$k^F_{l\to m}(t) = \frac{2(J_{lm}w_{lm})^2}{\hbar^2}\text{Re}\left[\int_0^t d\tau e^{i(\tilde{E}_l-\tilde{E}_m)\tau/\hbar}\left(e^{\mathcal{K}_{lm,lm}(\tau)}-1\right)\right]. \quad (2.104)$$

The FRET theory is reviewed in the next chapter. In the above rate expression, note that the divergent component that appears due to the neglect of the lifetime of the excited state has been subtracted. This regularization procedure is needed for super-Ohmic spectral density and amounts to including the average bath effects in the polaron picture within the system Hamiltonian.

As can be seen from Figure 2.3, the quantum coherence makes a significant contribution in the early phase of population dynamics, especially for weak and moderate system–bath couplings. It is also important to note that the assumption of sequential hopping dynamics, in general, underestimates the population transfer rate in such a coherent regime. It is also interesting to note that such coherent behaviour is fragile and quickly disappears as the system–bath coupling increases. Already at a value of $\eta = 5$, a reasonable agreement between the results of polaronic QME and hopping dynamics with FRET rate can be seen.

2.11 Path integral techniques

An alternative set of methods for the numerical evaluation of static and dynamic properties of quantum systems is the path integral techniques, which have found application in analytical studies right from their inception and have more recently also been applied to numerical investigations. In order to make the path integral approach computationally feasible a wide variety of approaches have been developed. Here, we are focusing on the quasi-adiabatic propagator path integral (QUAPI) that was originally developed by Makri and Makarov (Makarov and Makri, 1994; Makri and Makarov, 1995a, b) and which has been applied successfully to the problem of excitation transport in the presence of harmonic environments at finite temperatures (Nalbach and Thorwart, 2010; Thorwart *et al.*, 2009).

In the following we briefly outline the key points in the QUAPI approach that make it suitable for the study of excitation energy transfer in the presence of environments. For a Hamiltonian of the form,

$$H = W(x_1, \ldots, x_N) + \sum_{j=1}^{N} \frac{p_j^2}{2m_j}, \tag{2.105}$$

which describes an interacting quantum system of N degrees of freedom, the standard approach to the path integral splits it into a purely kinetic part and a potential part. Expanding the time evolution operator in small time-steps $\Delta t = t/n$, inserting the identity in the position representation and applying a second-order Trotter expansion of the time evolution operator one finds,

$$\langle x_{fin}|e^{-iHt/\hbar}|x_{ini}\rangle \cong \prod_{j=1}^{n} \sqrt{\frac{m_j n}{2\pi i \hbar t}} \int d^N x_1 \ldots \int d^N x_n$$
$$\exp\left(-\frac{i}{\hbar}\sum_{k=1}^{N}\left[\frac{t(W(x_k)+W(x_{k-1}))}{2n} - \sum_{j=1}^{N}\frac{m_j n}{2t}(x_{jk}-x_{jk-1})^2\right]\right), \tag{2.106}$$

where $\langle x_{fin}| \equiv \langle x_n|$ and $|x_{ini}\rangle \equiv |x_1\rangle$. This expression while being exact in the limit $\Delta t \to 0$ is of little use for numerical computations. Key problems are the fact that the integrand is oscillatory and not local. Hence the computation requires small time-steps and careful sampling of the integrand, requirements that generally lead to an effort that grows exponentially in the total integration time t. Furthermore, even for exactly solvable models such as the harmonic oscillator, evaluation of the integrals does not become simpler. The reason for these shortcomings lies in the non-optimal decomposition of the Hamiltonian into purely kinetic and purely potential parts. One innovation of QUAPI is the realization that there are many decompositions of the Hamiltonian that are all equivalent in the limit $\Delta t \to 0$, but which may lead to formulations of the problem that are computationally inequivalent. Indeed, one would normally aim to decompose the Hamiltonian into one part H_0 that includes the kinetic part of the Hamiltonian and as much as possible of the potential energy terms, while remaining either exactly solvable or at least efficiently computable. If the potential part that one includes in H_0 is confining the system to a finite spatial region, then one finds that in

$$\langle x_k|e^{-iH\Delta t/\hbar}|x_{k-1}\rangle \cong \langle x_k|e^{-iH_0\Delta t/\hbar}|x_{k-1}\rangle e^{-\frac{i\Delta t}{2\hbar}[(H-H_0)(x_k)-(H-H_0)(x_{k-1})]},$$

the first factor will now be local and decaying quite rapidly, thus greatly assisting the computation.

Further progress can be made by considering more specific problems. In the context of the exploration of quantum effects in biology we are particularly interested in the effect that an environment may have on the coherence properties of a system. Such environments are successfully described by sets of harmonic oscillators linearly coupled to the central system, leading to a Hamiltonian of the form,

$$H = \frac{p_s^2}{2m} + W(s) + \sum_{j=1}^{N}\left(\frac{p_j^2}{2m_j} + \frac{1}{2}m_j\omega_j^2\left(q_j - \frac{c_j s}{m_j\omega_j^2}\right)^2\right). \tag{2.107}$$

We discussed this model before in the context of EET: Equation (2.46). In this setting it is clearly natural to choose a decomposition of the total Hamiltonian into

$$H_0 = \frac{p_s^2}{2m} + W(s) \tag{2.108}$$

and

$$H_{int} = \sum_{j=1}^{N}\left(\frac{p_j^2}{2m_j} + \frac{1}{2}m_j\omega_j^2\left(q_j - \frac{c_j s}{m_j\omega_j^2}\right)^2\right). \tag{2.109}$$

Proceeding again as in the standard path integral approach, we now obtain a propagator,

$$\langle s_k, q_k | e^{-iH\Delta t/\hbar} | s_{k-1}, q_{k-1}\rangle \cong \langle s_k | e^{-iH_0\Delta t/\hbar} | s_{k-1}\rangle \times \prod_j \langle q_{jk} | e^{-iH_j(s_k)\Delta t/2\hbar} e^{-iH_j(s_{k-1})\Delta t/2\hbar} | q_{jk-1}\rangle.$$

The first factor in the propagator can usually be computed efficiently for small systems while the second part can be written out analytically. If the system and the environment are decoupled, the propagator is in fact exact. For finite coupling, all the computational effort is reserved for determining the perturbations around this exact solution and hence this is far more efficient than the standard approach.

As formulated here, we are still evaluating the full dynamics of system and environment. However, in many cases we are mainly interested in the dynamics of the system itself, i.e. the state of the reduced density matrix of the system alone,

$$\rho_S(r, s, t) = \mathrm{Tr}_B \langle r | e^{-iHt/\hbar} \rho_{SB}(0) e^{iHt/\hbar} | s \rangle, \tag{2.110}$$

where the initial state of system and environment is assumed to factor for simplicity: $\rho_{SB}(0) = \rho_S(0) \otimes \rho_B(0)$.

Now applying the path integral decomposition to both time evolution operators we find,

$$\rho_S(r_n, s_n, t) = \int dr_1 \ldots dr_{n-1} \int ds_1 \ldots ds_{n-1} \; \mathcal{A}(\{r_k\}, \Delta t) \langle r_1 | \rho_s(0) | s_1 \rangle \mathcal{B}(\{s_k\}, \Delta t) I(\{r_k\}, \{s_k\}, \Delta t),$$

where we abbreviate the propagators as,

$$\mathcal{A}(\{r_k\}, \Delta t) = \prod_{k=n}^{2} \langle r_k | e^{-iH_0\Delta t/\hbar} | r_{k-1} \rangle$$

$$\mathcal{B}(\{s_k\}, \Delta t) = \prod_{k=2}^{n} \langle s_{k-1} | e^{-iH_0\Delta t/\hbar} | s_k \rangle$$

and introduce the influence functional,

$$I(\{r_k\}, \{s_k\}, \Delta t) = \mathrm{Tr}_B \left[e^{-iH_{int} r_n/2\hbar} e^{-iH_{int} r_{n-1}/\hbar} \ldots e^{-iH_{int} r_1/2\hbar} \rho_B(0) \times e^{iH_{int} s_1/2\hbar} \ldots e^{iH_{int} s_{n-1}/\hbar} e^{iH_{int} s_n/2\hbar} \right].$$

Because we are considering an environment made up of independent harmonic oscillators, the influence functional takes the general form,

$$I(\{r_k\}, \{s_k\}, \Delta t) = \exp\left[-\frac{1}{\hbar} \sum_{k=1} n \sum_{l=1}^{k} (r_k - s_k)(\eta_{kl} r_l - \eta_{kl}^* s_l) \right], \qquad (2.111)$$

where the coefficients η_{kl} may be computed analytically for any given spectral density. We do not present the lengthy solution here, but note that for a smooth and slowly varying spectral density the influence functional decays quite swiftly in time, which is essential for efficient computation of the path integral. In fact, for any finite temperature, these correlations decay exponentially fast at asymptotic times, thereby setting the associated memory timescale. This timescale increases with decreasing temperature, which makes treatment of the low temperature limit computationally more challenging, even though temperatures not too far below the typical system frequencies are accessible (Thorwart *et al.*, 2000).

The QUAPI approach allows, with modest numerical effort, calculation of the dynamics for intermediate times extending to a few periods of oscillation of the system. Longer integration times are hard to achieve because the computation of multidimensional integrals requires an effort that grows exponentially with integration time. To overcome this limitation we need to make use of the finite correlation time of the influence functional, Equation (2.111). Indeed, if the influence functional decays sufficiently fast, this implies that we may neglect all terms for which $k - l$ differs by more than a value K. QUAPI then defines an object called the reduced density tensor, which lives on this memory time window and establishes an iteration scheme in order to extract the time evolution of this object. Within the memory time window, all correlations are included exactly over the finite memory time $\tau_{memory} = K\delta t$, and can safely be neglected for times beyond τ_{memory}. Then, the memory parameter K has to be increased, until convergence is found. The inclusion of all these tools permits treatment of the system–environment interaction outside of the perturbative regime. As with all approximative schemes that do not provide intrinsic error bounds, it has to be tested by either increasing computational effort until convergence has been reached or by comparison with other methods. Indeed, recently, the quasi-adiabatic propagator path integral method has been compared to the hierarchy methods described in the previous section, leading to matching results for a dimer subjected to a bath with a spectral density of a Brownian harmonic oscillator (Nalbach *et al.*, 2011).

2.12 DMRG based approaches

A third class of approach to the system–environment interaction preserves the full information about the environment state, and is able to treat arbitrary spectral densities and coupling strengths within one framework with known and controllable error. This method, which was developed in Prior *et al.* (2010) and Chin *et al.* (2010a, 2011), maps the spin-boson model exactly onto an effective 1D system, thus permitting deployment of the time-adaptive density matrix renormalization group (t-DMRG) technique (Daley *et al.*, 2004) to integrate the time evolution efficiently. Although the richly structured environments as they appear, for example, in photosynthetic complexes are perhaps the main application area, it should be emphasized that this approach is completely general and can be applied to any system linearly coupled to bosonic and even fermionic environments of arbitrary spectral density. Importantly, this method also provides complete information about the evolving state of the environment and opens the door to detailed studies of system–environment correlations, which in turn give rise to long-lasting coherences, entanglement and other novel effects.

Choosing a compromise between generality and transparency, here, we explain the basics of this method by considering a dimer molecule consisting of two pigments (henceforth referred to as 'sites') each subject to an interaction with a harmonic environment (see the upper part of Figure 2.4).[12] The internal dynamics of the dimer are described by a Hamiltonian,

$$H_S = \frac{\epsilon_1}{2}\sigma_{1z} + \frac{\epsilon_2}{2}\sigma_{2z} + J(\sigma_{1+}\sigma_{2-} + \sigma_{2+}\sigma_{1-}),$$

where $\sigma_{i+}, \sigma_{i-}, \sigma_{iz}$ are standard Pauli creation, annihilation and z matrices for site i of the dimer. The spin-down state $\sigma_{iz}|\downarrow_i\rangle = -|\downarrow_i\rangle$ represents the ground state of the site and the spin-up state represents a single local excitation which can hop between sites with a tunnelling amplitude J. Each site i interacts with its own continuous bath of harmonic oscillators, described by creation and annihilation operators $b_i^\dagger(k)$, $b_i(k)$ respectively. The environments are independent, and hence satisfy the continuum commutation relation $[b_i(k), b_j^\dagger(k')] = \delta_{ij}\delta(k - k')$. The environment Hamiltonian H_B can be written as

$$H_B = \sum_{i=1,2} \int_0^1 \omega_i(k) b_i^\dagger(k) b_i(k) dk, \tag{2.112}$$

while the site–environment interaction H_{SB} takes the form,

$$H_{SB} = \sum_{i=1,2} O_i \int_0^1 g_i(k)(b_i(k) + b_i^\dagger(k)) dk, \tag{2.113}$$

with an arbitrary operator O_i acting on the sites. Equations (2.112) and (2.113) describe a continuous version of the bosonic bath model given in Section 2.5. In

[12] The vast majority of environmental modes in a pigment–protein complex are to a very good approximation harmonic (Hayward and Gö, 1995).

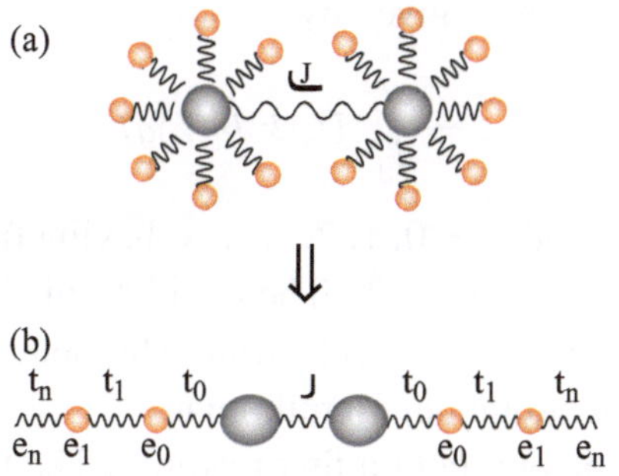

Figure 2.4 (a) The standard representation of the spin-boson model considers central systems (grey spheres) coupled individually to an environment of bosonic modes representing harmonic oscillators. (b) After an exact unitary transformation, solely affecting the bath degrees of freedom, the same number of bosonic modes are now arranged in a linear chain with nearest neighbour interactions only. The latter setting admits straightforward numerical description of the full dynamics with the time-dependent density renormalization group method.

the context of EET in photosynthetic complexes dephasing is the dominant source of noise which is described by $O_i = \frac{(1+\sigma_{iz})}{4}$.

We assume here that both environments have identical dispersions $\omega_i(k) = \omega(k)$ and coupling strengths $g_i(k) = g(k)$, though as will become evident, we can trivially accommodate different coupling structures on each site. While it is not essential, we assume that the spectrum of the environment frequencies is limited by a high-frequency cut-off γ that may be either sharp or realized by some rapidly decaying function, e.g. an exponential.

As such the spin-boson model is exceedingly hard to integrate in time because large amounts of entanglement are building up in the system. In order to make progress we will now need to perform a mathematical transformation that brings the spin-boson Hamiltonian into a form that then permits application of standard methods from condensed matter physics and quantum information theory. This is achieved by a unitary transformation of the configuration depicted in Figure 2.4(a) into the configuration shown in Figure 2.4(b), where the system couples to the first site of a linear chain which exhibits only nearest neighbour interaction. The Hamiltonian for the chain reads

$$H_C = \sum_{i=1,2} \sum_{n=0}^{\infty} \epsilon_{in} a_{in}^\dagger a_{in} + \left(t_{in} a_{in+1}^\dagger a_{in} + h.c.\right),$$

while the system–chain interaction is given by

$$H_{SC} = \sum_{i=1,2} \sqrt{\frac{\eta}{\pi}} \frac{(1+\sigma_{iz})}{4} (a_{i0} + a_{i0}^\dagger),$$

with the new modes of the chain given by

$$a_{in} = \int_0^1 U_n(k) b_i(k) dk,$$

with real functions $U_n(k)$ and $n = 0, 1, 2, \ldots, \infty$ having frequencies ϵ_{jn} and nearest neighbour couplings t_{jn}, $(j = 1, 2)$. The coupling of site j to the first member of its chain is determined by $\eta = \int \mathcal{J}(\omega)\, d\omega$. The determination of $U_n(k)$ then yields the desired nearest neighbour Hamiltonian.

(This paragraph may be omitted on first reading.) The key insight in constructing $U_n(k)$ is to realize that, as columns of a unitary operator, they are mutually orthogonal and may be interpreted as members of a family of orthonormal polynomials. Then we can write $U_n(k) = g(k) N_n^{-1} \pi_n(k)$, where N_n is a normalization constant, and $\pi_n(k)$ is the nth monic polynomial (that is $\pi_n(0) = 1$) of a sequence of orthogonal polynomials with inner product $\int_0^1 g^2(k)\pi_n(k)\pi_m(k) dk = \rho_n^2 \delta_{nm}$. The new modes obey bosonic commutation relation $[a_n, a_m^\dagger] = \delta_{nm}$ and for the natural choice $\omega(k) = \gamma k$ we find $g^2(k) = \pi^{-1}\gamma\mathcal{J}(\omega(k))$. The nearest neighbour chain structure is a direct consequence of using the basis of orthogonal polynomials and a linear dispersion $\omega(k)$. When the new operators are substituted into Equation (2.113), the effective coupling strength of the nth mode of the chain to the site is $\propto \int_0^1 g^2(k)\pi_n(k)\, dk$, which is only non-zero for $n = 0$, due to the orthogonality of $\pi_n(x)$ and the fact that $\pi_0(x) = 1$. Similarly, when the new modes are substituted into Equation (2.112), modes n and m couple to each other with a strength $\propto \int_0^1 g^2(k)\, k\, \pi_n(k)\pi_m(k)\, dk = \int_0^1 g^2(k)\, (\alpha_n \pi_n(k) + \beta_n \pi_{n-1}(k) + \pi_{n+1}(k))\pi_m(k)\, dk$, where we have used the standard three-term recurrence relation obeyed by monic orthogonal polynomials to replace the product $k\pi_n(k)$, which arises from our choice of a linear dispersion. For details see (Chin *et al.*, 2010a).

The transformation given by $U_n(k)$, which can be determined analytically for a wide variety of spectral densities (Chin *et al.*, 2010a), or can otherwise be obtained in a numerically stable and efficient way, acts only on the environment degrees of freedom. As a consequence, the dynamics of the system is unchanged. But the complexity of the simulation has decreased considerably as the structure (linear chain with nearest neighbour interactions) is such that it can now be efficiently integrated numerically, employing the time-dependent density matrix renormalization group algorithm (Daley *et al.*, 2004). The underlying principle of this approach is a clever parametrization of many-body quantum states in the form of so-called matrix product states (MPS) which takes account of the quantum correlations in the 1D chains in an efficient manner (see Figure 2.5 for an intuitive picture).

This approach has several advantages. Firstly, it works for arbitrary and possibly highly structured spectral densities of the environment without additional computational overheads. In particular, it is capable of including narrow features in general

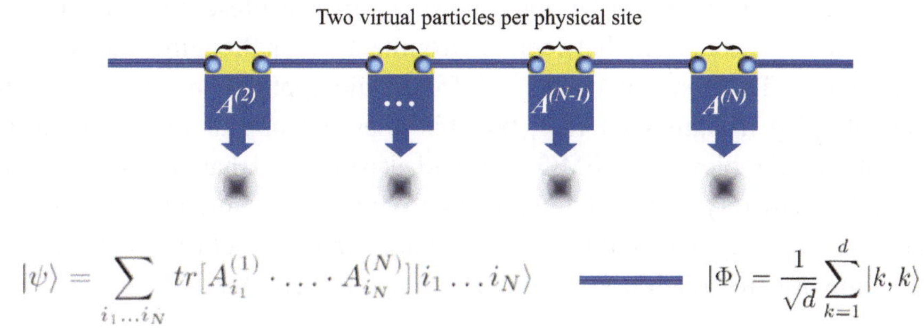

Figure 2.5 Matrix product states are a family of quantum states. For each physical site (blue sphere) we choose two virtual d-dimensional particles (red spheres) where neighbouring virtual particles in different sites are prepared in a maximally entangled state $\sum_{k=1}^{d} |k\rangle\langle k|$. The d^2-dimensional Hilbert-space of two virtual particles of one site is mapped onto the space of the physical particle by a linear map $\mathcal{M}_i$. Quantum states for N sites that are obtained in this way can be written as $\sum_{i_1 \ldots i_N} \mathrm{Tr}[A^{(1)}_{i_1} \cdots A^{(N)}_{i_N}]|i_1 \cdots i_N\rangle$, where the $A^{(k)}_{i_k}$ are $d \times d$ matrices that are obtained from the elements of the linear map $\mathcal{M}_k$. With each time-step the accurate description of the state requires an increasingly large dimension d but the time-dependent density matrix renormalization group approximates in each time-step the evolved state by the closest MPS with dimension fixed d. The error in this approximation can be bounded in each time-step. The computational effort scales linearly with the size of the environment and the simulation time and cubic in d.

backgrounds and spectral densities leading to power-law decay of the environment correlation function (Prior *et al.*, 2010) without sacrificing its efficiency. Secondly, the time-dependent density matrix renormalization group algorithm and hence the transformation approach can be made arbitrarily precise (at the cost of what is usually a polynomial increase in computation time) and, crucially, delivers in each time-step and hence also for the total evolution, an upper bound on the error committed. Hence the simulation delivers rigorous error bars from a single run. Finally, this approach provides the full information about system and environment and their dynamics and opens the possibility of studying system–environment interactions in which the environment is prepared in non-trivial initial states or in which knowledge about the state of the environment should be extracted. The versatility of this approach is demonstrated in Prior *et al.* (2010), by considering a structured environment and showing that both exponential decay and oscillatory behaviour can be treated within the same framework.

While the method presented here has been demonstrated for a dimer in contact with, zero temperature environments (Prior *et al.*, 2010; Chin *et al.*, 2010a, 2011), it is not restricted to this setting. It can be generalized to multi-site systems and finite temperatures at the expense of a moderate increase in computation time

(Chin *et al.*, 2013). It should be noted however, that at present the inclusion of spatial quantum correlations between environments is challenging and restricted to simple cases. This may be a drawback for certain applications in which spatial correlations play a significant role, but this does not appear to be the case in photosynthetic complexes like FMO. It should also be noted that the transformation approach immediately yields systematic approximation techniques (Woods *et al.*, 2014) by cutting short the chains that are obtained in the exact transformation and coupling the last site to a Markovian bath (Chin *et al.*, 2010a).

3

Generalized Förster resonance energy transfer

SEOGJOO JANG, HODA HOSSEIN-NEJAD AND GREGORY D. SCHOLES

3.1 Introduction

Decades of effort in structural biology and ultrafast spectroscopy have elucidated many details of natural photosynthesis (Blankenship, 2002; Hu *et al.*, 2002), and it is now a well-established fact that the initial light-harvesting process that leads up to the collection of energy at reaction centres, where charge transfer reaction occurs, is a process with almost perfect efficiency. The mechanism underlying the energy migration in a photosynthetic system is a fundamentally quantum-mechanical one, known as excitation energy transfer (EET) or resonance energy transfer (RET) (Silbey, 1976; Agranovich and Galanin, 1982; Scholes, 2003; May and Kühn, 2011; Olaya-Castro and Scholes, 2011).

Resonance energy transfer is ubiquitous, and had been observed as sensitized luminescence long before modern quantum-mechanical understanding of molecular systems was established (Agranovich and Galanin, 1982). Normally, when a molecule becomes excited electronically by absorbing a photon, it luminesces by emitting another photon, within about a nanosecond, if it is fluorescence, or much later for phosphorescence. However, when another molecule with similar excitation energy is present within a distance of tens of nanometres, it can swap its excitation with the molecule as follows:

$$D^* + A \rightarrow D + A^*, \tag{3.1}$$

where D^* (D) is the excited (ground) state donor of the energy and A (A^*) is the ground (excited) state acceptor. Thus, the excitation of D sensitizes that of A.

Clear and sensible understanding of the RET process had been beyond the reach of classical mechanics as had any other molecular processes involving matter–radiation interaction. In 1948, Förster (Förster, 1948) developed the first quantum-mechanical theory for RET and derived the following celebrated rate

Quantum Effects in Biology, ed. Masoud Mohseni, Yasser Omar, Gregory S. Engel and Martin B. Plenio. Published by Cambridge University Press.

expression:

$$k_F = \frac{9000(ln10)\kappa^2}{128\pi^5 N_A \tau_D n_r^4 R^6} \left(\int d\tilde{\nu} \frac{f_D(\tilde{\nu})\epsilon_A(\tilde{\nu})}{\tilde{\nu}^4} \right), \tag{3.2}$$

where $f_D(\tilde{\nu})$ is the normalized emission spectrum of the donor, $\epsilon_A(\tilde{\nu})$ is the molar extinction coefficient of the acceptor, R is the distance between the donor and the acceptor, n_r is the refractive index of the medium, N_A is Avogadro's number, τ_D is the spontaneous decay lifetime of the excited D and κ^2 is the orientational factor, which is usually taken to be $2/3$.

Förster's rate expression, Equation (3.2), has made a profound contribution to the fields of chemistry, physics and biology involving luminescence properties (Birks, 1976; Silbey, 1976; Agranovich and Galanin, 1982; Agranovich and Hochstrasser, 1982; Scholes, 2003; Olaya-Castro and Scholes, 2011). It made it possible to understand and describe RET processes (Birks, 1976; Agranovich and Galanin, 1982; Andrews and Demidov, 1999) at a semi-quantitative level long before the advance of modern experimental (Ha *et al.*, 1996; Ranasinghe, M. *et al.*, 2003; Kleiman *et al.*, 2001; Varnavski *et al.*, 2002; Lipman *et al.*, 2003; Scholes, 2003; Gaab and Bardeen, 2004b) and computational techniques (Scholes, 2003; Thompson *et al.*, 2004; Wong *et al.*, 2004; Beljonne *et al.*, 2005; Russo *et al.*, 2007), which were later able to provide more direct confirmation or reliable application of the theory. Equation (3.2) has also been used for the modelling of complex spectroscopic and kinetic data of energy transfer processes (Hsiao *et al.*, 1996; Yang *et al.*, 1998; Nguyen *et al.*, 2000; Kleiman *et al.*, 2001; Gronheid *et al.*, 2002; Serin *et al.*, 2002; Varnavski *et al.*, 2002; Aratani *et al.*, 2003; Fan *et al.*, 2003; Gaab and Bardeen, 2004b; Hindin *et al.*, 2004; Chen *et al.*, 2005; Schlosser and Lochbrunner, 2006), or determination of nanoscale distances through detection of fluorescence signals (Stryer and Haugland, 1967; Haugland *et al.*, 1969; Selvin, 2000; Hillisch *et al.*, 2001; Heyduk, 2002). This latter application, referred to as FRET (fluorescence, or Förster, resonance energy transfer) in the biophysics community, is now well established as a powerful tool for probing both in vivo and in vitro structural information of biological molecules (Ha *et al.*, 1996; Selvin, 2000; Hillisch *et al.*, 2001; Heyduk, 2002; Lipman *et al.*, 2003; Koushik *et al.*, 2006) and synthetic macromolecular systems (Nguyen *et al.*, 2000; Serin *et al.*, 2002; Kim and Osuka, 2004).

While Förster's theory laid down the fundamental basis for understanding the RET process in general, it is a theory based on rather simple assumptions that do not reflect the complex nature of many molecular systems. This is true in particular for photosynthetic light-harvesting complexes (Blankenship, 2002; Hu *et al.*, 2002; Scholes *et al.*, 2011), where tens or hundreds of pigment molecules are embedded in soft protein environments with nearest neighbour distances typically on the order of

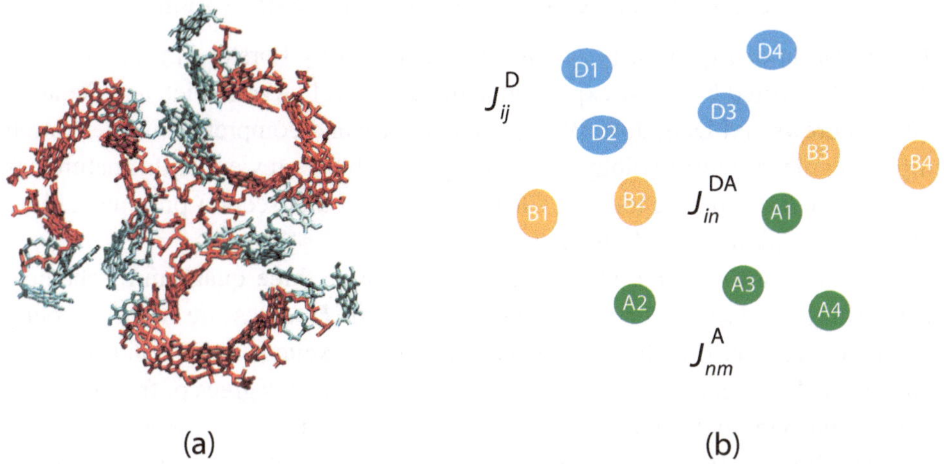

Figure 3.1 (a) Chromophores in the LHCII antenna complex with protein scaffolds removed. (b) Schematic depiction of a prototype aggregate comprised of several donor (D) and acceptor (A) molecules, and a set of bridge molecules (B). Couplings between donors (acceptors) are denoted as J_{ij}^{D} (J_{nm}^{A}). There is also coupling among the bridge molecules, as well as between the bridge and donor/acceptor sets, which is not represented in the diagram.

nanometers. As an example, Figure 3.1(a) shows the chromophores (chlorophyll *a* and chlorophyll *b*) of the light-harvesting complex II (LHCII), an antenna complex of higher plants and green algae. It is obvious that several molecules are clustered together at separations comparable to their size. Figure 3.1(b) is a schematic representation of a generic aggregate where groups of molecules constitute the donor and the acceptor. For these cases, the approximations underlying Equation (3.2) become unreliable in many ways. More often than not, the excitation is coherently shared by a cluster of proximate pigment molecules, and transfer between groups of delocalized exciton states needs to be considered. Furthermore, the soft nature of protein environments may incur the possibility of additional inelastic process where quantum-mechanical exchange between transferring electronic excitation energy and the vibrational energy of molecules can occur. In recent decades, theoretical and computational efforts (Sumi, 1999; Jang *et al.*, 2002c; Scholes, 2003; Jang *et al.*, 2004; Jang, 2007; Renger, 2009; Olaya-Castro and Scholes, 2011; Jang and Cheng, 2013) demonstrate that these issues can be addressed by following in the footsteps of Förster's theory, but with his original assumptions relaxed for more general situations. Thus, so-called generalizations (Sumi, 1999; Jang *et al.*, 2002c; Scholes, 2003; Jang *et al.*, 2004; Jang, 2007) of Förster's theory have been developed. The scope of this chapter is to introduce these new theoretical advances. As a preliminary step for this, a complete and detailed derivation of Förster's theory is presented in the next section.

3.2 Förster's rate expression: a complete derivation

The derivation of Equation (3.2) has been reviewed by Förster himself (Förster, 1965) and revisited by a few experts at a different level (Agranovich and Galanin, 1982; Andrews and Demidov, 1999), but a concise and comprehensive derivation employing modern terminology is not easily available. A review of this nature has been provided recently (Jang and Cheng, 2013) and this section presents a more detailed exposition of the derivation.

The description of RET at the minimum requires three quantum-mechanical electronic states. Let us denote the state where both D and A are in their ground electronic states as $|g\rangle$, the state where only D is excited, as $|D\rangle$ and the state where only A is excited, as $|A\rangle$. The rest of the dynamical degrees of freedom such as molecular vibrations and solvent degrees of freedom are represented as a bath with Hamiltonian H_b. If the spontaneous decay of excitation occurs much more slowly than the energy transfer dynamics, which is typical, the ground electronic state $|g\rangle$ can be removed effectively from the description. Thus, the following total Hamiltonian can be used:

$$H = H_0 + H_{DA} + H_{eb}, \tag{3.3}$$

where

$$H_0 = E_D|D\rangle\langle D| + E_A|A\rangle\langle A| + H_b, \tag{3.4}$$

$$H_{DA} = J(|D\rangle\langle A| + |A\rangle\langle D|), \tag{3.5}$$

$$H_{eb} = B_D|D\rangle\langle D| + B_A|A\rangle\langle A|. \tag{3.6}$$

In the above expressions, the zeroth order Hamiltonian H_0 represents the system states $|D\rangle$ and $|A\rangle$ and the bath degrees of freedom that do not interact with each other. H_{DA} represents the interaction between $|D\rangle$ and $|A\rangle$, with electronic coupling J. Förster assumed this to be a constant parameter in deriving the rate expression, but it is possible to treat them as operators, as will be shown in the next section. H_{eb} is the exciton–bath Hamiltonian, and B_D and B_A are bath operators.

3.2.1 Fermi's golden rule

Let us consider the simplest situation, where an impulsive excitation prepares the state $|D\rangle$ at time $t = 0$, while the bath remains in canonical equilibrium with respect to the ground electronic state $|g\rangle$. This corresponds to the following initial condition of the total density operator:

$$\rho(0) = |D\rangle\langle D|e^{-\beta H_b}/Z_b, \tag{3.7}$$

where $\beta = 1/k_B T$ and $Z_b = Tr_b\{e^{-\beta H_b}\}$. Then, the probability of finding A^* at time t is

$$P_A(t) = Tr_b\left\{\langle A|e^{-iHt/\hbar}\rho(0)e^{iHt/\hbar}|A\rangle\right\}. \tag{3.8}$$

For sufficiently short time t compared to $\hbar/J$, perturbative expansion of $P_A(t)$ with respect to H_{DA} is well justified. Inserting the first-order approximation of the time evolution operator $e^{-iHt/\hbar}$ with respect to H_{DA} and its complex conjugate into Equation (3.8), we obtain the following expression valid up to the second order of J:

$$P_A(t) \approx \frac{J^2}{\hbar^2}\int_0^t dt' \int_0^t dt'' e^{i(E_A-E_D)(t'-t'')/\hbar}$$
$$\times \frac{1}{Z_b} Tr_b\{e^{i(H_b+B_A)(t'-t'')/\hbar}e^{-i(H_b+B_D)t'/\hbar}e^{-\beta H_b}e^{i(H_b+B_D)t''/\hbar}\}. \tag{3.9}$$

The time-dependent rate of RET can be defined as the time derivative of $P_A(t)$ as follows:

$$k(t) \equiv \frac{d}{dt}P_A(t) \approx \frac{2J^2}{\hbar^2}\text{Re}\left[\int_0^t dt' e^{i(E_D-E_A)t'/\hbar}\right.$$
$$\left.\times \frac{1}{Z_b} Tr_b\{e^{i(H_b+B_D)t/\hbar}e^{-i(H_b+B_A)t'/\hbar}e^{-i(H_b+B_D)(t-t')/\hbar}e^{-\beta H_b}\}\right]. \tag{3.10}$$

Let us assume that the bath Hamiltonian is an independent sum of donor and acceptor baths as follows:

$$H_b = H_{b_D} + H_{b_A}. \tag{3.11}$$

Thus,

$$\frac{1}{Z_b}e^{-\beta H_b} = \rho^g_{b_D}\rho^g_{b_A}, \tag{3.12}$$

where $\rho^g_{b_D} = e^{-\beta H_{b_D}}/Tr_{b_D}\{e^{-\beta H_{b_D}}\}$ and $\rho^g_{b_A} = e^{-\beta H_{b_A}}/Tr_{b_A}\{e^{-\beta H_{b_A}}\}$. In addition, assume that B_D (B_A) commutes with H_{b_A} (H_{b_D}). Then, the trace over the bath in Equation (3.10) can be decoupled into those for donor and acceptor baths as follows:

$$k(t) = \frac{2J^2}{\hbar^2}\text{Re}\left[\int_0^t dt' e^{i(E_D-E_A)t'/\hbar}\right.$$
$$\times Tr_{b_A}\left\{e^{iH_{b_A}t/\hbar}e^{-i(H_{b_A}+B_A)t'/\hbar}e^{-iH_{bA}(t-t')/\hbar}\rho^g_{bA}\right\}$$
$$\left.\times Tr_{b_D}\left\{e^{i(H_{b_D}+B_D)t/\hbar}e^{-iH_{b_D}t'/\hbar}e^{-i(H_{b_D}+B_D)(t-t')/\hbar}\rho^g_{bD}\right\}\right]. \tag{3.13}$$

Assuming that the bath for the excited donor relaxes fast and reaches the stationary limit before significant progress of RET occurs, the following approximation can be made:

$$e^{-i(H_{b_D}+B_D)(t-t')/\hbar}\rho^g_{b_D}e^{i(H_{b_D}+B_D)(t-t')/\hbar}\approx\frac{e^{-\beta(H_{b_D}+B_D)}}{Z'_{b_D}}\equiv\rho^e_{b_D}, \quad (3.14)$$

where $Z'_{b_D}=Tr_b\{e^{-\beta(H_{b_D}+B_D)}\}$. Inserting the above approximation into Equation (3.13) and going to the limit of $t=\infty$, we obtain the following steady-state expression:

$$\begin{aligned} k(\infty)=&\frac{2J^2}{\hbar^2}\mathrm{Re}\Bigg[\int_0^\infty dt' e^{i(E_D-E_A)t'/\hbar}\\ &\times\frac{1}{Z_{bA}}Tr_{bA}\{e^{iH_{bA}t'/\hbar}e^{-i(H_{bA}+B_A)t'/\hbar}e^{-\beta H_{bA}}\}\\ &\times\frac{1}{Z'_{bD}}Tr_{bD}\left\{e^{i(H_{bD}+B_D)t'/\hbar}e^{-iH_{bD}t'/\hbar}e^{-\beta(H_{bD}+B_D)}\right\}\Bigg]\\ =&\,k_{FG}. \end{aligned} \quad (3.15)$$

As the second equality in the above equation suggests, the above expression is nothing but the Fermi golden rule (FGR) expression for RET, starting from the relaxed initial state given by Equation (3.14).

Now let us introduce the following lineshape functions of the donor and the acceptor:

$$L_D(\omega)=\int_{-\infty}^{\infty}dte^{-i\omega t+iE_Dt/\hbar}Tr_{bD}\left\{e^{i(H_{bD}+B_D)t/\hbar}e^{-iH_{bD}t/\hbar}\rho^e_{b_D}\right\}, \quad (3.16)$$

$$I_A(\omega)=\int_{-\infty}^{\infty}dt\, e^{i\omega t-iE_At/\hbar}Tr_{bA}\left\{e^{iH_{bA}t/\hbar}e^{-i(H_{bA}+B_A)t/\hbar}\rho^g_{b_A}\right\}. \quad (3.17)$$

Then, Equation (3.15) can be expressed as

$$\begin{aligned} k_{FG}&=\frac{2J^2}{4\pi^2\hbar^2}\mathrm{Re}\left[\int_0^\infty dt\int_{-\infty}^{\infty}d\omega\int_{-\infty}^{\infty}d\omega' e^{i(\omega-\omega')t}L_D(\omega)I_A(\omega')\right]\\ &=\frac{J^2}{2\pi\hbar^2}\int_{-\infty}^{\infty}d\omega\int_{-\infty}^{\infty}d\omega'\delta(\omega-\omega')L_D(\omega)I_A(\omega')\\ &=\frac{J^2}{2\pi\hbar^2}\int_{-\infty}^{\infty}d\omega L_D(\omega)I_A(\omega). \end{aligned} \quad (3.18)$$

This FGR expression can be shown to be equivalent to Equation (3.2). First, the integration over ω can be converted into that over $\tilde{\nu} = \omega/(2\pi c)$, as follows:

$$k_{FG} = \frac{c}{\hbar^2} J^2 \int_{-\infty}^{\infty} d\tilde{\nu} L_D(2\pi c\tilde{\nu}) I_A(2\pi c\tilde{\nu}). \tag{3.19}$$

Second, employing the standard theory of emission and absorption, we can express $L_D(\omega)$ in terms of $f_D(\tilde{\nu})$ and $I_A(\omega)$ in terms of $\epsilon_A(\tilde{\nu})$. Detailed derivation of these relations based on the quantum electrodynamics formalism is provided below.

3.2.2 The relation between $L_D(\omega)$ and $f_D(\tilde{\nu})$

Let us consider only the donor and its bath, for which the total Hamiltonian is $(E_D + B_D)|D\rangle\langle D| + H_{b_D}$. The initial condition corresponds to an excited donor (D^*) with fully equilibrated bath, which can be represented by the following density operator:

$$\rho_D^e = |D\rangle\langle D|\rho_{b_D}^e. \tag{3.20}$$

In the formalism of second quantization (Craig and Thirunamachandran, 1984), the interaction Hamiltonian between matter and radiation has the following form:

$$H_{int} = -i \sum_{\mathbf{k},\lambda} \left(\frac{2\pi\hbar\omega}{n_r^2 V}\right)^{1/2} \left\{\boldsymbol{\mu}_D \cdot \mathbf{e}^{(\lambda)}(\mathbf{k}) a^{(\lambda)}(\mathbf{k}) e^{i\mathbf{k}\cdot\mathbf{r}} - \boldsymbol{\mu}_D \cdot \mathbf{e}^{(\lambda)}(\mathbf{k}) a^{(\lambda)\dagger}(\mathbf{k}) e^{-i\mathbf{k}\cdot\mathbf{r}}\right\}, \tag{3.21}$$

where V is the volume of the medium, ω is the (angular) frequency of the radiation, $\mathbf{e}^{(\lambda)}(\mathbf{k})$ is the polarization of the radiation, $a^{(\lambda)}(\mathbf{k})$ is the annihilation operator of the photon, $a^{(\lambda)\dagger}(\mathbf{k})$ is the creation operator of the photon, and $\boldsymbol{\mu}_D$ is the transition dipole moment of D^*. No local field effect is assumed here and the refractive index n_r is the same as that of the bulk medium.

Let us introduce $G_\lambda^e(\mathbf{k})$ as the rate of photon emission from D^* with wave vector $\mathbf{k}$ and polarization λ. It can be calculated according to the FGR, starting from the initial condition of Equation (3.20), for the matter and vacuum state for the radiation, and employing Equation (3.21) as the matter–radiation interaction Hamiltonian responsible for the transition. The resulting expression is as follows:

$$G_\lambda^e(\mathbf{k}) = 2\pi \frac{|\boldsymbol{\mu}_D \cdot \mathbf{e}^{(\lambda)}(\mathbf{k})|^2}{\hbar n_r^2 V} \omega L_D(\omega), \tag{3.22}$$

where $L_D(\omega)$ has been defined by Equation (3.16). The average of Equation (3.22) over all possible orientations of radiation results in

$$\langle G_\lambda^e(\mathbf{k})\rangle_{or} = \frac{2\pi \mu_D^2 \omega}{3\hbar n_r^2 V} L_D(\omega). \tag{3.23}$$

Let us also introduce $d\Gamma_k^e$ as the rate of photon emission with a magnitude of wave vector in the range k to $k + dk$. Then,

$$d\Gamma_k^e = \langle G_\lambda^e(\mathbf{k})\rangle_{or} \frac{V}{4\pi^3} 4\pi k^2 dk = L_D(\omega) \frac{2n_r \mu_D^2 \omega^3}{3\hbar\pi c^3} d\omega, \tag{3.24}$$

where the sum over two polarization directions have been made and the fact that $\omega = ck/n_r$ has been used. From Equation (3.24), the following expression for the density of emission rate per unit frequency is obtained:

$$\frac{d\Gamma_k^e}{d\omega} = \frac{2n_r \mu_D^2 \omega^3}{3\hbar\pi c^3} L_D(\omega). \tag{3.25}$$

Then, the spontaneous decay lifetime of D^* can be related to the total rate of emission as follows:

$$\begin{aligned} \frac{1}{\tau_D} &= \frac{2n_r \mu_D^2}{3\hbar\pi c^3} \left(\int d\omega\, \omega^3 L_D(\omega) \right) \\ &= \frac{2n_r \mu_D^2 (2\pi c)^4}{3\hbar\pi c^3} \left(\int d\tilde{\nu}\, \tilde{\nu}^3 L_D(2\pi c\tilde{\nu}) \right). \end{aligned} \tag{3.26}$$

The above relation can be used to replace the integration of $L_D(2\pi c\tilde{\nu})$ in the normalized emission spectrum as follows:

$$f_D(\tilde{\nu}) = \frac{\tilde{\nu}^3 L_D(2\pi c\tilde{\nu})}{\int d\tilde{\nu}\tilde{\nu}^3 L_D(2\pi c\tilde{\nu})} = \tau_D \frac{2^5 \pi^3 n_r \mu_D^2 c}{3\hbar} \tilde{\nu}^3 L_D(2\pi c\tilde{\nu}). \tag{3.27}$$

Inserting the above expression into Equation (3.19), we find that

$$k_{FG} = \frac{3}{32\pi^3 \hbar \tau_D n_r \mu_D^2} J^2 \int_{-\infty}^{\infty} d\tilde{\nu} \frac{f_D(\tilde{\nu}) I_A(2\pi c\tilde{\nu})}{\tilde{\nu}^3}. \tag{3.28}$$

3.2.3 The relation between $I_A(\omega)$ and $\epsilon_A(\tilde{\nu})$

Consider an absorption process where A and the bath are prepared in equilibrium. Assume that the incident light is in the state $|N_p(\mathbf{k}, \lambda)\rangle$, where N_p is the number of photons with wave vector $\mathbf{k}$ and polarization λ. Let us introduce $G_\lambda^a(\mathbf{k})$ as the absorption rate of the corresponding photon by A. Then, according to the FGR, starting from the initial states described above and employing a matter–radiation interaction term similar to Equation (3.21), we can obtain the following expression:

$$G_\lambda^a(\mathbf{k}) = \frac{|\boldsymbol{\mu}_A \cdot \mathbf{e}^{(\lambda)}(\mathbf{k})|^2}{\hbar^2} \frac{2\pi N_p \hbar\omega}{n_r^2 V} I_A(\omega). \tag{3.29}$$

For N molecules with random orientation, the average of the above expression over all possible angular distribution results in

$$\langle G_\lambda^a(\mathbf{k})\rangle_{or} = N\frac{\mu_A^2}{3\hbar^2}\frac{2\pi N_p\hbar\omega}{n_r^2 V}I_A(\omega). \tag{3.30}$$

Let us denote the radiant energy density per unit frequency as $\mathcal{J}(\nu)$. Then, the number of radiation states within ω and $\omega + d\omega$ can be expressed as

$$\frac{\mathcal{J}(\nu)}{N_p\hbar\omega}V\frac{d\omega}{2\pi} = \frac{I_R(\nu)n_r}{N_p\hbar\omega c}V\frac{d\omega}{2\pi}, \tag{3.31}$$

where $I_R(\nu) = c\mathcal{J}(\nu)/n_r$ is defined as the irradiance per unit frequency and is equal to the radiant energy flux per unit area and frequency interval. Let us define $d\Gamma_\omega^a$ as the absorption rate of a photon with frequency in the range ω to $\omega + d\omega$. Then, the density of absorption rate per unit frequency is

$$\frac{d\Gamma_\omega^a}{d\omega} = \langle G_\lambda^a(\mathbf{k})\rangle\frac{I_R(\nu)n_r}{N_p\hbar\omega c}V\frac{d\omega}{2\pi}\frac{1}{d\omega} = N\frac{\mu_A^2}{3\hbar^2 n_r c}I_A(\omega)I_R(\nu). \tag{3.32}$$

Given that the length of the sample along the direction of the radiation is l, the irradiance (in the unit of wavenumbers) going out of the sample is equal to

$$\tilde{I}_R(\tilde{\nu}) = \tilde{I}_{R,0}(\tilde{\nu})10^{-\epsilon_A(\tilde{\nu})Ml}, \tag{3.33}$$

where $\tilde{I}_R(\tilde{\nu}) = cI_R(\nu)$, $\tilde{I}_{R,0}(\tilde{\nu})$ is the irradiance incident on the sample, $\epsilon_A(\tilde{\nu})$ is the molar extinction coefficient, and M is the molarity of the sample. Assuming that the difference between $\tilde{I}_R(\tilde{\nu})$ and $\tilde{I}_{R,0}(\tilde{\nu})$ is infinitesimally small, the number of photons lost during passage through a sample of length l can be approximated as

$$\tilde{I}_{R,0}(\tilde{\nu})(\ln 10)\epsilon_A(\tilde{\nu})Ml\frac{d\omega}{2\pi c}\frac{V}{c}. \tag{3.34}$$

On the other hand, the number of photons being absorbed into the acceptor molecules is

$$\frac{N_A M}{10^3}V\frac{\mu_A^2}{3\hbar^2 n_r c}I_A(\omega)\frac{\tilde{I}_{R,0}(\tilde{\nu})}{c}d\omega\frac{l}{c}\hbar\omega. \tag{3.35}$$

Equating (3.34) and (3.35), we find that

$$I_A(\omega) = \frac{3000(\ln 10)n_r\hbar}{(2\pi)^2 N_A\mu_A^2\tilde{\nu}}\epsilon_A(\tilde{\nu}). \tag{3.36}$$

Inserting this expression into Equation (3.28), we find the following rate expression:

$$k_{FG} = \frac{9000(\ln 10)}{128\pi^5 N_A\tau_D}\frac{J^2}{\mu_D^2\mu_A^2}\int_{-\infty}^{\infty}d\tilde{\nu}\frac{f_D(\tilde{\nu})\epsilon_A(\tilde{\nu})}{\tilde{\nu}^4}. \tag{3.37}$$

3.2.4 Transition dipole interaction

Equation (3.37) is more general than Equation (3.2) in the sense that no specification has been made for the electronic coupling constant, J. This also means that Equation (3.37) may be used to determine the rate for any mechanism of electronic coupling, as long as the value of J can be calculated directly by a computational method. For example, the transition density cube (TDC) method (Krueger *et al.*, 1998), which will be described in detail in the next section, or evaluation within the density functional theory (DFT) formalism (Iozzi *et al.*, 2004), can be used for this purpose. At the moment, let us focus on completing the derivation of Förster's expression employing the transition dipole approximation.

If the distance between D and A is much larger than the length scales characteristic of their transition dipoles, the following dipole approximation for the electron coupling can be made:

$$J = \frac{\boldsymbol{\mu}_D \cdot \boldsymbol{\mu}_A - 3(\boldsymbol{\mu}_D \cdot \hat{\mathbf{R}})(\boldsymbol{\mu}_A \cdot \hat{\mathbf{R}})}{n_r^2 R^3}, \tag{3.38}$$

where R is the distance between the donor and the acceptor and $\hat{\mathbf{R}}$ is the corresponding unit vector. Then,

$$\frac{J^2}{\mu_D^2 \mu_A^2} = \frac{\kappa^2}{n_r^4 R^6}, \tag{3.39}$$

where

$$\kappa = \frac{\boldsymbol{\mu}_D \cdot \boldsymbol{\mu}_A - 3(\boldsymbol{\mu}_D \cdot \hat{\mathbf{R}})(\boldsymbol{\mu}_A \cdot \hat{\mathbf{R}})}{\mu_D \mu_A}. \tag{3.40}$$

Then, Equation (3.2), can be recovered by inserting Equation (3.39) into Equation (3.37). This completes the derivation of the original Förster expression.

Throughout the derivation described above, it was assumed that the dielectric response near the chromophores is the same as that of the bulk. This simplification is reasonable (Knox and van Amerongen, 2002) in many cases and its error is often less significant than those due to the orientational factor, κ. However, as progress is made in accuracy of measurement and theoretical calculation, more rigorous and careful examination of the effect of the local field effect may become necessary in the future. For this, consideration within a full quantum electrodynamics formalism (Juzeliūnas and Andrews, 1994) is necessary.

3.3 Transition density cube method

The transition dipole approximation becomes inaccurate if the distance between the donor and the acceptor is comparable to the size of the molecules, which is typically the case for less than 2 nm between moderate-sized molecules. For this situation, Equation (3.37) or its generalizations need to be used along with a more general evaluation of the electronic coupling J. As long as the contributions of exchange interaction (Dexter, 1953) or direct orbital overlap are negligible, the TDC method (Krueger *et al.*, 1998) described below can serve as a general and reliable approach to calculate the electronic coupling.

Given that intra-molecular non-adiabatic effects within the donor and the acceptor molecule can be disregarded, the microscopic interaction Hamiltonian responsible for the resonance interaction is the following Coulomb interaction between electrons:

$$H_c = \frac{e^2}{4\pi\epsilon} \sum_{kl} \frac{1}{|\mathbf{r}_{D_k} - \mathbf{r}_{A_l}|}, \tag{3.41}$$

where ϵ is the dielectric constant of the medium. The electronic coupling J due to the above Coulomb interaction is given by

$$J = \langle D|H_c|A\rangle = \frac{e^2}{4\pi\epsilon} \langle D| \sum_{kl} \frac{1}{|\mathbf{r}_{D_k} - \mathbf{r}_{A_l}|} |A\rangle. \tag{3.42}$$

Let us assume that $|D\rangle$ and $|A\rangle$ can be approximated as the direct products of independent donor and acceptor electronic states as follows: $|D\rangle = |e_D\rangle \otimes |g_A\rangle$; $|A\rangle = |g_D\rangle \otimes |e_A\rangle$. Here, $|e_D\rangle$ is the excited state of D, $|g_A\rangle$ the ground state of A, $|g_D\rangle$ the ground state of D, and $|e_A\rangle$ the excited state of A. These are electronic states of the single donor or acceptor molecules. Inserting these into Equation (3.42) and employing the completeness relation in the configuration space of electrons, we obtain the following expression:

$$J = \frac{e^2}{4\pi\epsilon} \int \left(\prod_k d\mathbf{r}_{D_k}\right) \left(\prod_l d\mathbf{r}_{A_l}\right) \langle e_D| \cdots \mathbf{r}_{D_k} \cdots \rangle \langle \cdots \mathbf{r}_{D_k} \cdots |g_D\rangle$$
$$\times \left(\sum_{kl} \frac{1}{|\mathbf{r}_{D_k} - \mathbf{r}_{A_l}|}\right) \langle g_A| \cdots \mathbf{r}_{A_l} \cdots \rangle \langle \cdots \mathbf{r}_{A_l} \cdots |e_A\rangle, \tag{3.43}$$

where $\langle e_D| \cdots \mathbf{r}_{D_k} \cdots \rangle$ is the electron wavefunction of the excited donor in the full coordinate space, and others are defined in a similar manner. Then, the following transition densities can be defined:

$$\rho_D^{tr}(\mathbf{r}_D) = \int \left(\prod_k d\mathbf{r}_{D_k}\right) \left(\sum_k \delta(\mathbf{r}_D - \mathbf{r}_{D_k})\right)$$
$$\times \langle e_D| \cdots \mathbf{r}_{D_k} \cdots \rangle \langle \cdots \mathbf{r}_{D_k} \cdots |g_D\rangle, \tag{3.44}$$

$$\rho_A^{tr}(\mathbf{r}_A) = \int \left(\prod_l d\mathbf{r}_{A_l}\right) \left(\sum_l \delta(\mathbf{r}_A - \mathbf{r}_{A_l})\right)$$
$$\times \langle e_A| \cdots \mathbf{r}_{A_l} \cdots \rangle \langle \cdots \mathbf{r}_{A_l} \cdots |g_A\rangle. \tag{3.45}$$

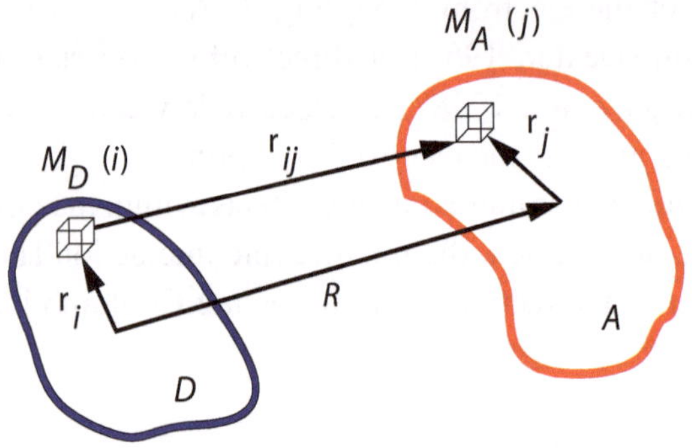

Figure 3.2 Depiction of two cells in the arbitrary donor (D) and acceptor (A) transition densities. The positions of the TDC elements are defined by the $\mathbf{r}_i$ and $\mathbf{r}_j$ elements respectively, relative to the centre of D and A. Diagram taken with permission from (Krueger *et al.*, 1998).

Utilizing the fact that all the electrons in the donor molecule are equivalent, as are those in the acceptor molecule, Equation (3.43) can be expressed in terms of the above transition densities as follows:

$$J = \frac{e^2}{4\pi\epsilon} \int d\mathbf{r}_D d\mathbf{r}_A \, \rho_D^{tr}(\mathbf{r}_D) \frac{1}{|\mathbf{r}_D - \mathbf{r}_A|} \rho_A^{tr}(\mathbf{r}_A). \tag{3.46}$$

This is referred to as Coulomb coupling, in order to distinguish it from the exchange interaction (Dexter, 1953) between the donor and the acceptor electrons, which is neglected here. Except for the fact that Equation (3.46) involves quantum-mechanical transition densities, it is mathematically equivalent to the Coulomb interaction between charge densities in classical electrodynamics. Thus, multipolar expansion of Equation (3.46) can be done by employing a Taylor expansion of the $1/r$ potential, and the first term in this series is the transition dipole interaction given by Equation (3.38).

The TDC method is to evaluate Equation (3.46) numerically by computing the transition densities over the three-dimensional grid, as demonstrated in Figure 3.2. First, the transition density cube elements can be defined for the donor and the acceptor transitions and approximated as follows:

$$M_D(i) = \int_{\Delta V_i} d\mathbf{r}_D \, \rho_D^{tr}(\mathbf{r}_D) \approx \Delta V_i \rho_D^{tr}(\mathbf{r}_i), \tag{3.47}$$

$$M_A(j) = \int_{\Delta V_j} d\mathbf{r}_A \, \rho_A^{tr}(\mathbf{r}_A) \approx \Delta V_j \rho_A^{tr}(\mathbf{r}_j), \tag{3.48}$$

where ΔV_i and ΔV_j are volume elements of the cubes shown in Figure 3.1. Then, Equation (3.46) can be approximated as

$$J \approx \frac{e^2}{4\pi\epsilon} \sum_{ij} \frac{M_D(i) M_A(j)}{|\mathbf{r}_i - \mathbf{r}_j|}. \tag{3.49}$$

The only approximation involved here, apart from those in any electronic structure calculation, is the discretization of the integral over the space. Compared to multipolar expansion, the TDC method is valid at all molecular separations. The transition dipole for D can be obtained from the cube elements as follows:

$$\boldsymbol{\mu}_D \simeq e \sum_i \mathbf{r}_i M_D(i), \tag{3.50}$$

which is independent of origin, because the integration of the transition density is zero due to the orthogonality of the two states. The transition dipole for A can be obtained in a similar manner. In actual applications, these approximations for transition dipoles can be compared with experimental values, in order to estimate the accuracy of the discretization error and to introduce a scaling factor if necessary.

3.4 Generalized Förster theories

As addressed in Section 3.2, there are three major assumptions implicit in Förster's theory, assuming that the underlying time-dependent perturbation theory remains valid. (i) The excited donor molecule and its environments are assumed to be fully equilibrated before the energy transfer occurs. (ii) It is based on the so-called Condon approximation, where the electronic coupling constant J is assumed to be independent of any nuclear motion. (iii) It assumes that the donor and the acceptor consist of single chromophores. Theories (Sumi, 1999; Scholes and Fleming, 2000; Jang *et al.*, 2002c, 2004; Jang, 2007) generalizing these assumptions are described in this section.

3.4.1 Non-equilibrium generalization

If the energy transfer dynamics is fast enough to occur before equilibration in the excited state becomes complete, the assumption of Equation (3.14) cannot be justified. In this situation, a more general rate expression, Equation (3.13), needs to be used. Inserting the inverse Fourier transform of $I_A(\omega)$ defined by Equation (3.17) directly into Equation (3.13), we obtain the following expression:

$$k(t) = \frac{J^2}{\pi\hbar^2}\int_{-\infty}^{\infty} d\omega\, I_A(\omega)\mathrm{Re}\Bigg[\int_0^t dt' e^{-i\omega t' + iE_D t'/\hbar}\frac{1}{Z_{bD}} Tr_{bD} \times \left\{ e^{i(H_{bD}+B_D)t/\hbar} e^{-iH_{bD}t'/\hbar} e^{-i(H_{bD}+B_D)(t-t')/\hbar} e^{-\beta H_{bD}} \right\}\Bigg]. \tag{3.51}$$

In the above expression, the time-integration with respect to t' can be expressed as the time-dependent emission profile of D^*, as described below.

Let us consider the situation where D is excited at time zero by an impulsive pulse, as assumed in Section 2.1. The corresponding initial density operator at

$t = 0$ is given by $\rho(0) = |D\rangle\langle D|\rho^g_{b_D}$. Assuming that there is no A and its bath, the total matter Hamiltonian governing the dynamics starting from this initial condition is $H_D = (E_D + B_D)|D\rangle\langle D| + H_{b_D}$, and the matter–radiation interaction Hamiltonian is the same as Equation (3.21). Employing the same time-dependent perturbation theory with respect to the matter–radiation interaction Hamiltonian, we find that the probability for the donor to emit a photon and return to the ground electronic state at time t can be expressed as

$$P_g(t) = |\boldsymbol{\mu}_D \cdot \hat{\mathbf{e}}|^2 \int_0^t dt' \int_0^t dt''\, e^{i\omega(t'-t'')-iE_D(t'-t'')/\hbar} \frac{1}{Z_{b_D}} Tr_{b_D}$$
$$\times \left\{ e^{-iH_{b_D}(t-t')/\hbar} e^{-i(B_D+H_{b_D})t'/\hbar} e^{-\beta H_{b_D}} e^{i(B_D+H_{b_D})t''/\hbar} e^{iH_{b_D}(t-t'')/\hbar} \right\}. \quad (3.52)$$

The time-dependent emission profile is defined as the time derivative of this probability as follows:

$$L_D(t,\omega) \equiv \frac{1}{|\boldsymbol{\mu}_D \cdot \hat{\mathbf{e}}|^2} \frac{d}{dt} P_g(t)$$
$$= 2\mathrm{Re}\Bigg[\int_0^t dt' e^{-i\omega t'+iE_D t'/\hbar} \frac{1}{Z_{b_D}} Tr_{b_D}$$
$$\times \left\{ e^{i(H_{b_D}+B_D)t/\hbar} e^{-iH_{b_D}t'/\hbar} e^{-i(H_{b_D}+B_D)(t-t')/\hbar} e^{-\beta H_{b_D}} \right\} \Bigg]. \quad (3.53)$$

Inserting Equation (3.53) into Equation (3.51), we find the following rate expression for the non-equilibrium FRET (N-FRET) (Jang *et al.*, 2002c):

$$k(t) = \frac{J^2}{2\pi\hbar^2} \int_{-\infty}^{\infty} d\omega I_A(\omega) L_D(t,\omega). \quad (3.54)$$

In the limit where $t \to \infty$, this expression approaches Equation (3.18), given that $L_D(\infty, \omega)$ approaches the emission lineshape in the stationary limit. An expression analogous to Equation (3.37) can be obtained by employing Equation (3.36) and an analogue of Equation (3.27) for the time-dependent stimulated emission of D.

3.4.2 Inelastic generalization

The assumption that the electronic coupling J in H_{DA} of Equation (3.5) is a parameter is reasonable if the donor–acceptor configuration remains rigid or fluctuates in a way independent of the electronic excitation dynamics. In the latter case, either the population or the rate can be averaged over the sample of fluctuating J. However, if the donor and acceptor are connected by a bridge molecule or locked in soft environments, significant quantum-mechanical modulation of the electronic coupling J can occur. Exchange of energy between the electronic excitation and nuclear degrees of freedom is possible in this situation, and Förster's rate expression needs to be modified accordingly.

Let us assume that the bath Hamiltonian H_b can be decomposed into three components as follows:

$$H_b = H_{b_D} + H_{b_A} + H_{b_J}, \tag{3.55}$$

where H_{b_J} is the bath Hamiltonian governing the dynamics of J. The original Förster assumption that donor and acceptor baths are independent of each other remains the same. In addition, it is assumed that H_{b_J} is independent of H_{b_D} and H_{b_A}, and satisfies the following two conditions:

$$[H_{b_J}, B_D] = [H_{b_J}, H_{b_D}] = 0, \tag{3.56}$$

$$[H_{b_J}, B_A] = [H_{b_J}, H_{b_A}] = 0. \tag{3.57}$$

The starting expression for the population of A at time t is the same as Equation (3.8). Inserting the first-order approximation for the time evolution operator $e^{-iHt/\hbar}$ with respect to H_{DA} and its complex conjugated into Equation (3.8), and taking the fact that J is an operator into consideration now, we obtain the following perturbative approximation:

$$P_A(t) \approx \frac{1}{\hbar^2}\int_0^t dt' \int_0^t dt'' e^{i(E_A - E_D)(t'-t'')/\hbar}$$
$$\times \frac{1}{Z_b} Tr_b\{e^{i(H_b+B_A)(t'-t'')/\hbar} J e^{-i(H_b+B_D)t'/\hbar} e^{-\beta H_b} e^{i(H_b+B_D)t''/\hbar} J\}. \tag{3.58}$$

The time-dependent rate $k(t)$, which is defined as the time derivative of $P_A(t)$, can be expressed as

$$k(t) \approx \frac{2}{\hbar^2}\mathrm{Re}\int_0^t dt' e^{i(E_D-E_A)t'/\hbar} Tr_b\{e^{i(H_b+B_D)t'/\hbar} J e^{-i(H_b+B_A)t'/\hbar} J\rho_b(t-t')\}, \tag{3.59}$$

where

$$\rho_b(t) = e^{-i(H_b+B_D)t/\hbar} \frac{e^{-\beta H_b}}{Z_b} e^{i(H_b+B_D)t/\hbar}. \tag{3.60}$$

The assumption that the three bath terms H_{b_D}, H_{b_A}, and H_{b_J} are independent of each other makes it possible to decouple the bath density operator, Equation (3.60), as follows:

$$\frac{e^{-\beta H_b}}{Z_b} = \rho^g_{b_D}\rho^g_{b_A}\rho_{b_J}, \tag{3.61}$$

where $\rho^g_{b_D}$ and $\rho^g_{b_A}$ have been defined below Equation (3.12) and $\rho_{b_J} = e^{-\beta H_{b_J}}/Tr_{b_J}\{e^{-\beta H_{b_J}}\}$. Inserting Equation (3.61) into Equation (3.60), we find that

$$\rho_b(t) = \rho_{b_D}(t)\rho^g_{b_A}\rho_{b_J}, \tag{3.62}$$

where

$$\rho_{b_D}(t) = e^{-i(H_b+B_D)t/\hbar}\rho^g_{b_D}e^{i(H_b+B_D)t/\hbar}. \tag{3.63}$$

Inserting Equation (3.62) into Equation (3.59),

$$\begin{aligned} k(t) = \frac{2}{\hbar^2}\text{Re}\int_0^t dt' e^{i(E_D-E_A)t'/\hbar} Tr_{b_J}\left\{e^{iH_{b_J}t'/\hbar} J e^{-iH_{b_J}t'/\hbar} J\rho_{b_J}\right\} \\ \times Tr_{b_A}\left\{e^{iH_{b_A}t'/\hbar}e^{-i(H_{b_A}+B_A)t'/\hbar}\rho^g_{b_A}\right\} \\ \times Tr_{b_D}\left\{e^{i(H_{b_D}+B_D)t'/\hbar}e^{-iH_{b_D}t'/\hbar}\rho_{b_D}(t-t')\right\}. \end{aligned} \tag{3.64}$$

Given that the assumption of ergodicity holds in the excited donor space, Equation (3.64) approaches the following steady state limit:

$$\begin{aligned} k_{IF} = \frac{2}{\hbar^2}\text{Re}\int_0^\infty dt\, e^{i(E_D-E_A)t/\hbar} Tr_{b_J}\left\{e^{iH_{bJ}t/\hbar} J e^{-iH_{bJ}t/\hbar} J\rho_{b_J}\right\} \\ \times Tr_{b_A}\left\{e^{iH_{bA}t/\hbar}e^{-i(H_{bA}+B_A)t/\hbar}\rho^g_{b_A}\right\} \\ \times Tr_{b_D}\left\{e^{i(H_{bD}+B_D)t/\hbar}e^{-iH_{bD}t/\hbar}\rho^e_{b_D}\right\}. \end{aligned} \tag{3.65}$$

Inserting the inverse Fourier transforms of $I_A(\omega)$ and $L_D(\omega)$, defined by Equations (3.16) and (3.17), into the above expression, we find that

$$k_{IF} = \frac{1}{2\pi\hbar^2}\int_{-\infty}^\infty d\omega \int_{-\infty}^\infty d\omega' L_D(\omega)I_A(\omega')K_J(\omega-\omega'), \tag{3.66}$$

where

$$K_J(\omega) = \frac{1}{\pi}\text{Re}\int_0^\infty dt\, e^{i\omega t} Tr_{b_J}\left\{e^{iH_{b_J}t/\hbar} J e^{-iH_{b_J}t/\hbar} J\rho_{b_J}\right\}. \tag{3.67}$$

The form of Equation (3.66) is generic for inelastic processes where the exchange of energy between transferring excitation and the modulating degrees of freedom is possible. Introducing,

$$\tilde{K}_J(\tilde{\nu}) = 2\pi c K_J(2\pi c\tilde{\nu}), \tag{3.68}$$

and inserting Equations (3.27) and (3.36) into Equation (3.66), we find the following expression for the inelastic FRET (I-FRET) (Jang, 2007):

$$k_{IF} = \frac{9000(\ln 10)}{128\pi^5 N_A \tau_D \mu_D^2 \mu_A^2}\int d\tilde{\nu}\int d\tilde{\nu}' \frac{f_D(\tilde{\nu})\epsilon_A(\tilde{\nu}')}{\tilde{\nu}^3\tilde{\nu}'}\tilde{K}_J(\tilde{\nu}-\tilde{\nu}'). \tag{3.69}$$

Equation (3.2) corresponds to the limit of the above expression, where $\tilde{K}_J(\tilde{\nu})$ approaches the delta function and J is given by transition–dipole interaction, i.e. $\tilde{K}_J(\tilde{\nu}-\tilde{\nu}') \approx \mu_D^2\mu_A^2\kappa^2\delta(\tilde{\nu}-\tilde{\nu}')/(n_r^4 R^6)$.

3.4.3 Multichromophoric generalization

In natural light-harvesting complexes or synthetic multichromophore systems, it is common to find transfer of excitons delocalized over multiple chromophores or sites. As long as the group of donor molecules in these systems is well separated from that of acceptor molecules, Förster's theory can be extended to include this situation as described below.

Assume that the system consists of two distinctive sets of chromophores, donors (D_j, $j = 1, \ldots, N_D$), and acceptors (A_k, $k = 1, \ldots, N_A$). The state where all the D_j's and A_k's are in their ground electronic states is denoted as $|g\rangle$. The state where D_j is excited while all others remain in the ground electronic state is denoted as $|D_j\rangle$. The state $|A_k\rangle$ is defined similarly. All the remaining degrees of freedom such as molecular vibrations and solvation coordinates are termed as bath. The bath Hamiltonian is assumed to be $H_b = H_{b,D} + H_{b,A}$, where the subscripts D and A respectively denote the components coupled to the set of donors and acceptors.

The dynamics of an exciton is assumed to be much faster than its spontaneous decay rate. Thus, the latter process is ignored. The zeroth order Hamiltonian describing the interaction-free dynamics can be written as,

$$H_0 = H_{e,D} + H_{e,A} + H_b, \tag{3.70}$$

$$H_{e,D} = \sum_{j=1}^{N_D} E_{Dj}|D_j\rangle\langle D_j| + \sum_{j,j'} \Delta_{D,jj'}(|D_j\rangle\langle D_{j'}| + |D_{j'}\rangle\langle D_j|), \tag{3.71}$$

$$H_{e,A} = \sum_{k=1}^{N_A} E_{Ak}|A_k\rangle\langle A_k| + \sum_{k,k'} \Delta_{A,kk'}(|A_k\rangle\langle A_{k'}| + |A_{k'}\rangle\langle A_k|). \tag{3.72}$$

The resonance interaction between $|D_j\rangle$ and $|A_k\rangle$ is represented by

$$H_{DA} = \sum_{j=1}^{N_D}\sum_{k=1}^{N_A} J_{jk}(|D_j\rangle\langle A_k| + |A_k\rangle\langle D_j|), \tag{3.73}$$

where J_{jk} is assumed to be independent of any bath coordinates.

The excitation–bath coupling is assumed to be diagonal in the site excitation basis as follows:

$$\begin{aligned} H_{eb} &= \sum_{j=1}^{N_D} B_{Dj}|D_j\rangle\langle D_j| + \sum_{k=1}^{N_A} B_{Ak}|A_k\rangle\langle A_k| \\ &\equiv H_{eb,D} + H_{eb,A}, \end{aligned} \tag{3.74}$$

where B_{Dj} and B_{Ak} are bath operators coupled to $|D_j\rangle$ and $|A_k\rangle$ respectively. These operators and the bath Hamiltonian can be arbitrary, except that they satisfy the following conditions:

$$[H_{bD}, H_{bA}] = [H_{bD}, B_{Ak}] = [H_{bA}, B_{Dj}] = 0, \tag{3.75}$$

which implies that the sets of $|D_j\rangle$'s and $|A_k\rangle$'s are coupled to independent bath modes. However, it is important to note that there can still be common modes within the respective sets of donors and acceptors. Summing up Equations (3.70) to (3.74), the total Hamiltonian is given by

$$H = H_0 + H_{DA} + H_{eb} = H_D + H_A + H_{DA}, \tag{3.76}$$

where

$$H_D \equiv H_{e,D} + H_{b,D} + H_{eb,D}, \tag{3.77}$$

$$H_A \equiv H_{e,A} + H_{b,A} + H_{eb,A}. \tag{3.78}$$

Let us consider the situation where excited donor molecules are populated by an impulsive pulse at $t = 0$, while the bath remains in thermal equilibrium with respect to the ground electronic state. As in the single chromophoric case, the duration of pulse is assumed to be short and can be approximated well by a delta function. Thus, the density operator of the system plus the bath after excitation by the pulse can be approximated as

$$\rho(0) = |D_{\hat{\mathbf{e}}}\rangle\langle D_{\hat{\mathbf{e}}}|\rho_b^g = |D_{\hat{\mathbf{e}}}\rangle\langle D_{\hat{\mathbf{e}}}|\rho_{bD}^g \rho_{bA}^g, \tag{3.79}$$

where $|D_{\hat{\mathbf{e}}}\rangle = \hat{\mathbf{e}} \cdot \sum_j \boldsymbol{\mu}_j |D_j\rangle$, with $\hat{\mathbf{e}}$ being the polarization of the impulsive pulse and $\boldsymbol{\mu}_j$ the transition dipole moment vector of D_j and $\rho_b^g = e^{-\beta H_b}/Tr_b\{e^{-\beta H_b}\}$. ρ_{bD}^g and ρ_{bA}^g are also defined accordingly as canonical density operators for H_D and H_A.

The probability for any of the acceptor molecules to be excited at time t is then given by

$$P_A(t) = \sum_{k=1}^{N_A} Tr_b\left\{\langle A_k|e^{-iHt/\hbar}\rho(0)e^{iHt/\hbar}|A_k\rangle\right\}. \tag{3.80}$$

For sufficiently short time compared to $\hbar/\max(J_{jk})$, a perturbative expansion of $P_A(t)$ with respect to H_{DA} is well justified. Inserting the first-order approximation of the time evolution operator $e^{-iHt/\hbar}$ and its complex conjugate into Equation (3.80), we obtain the following expression valid up to the second order of J_{jk}:

$$P_A(t) \approx \sum_{j'j''}\sum_{k'k''}\sum_{k} \frac{J_{j'k'}J_{j''k''}}{\hbar^2}\int_0^t dt' \int_0^t dt'' Tr_b\left\{\langle A_k|e^{-i(H_0+H_{eb})(t-t'')/\hbar}|A_{k''}\rangle \right.$$
$$\left. \times \langle D_{j''}|e^{-i(H_0+H_{eb})t''/\hbar}\rho(0)e^{i(H_0+H_{eb})t'/\hbar}|D_{j'}\rangle \, \langle A_{k'}|e^{i(H_0+H_{eb})(t-t')/\hbar}|A_k\rangle\right\}. \tag{3.81}$$

The time-dependent RET rate is defined as the derivative of this acceptor probability. Thus, taking the time derivative of Equation (3.81), we obtain

$$k(t) = \sum_{j'j''}\sum_{k'k''} \frac{2J_{j'k'}J_{j''k''}}{\hbar^2}\mathrm{Re}\left[\int_0^t dt' Tr_b\{\langle A_{k'}|e^{-i(H_0+H_{eb})(t-t')/\hbar}|A_{k''}\rangle\right.$$
$$\left.\times \langle D_{j''}|e^{-i(H_0+H_{eb})t'/\hbar}\rho(0)e^{i(H_0+H_{eb})t/\hbar}|D_{j'}\rangle\}\right]. \quad (3.82)$$

In obtaining the above expression, the following identity has been used:

$$\sum_k Tr_b\left\{\langle A_{k'}|e^{i(H_0+H_{eb})(t-t')/\hbar}[|A_k\rangle\langle A_k|, H_0+H_{eb}]e^{-i(H_0+H_{eb})(t-t'')/\hbar}|A_{k''}\rangle\right.$$
$$\left.\times\langle D_{j''}|e^{-i(H_0+H_{eb})t''/\hbar}\rho(0)e^{i(H_0+H_{eb})t'/\hbar}|D_{j'}\rangle\right\} = 0. \quad (3.83)$$

Under the assumption of Equation (3.75), the trace over the bath degrees of freedom comprising H_{b_D} and H_{b_A} can be decoupled from each other. Changing the integration variable t' to $t-t'$ in Equation (3.82) and decoupling the trace over the bath degrees of freedom, we obtain the following expression:

$$k(t) = \sum_{j'j''}\sum_{k'k''} \frac{2J_{j'k'}J_{j''k''}}{\hbar^2}\mathrm{Re}\left[\int_0^t dt' Tr_{b_A}\{e^{iH_{b,A}t'/\hbar}\langle A_{k'}|e^{-iH_At'/\hbar}|A_{k''}\rangle\rho^g_{b_A}\}\right.$$
$$\left.\times Tr_{b_D}\{e^{-iH_{b,D}t'/\hbar}\langle D_{j''}|e^{-iH_D(t-t')/\hbar}|D_{\hat{e}}\rangle\rho^g_{b_D}\langle D_{\hat{e}}|e^{iH_Dt/\hbar}|D_{j'}\rangle\}\right]. \quad (3.84)$$

Equivalently, this can be expressed as the sum of spectral overlaps as follows (Jang *et al.*, 2004):

$$k(t) = \sum_{j'j''}\sum_{k'k''} \frac{J_{j'k'}J_{j''k''}}{2\pi\hbar^2}\int_{-\infty}^{\infty} d\omega I_A^{k'k''}(\omega)L_D^{j''j'}(t,\omega), \quad (3.85)$$

where $I_A^{k'k''}(\omega)$ and $L_D^{j''j'}(t;\omega)$ represent the absorption of the acceptors and the stimulated emission of the donors, and are defined as

$$I_A^{k'k''}(\omega) \equiv \int_{-\infty}^{\infty} dt e^{i\omega t} Tr_{bA}\left\{\langle A_{k'}|e^{iH_{b,A}t'/\hbar}e^{-iH_At'/\hbar}\rho^g_{b_A}|A_{k''}\rangle\right\}, \quad (3.86)$$

$$L_D^{j''j'}(t;\omega) \equiv 2\mathrm{Re}\left[\int_0^t dt' e^{-i\omega t'} Tr_{bD}\left\{\langle D_{j''}|e^{-iH_{b,D}t'/\hbar}e^{-iH_D(t-t')/\hbar}\right.\right.$$
$$\left.\left.\times |D_{\hat{e}}\rangle\langle D_{\hat{e}}|\rho^g_{b_D}e^{iH_Dt/\hbar}|D_{j'}\rangle\right\}\right]. \quad (3.87)$$

In the stationary limit of $t\to\infty$, $k(t)$ approaches the following limit:

$$k_s = \sum_{j'j''}\sum_{k'k''} \frac{J_{j'k'}J_{j''k''}}{2\pi\hbar^2}\int_{-\infty}^{\infty} d\omega I_A^{k'k''}(\omega)L_D^{j''j'}(\infty,\omega). \quad (3.88)$$

We call the above a multichromophoric FRET (MC-FRET) rate expression (Jang *et al.*, 2004). Under the assumption that the dynamics in the donor exciton manifold is ergodic, we find that

$$L_D^{j''j'}(\infty,\omega) = 2\text{Re}\int_0^{\infty} dt e^{-i\omega t} Tr_{bD}\left\{\langle D_{j''}|e^{-iH_{b,D}t}\rho_{D,b}^{e}e^{iH_{D}t}|D_{j'}\rangle\right\}, \quad (3.89)$$

where $\rho_{D,b}^{e} = e^{-\beta H_D}/Tr\{e^{-\beta H_D}\}$ is the canonical density operator of the donors and the bath in the single exciton manifold. Sumi's starting expression in his theory (Sumi, 1999) is equivalent to Equation (3.88), with the above expression of emission lineshape.

Equation (3.88) is presented in a tensor product form and is invariant with respect to any unitary transformation. Thus, when all the matrix elements and the lineshape functions are transformed to those in the exciton basis, the form of the rate expression remains the same. Thus, the same expression can be used when the indices represent delocalized exciton states. As a further simplification, it is useful to consider the limit where the lineshape functions become diagonal; where Equation (3.88) reduces to

$$k_s \approx \sum_j \sum_k \frac{J_{jk}^2}{2\pi\hbar^2}\int_{-\infty}^{\infty} d\omega I_A^{kk}(\omega) L_D^{jj}(\infty,\omega). \quad (3.90)$$

Thus, the rate expression is an independent sum of each pairwise contribution. If j and k represent site-local excitation states, the above approximation represents the situation where each exciton is completely localized at each chromophore, and amounts to an independent sum of Förster's rates. If j and k represent delocalized exciton states, the above expression represents the limit of weak exciton–bath coupling, where the bath-mediated inter-exciton coupling can be neglected. This latter limit is often referred to as the generalized Förster theory (Ritz *et al.*, 2001; Scholes, 2003; Renger, 2009).

3.5 Important computational issues in an actual application

3.5.1 Identification of donors and acceptors

In systems with multiple chromophores, identification of appropriate groups of donors and acceptors is the first important step. Although such identification can be somewhat arbitrary, as depicted in the schematics of Figure 3.1b, the classification of different groups is possible, based on observation of the physical proximity of molecules. Thus, in general, the electronic couplings can be put into three distinct categories: intra-donor couplings, intra-acceptor couplings, and donor–acceptor couplings. In the limit where all the coupling constants are small, the Förster's theory can be applied to the energy transfer between each pair. However, if some of the coupling constants are large, to the extent that the spectral line-shapes are altered significantly compared with those of independent single molecules, judicious definition of donor groups and acceptor groups has to be made. In addition, as shown in Figure 3.1b, it is sometimes necessary to introduce an additional set of bridge states, which are intermediate molecules or connecting bonds that can indirectly influence the transfer rate, but are far off resonance. It is possible to include the effects of these couplings through bridges in the form of effective couplings between the donors and the acceptors, as will be shown below.

Given all the numerical values of the coupling constants calculated by the TDC method or any other approaches, an informed decision to partition the electronic Hamiltonian into contributions from the donor, the acceptor and the bridge states can be made. Partitioning of the Hamiltonian can be carried out on the basis of spectroscopic and structural information. In other words, donors and acceptors can often be identified from the information on the direction of the energy flow. For the interpretation of spectroscopic data, an 'optical partitioning' of the Hamiltonian has proven useful. The sequence of events that determines the Hamiltonian partitioning is as follows: (i) the system is initially prepared by a short optical pulse; (ii) population relaxation leads to an initial thermalized population on a set of molecules, which are identified as the donor states; (iii) the complementary set of molecules is identified as the acceptor states; and (iv) any remaining molecules that may indirectly influence the transfer process but are far off resonance are subsequently identified as the bridge states. This type of protocol is reasonable if the timescales associated with the initial two processes, (i) and (ii), are distinct from those of (iii) and (iv).

The issue of partitioning a multi-state Hamiltonian was addressed by Löwdin (Löwdin, 1951) and has emerged in many fields treating transitions between quantum states in a perturbative manner. Similarly, we can apply Löwdin partitioning to the multichromophore exciton Hamiltonian. This provides a natural way to include superexchange transfer mechanisms, which were first used in electronic transfer processes (McConnell, 1961). Further theoretical developments that combine superexchange with a coherent hopping mechanism have been shown more recently (Ratner, 1990; Sumi and Kakitani, 2001; Kimura, 2009; Saito and Sumi, 2009).

For the simple case where there are three states, $|D\rangle$, $|A\rangle$ and $|B\rangle$, the exciton Hamiltonian in the basis of these states can be represented by the following matrix:

$$H = \begin{pmatrix} E_D & J_{DA} & J_{DB} \\ J_{AD} & E_A & J_{AB} \\ J_{BD} & J_{BA} & E_B \end{pmatrix}. \tag{3.91}$$

Incorporating the off-diagonal bridge coupling matrix elements in either the donor or acceptor components, we can divide the above exciton Hamiltonian into effective donor and acceptor Hamiltonians, as follows:

$$H_D = \begin{pmatrix} E_D & 0 & J_{DB} \\ 0 & 0 & 0 \\ J_{BD} & 0 & E_B \end{pmatrix}, \tag{3.92}$$

$$H_A = \begin{pmatrix} 0 & 0 & 0 \\ 0 & E_A & J_{AB} \\ 0 & J_{BA} & E_B \end{pmatrix}. \tag{3.93}$$

H_D and H_A can be diagonalized separately to obtain a set of effective donor and acceptor states. Among these, the states that have dominant characteristics of the bridge can be neglected, and the effective donor and acceptor states can be defined as follows:

$$|D_{eff}\rangle = \frac{1}{\sqrt{N_D}}(|D\rangle + \delta_D|B\rangle), \tag{3.94}$$

$$|A_{eff}\rangle = \frac{1}{\sqrt{N_A}}(|A\rangle + \delta_A|B\rangle), \tag{3.95}$$

where $|A\rangle$, $|D\rangle$ and $|B\rangle$ are the unperturbed acceptor, donor and bridge states respectively, N_D and N_A are the normalization constants and the coefficients δ_D and δ_A, which are much smaller than unity, are determined from diagonalization of H_D and H_A. It is important to note that the two states $|D_{eff}\rangle$ and $|A_{eff}\rangle$ are not orthogonal to each other. Then, the effective donor–acceptor coupling can be calculated as follows:

$$\langle D_{eff}|H|A_{eff}\rangle = \frac{1}{\sqrt{N_D N_A}}\{J_{DA} + \delta_D J_{DB} + \delta_A J_{BA} + \delta_D \delta_A E_B\}. \tag{3.96}$$

For the general case where there are multiple states in each group, each matrix element of Equations (3.91), (3.92) and (3.93) itself becomes a matrix. Thus, the diagonalization of H_D and H_A leads to multiple effective donor and acceptor states, and the effective MC donor–acceptor coupling elements can be calculated as in Equation (3.96).

3.5.2 Evaluation of spectral overlap and averaging over inhomogeneous distribution

Reliable calculation of spectral overlap is crucial for any of the rate theories presented in this chapter. For disordered and condensed phase systems, this can be a difficult task. The experimental spectral lineshape of any molecule is a convolution of various sources of line broadening mechanisms with different timescales; and not all of them contribute to the lineshape functions that enter the rate expression. Loosely speaking, any ensemble spectral lineshape consists of homogeneous and inhomogeneous contributions. The former reflect the dynamics of an individual molecule and enter the rate expression, whereas the latter result from static or quasistatic disorder and simply broaden ensemble or single molecule spectroscopy lineshapes. It is often assumed that homogeneous broadening is identical for all molecules of the same kind which, however, may not be true for pigment molecules embedded in proteins or any other disordered media. In this case, there can be heterogeneity even in the theoretical lineshape of individual pigment molecules.

As a simple case, let us consider an ensemble of single chromophoric donor and acceptor pairs with fixed relative distance and orientation. All of the conditions for the FRET theory are assumed to be valid. It is also assumed that inhomogeneous line broadening originates only from disorder in the excitation energies. The ensemble emission lineshape of the donor and the absorption lineshape of the acceptor can be expressed as

$$L_D(\omega) = \int d\omega_0 \, L_D^h(\omega - \omega_0) G_D(\omega_0), \tag{3.97}$$

$$I_A(\omega) = \int d\omega_0' \, I_A^h(\omega - \omega_0') G_A(\omega_0'), \tag{3.98}$$

where $G_D(\omega_0)$ and $G_A(\omega_0')$ are distributions of transition frequencies of the donor and the acceptor, ω_0 and ω_0'. The FRET rate for each pair with excitation frequencies ω_0 and ω_0' is

$$k_F^h(\omega_0, \omega_0') = \frac{J^2}{2\pi\hbar^2} \int d\omega L_D^h(\omega - \omega_0) I_A^h(\omega - \omega_0'). \tag{3.99}$$

Given that the distributions of the donor and the acceptor remain the same in the ensemble of the pairs of the donor and the acceptor, the average rate becomes

$$\begin{aligned} \langle k_F \rangle &= \int d\omega_0 \int d\omega_0' G_D(\omega_0) G_A(\omega_0') k_F^h(\omega_0, \omega_0') \\ &= \frac{J^2}{2\pi\hbar^2} \int d\omega L_D(\omega) I_A(\omega), \end{aligned} \tag{3.100}$$

which amounts to using ensemble lineshapes for calculation of the rate.

While the above average rate can serve as a useful estimate for the timescale of the RET dynamics, there are a number of possibilities where it does not represent a genuine RET dynamics of the whole ensemble. Firstly, for the case where the inhomogeneous distribution is broad, there can be a significant discrepancy

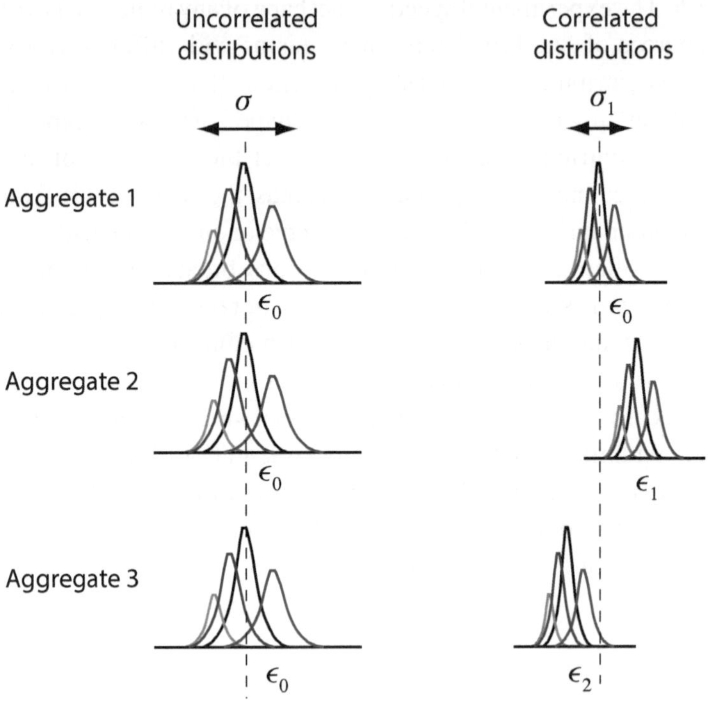

Figure 3.3 Illustration of the inhomogeneously broadened absorption spectra of molecular aggregates. The disorder in the site energies of the aggregates is uncorrelated in the left panel and correlated in the right panel. In the latter scenario, the energy offsets ($\epsilon_i - \epsilon_0$) form a distribution of width Σ. Both cases have the same total disorder, σ for the left and $\sigma = \sqrt{\Sigma^2 + \sigma_1^2}$ for the right.

between the dynamics represented by the average rate and the average of the time-dependent population dynamics. Secondly, any inhomogeneity in J or in the homogeneous lineshape invalidates Equation (3.100), because the total average cannot be expressed as the product of two average quantities in general. In spite of these issues, the use of Equation (3.100) can provide a reasonable first-order estimate of the RET timescale, as long as the inhomogeneity is not substantially larger than the homogeneous contribution.

Further complications arise for aggregates of multiple chromophores. For example, the correlation between intra-aggregate and inter-aggregate disorder can lead to subtle but important effects. Figure 3.3 provides a pictorial representation of two limiting scenarios of the distribution of disorder: one in which the disorder is dominated by inter-aggregate offsets Σ (right panel), and the other in which the intra-aggregate offsets σ are dominant. Both cases have the same total static disorder. Thus, the resulting ensemble lineshapes should be identical although the distributions of rates will be significantly different.

A more fundamental issue that arises in the application of MC-FRET is that none of the tensor elements of the lineshape functions in Equation (3.88) can be obtained easily without full quantum-dynamics calculations. Thus, instead of the full expression, the expression of Equation (3.90) can be used along with an additional approximation to model each exciton peak with pre-determined independent lineshape functions. The majority of applications of MC-FRET to light-harvesting systems employ this approach which seems to result in a reasonable estimate for the average rate in various light-harvesting systems (Mukai *et al.*, 1999; Scholes and Fleming, 2000; Ritz *et al.*, 2001; Scholes, 2003; Renger, 2009; Hossein-Nejad *et al.*, 2011).

3.6 Applications of MC-FRET

Natural photosynthetic antenna complexes (Blankenship, 2002) exhibit a remarkable degree of diversity, and it is believed that their spatial and spectral cross-sections have been tailored to maximize their performance under different conditions. The detailed dynamics of the RET processes in these systems differ widely in different antenna complexes, some of which may involve substantially coherent exciton dynamics. However, in many of these systems, rate processes between groups of coherent delocalized excitons play an important role. Application of MC-FRET theory has elucidated the main mechanisms of robust and efficient energy transfer mechanisms in some systems and helped to identify the rough timescales of energy flow dynamics in others.

3.6.1 Light-harvesting 2 (LH2) complex of purple bacteria

Light-harvesting 2 (LH2) complex is the major antenna complex found in purple bacteria, and mostly consists of 24 or 27 bacteriochlorophylls (BChls) and 16 and 18 protein units that serve as scaffolds. The structure was first determined (McDermott *et al.*, 1995) in 1995 and detailed structural and dynamical information is available in various reviews (Blankenship, 2002; Hu *et al.*, 2002; Cogdell *et al.*, 2006b; Hunter *et al.*, 2009). Figure 3.4 shows the LH2 of *Rps. Acidophila* with all the proteins stripped off. The densely packed ring of 18 BChls absorb photons at around 850 nm and the loosely packed ring of 9 BChls absorbs photons at around 800 nm. Thus, the former is called B850 and the latter B800. Pump probe spectroscopy has confirmed that photons absorbed by the B800 unit transfer quickly to B850 in about 0.7 ps at room temperature (Jimenez *et al.*, 1996) and about 1.5 ps in the low temperature limit of 4 K (Pullerits *et al.*, 1997; Sundstrom *et al.*, 1999). On the other hand, the estimate based on FRET theory was found to be at least a factor of five slower (Jimenez *et al.*, 1996).

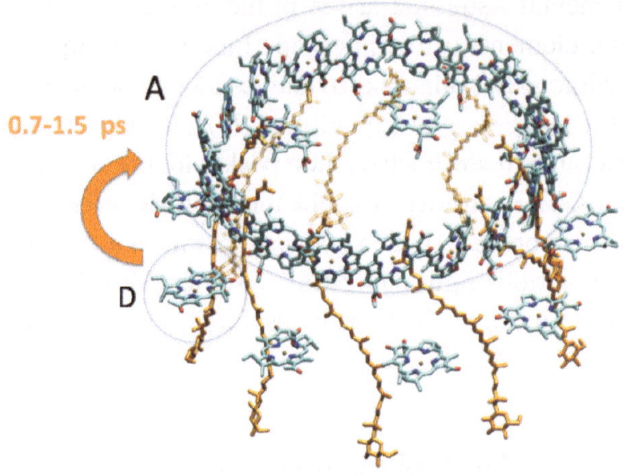

Figure 3.4 Light-harvesting 2 (LH2) of purple bacterium, *Rps. Acidophila*. Not all the proteins are shown, for clarity. The upper ring of densely packed 18 BChls is called B850 and the lower ring of loosely packed 9 BChls is called B800.

The MC-FRET theory was applied to this system (Jang *et al.*, 2007) representing the B800 as a single BChl and treating the entire BChls of the B850 as the acceptor. The electronic coupling Hamiltonian H_{DA} is given by

$$H_{DA} = \sum_{n=1}^{18} J_n(|D\rangle\langle n| + |n\rangle\langle D|), \tag{3.101}$$

where $|D\rangle$ represents the excitation of the B800 BChl and $|n\rangle$ the excitation of the nth BChl in B850. J_n is the transition dipole interaction between $|D\rangle$ and $|n\rangle$, and has the following form:

$$J_n = \frac{\boldsymbol{\mu}_D \cdot \boldsymbol{\mu}_n - 3(\boldsymbol{\mu}_D \cdot \hat{\mathbf{R}}_n)(\boldsymbol{\mu}_n \cdot \hat{\mathbf{R}}_n)}{n_r^2 R_n^3}. \tag{3.102}$$

The MC-FRET rate expression, Equation (3.88), for this system has the following form:

$$k_{B800 \to B850} = \sum_{n,n'=1}^{18} \frac{J_n J_{n'}}{2\pi\hbar^2} \int_{-\infty}^{\infty} d\omega I_{A,nn'}(\omega) L_D(\omega). \tag{3.103}$$

Introducing the following linear combination of the acceptor states weighted by J_n:

$$|J\rangle = \sum_{n=1}^{18} J_n |n\rangle, \tag{3.104}$$

and based on a non-Markovian Redfield type approach, we find that the double summation over n and n' in Equation (3.103) can be approximated as

$$\sum_{n,n'=1}^{18} J_n J_{n'} I_{A,nn'}(\omega) \approx -\frac{1}{\pi}\mathrm{Im}Tr_A\left\{\frac{|J\rangle\langle J|}{\omega + (E_g - H_A^0) + i\mathcal{K}(\omega)}\right\} \equiv J_A(\omega), \tag{3.105}$$

where $\mathcal{K}(\omega)$ represents the effects of the bath on the excitons and is a complex matrix defined in the exciton space. With this, Equation (3.103) can be expressed as

$$k_{B800\to B850} \approx \frac{1}{2\pi\hbar^2}\int_{-\infty}^{\infty} J_A(\omega) L_D(\omega). \tag{3.106}$$

The rate expression, Equation (3.106), has been calculated over an ensemble of disorder that reproduces the experimental ensemble lineshape at a low temperature limit of $k_B T = 10\ \mathrm{cm}^{-1}$. A broad distribution of MC-FRET rates has been found. One of the most important results obtained from this calculation is that the average MC-FRET is about a factor of five times larger than the FRET rate, which is consistent with experimental finding and previous works based on phenomenological lineshape functions (Mukai *et al.*, 1999; Scholes and Fleming, 2000). It was also confirmed that the excitation energies of B800–BChls to those of B850–BChls have been optimized so as to maximize the energy transfer from B800 to B850. Similar work, including the coherent delocalization of B800, also found subtle but important effects (Cheng and Silbey, 2006).

3.6.2 Antenna complexes in LHCII

In higher plants and algae, the majority of light harvesting occurs at LHCII, which absorbs light across a broad spectral range and directs the excitation to the exit sites. Figure 3.5 is a depiction of the LHCII complex. LHCII is not directly linked to the reaction centre of photosystem II (PSII), but the excitation energy that leaves its exit sites can ultimately be delivered to the reaction centres of photosystem II through two or three intervening large protein–pigment complexes. How the overall long-range energy flow occurs with such efficiency and robustness remains to be understood at a quantitative level.

The structure of LHCII was determined in 2005 with a resolution of 2.5 Å, which allowed unambiguous determination of chlorophyll (Chl) species and chromophore orientations (Standfuss *et al.*, 2005). Following this discovery, two-dimensional electronic spectroscopy was used to investigate the real-time energy flow dynamics within the complex (Schlau-Cohen *et al.*, 2009). Two-dimensional spectra provide

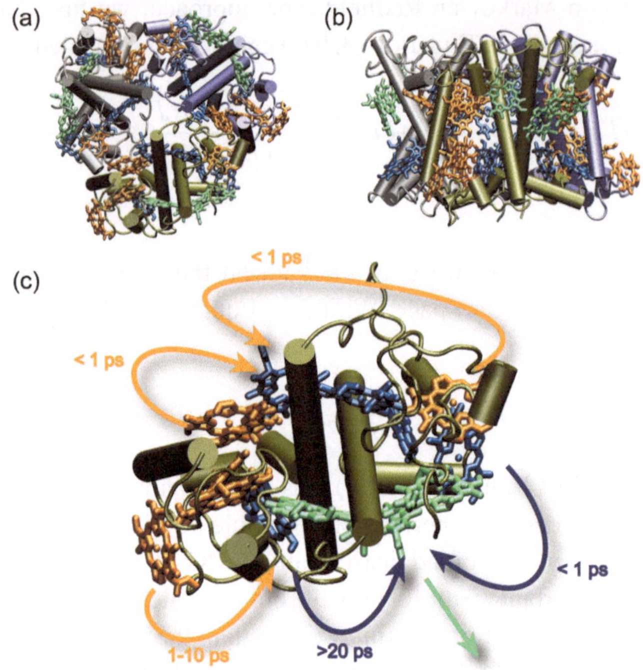

Figure 3.5 Illustration of the structure of plant light-harvesting complex of photosystem II (LHC-II), a trimeric pigment–protein complex that carries over 50% of plant chlorophylls (Chl-a molecules contributing to the final energy acceptor state in green, other Chl-a in blue, Chl-b in orange). (a) Top view from the stromal side. (b) Side view. (c) Schematic picture of the light-harvesting funnel in a single monomer of LHC-II, indicating the approximate timescales for energy transfer (Schlau-Cohen *et al.*, 2009). Figure taken with permission from Andrews *et al.* (2011).

a more detailed reconstruction of the energy transfer channels and their associated timescales, and have unveiled the crucial role played by Chl types and distinct binding sites in determining energy transfer pathways.

LHCII is a trimer, consisting of three monomeric units. Each monomeric unit contains a set of chromophores embedded within the protein matrix, located in the thylakoid membrane of the cell. The shortest inter-pigment distances are of the order of 4 or 5 Å. The monomer subunits contain eight molecules of Chl a, six molecules of Chl b, four carotenoids and five α-helices. Chl b has an extra carbonyl group and therefore a higher excitation energy. Energy transfer within the monomeric subunits consists of fast relaxation ($\sim$100 fs) within the exciton bands of dimeric or trimeric units and slower transfer ($\sim$1 ps) between the exciton bands. Overall, the coherent exciton dynamics within each monomer can be well described by the Redfield equation approach. However, this is not the case for the exciton dynamics between different monomers.

After equilibration within each monomeric subunit, the dynamics becomes dominated by transfer between the Chl *a* clusters of adjacent subunits. The weak coupling between the monomers implies that the MC-FRET theory provides a better description of the transfer dynamics, and that the Redfield theory would overestimate transfer rates. For a comparison of the two results in LHCII, readers are referred to Andrews *et al.* (2011), Novoderezhkin *et al.* (2011a) and Renger *et al.* (2011).

3.7 Summary

This chapter has provided a detailed and self-contained review of Förster's theory and its generalizations. The main objective is to help readers examine all the major steps of the derivation and understand the physical assumptions behind them. This is crucial for clarifying the capability and limitation of the theories when applied to complex systems, such as photosynthetic light-harvesting complexes. In an effort to elucidate the design principles behind the extraordinary efficiency of photosynthetic light-harvesting complexes, this is a critical first step that provides a solid basis for quantitative analysis. In the case of MC-FRET theory, it has already been demonstrated that the theory can explain the hidden role of coherent quantum-mechanical delocalization of excitons for efficient and robust energy flow dynamics. Further investigation of this issue for other light-harvesting complexes and new synthetic systems is a promising area of theoretical research. Considering the softness of these systems, the prospect that inelastic effects play a significant role seems high. This is still a largely unexplored issue but has potential importance.

4

Principles of multi-dimensional electronic spectroscopy

TOMÁŠ MANČAL

4.1 Photo-induced dynamics of molecular systems

Life is intimately connected with the light from the sun, and many properties of living matter are tuned to its spectrum. Many biological functions, be it for harnessing the sun's power or for protection from its harmful rays, rely on interaction of specific molecules with ultraviolet (UV), visible or infrared (IR) light. In this chapter, we will develop a theoretical description of this interaction, which forms the basis both of our current understanding of photo-induced processes and the design of experimental methods used to investigate their details.

Among the biologically relevant photo-induced processes, photosynthesis probably occupies the most prominent place. Essentially all of the energy that is used by the biosphere is a result of photosynthesis, and the prospect of connecting humanity to the same energy source has recently motivated many interdisciplinary research lines. Insight about the design and function of natural photosynthetic systems is likely to play an important role in future research. Within the limited scope of this chapter it is therefore reasonable to limit ourselves to photosynthesis motivated molecular systems. Much of the theory presented in the following sections applies even beyond photosynthesis, e.g. in experiments on molecules related to vision, DNA photo-protection, or in IR spectroscopy of molecular vibrations. However, one has to keep in mind that approximations and models applicable in one field might not work well in another. In this chapter we will introduce and discuss concepts that, together with a good knowledge of a particular physical problem, should enable the reader to develop a theory in order to understand certain important classes of spectroscopic experiments on biomolecules.

To understand light–matter interaction as it plays out in modern spectroscopic experiments, one has to span several fields of classical and quantum physics. Light,

Quantum Effects in Biology, ed. Masoud Mohseni, Yasser Omar, Gregory S. Engel and Martin B. Plenio. Published by Cambridge University Press.

as used in laser spectroscopy, has the convenient property that it can often be described classically. The equations that describe the absorption and non-linear signal generation processes in macroscopic samples are the well-known Maxwell equations of electromagnetism. However, light interacts with individual molecules, which have to be described quantum mechanically. The theory of spectroscopy therefore lies on the crossing point of the macroscopic and microscopic (classical and quantum) theories of light–matter interaction, the theory of open quantum systems and the theory of molecular structure from which it derives its models.

4.1.1 Semi-classical description of light–matter interaction

Light–matter interaction can in principle be included quantum mechanically into the Hamiltonian H, which would then remain time-independent. Such a quantum-mechanical treatment is indeed required in cases where non-classical properties of light play an important role, e.g. in spontaneous emission. Many spectroscopic techniques are, however, well described by the electric field of the light entering the Hamiltonian as an external (classical) parameter. A complete derivation of the semi-classical Hamiltonian is beyond the scope of this chapter. A detailed discussion can be found e.g. in Mukamel (1995) and Grynberg *et al.* (2010). The Hamiltonian which describes the matter and its interaction with the classical field is derived there in a form:

$$\hat{H} \equiv \hat{H}_{sc} = \hat{H}_{mol} + \hat{V}_{inter} - \int d\boldsymbol{r} \hat{\boldsymbol{P}}(\boldsymbol{r}) \cdot \boldsymbol{E}^{\perp}(\boldsymbol{r}, t). \tag{4.1}$$

The Hamiltonian $\hat{H}_{mol}$ is that of isolated molecules (their electronic and nuclear degrees of freedom (DOF)), $\hat{V}_{inter}$ represents the electrostatic interaction between molecules, $\boldsymbol{E}^{\perp}(\boldsymbol{r}, t)$ is the external transverse (radiative) electric field acting on the molecules and $\hat{\boldsymbol{P}}(\boldsymbol{r})$ is the polarization operator. The field $\boldsymbol{E}^{\perp}(\boldsymbol{r}, t)$ is coupled to material polarization $\boldsymbol{P}(\boldsymbol{r}, t) = \mathrm{tr}\{\hat{\boldsymbol{P}}(\boldsymbol{r})\hat{W}(t)\}$ which it generates in the sample through the Maxwell equations. We assume that the wavelength of the light is much larger than the typical dimension of the molecular systems under study, and the interaction of a molecule with light is dominated by its transition dipole moment. Thus we have,

$$\hat{\boldsymbol{P}}(\boldsymbol{r}) = \sum_m \hat{\boldsymbol{\mu}}_m \delta(\boldsymbol{r} - \boldsymbol{R}_m), \tag{4.2}$$

where $\boldsymbol{R}_m$ is the position of the mth molecule in the sample, and $\hat{\boldsymbol{\mu}}_m$ is the transition dipole moment operator of the molecule, given by the charges q_α of all charged particles distributed at positions $\boldsymbol{r}_\alpha$ in the molecule $\hat{\boldsymbol{\mu}}_m = \sum_\alpha q_\alpha(\hat{\boldsymbol{r}} - \boldsymbol{r}_\alpha)$.

Often, molecules carrying the relevant DOF can be grouped into complexes between which interaction through $\hat{V}_{inter}$ is negligible. The problem of N such complexes interacting with their environment through corresponding terms in $\hat{V}_{inter}$ can then be replaced by N problems of a single complex embedded in its environment. Integration over the coordinate $\boldsymbol{r}$ in the light–matter interaction Hamiltonian (third term of Equation (4.1)) leads to $\hat{H}_{int} = -\sum_m \hat{\boldsymbol{\mu}}_m \cdot \boldsymbol{E}^{\perp}(\boldsymbol{R}_m, t)$. One can then choose the centre of a representative complex as the origin of the coordinates. Because the typical dimension L of the complex satisfies $L \ll \lambda$, where λ is a typical wavelength, one can set $\boldsymbol{E}^{\perp}(\boldsymbol{R}_m, t) \approx \boldsymbol{E}^{\perp}(0, t) \equiv \boldsymbol{E}(t)$. This enables us to write the light–molecular complex interaction by means of the total dipole moment operator $\hat{\boldsymbol{\mu}} = \sum_m \hat{\boldsymbol{\mu}}_m$ of the complex, i.e. $\hat{H}_{int} = -\hat{\boldsymbol{\mu}} \cdot \boldsymbol{E}(t)$. The Hamiltonian of the complex in the absence of external light will be denoted $\hat{H}_0 = \hat{H}_{mol} + \hat{V}_{inter}$, and the equation of motion for the molecular system's statistical operator (or density operator) $\hat{W}(t)$ has the following form (May and Kühn, 2011):

$$\frac{\partial}{\partial t}\hat{W}(t) = -\frac{i}{\hbar}[\hat{H}_0, \hat{W}(t)]_- + \frac{i}{\hbar}[\hat{\boldsymbol{\mu}} \cdot \boldsymbol{E}(t), \hat{W}(t)]_-. \tag{4.3}$$

The complexes we have just introduced are the natural entities to work with in spectroscopy. The light is resonantly exciting their *eigenstates* instead of the eigenstates of the individual molecules, so that we cannot investigate easily the properties of individual molecules in the complex. On the other hand, molecules usually have a fixed position and orientation in the complex on the timescale of the experiments we will be discussing, and their spatial relations are reflected in the properties of the optical spectra of the complex. We will assume that the complexes are distributed homogeneously within the sample. This allows us to assume that all points in the sample are essentially equivalent, except for the differences in arrival times of the electromagnetic waves of the light (i.e. in the phase factor $e^{i\boldsymbol{k}\cdot\boldsymbol{R}}$) determined by the position $\mathbf{R}$ of the molecule in the sample. There is no communication between different points of the sample, except for the light waves travelling through it. The position of the complex in the sample bears no importance for the quantum-mechanical calculation, as long as we are able to assign the correct spatial phase to the resulting response.

4.1.2 Response functions

In a non-linear experiment, the field comes in the form of pulses of various durations. It is a matter of experimental evidence that spectroscopic methods can be classified according to the dependency of the signal intensity on the input intensity. We thus have first-order (linear), second- and higher-order (non-linear) methods, and somewhere in our description of the light–matter interaction this notion of orders has to be hidden. Already now, before we connect Equation (4.3) with macroscopic spectroscopic signals, we can demonstrate how the total density operator $\hat{W}(t)$ can naturally be expressed in terms of contributions of different orders in $\boldsymbol{E}(t)$. To simplify the treatment of Equation (4.3), we will take the polarization of the field to be given by a unity vector $\boldsymbol{e}$, so that $\boldsymbol{E}(t) = \boldsymbol{e}E(t)$, and we will introduce a superoperator notation. The commutator with a Hamiltonian $\hat{H}_x$, where x is an arbitrary index, will be denoted as $\hbar\mathcal{L}_x$, i.e. $\hbar\mathcal{L}_x\hat{A} \equiv \hat{H}_x\hat{A} - \hat{A}\hat{H}_x \equiv [\hat{H}_x, \hat{A}]$, for an arbitrary operator $\hat{A}$. Similarly, we will denote $\hbar\mathcal{V}$ the commutator with the scalar transition dipole moment operator $\hat{\mu} = \boldsymbol{e} \cdot \hat{\boldsymbol{\mu}}$. We thus have Equation (4.3) in a form,

$$\frac{\partial}{\partial t}\hat{W}(t) = -i\mathcal{L}_0\hat{W}(t) + i\mathcal{V}\hat{W}(t)E(t). \tag{4.4}$$

To abbreviate this equation we turn to the *interaction picture*. We define $\hat{W}^{(I)}(t) = \mathcal{U}_0^\dagger(t)\hat{W}(t) \equiv \hat{U}_0^\dagger(t)\hat{W}(t)\hat{U}_0(t)$, where ordinary evolution operators are $\hat{U}_x(t) \equiv \exp\{-\frac{i}{\hbar}\hat{H}_x t\}$ (for an arbitrary index x), and the evolution superoperators $\mathcal{U}_x(t)$ are defined similarly as $\mathcal{U}_x(t) \equiv \exp\{-i\mathcal{L}_x t\}$. By definition, the evolution (super)operators satisfy Equation (4.4) with $E(t) \equiv 0$, and can be used to write the density operator at a time t knowing its initial condition at time $t_0 < t$ as

$$\hat{W}(t) = \mathcal{U}_0(t - t_0)\hat{W}(t_0) \equiv \hat{U}_0(t - t_0)\hat{W}(t_0)\hat{U}_0^\dagger(t - t_0). \tag{4.5}$$

The equation of motion for the operator $\hat{W}^{(I)}(t)$ reads

$$\frac{\partial}{\partial t}\hat{W}^{(I)}(t) = i\mathcal{V}(t)\hat{W}^{(I)}(t)E(t), \tag{4.6}$$

where

$$\mathcal{V}(t) \equiv \mathcal{V}^{(I)}(t) = \mathcal{U}_0^\dagger(t - t_0)\mathcal{V}\mathcal{U}_0(t - t_0). \tag{4.7}$$

The initial condition for Equation (4.6) should be specified at $t = t_0$ where $\hat{W}^{(I)}(t_0) = \hat{W}(t_0)$.

A formal solution of Equation (4.6) can be written as the time-ordered exponential

$$\hat{W}^{(I)}(t) = \exp_+\Big\{-i\int_{t_0}^{t} d\tau\,\mathcal{V}(\tau)E(\tau)\Big\}\hat{W}(t_0), \tag{4.8}$$

which is equivalent to an infinite series $\hat{W}^{(I)}(t) = \sum_{n=0}^{\infty}\hat{W}^{(n)}(t)$, where the nth order contribution reads,

$$\hat{W}^{(n)}(t) = i^n\int_{t_0}^{t} d\tau_n\int_{t_0}^{\tau_n} d\tau_{n-1}\cdots\int_{t_0}^{\tau_2} d\tau_1\,\mathcal{V}(\tau_n)\cdots\mathcal{V}(\tau_1)\hat{W}(t_0)E(\tau_n)\cdots E(\tau_1). \tag{4.9}$$

Expectation value $\langle \hat{A} \rangle$ of an arbitrary operator $\hat{A}$ can be expressed in orders of the electric field simply by returning back to the Schrödinger picture and applying the trace to the expansion, Equation (4.8). In this chapter, we will deal exclusively with polarization, i.e. with the expectation value of the transition dipole moment. This reads, $\langle \hat{\mu} \rangle = \sum_n \text{tr}\{\hat{\mu}\mathcal{U}_0(t-t_0)\hat{W}_n(t)\}$. For the nth order we therefore have

$$\langle \hat{\mu} \rangle^{(n)} = i^n \int_{t_0}^{t} d\tau_n \int_{t_0}^{\tau_n} d\tau_{n-1} \cdots \int_{t_0}^{\tau_2} d\tau_1 \text{tr}\{\hat{\mu}\mathcal{U}_0(t-t_0)\mathcal{V}(\tau_n)\cdots\mathcal{V}(\tau_1)\hat{W}(t_0)\}$$
$$\times E(\tau_n)\cdots E(\tau_1). \tag{4.10}$$

Using the properties of evolution operators and the definition of $\mathcal{V}(t)$, Equation (4.7), we can rewrite the trace in Equation (4.10) in terms of the differences $t_a = \tau_{a+1} - \tau_a$ ($\tau_{n+1} = t$), where $a = 1, \ldots, n$. Because the initial state of the system $\hat{W}(t_0)$ should correspond to a thermal equilibrium, it does not evolve under the evolution superoperator $\mathcal{U}_0(t)$. The first factor from the right-hand side inside the trace therefore reads, $\mathcal{U}_0(\tau_1 - t_0)\hat{W}(t_0) = \hat{W}(t_0)$. After substituting t_as in Equation (4.10) we obtain

$$\langle \hat{\mu} \rangle^{(n)} = \int_0^{\infty} dt_n \int_0^{\infty} dt_{n-1} \cdots \int_0^{\infty} dt_1 S_{\mu}^{(n)}(t_n, \ldots, t_1)$$
$$\times E(t-t_n)E(t-t_n-t_{n-1})\cdots E(t-t_n-t_{n-1}\cdots-t_1), \tag{4.11}$$

where we define the nth order *response function* $S_{\mu}^{(n)}$ as

$$S_{\mu}^{(n)}(t_n, \ldots, t_1) = \Theta(t_n)\cdots\Theta(t_1)$$
$$\times i^n \text{tr}\{\hat{\mu}\mathcal{U}_0(t_n)\mathcal{V}\mathcal{U}_0(t_{n-1})\cdots\mathcal{U}_0(t_1)\mathcal{V}\hat{W}(-\infty)\}. \tag{4.12}$$

We have also sent the initial condition formally to negative infinity, $t_0 \to -\infty$, and introduced the Heaviside step function, $\Theta(t)$, to make sure the delays to t_n are always positive.

4.1.3 Meaning of response functions in time and frequency domains

The response functions represent a key to linear and non-linear spectroscopy. They will allow us to establish a link between microscopic molecular dynamics and macroscopic fields. The expectation value of the transition dipole operator can be identified with the transverse polarization $P(t) = \langle \hat{\mu} \rangle_t$, and different orders of polarization can be calculated by the response functions of the corresponding order,

Equation (4.12). For example, in the linear case we can write

$$P^{(1)}(t) = \int_{-\infty}^{\infty} \mathrm{d}\tau\, S_{\mu}^{(1)}(\tau) E(t-\tau), \tag{4.13}$$

$$\tilde{P}^{(1)}(\omega) = \epsilon_0 \chi^{(1)}(\omega) \tilde{E}(\omega). \tag{4.14}$$

Here, we denote Fourier transforms of the polarization and electric field by a tilde, and we define the Fourier transform of the first-order response function as

$$\chi^{(1)}(\omega) = \frac{1}{\epsilon_0} \int_{-\infty}^{\infty} \mathrm{d}t\, S_{\mu}^{(1)}(t) e^{i\omega t}. \tag{4.15}$$

We have used the fact that Equation (4.13) is a convolution, and the Fourier transform of a convolution leads to a product of Fourier transforms. In Equation (4.14), we recognize $\chi^{(1)}$ as the linear susceptibility, which is well known to be related to the effects of absorption and refraction. In a similar manner, higher-order response functions (and susceptibilities) are related to higher-order spectroscopic signals.

The susceptibility or, in other words, the frequency domain response function, can be interpreted as a response of the system to a single frequency component of the light. We define $E_{\omega}^{(+)}(t) = E_0 e^{-i\omega t}$, $E_{\omega}^{(-)}(t) = E_0^* e^{i\omega t}$, so that the real monochromatic electric field is $E_{\omega}(t) = E_{\omega}^{(+)}(t) + E_{\omega}^{(-)}(t)$. We calculate the linear polarization resulting from such a field,

$$P^{(1)}(t) = \int_{-\infty}^{\infty} \mathrm{d}\tau\, S_{\mu}^{(1)}(\tau) e^{i\omega\tau} E_0 e^{-i\omega t} + \int_{-\infty}^{\infty} \mathrm{d}\tau\, S_{\mu}^{(1)}(\tau) e^{-i\omega\tau} E_0^* e^{i\omega t}, \tag{4.16}$$

which yields

$$P^{(1)}(t) = P_{\omega}^{(+)}(t) + P_{\omega}^{(-)}(t) = \epsilon_0 \chi^{(1)}(\omega) E_{\omega}^{(+)}(t) + \epsilon_0 \chi^{(1)}(-\omega) E_{\omega}^{(-)}(t). \tag{4.17}$$

This means that the polarization can also be split into positive and negative frequency components, and $\chi(\omega)$ relates them to the electric field.

Another interpretation can be given to the response function in the time domain. To this end we will consider the field in the form of an ultra short pulse,

$$E^{(+)}(t) = \mathcal{E}(t) e^{-i\omega t}, \tag{4.18}$$

and $\mathcal{E}(t) = E_0 \delta(t)$. Now, Equation (4.13) yields

$$P^{(1)(+)}(t) = S_{\mu}^{(1)}(t) E_0, \tag{4.19}$$

so the time domain response function can be interpreted as the polarization induced by a Dirac δ-pulse of unit intensity.

4.1.4 Macroscopic polarization and the spectroscopic signal

The quantity that connects microscopic photo-induced processes in the studied spectroscopic sample with the detectable macroscopic signal is the polarization $\mathbf{P}(\mathbf{r}, t)$. It is induced by perturbing the state of the system by light; and in the previous section we introduced response functions by which it can be calculated. The Maxwell wave equation describes how the light that enters the sample is modulated by the polarization it induces.

Non-linear spectroscopic methods discussed in this chapter belong to those utilizing the so-called N-wave mixing process, where light in the form of $N-1$ fields,

$$\mathbf{E}(\mathbf{r}, t) = \sum_{n=1}^{N-1} \mathbf{e}_n \mathcal{E}_n(\mathbf{r}, t) e^{i\mathbf{k}_n \cdot \mathbf{r} - i\omega_n t} + c.c., \tag{4.20}$$

generates a field in a direction given by a combination of the incoming wave vectors $\mathbf{k}_s = \pm\mathbf{k}_1 \cdots \pm \mathbf{k}_{N-1}$ (Mukamel, 1995). Here $c.c.$ denotes the complex conjugated term. The most widely used methods belong to the *four-wave mixing* (FWM) type, and the direction $-\mathbf{k}_1 + \mathbf{k}_2 + \mathbf{k}_3$, where the index denotes the order in which the pulses enter the sample, corresponds to the most prominent of the FWM methods, namely the *photon echo.*

Applying Maxwell theory to the problem of propagation of light through a simple slab of width h (Mukamel, 1995), we can show that the light generated in the third-order of the perturbation is proportional to the third-order polarization. Considering the light and polarization in the form of a slowly oscillating envelope and a carrier frequency ω,

$$E^{(3)}(t) = \mathcal{E}^{(3)}(t) e^{-i\omega t} + c.c., \quad P^{(3)}(t) = \mathcal{P}^{(3)}(t) e^{-i\omega t} + c.c., \tag{4.21}$$

we can write

$$\mathcal{E}^{(3)}(t) = i \frac{\omega}{n(\omega)\epsilon_0 c} \mathcal{P}^{(3)}(t) h. \tag{4.22}$$

Here, $n(\omega)$ is the refractive index, ϵ_0 is the permittivity of the vacuum and c is the speed of light. The light emerges from the sample with sharp intensity maxima in the directions of the combination wave vectors corresponding to its order, i.e. in $\mathbf{k}_s = \pm\mathbf{k}_1 \pm \mathbf{k}_2 \pm \mathbf{k}_3$, in the third order (Mukamel, 1995).

A non-linear method of the third order can also use just two directions. The *pump–probe* experiment (Cho *et al.*, 1992) is described by the same general theory as the photon echo, only with $\mathbf{k}_1 \equiv \mathbf{k}_2$. The field generated by the third-order polarization mixes with the incoming field $E_{out}(t) = E_{in}(t) + E^{(3)}(t)$, and the mixing leads to a decrease in the outgoing intensity $I_{out}(t) \approx |E_{out}(t)|^2 \approx |E_{in}(t)|^2 + 2\mathrm{Re}\, E^*(t)E^{(3)}(t)$. Equation (4.22) yields $2\mathrm{Re}\, E(t)E^{(3)}(t) = -2\frac{\omega}{n(\omega)\epsilon_0 c} Im[E^*(t) P^{(3)}(t)]h$. This relation between the absorption, the generating field and the non-linear polarization enables us to define 'absorption' also in cases where the field $E^{(3)}(t)$ travels in a direction where it does not mix with any of its generating fields. By mixing it artificially with a so-called local oscillator field $E_{LO}(t)$ which has the same properties as the generating fields, and comparing such a mixed signal with the real pump–probe measurement, we can define an 'absorptive' part of the non-linear signal. This technique, called *heterodyne detection*, is used e.g. in two-dimensional (2D) coherent spectroscopy (Jonas, 2003; Brixner *et al.*, 2004), which we will introduce in Section 4.4.2

4.2 Non-linear response of multi-state systems

In this section, we will specify a molecular (the three-band) model that will be treated in the following sections. The resonance condition in the light–matter interaction enables us to disregard off-resonant transitions, and the three-band model thus encompasses a broad class of real-world molecular systems.

4.2.1 Two- and three-band molecules

Let us start right away with the notion of bands. Let us assume that the (pulsed) light we use to investigate matter has a carrier frequency Ω, which is similar to the transition frequency between the electronic ground state of matter and a whole band of its excited states (see Figure 4.1). Let us denote the ground state by g and the first excited band by e. Furthermore, once the system is excited in some of the e-band states, it can be promoted by light to another band f lying at an energy $\hbar\Omega$ higher.

An important example of a generic molecular system which fits the above description are the photosynthetic excitonic complexes. In these complexes some relatively simple molecules (chromophores) are bound together non-covalently, and they are held in well-defined positions by a protein scaffold. Thanks to mutual electrostatic interaction between the chromophores, the complex has spectroscopic properties different from a simple sum of the individual chromophores (e.g. in solution). The states of the complex can be well classified using the states of individual chromophores, however. The complex with N chromophores is said to be in its ground state $|g\rangle$, if all its constituting chromophores are in their ground states $|g_i\rangle$. Thus, the complex ground state might be represented as

$$|g\rangle = |g_1\rangle \cdots |g_N\rangle. \tag{4.23}$$

If the complex is formed of similar chromophores with transition frequencies between their ground state and some excited states $|\bar{e}_i\rangle$ equal approximately to

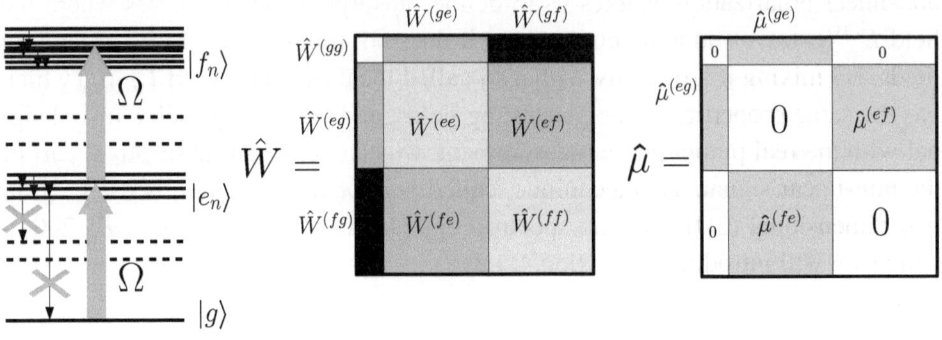

Figure 4.1 Energy level structure of a three-electronic-band system, and the block structure of its statistical operator $\hat{W}$ and the transition dipole operator $\hat{\boldsymbol{\mu}}$. The three-band model consist of the ground state $|g\rangle$ and two bands (e and f) of excited states $|e_n\rangle$ and $|f_n\rangle$. Energy relaxation is allowed within the band, but not between the bands or to other states with non-resonant transitions (dashed lines). The statistical operator $\hat{W}$ has a block structure (centre of the figure) with six independent blocks (plus three obtained by Hermitian conjugation) connected by optical coupling only.

some Ω, then the whole complex will have N excited states of a form,

$$|e_k\rangle = |g_1\rangle \cdots |g_{k-1}\rangle|\bar{e}_k\rangle|g_{k+1}\rangle \cdots |g_N\rangle \tag{4.24}$$

with this transition frequency. In the same way, we also obtain a band of excited states that have an energy $2\hbar\Omega$ larger than the ground state $|g\rangle$. There must be two excitations present in the complex to achieve this, and the two-excitation states read as

$$|f_{kl}\rangle = |g_1\rangle \cdots |g_{k-1}\rangle|\bar{e}_k\rangle|g_{k+1}\rangle \cdots |g_{l-1}\rangle|\bar{e}_l\rangle|g_{kl+1}\rangle \cdots |g_N\rangle. \tag{4.25}$$

The states $|e_k\rangle$ and $|f_{kl}\rangle$ are not the eigenstates of the complex Hamiltonian, but they can be used as a basis to construct a suitable Hilbert space in which the eigenstates can be found by diagonalization of the Hamiltonian. The eigenstates still represent three well-defined bands, as shown in Figure 4.1. Photosynthetic antennae of bacteria and plants belong to this class of molecular complexes. Considering their functions as light energy harvesting and transport units, it is a reasonable expectation that the individual molecules do not de-excite non-radiatively on their 'operational' timescale, i.e. on the timescale of the energy transfer. Indeed, the chlorophyll excited state lifetime is several orders of magnitude longer that the energy transfer timescale. Consequently, non-radiative relaxation between the bands can be assumed to be negligible (this is also denoted in Figure 4.1).

The evolution superoperator elements can be organized into blocks according to the bands, too. The density matrix breaks up into nine blocks (see Figure 4.1) to which we can assign corresponding blocks in the evolution superoperator. Because our model assumes no direct relaxation between the bands, all evolution superoperator blocks connecting two bands are zero. Thus, e.g. $\mathcal{U}^{(eeee)}(t)$ will be the block of the evolution superoperator governing evolution of the populations and intra-band coherences in the e band, which is in turn described by the density operator block $\hat{W}^{(ee)}$. Whenever we need to evaluate the evolution of the density operator block in detail, we can go back to the indices of the individual states, so we can write, e.g.

$$W_{mn}^{(ee)}(t+t_0) = \sum_{k,l=1}^{N} \mathcal{U}_{mn,kl}^{(eeee)}(t) W_{kl}^{(ee)}(t_0), \ n, m = 1, \ldots, N \tag{4.26}$$

or

$$\hat{W}^{(ee)}(t+t_0) = \mathcal{U}^{(eeee)}(t)\hat{W}^{(ee)}(t_0), \tag{4.27}$$

with the same meaning.

One important property of the statistical operator is that the elements of the off-diagonal blocks, e.g. $\hat{W}^{(ge)}(t)$ or $\hat{W}^{(ef)}(t)$, oscillate with frequencies close to Ω. The block $\hat{W}^{(fg)}(t)$ oscillates with 2Ω. The evolution superoperator has to reflect this fact, and corresponding frequencies are indeed imprinted in its time dependence. For further manipulations with its blocks, we will explicitly subtract this fast frequency, and introduce a slowly varying envelope, as in the following example,

$$\mathcal{U}^{(egeg)}(t) = \tilde{\mathcal{U}}^{(egeg)}(t)e^{-i\Omega t}, \ \mathcal{U}^{(fgfg)}(t) = \tilde{\mathcal{U}}^{(fgfg)}(t)e^{-i2\Omega t}. \tag{4.28}$$

From now on in this chapter, tilde will denote these slow envelopes.

The second quantity that enters the response functions in both operator and superoperator form is the transition dipole moment. Here, we assume that this operator mediates transitions between bands. Consequently, the dipole operator has a block form, as in Figure 4.1. Matrices can be multiplied in blocks, and thus it is easy to verify that the action of the transition dipole moment on the density operator transforms one block into another. With the transition dipole moment operator,

$$\hat{\mu} = \hat{\mu}^{(eg)}|e\rangle\langle g| + \hat{\mu}^{(fe)}|f\rangle\langle e| + h.c., \tag{4.29}$$

where $h.c.$ stands for Hermitian conjugate, we have e.g. $\hat{\mu}\hat{W}^{(gg)}|g\rangle\langle g| = \hat{\mu}^{(eg)}\hat{W}^{(gg)}|e\rangle\langle g|$. The action of the operator $\hat{\mu}$ can equally well be written using the corresponding superoperator. Formally, the difference is that, in the case of a superoperator, its action changes both of the operator (block) indices at the same time (the superoperator acts on both the ket and the bra of the operator at the same time). We denote this by writing out both the initial and the final pairs of the operator (block) indices in the acting superoperator element. Thus we write $\hat{\mu}^{(eg)}\hat{W}^{(gg)} = \mathcal{V}^{(eggg)}\hat{W}^{(gg)}$. We can see that the action of the dipole (super)operator transforms the diagonal elements of the density operator to the off-diagonal ones; and similarly we can show that off diagonal blocks are transformed into diagonal ones by the action of the dipole moment operator from the left or right-hand sides. For the action of the transition dipole moment superoperator we have

$$\mathcal{V}\hat{W}^{(gg)}(t)|g\rangle\langle g| = \mathcal{V}^{(eggg)}\hat{W}^{(gg)}(t)|e\rangle\langle g| - \mathcal{V}^{(gegg)}\hat{W}^{(gg)}(t)|g\rangle\langle e|. \tag{4.30}$$

The action of a transition dipole moment superoperator on a block of the statistical operator yields contributions to its two different blocks.

4.2.2 Liouville pathways

Response functions consist of many such actions of transition dipole operators and propagations by evolution superoperators, with a trace over everything at the end. In optical spectroscopy, we always start with only the electronic ground state populated (thermal energy is much smaller than the optical energy gap, $\hbar\Omega \gg k_B T$), and the tracing requires that after all actions of the dipole moment operators, we end up on the diagonal block. This means that not all terms in the response function, Equation (4.12), are non-zero. There are only certain *pathways* through the bands of the molecular system that can contribute. These pathways occur in the so-called *Liouville space* of projectors $|a\rangle\langle b|$, where a and b denote bands (or individual energy levels). To visualize and organize these so-called *Liouville pathways*, we can use the diagram in Figure 4.2.

We denote the projector corresponding to the non-zero density matrix element by a circle with its band indices. The action of a dipole moment operator from the left-hand side changes the left band index, and similarly, action from the right-hand side changes the right band index. In the diagram we have denoted the left action of the dipole operator by a downward pointing arrow, and a corresponding change of the left index, and the right action by a right arrow and a change of the right index. Each commutator in Equation (4.12) corresponds to two arrows. In a three-band system, some actions of the dipole moment operator correspond to two possible changes of indices ($|e\rangle \rightarrow |g\rangle$ and $|e\rangle \rightarrow |f\rangle$), so we keep both alternatives in the diagram. Thanks to the cyclic invariance of the operators under the trace, the dipole moment on the furthest left in Equation (4.12) can be equivalently written on the furthest right of the expression. It can therefore correspond either to the downward or the right arrow.

Figure 4.2 shows that there are sixteen possible Liouville pathways between the three bands that can be travelled by three actions of the dipole moment. The projectors $|a\rangle\langle b|$ and $|b\rangle\langle a|$ correspond to mutually Hermite conjugated blocks of the density matrix. The pathways that are mirrors of each other with respect

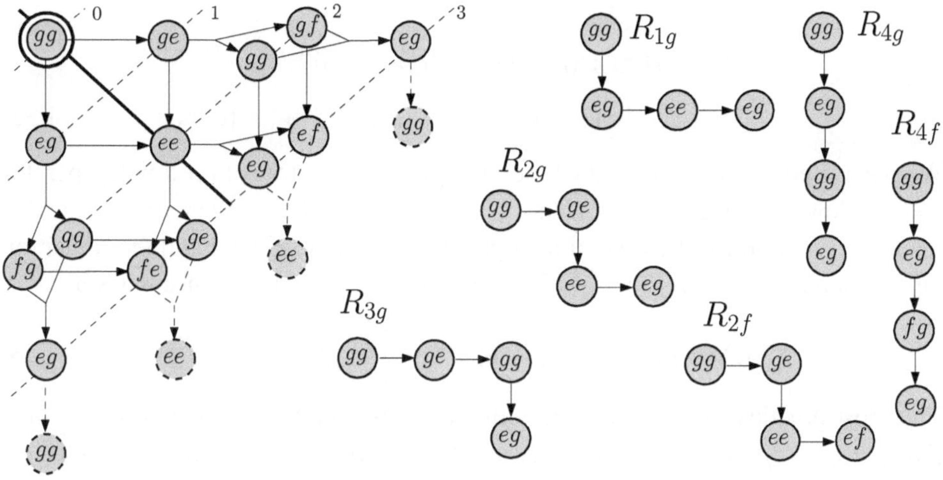

Figure 4.2 Liouville pathways: The left side of the figure shows possible pathways, starting from the electronic ground state $|g\rangle$. The dashed lines connect the states after zero, one, two and three perturbative interactions with the field. Pathways which are mirror images of each other with respect to the thick full line are complex conjugated to each other. Examples of individual Liouville pathways are given on the right-hand part of the figure.

to the diagonal line of the diagram, therefore have mutually complex conjugated contributions. We have eight independent contributions, four of which contain only bands g and e, and four of which also include the f band. Moreover, those pathways that involve an odd number of dipole operator actions from the right, carry a minus sign, which originates from the second term of a commutator. There are four pairs of pathways with a plus sign that correspond to four different orderings of left and right actions of the dipole moment, and which differ by exchanging g and f. We denote their contributions to the total response function by the letter R with a lower index $1, \ldots, 4$, denoting the four independent orderings of interactions from the left and right. Consequently, there are four independent shapes of Liouville pathways in Figure 4.2. We will introduce a second index, g or f, to distinguish pathways which reach the f-band from those that do not. Thus, the total third-order response function reads as

$$S_{\mu}^{(3)}(t_3, t_2, t_1) = i^3 \Theta(t_3)\Theta(t_2)\Theta(t_1) \sum_{n=1}^{4} \sum_{a=g,f} [R_{na}(t_3, t_2, t_1) - R_{na}^{*}(t_3, t_2, t_1)]. \tag{4.31}$$

Now it is possible to construct individual Liouville pathways and evaluate them using the evolution superoperators.

Before we do this it is a good exercise to follow the same approach on a simpler problem of linear responses. Let us start with the linear response function written as

$$S_{\mu}^{(1)}(t) = i\Theta(t)\mathrm{tr}\{\hat{\mu}^{(ge)}\tilde{\mathcal{U}}^{(egeg)}(t)\mathcal{V}^{(eggg)}W^{(gg)}\}e^{-i\Omega t} - i\Theta(t)\mathrm{tr}\{\hat{\mu}^{(eg)}\tilde{\mathcal{U}}^{(gege)}(t)\mathcal{V}^{(gegg)}W^{(gg)}\}e^{i\Omega t}. \quad (4.32)$$

We have used the superoperator block notation for the transition dipole operator and the slow envelopes for the evolution superoperator blocks. We will evaluate this response function for the case of a single level $|e\rangle$ in the e band and the transition dipole moment $d_{eg} = \langle e|\hat{\mu}^{(eg)}|g\rangle$. The two-level first-order response reads as

$$S_{\mu}^{(1)}(t) = \frac{i}{\hbar}|d_{eg}|^2\tilde{\mathcal{U}}_{egeg}^{(egeg)}(t)e^{-i\Omega t} - \frac{i}{\hbar}|d_{eg}|^2\tilde{\mathcal{U}}_{gege}^{(gege)}(t)e^{i\Omega t}. \quad (4.33)$$

The most practical way of dealing with a two-level system is to choose $\Omega = \omega_{eg}$ ($\hbar\omega_{eg} = \epsilon_e - \epsilon_g$, where ϵ_e and ϵ_g are the energy of the state $|e\rangle$ and $|g\rangle$, respectively). With this choice, $\tilde{\mathcal{U}}^{(egeg)}(t)$ would contain just the dephasing of the $|g\rangle \rightarrow |e\rangle$ transition. For our purpose here, we can assume it in a form $\tilde{\mathcal{U}}^{(egeg)}(t) = \tilde{\mathcal{U}}^{(gege)}(t) = e^{-\Gamma t}$, with some dephasing rate Γ. The absorption coefficient can be obtained by Fourier transform of the $S_{\mu}^{(1)}$ (Mukamel, 1995). Because the Fourier transform is performed only over positive frequencies, the second term in Equation (4.33) will contribute much less then the first, and we can neglect it. The Fourier transform yields the real and imaginary part of the susceptibility $\chi^{(1)} = \chi' + i\chi''$,

$$\chi'(\omega) = \frac{|d_{eg}|^2}{\epsilon_0\hbar}\frac{(\omega_{eg} - \omega)}{(\omega_{eg} - \omega)^2 + \Gamma^2}, \quad \chi''(\omega) = \frac{|d_{eg}|^2}{\epsilon_0\hbar}\frac{\Gamma}{(\omega_{eg} - \omega)^2 + \Gamma^2}. \quad (4.34)$$

The latter gives the well-known Lorentz absorption line-shape.

Let us construct the contributions to the third-order signal. Following Figure 4.2 we can write for R_1 and R_2,

$$R_{1g}(t_3, t_2, t_1) = \mathrm{tr}\{\hat{\mu}^{(ge)}\tilde{\mathcal{U}}^{(egeg)}(t_3)\mathcal{V}^{(egee)}\tilde{\mathcal{U}}^{(eeee)}(t_2) \times \mathcal{V}^{(eeeg)}\tilde{\mathcal{U}}^{(egeg)}(t_1)\mathcal{V}^{(eggg)}\hat{W}^{(gg)}\}e^{-i\Omega(t_1+t_3)}, \quad (4.35)$$

$$R_{2g}(t_3, t_2, t_1) = \mathrm{tr}\{\hat{\mu}^{(ge)}\tilde{\mathcal{U}}^{(egeg)}(t_3)\mathcal{V}^{(egee)}\tilde{\mathcal{U}}^{(eeee)}(t_2) \times \mathcal{V}^{(eege)}\tilde{\mathcal{U}}^{(gege)}(t_1)\mathcal{V}^{(gegg)}\hat{W}^{(gg)}\}e^{-i\Omega(t_3-t_1)}, \quad (4.36)$$

$$R_{1f}(t_3, t_2, t_1) = \mathrm{tr}\{\hat{\mu}^{(fe)}\tilde{\mathcal{U}}^{(efef)}(t_3)\mathcal{V}^{(efee)}\tilde{\mathcal{U}}^{(eeee)}(t_2) \times \mathcal{V}^{(eeeg)}\tilde{\mathcal{U}}^{(egeg)}(t_1)\mathcal{V}^{(eggg)}\hat{W}^{(gg)}\}e^{-i\Omega(t_1-t_3)}, \quad (4.37)$$

$$R_{2f}(t_3, t_2, t_1) = \mathrm{tr}\{\hat{\mu}^{(fe)}\tilde{\mathcal{U}}^{(efef)}(t_3)\mathcal{V}^{(efee)}\tilde{\mathcal{U}}^{(eeee)}(t_2) \times \mathcal{V}^{(eege)}\tilde{\mathcal{U}}^{(gege)}(t_1)\mathcal{V}^{(gegg)}\hat{W}^{(gg)}\}e^{i\Omega(t_1+t_3)}, \quad (4.38)$$

and the rest of the expressions can be obtained in the same way. The eight functions $R_{1g}, \ldots, R_{4g}$ and $R_{1f}, \ldots, R_{4f}$ completely determine the third-order response of a

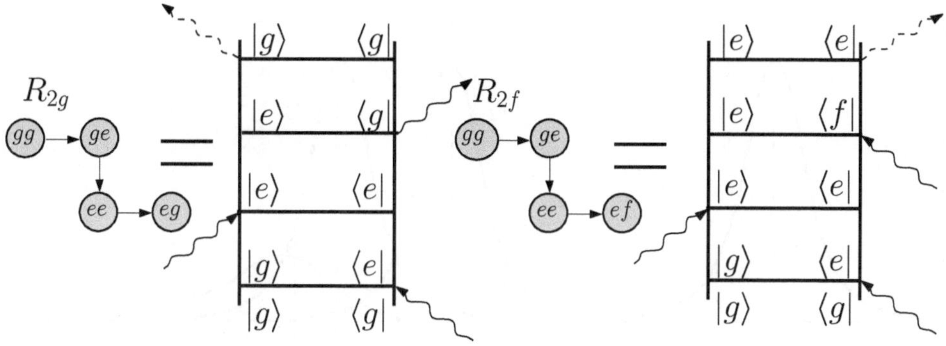

Figure 4.3 Two examples of the double-sided Feynman diagram. Liouville pathways R_{2g} and R_{2f} are translated from the scheme of Figure 4.2 into Feynman diagrams. Arrows denote absorption and emission events, the dashed arrow represents the signal emission.

three-band system. Each of the functions involve a sum over a possibly large number of 'subpathways' involving individual levels of bands. The evolution superoperator $\tilde{\mathcal{U}}^{(eeee)}$ includes the effects of population transfer between different levels of the excited state band. Very often these are the processes which interest us the most. The time variables t_1 and t_3 are often referred to as *coherence times*, because they appear only in the blocks of evolution superoperators that describe optical coherences. The time variable t_2 is always associated with the propagation of diagonal block(s) of the statistical operator, and it is therefore referred to as the *population time*.

At this point, it is useful to comment on a frequent and powerful way of representing individual contributions to the response function, namely so-called *double-sided Feynman diagrams*. A double-sided Feynman diagram carries at least the same information as the diagrams on the right-hand side of Figure 4.2. Depending on notation, additional information can be included, which completely determines the Liouville pathway's characteristics (Mukamel, 1995). Figure 4.3 shows two Feynman diagrams compared with the Liouville pathway diagram from Figure 4.2. The two vertical lines denote the time evolution of the ket and bra of the system's statistical operator (time is running upwards). Each horizontal bar denotes the time when an interaction with the external field (i.e. multiplication of the statistical operator from left or right by the transition dipole operator) occurs. The ket and bra after each interaction are written above the bar.

4.2.3 Third-order polarization in a rotating wave approximation

So far we have evaluated the response functions separately from Equation (4.11) which connects them with the third-order polarization. In Equation (4.11), the

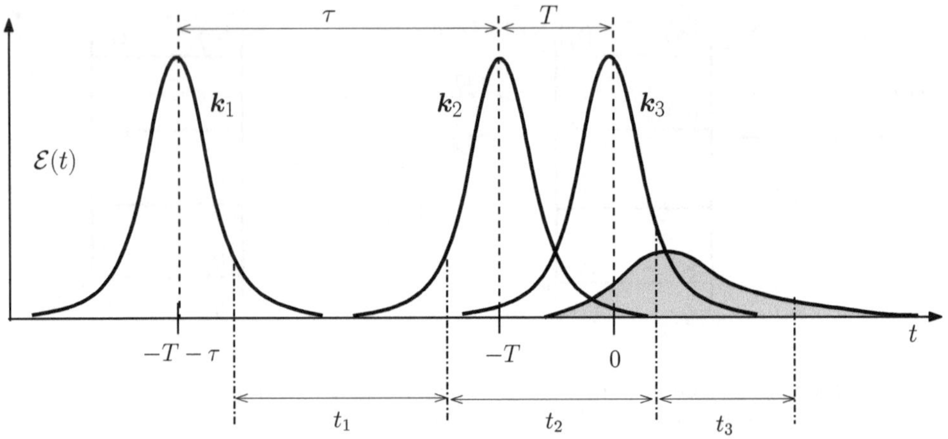

Figure 4.4 Pulse scheme of the four-wave mixing experiment. The excitation pulses are centred at times $-T-\tau$, $-\tau$ and 0. The third-order signal at time t is then an integral over contributions from all possible combinations of positive delays t_1, t_2 and t_3.

electric field $E(t)$ carries phase factors not dissimilar to those of Liouville pathways. The polarization is obtained by integrating the product of the response functions and the fields. Because the optical frequency is very fast with respect to evolution of the pulse and response envelopes, it is expected that terms where optical frequencies cancel out will yield a much larger contribution to the integral than those where some optical frequency remains uncancelled.

To put these words into equations, let us assume the incoming electric field to be formed of three pulses with slowly varying envelopes, depicted in Figure 4.4, i.e.

$$E(\boldsymbol{r},t) = \mathcal{E}_1(t+T+\tau)e^{-i\Omega_1(t+T+\tau)+i\boldsymbol{k}_1\cdot\boldsymbol{r}} + \mathcal{E}_2(t+T)e^{-i\Omega_2(t+\tau)+i\boldsymbol{k}_2\cdot\boldsymbol{r}} + \mathcal{E}_3(t)e^{-i\Omega_3 t+i\boldsymbol{k}_3\cdot\boldsymbol{r}} + c.c. \quad (4.39)$$

Here $c.c.$ denotes the complex conjugated term. In the product $E(\boldsymbol{r},t-t_3-t_2-t_1)E(\boldsymbol{r},t-t_3-t_2)E(\boldsymbol{r},t-t_3)$ there are six terms with spatial phase factor $e^{i(-\boldsymbol{k}_1+\boldsymbol{k}_2+\boldsymbol{k}_3)\cdot\boldsymbol{r}}$. Each of these field terms also carries a phase factor in t_3, t_2 and t_1 over which we integrate (and phase factors in t, T and τ which can be taken out of the integration). The differences between two laser frequencies are assumed to be small in comparison with optical frequencies, i.e. $0 \approx \Omega_n - \Omega_m \ll \Omega_n \approx \Omega, n, m = 1, 2, 3$, where Ω is the average carrier frequency of the excitation pulses. Thus the phase factors are of three possible kinds, $e^{i\Omega(t_3-t_1)}$, $e^{i\Omega(t_3+t_1)}$ and $e^{i\Omega(t_1+2t_2+t_3)}$. Looking at the expressions for the Liouville pathways, e.g. Equation (4.35), we see that they have similar phase factors in variables t_3, t_2 and t_1. If a given Liouville pathway does not have a phase factor which cancels some phase factor of the field, its contribution to the integral will be neglected. This is a version of the frequently used *rotating wave approximation* (RWA).

For the third-order polarization in RWA we can now write the following integral expression (Brixner *et al.*, 2004):

$$
\begin{aligned}
&P^{(3)}_{RWA}(t, T, \tau) \\
&\quad \approx e^{i(\Omega_1-\Omega_2-\Omega_3)t+i(\Omega_1-\Omega_2)T+i\Omega_1\tau} \\
&\quad \times \int_0^\infty dt_3 \int_0^\infty dt_2 \int_0^\infty dt_1 \Big\{ S_R(t_3, t_2, t_1)\mathcal{E}_1^*(t+T+\tau-t_3-t_2-t_1) \\
&\quad \times \big[\mathcal{E}_2(t+T-t_3-t_2)\mathcal{E}_3(t-t_3)e^{-i(\Omega_1-\Omega_2-\Omega_3)t_3-i(\Omega_1-\Omega_2)t_2-i\Omega_3 t_1} \\
&\quad + \mathcal{E}_3(t-t_3-t_2)\mathcal{E}_2(t+T-t_3)e^{-i(\Omega_1-\Omega_2-\Omega_3)t_3-i(\Omega_1-\Omega_3)t_2-i\Omega_3 t_1}\big] \\
&\quad + S_{NR}(t_3, t_2, t_1)\mathcal{E}_1^*(t+T+\tau-t_3-t_2) \\
&\quad \times \big[\mathcal{E}_2(t+T+\tau-t_3-t_2-t_1)\mathcal{E}_3(t-t_3)e^{-i(\Omega_1-\Omega_2-\Omega_3)t_3-i(\Omega_1-\Omega_2)t_2+i\Omega_2 t_1} \\
&\quad + \mathcal{E}_3(t-t_3-t_2-t_1)\mathcal{E}_2(t+T-t_3)e^{-i(\Omega_1-\Omega_2-\Omega_3)t_3-i(\Omega_1-\Omega_3)t_2+i\Omega_3 t_1}\big] \\
&\quad + S_{DC}(t_3, t_2, t_1)\mathcal{E}_1^*(t+T+\tau-t_3) \\
&\quad \times \big[\mathcal{E}_2(t+T-t_3-t_2-t_1)\mathcal{E}_3(t-t_3-t_2)e^{-i(\Omega_1-\Omega_2-\Omega_3)t_3+i(\Omega_2+\Omega_3)t_2+i\Omega_2 t_1} \\
&\quad + \mathcal{E}_3(t-t_3-t_2-t_1)\mathcal{E}_2(t+T-t_3-t_2)e^{-i(\Omega_1-\Omega_2-\Omega_3)t_3+i(\Omega_2+\Omega_3)t_2+i\Omega_3 t_1}\big]\Big\}.
\end{aligned}
\tag{4.40}
$$

Here we have collected the Liouville pathways into groups according to their characteristic oscillating phase factors. Thus we have *rephasing pathways*,

$$S_R(t_3, t_2, t_1) = R_{2g}(t_3, t_2, t_1) + R_{3g}(t_3, t_2, t_1) - R^*_{1f}(t_3, t_2, t_1), \tag{4.41}$$

non-rephasing pathways,

$$S_{NR}(t_3, t_2, t_1) = R_{1g}(t_3, t_2, t_1) + R_{4g}(t_3, t_2, t_1) - R^*_{2f}(t_3, t_2, t_1), \tag{4.42}$$

and *double coherence pathways*,

$$S_{DC}(t_3, t_2, t_1) = R_{4f}(t_3, t_2, t_1) - R^*_{3f}(t_3, t_2, t_1). \tag{4.43}$$

Equation (4.40) enables us to calculate the signal for any third-order non-linear experiment employing a three-pulse sequence with arbitrary pulse shapes.

4.2.4 Third-order polarization in the impulsive limit

Laser pulses that are used in the experiment to excite third-order polarization in the sample may be very short, sometimes with just a few cycles of optical oscillation within the pulse envelope. It is therefore often possible to simulate the non-linear response assuming that it was excited by an infinitely short pulse of the form,

$$E(t) = E_0\delta(t)e^{-i\Omega t} + c.c.. \tag{4.44}$$

The above delta function must be understood as itself 'slowly' varying with respect to the optical frequency, and it is allowed to have Dirac delta function properties only with respect to the slow envelopes of the response functions. Such a delta function is often referred to as a *physical delta function* (see e.g. the discussion in Mukamel (1995)).

Equation (4.40) can be simplified significantly by the assumption, Equation (4.44). Most importantly, one can immediately see that for different pulse orderings, different types of Liouville pathways contribute to the signal in any chosen direction. Let us choose $\mathbf{k}_s = -\mathbf{k}_1 + \mathbf{k}_2 + \mathbf{k}_3$ as the investigated direction, and let us first assume that the pulses arrive in the order $\boldsymbol{k}_1$, $\boldsymbol{k}_2$, $\boldsymbol{k}_3$ (we will denote this order as $1-2-3$), and that they are not overlapping. We have six combinations of pulse envelopes in Equation (4.40), but only one of them yields a non-zero integration. Let us examine the first term, i.e.

$$\begin{aligned}&\mathcal{E}_1^*(t+T+\tau-t_3-t_2-t_1)\mathcal{E}_2(t+T-t_3-t_2)\mathcal{E}_3(t-t_3)\\&\quad\approx\delta(t+T+\tau-t_3-t_2-t_1)\delta(t+T-t_3-t_2)\delta(t-t_3).\end{aligned} \tag{4.45}$$

This term yields the conditions, $t+T+\tau-t_3-t_2-t_1=0, t+T-t_3-t_2=0$ and $t-t_3=0$, for the integral in Equation (4.40) to give a non-zero contribution. They can easily be satisfied by $t_3=t$, $t_2=T$ and $t_1=\tau$, and thus the contribution to the polarization yields $S_R(t,T,\tau)E_0^3$. The second term in Equation (4.40), which corresponds with the roles of the second and third pulses switched, similarly yields the conditions $t+T+\tau-t_3-t_2-t_1=0$, $t-t_3-t_2=0$ and $t+T-t_3=0$. This can be satisfied with $t_3=t+T$, $t_2=-T$ and $t_1=T+\tau$, but for negative t_2 ($T>0$), the response functions are zero. A similar conclusion will be reached for all other integrals, and thus the impulsive signal from the $(1-2-3)$ order of pulses consists of one contribution only, namely,

$$P^{(3)}_{RWA}(t,T,\tau) \approx S_R(t,T,\tau)E_0^3. \tag{4.46}$$

Switching the order of pulses into $(1-3-2)$ can be achieved formally by making T negative and keeping $|T|<\tau$ (see Figure 4.4). The time should be measured from the last arriving pulse, which is the $\boldsymbol{k}_2$-pulse centred at $|T|$. We define a new time $t' = t - |T|$, a new time $T' = |T|$ separating the second and the last arriving pulses and a new time $\tau' = \tau - |T|$. These new 'natural' variables t', T' and τ' again measure the delays between the pulses by positive values. The third-order polarization in these new variables reads as

$$P^{(3)}_{RWA}(t', T', \tau') \approx S_R(t', T', \tau')E_0^3. \tag{4.47}$$

We can therefore conclude that for both orderings $(1-2-3)$ and $(1-3-2)$, only the rephasing Liouville pathways contribute to the signal.

Let us now switch the order of the first two pulses to yield the order $(2-1-3)$. This corresponds to $\tau < 0$. In addition $|\tau| < T$, so that pulse $\boldsymbol{k}_3$ does not precede pulse $\boldsymbol{k}_1$. The corresponding delta functions yield conditions leading to $t_3 = t$, $t_2 = T + \tau$ and $t_1 = -\tau$. The natural time variables in this order are $t'' = t_3$, $T'' = T + |\tau|$ and $\tau'' = |\tau|$. Now however, it is the third term in Equation (4.40) which yields a non-zero contribution and thus,

$$P^{(3)}_{RWA}(t'', T'', \tau'') \approx S_{NR}(t'', T'', \tau'')E_0^3. \tag{4.48}$$

The very same conclusion is reached for pulse order $(3-1-2)$ in its natural delay variables.

To conclude, by measuring in the $-\boldsymbol{k}_1 + \boldsymbol{k}_2 + \boldsymbol{k}_3$ direction one can selectively probe the rephasing or non-rephasing groups of Liouville pathways by selecting which of the pulses $\boldsymbol{k}_1$ and $\boldsymbol{k}_2$ will act first. In an experimental arrangement, instead of exchanging the order of pulses in time and keeping the same detection direction, we may look into a direction with the role of pulses 1 and 2 exchanged $(-\boldsymbol{k}_2 + \boldsymbol{k}_1 + \boldsymbol{k}_3)$, while keeping the original time order $(1-2-3)$. Thus we can measure S_R and S_{NR} simultaneously by collecting the signal from two directions.

The suspicious looking pathways S_{DC} contribute to the signal only when the order of the pulses is $(2-3-1)$ or $(3-2-1)$. The S_{DC} pathways contain a fast oscillating term, cancelling the 2Ω term in the field factor which appears in the orderings $(2-3-1)$ and $(3-2-1)$. Although the oscillations over the population interval t_2 are very fast, they also survive RWA. This signal is emitted only if the last of the interacting pulses is $\boldsymbol{k}_1$.

4.3 Cumulant expansion of a non-linear response

The results discussed in previous sections are valid for a very broad class of two- and three-band molecules. They are very general, and they can provide a starting point for calculation of the non-linear response from equations of motion for the (reduced) density operator. For few electronic level systems, one can go into greater depth analytically, as we will demonstrate below.

In this section, we will reduce our original three-band model to just a single electronic level per band. We will assume that this three-level system interacts with a bath of harmonic oscillators. For this problem, the response functions can be evaluated using so-called cumulant expansion, which is exact for a bath of harmonic oscillators linearly coupled to the electronic system.

4.3.1 Energy gap correlation function

Let us consider a three-level electronic system with the following Hamiltonian:

$$\hat{H} = \hat{T} + (\varepsilon_g + \hat{V}_g)|g\rangle\langle g| + (\varepsilon_e + \hat{V}_e)|e\rangle\langle e| + (\varepsilon_f + \hat{V}_f)|f\rangle\langle f|. \tag{4.49}$$

This Hamiltonian represents a system with electronic ground state $|g\rangle$, ground state electronic energy ε_g, nuclear kinetic energy operator $\hat{T}$ and ground state nuclear potential $\hat{V}_g$. When the system is in its electronically excited state $|e\rangle$ ($|f\rangle$), the electronic energy is ε_e (ε_f), and the nuclei feel potential energy $\hat{V}_e$ ($\hat{V}_f$). This model does not include any terms allowing a non-radiative transition between electronic states. Our simple Hamiltonian has to be augmented, if we want to include these processes into our considerations. Let us consider the Hamiltonian, Equation (4.49), as fully describing our system of interest. We can proceed by splitting it into the usual system–bath form, $\hat{H} = \hat{H}_B + \hat{H}_S + \hat{H}_{S-B}$. Because we are dealing with optical transitions and we assume room temperature (and lower), the equilibrium density matrix can be written as

$$\hat{W}_{eq} = \hat{W}_{eq}^{(\text{bath})}|g\rangle\langle g|. \tag{4.50}$$

From this it follows that the Hamiltonian of the bath can be identified with the nuclear Hamiltonian of the electronic ground state. Thus we have

$$\hat{H}_B = \hat{T} + \hat{V}_g. \tag{4.51}$$

To split Equation (4.49) into purely electronic and purely bath parts, we will use the electronic completeness relation in the form, $1_{el} = |g\rangle\langle g| + |e\rangle\langle e| + |f\rangle\langle f|$. We can write

$$\begin{aligned}\hat{H} = \hat{H}_B \otimes \hat{1}_{el} + \hat{1}_{bath} \otimes \left(\varepsilon_g|g\rangle\langle g| + \varepsilon_e|e\rangle\langle e| + \varepsilon_f|f\rangle\langle f|\right)\\ + (\hat{V}_e - \hat{V}_g)|e\rangle\langle e| + (\hat{V}_f - \hat{V}_g)|f\rangle\langle f|.\end{aligned} \tag{4.52}$$

Here, we have introduced a unity operator $\hat{1}_{bath}$ on the bath Hilbert space. We have thus split the Hamiltonian as required, and we can identify $\hat{H}_{S-B}$ with the last term of Equation (4.52). It is advantageous to add a condition that the bath part of the interaction Hamiltonian has an equilibrium expectation value equal to zero, i.e.

$\mathrm{tr}_{\mathrm{bath}}\{\hat{H}_{S-B}\hat{W}_{eq}^{(\mathrm{bath})}\} = 0$. To achieve this we define so-called *energy gap operators*,

$$\Delta\hat{V}_{eg} = \hat{V}_e - \hat{V}_g - \mathrm{tr}_{\mathrm{bath}}\{(\hat{V}_e - \hat{V}_g)\hat{W}_{eq}^{(bath)}\}, \tag{4.53}$$

$$\Delta\hat{V}_{fg} = \hat{V}_f - \hat{V}_g - \mathrm{tr}_{\mathrm{bath}}\{(\hat{V}_f - \hat{V}_g)\hat{W}_{eq}^{(bath)}\}, \tag{4.54}$$

so that

$$\hat{H}_{S-B} = \Delta\hat{V}_{eg}|e\rangle\langle e| + \Delta\hat{V}_{ef}|f\rangle\langle f|. \tag{4.55}$$

The values that we subtracted from the last two terms of Equation (4.52) are added in the second one by defining new electronic energies,

$$\begin{aligned}\epsilon_g &= \varepsilon_g, \epsilon_e = \varepsilon_e + \mathrm{tr}_{bath}\{(\hat{V}_e - \hat{V}_g)\hat{W}_{eq}\},\\ \epsilon_f &= \varepsilon_f + \mathrm{tr}_{bath}\{(\hat{V}_f - \hat{V}_g)\hat{W}_{eq}\}.\end{aligned} \tag{4.56}$$

The electronic Hamiltonian thus has a very simple form,

$$\hat{H}_S = \epsilon_g|g\rangle\langle g| + \epsilon_e|e\rangle\langle e| + \epsilon_f|f\rangle\langle f|. \tag{4.57}$$

Let us now evaluate the first-order response function for this system. We will use the so-called Condon approximation, which states that the transition dipole moment does not depend on the nuclear coordinates, and it can thus be represented by a rather simple operator,

$$\hat{\mu} = d_{eg}|e\rangle\langle g| + d_{ge}|g\rangle\langle e|. \tag{4.58}$$

Here, $d_{eg} = (d_{ge})^*$ are complex numbers.

Let us first demonstrate evaluation of the material response on the first-order response functions. The first-order response function $S_\mu^{(1)}$, Equation (4.32), is composed of two short Liouville pathways, $J(t)$ and $J^*(t)$,

$$S_\mu^{(1)}(t) = \frac{i}{\hbar}|d_{eg}|^2\Theta(t)[J(t) - J^*(t)], \tag{4.59}$$

where

$$J(t) = \mathrm{tr}\{\hat{U}_g^\dagger(t)\hat{U}_e(t)\hat{W}_{eq}^{(\mathrm{bath})}\}. \tag{4.60}$$

Here, $U_g(t)$ and $U_e(t)$ are evolution operators of the molecular system provided it is in electronic state $|g\rangle$ and $|e\rangle$, respectively. Evaluation of Equation (4.60) is done in the second cumulant approximation (see e.g. (Mukamel, 1995)). First, we realize that

$$\hat{U}_g^\dagger(t)\hat{U}_e(t) \equiv \mathcal{G}_{ge}^{(+)}(t) = \exp_+\left\{\frac{i}{\hbar}\int_0^t \mathrm{d}\tau \hat{U}_g^\dagger(\tau)\Delta\hat{V}_{eg}\hat{U}_g(\tau)\right\}. \tag{4.61}$$

Here, we have defined the so-called *coherence Green's function* $\mathcal{G}_{ge}(t)$ which will be of use later. Equation (4.61) can be verified by taking the time derivative of both sides. One can see that they satisfy the same differential equation. The idea of the second cumulant approximation is to replace the equilibrium average of the ordered exponential of an operator, Equation (4.60), by an ordinary exponential of some suitable equilibrium averages. We do this by comparing two Taylor expansions of the same function A, which we expect to have an exponential form $A = \exp\{F\}$. An ordinary Taylor expansion yields

$$A(\lambda) = 1 + A_1\lambda + A_2\lambda^2 + \cdots , \tag{4.62}$$

and a Taylor expansion in the exponent gives

$$\begin{aligned} A(\lambda) &= \exp\left\{F_1\lambda + F_2\lambda^2 + \cdots\right\} \\ &= 1 + (F_1\lambda + F_2\lambda^2 + \cdots) + \frac{1}{2}(F_1\lambda + F_2\lambda^2 + \cdots)^2 + \cdots . \end{aligned} \tag{4.63}$$

Here, A_1, A_2, F_1 and F_2 are the coefficients of the corresponding Taylor expansions in parameter λ. Comparing the two expansions to the second order in λ yields

$$F_1 = A_1, \ F_2 = A_2 - \frac{1}{2}A_1^2. \tag{4.64}$$

Consequently, starting from an ordinary Taylor expansion, Equation (4.62), we can find coefficients F_1 and F_2, which would enable us to write A as an exponential, and thus partially sum the expansion terms up to infinity. Equation (4.60) can easily be expanded to the second order in $\Delta\hat{V}_{eg}$ and the coefficients A_1 and A_2 yield

$$A_1 = \frac{i}{\hbar}\int_0^t \mathrm{d}\tau \mathrm{tr}\{\hat{U}_g^\dagger(\tau)\Delta\hat{V}_{eg}\hat{U}_g(\tau)\hat{W}_{eq}^{(\mathrm{bath})}\} = 0, \tag{4.65}$$

and

$$A_2 = -\frac{1}{\hbar^2}\int_0^t \mathrm{d}\tau \int_0^\tau \mathrm{d}\tau' \mathrm{tr}\{\hat{U}_g^\dagger(\tau-\tau')\Delta\hat{V}_{eg}\hat{U}_g(\tau-\tau')\Delta\hat{V}_{eg}\hat{W}_{eq}^{(\mathrm{bath})}\}. \tag{4.66}$$

In both equations we used the fact that $\hat{W}_{eq}^{(\mathrm{bath})}$ does not evolve under $\hat{U}_g(t)$. Introducing the *energy gap correlation function* (EGCF) $C_{ee}(t)$ as

$$C_{ee}(t) = \mathrm{tr}\{\hat{U}_g^\dagger(t)\Delta\hat{V}_{eg}\hat{U}_g(t)\Delta\hat{V}_{eg}\hat{W}_{eq}^{(\mathrm{bath})}\}, \tag{4.67}$$

and the *line broadening function* $g_{ee}(t)$ as

$$g_{ee}(t) = \frac{1}{\hbar^2}\int_0^t \mathrm{d}\tau \int_0^\tau \mathrm{d}\tau' C_{ee}(\tau'), \tag{4.68}$$

we can write the first-order Liouville pathway $J(t)$ in a simple form,

$$J(t) = e^{-g_{ee}(t)-i\omega_{eg}t}. \tag{4.69}$$

The two indices ee denote the energy gaps of which the $C_{ee}(t)$ is composed. In more general problems, various cross-correlation functions such as $C_{ef}(t)$ may appear.

The EGCF is a central quantity of the response function description of linear and non-linear spectroscopy. It determines not only the lineshape, but also the dephasing lifetime of coherences, relaxation dynamics of excitons etc. (May and Kühn, 2011). The energy gap correlation function and its Fourier transform satisfy several general conditions, such as $C^*_{ee}(t) = C_{ee}(-t)$, and many models of the $C_{ee}(t)$ function exist (Mukamel, 1995; May and Kühn, 2011; Renger and Marcus, 2002; Jang and Silbey, 2003) which allow its application to a variety of problems. One of the most useful and intuitive models, the Brownian oscillator, is discussed in detail in Mukamel (1995).

4.3.2 Energetic disorder

The situation where all molecules in the sample are exactly the same is rather rare. Quantities such as energy of transition might depend, for example, on some very slowly changing parameter of the environment. This is referred to as *disorder* and leads to the so-called *inhomogeneous broadening* of the spectra. The word inhomogeneous refers to the fact that during the spectroscopic experiment individual molecules exhibit different parameters from each other, and it is in contrast to *homogeneous broadening*, which stems from interaction of the molecule with some component of its environment which is faster than the typical timescale of the experiment. These fast fluctuations average out quickly so that they are essentially the same for all molecules. Our averaging that leads to the line broadening function was of this latter type.

For simplicity, let us assume that the disorder concerns only the energy levels of the system in question, and that the fast fluctuations of the environment responsible for the EGCF are independent of the slow fluctuations that cause the disorder. This enables us to perform two averaging procedures, one over fast and the other over slow fluctuations, independently. We might assume that to a good approximation, the total density matrix reads as $\hat{W}_{eq} = \hat{W}^{(fast)}_{eq}\hat{W}^{(slow)}_{eq}$, and the evolution operator corresponding to the slow bath is $\hat{U}^{(slow)}_{B}(t) \approx 1$. As the origin of the two types of fluctuations is essentially the same, we may add the slow component to the energy gap operator, Equation (4.53), and write for its time dependence,

$$\Delta V_{eg}(t) = \Delta V^{(slow)}_{eg} + \Delta V^{(fast)}_{eg}(t). \tag{4.70}$$

The response functions can now be constructed in the same way as in Section 4.3.1, i.e. by cumulant expansion. We assume only the term $\text{tr}\{\Delta V_{eg}^{(slow)} \Delta V_{eg}^{(slow)} W_{eq}^{(slow)}\} \equiv \hbar^2 \Delta^2$ to be non-zero. Because of the independence of the two types of fluctuation, the response functions can be written as products of the slow and fast parts, $R_n(t_3, t_2, t_1) = R_n^{(slow)}(t_3, t_2, t_1) R_n^{(fast)}(t_3, t_2, t_1)$, and the fast and slow response functions differ only in their corresponding EGCF. The slow EGCF and the lineshape function read as

$$C_{ee}^{(slow)}(t) = \hbar^2 \Delta^2, \; g_{ee}^{(slow)}(t) = \frac{1}{2} \Delta^2 t^2. \tag{4.71}$$

Interestingly, the slow part of the response is the same for all rephasing pathways,

$$R_{2g}^{(slow)}(t_3, t_2, t_1) = R_{3g}^{(slow)}(t_3, t_2, t_1) = I_{inh}(t_3 - t_1) = e^{-\frac{\Delta^2}{2}(t_3 - t_1)^2}, \tag{4.72}$$

and for all non-rephasing pathways,

$$R_{1g}^{(slow)}(t_3, t_2, t_1) = R_{4g}^{(slow)}(t_3, t_2, t_1) = I_{inh}(t_3 + t_1) = e^{-\frac{\Delta^2}{2}(t_3 + t_1)^2}. \tag{4.73}$$

We have introduced the inhomogeneity factor $I_{inh}(t)$ here, which will be useful in Section 4.4. This form of the response function contains the famous *photon echo effect*. The inhomogeneity factor $I_{inh}(t_3 - t_1)$ is equal to one at $t_1 = t_3$, and thus the rephasing responses are equal to their fast (homogeneous) part at this time point. For a given time t_1 this leads to a recovery in the macroscopic experimental signal from some value $S_0 e^{-\Delta^2 t_1^2}$ to S_0. The non-rephasing responses and their corresponding spectroscopic signal, however, continue to decay with t_3, and would never rephase.

4.3.3 Response functions of a three-level system

Third-order response functions can be expressed using the same line broadening function $g_{ee}(t)$ as for linear absorption. We deal with three levels and two transitions with transition dipole moment d_{eg} and d_{ef}. For this simple case, response functions corresponding to Liouville pathways from Section 4.2.2 can be written in terms of a single function (see e.g. (Zhang *et al.*, 1998; Brixner *et al.*, 2004)),

$$F_\beta(\tau_1, \tau_2, \tau_3, \tau_4) = |d_{eg}|^2 |d_{\beta e}|^2 \text{tr}\big\{\mathcal{G}_{ge}^{(+)}(\tau_1 - \tau_2) \mathcal{G}_{\beta g}^{(-)}(\tau_2 - \tau_3) \times \mathcal{G}_{ge}^{(-)}(\tau_2 - \tau_3) \mathcal{G}_{eg}^{(-)}(\tau_2 - \tau_4) W_{eq}^{(\text{bath})}\big\}, \tag{4.74}$$

where $\beta = g, f$, and we define another coherence Green's function,

$$\mathcal{G}_{ab}^{(-)}(t) = U_a(t) U_b^\dagger(t). \tag{4.75}$$

Individual pathways read,

$$R_{1\beta}(t_3, t_2, t_1) = F_\beta(t_1, t_1 + t_2, t_1 + t_2 + t_3, 0), \tag{4.76}$$

$$R_{2\beta}(t_3, t_2, t_1) = F_\beta(0, t_1 + t_2, t_1 + t_2 + t_3, t_1), \tag{4.77}$$

$$R_{3\beta}(t_3, t_2, t_1) = F_\beta(0, t_1, t_1 + t_2, t_1 + t_2 + t_3, t_1), \tag{4.78}$$

$$R_{4\beta}(t_3, t_2, t_1) = F_\beta(t_1 + t_2 + t_3, t_1 + t_2, t_1, 0). \tag{4.79}$$

The second cumulant evaluation of F_β proceeds as follows. First we expand all coherence Green's functions (see Equations (4.75) and (4.61)) to the second order in their respective energy gap operator $\Delta\hat{V}$. Then we perform the product of the second-order expansions, and collect separately the first and the second-order terms in $\Delta\hat{V}$. The first-order terms become zero after the trace, similarly to Equation (4.65). The second-order terms are then all rewritten using the line broadening functions, Equation (4.68). The calculation, details of which can be found e.g. in Mukamel (1995) and Zhang *et al.* (1998), is rather tedious but straightforward. The method yields

$$F_\beta(\tau_1, \tau_2, \tau_3, \tau_4) = |d_{eg}|^2 |d_{\beta e}|^2 e^{-i\omega_{eg}(\tau_1-\tau_4)+i\omega_{e\beta}(\tau_2-\tau_3)} \exp\{-h_\beta(\tau_1, \tau_2, \tau_3, \tau_4)\}, \tag{4.80}$$

where

$$\begin{aligned} h_\beta(\tau_1, \tau_2, \tau_3, \tau_4) = {} & g_{ee}(\tau_1 - \tau_2) + g_{\beta\beta}(\tau_2 - \tau_3) + g_{ee}(\tau_3 - \tau_4) - g_{\beta e}(\tau_1 - \tau_2) \\ & + g_{\beta e}(\tau_1 - \tau_3) - g_{e\beta}(\tau_2 - \tau_3) - g_{ee}(\tau_1 - \tau_3) + g_{ee}(\tau_2 - \tau_3) \\ & + g_{ee}(\tau_1 - \tau_4) - g_{ee}(\tau_2 - \tau_4) - g_{\beta e}(\tau_2 - \tau_3) + g_{e\beta}(\tau_2 - \tau_4) \\ & - g_{e\beta}(\tau_3 - \tau_4). \end{aligned} \tag{4.81}$$

For $\beta = g$, the expression simplifies because $g_{\beta g}(t) = g_{g\beta}(t) = g_{gg}(t) = 0$. The rephasing Liouville pathways read as

$$\begin{aligned} R_{2g}(t_3, t_2, t_1) = {} & |d_{eg}|^4 e^{-g^*_{ee}(t_3)-g^*_{ee}(t_1)+g_{ee}(t_2)-g_{ee}(t_2+t_3)} \\ & \times e^{-g^*_{ee}(t_1+t_2)+g^*_{ee}(t_1+t_2+t_3)} e^{-i\omega_{eg}(t_3-t_1)}, \end{aligned} \tag{4.82}$$

$$\begin{aligned} R_{3g}(t_3, t_2, t_1) = {} & |d_{eg}|^4 e^{-g_{ee}(t_3)-g^*_{ee}(t_1)+g^*_{ee}(t_2)-g^*_{ee}(t_2+t_3)} \\ & \times e^{-g^*_{ee}(t_1+t_2)+g^*_{ee}(t_1+t_2+t_3)} e^{-i\omega_{eg}(t_3-t_1)}, \end{aligned} \tag{4.83}$$

$$\begin{aligned} R_{1f}(t_3, t_2, t_1) = {} & |d_{eg}|^2 |d_{ef}|^2 e^{-g^*_{ff}(t_3)-g_{ee}(t_1)-g_{ee}(t_1+t_2+t_3)+g_{ef}(t_1+t_2+t_3)} \\ & \times e^{-g_{ef}(t_1+t_2)+g_{ee}(t_1+t_2)+g^*_{fe}(t_3)-g^*_{ee}(t_3)+g^*_{ee}(t_2+t_3)+g^*_{ef}(t_3)} \\ & \times e^{-g^*_{ef}(t_2+t_3)+g^*_{ef}(t_2)-g^*_{ee}(t_2)} e^{-i\omega_{fe}t_3-i\omega_{eg}t_1}. \end{aligned} \tag{4.84}$$

All other pathways can be obtained analogically. The non-linear spectroscopy of a three-level system is fully characterized by the eight response functions. They enable us to calculate the complete three-level system response, based on three correlation functions $C_{ee}(t)$, $C_{ff}(t)$ and $C_{ef}(t)$.

4.4 Selected non-linear spectroscopic methods

Non-linear interaction between light and matter enables us to devise a variety of spectroscopic methods (Parson, 2007; Mukamel, 1995) suited to extract different types of information from the studied molecular systems. Some of the most fundamental information about photon induced time evolution of excited states, e.g. about excitation energy processes in photosynthesis and other biologically relevant processes, can be obtained from so-called *pump–probe spectroscopy* (Sundstrom *et al.*, 1999). This is one of the most widely used time-resolved spectroscopic methods, and nowadays it still represents a certain reality check for more involved non-linear methods. A description of pump–probe spectroscopy is possible using theory presented in previous sections (Yan *et al.*, 1989; Mukamel, 1995; Parson, 2007).

In this section, we will discuss in some detail another two non-linear methods and the information they can reveal about molecular systems. Firstly, we will consider an example of a non-linear method sensitive to system–bath interaction, the *photon-echo peakshift*. We will show how its measurement reveals the shape of the energy gap correlation function. Secondly, we will see how a 2D coherent spectrum uncovers coherence in a system of excitons, and how this method is sensitive to resonance coupling between molecules in an excitonic complex. The same method can also reveal coherent vibrational motion.

4.4.1 Photon echo: learning about system–bath interactions

Linear and non-linear spectra provide us with information about both the electronic properties of the studied molecules (positions of the absorption band) and the properties of the molecular environment (lineshapes). The latter is carried by the correlation function $C_{ee}(t)$ of the energy gap. The so-called *photon-echo peakshift* experiment is one from which an energy gap correlation function can be extracted. In the analysis of the photon-echo peakshift, we will demonstrate that even in the theory of non-linear spectroscopy, one can obtain highly interesting and relevant analytical results. This is achieved here by a short time expansion of the exponent in the non-linear response functions.

In the photon-echo peakshift experiment, a third-order non-linear signal in the direction $-\boldsymbol{k}_1 + \boldsymbol{k}_2 + \boldsymbol{k}_2$ (photon echo) is measured by a slow detector in a FWM set-up, with delays $\tau \geq 0$ and $T \geq 0$ between the three pulses. The detector integrates the signal intensity over time t. Thus we measure

$$S_{PE}(T, \tau) = \int_0^\infty \mathrm{d}t |S^{(3)}(t, T, \tau)|^2. \tag{4.85}$$

In S_{PE} we look for the value $\tau^*(T)$ of the delay τ for which S_{PE} is maximal at a given population time T. This value is called the *photon-echo peakshift* and can be formally defined by the relation

$$\frac{\partial}{\partial \tau} S_{PE}(T, \tau)|_{\tau=\tau^*(T)} = 0. \tag{4.86}$$

In the impulsive limit, the third-order signal reads as

$$|S^{(3)}(t, T, \tau)|^2 \approx I_{inh}^2(t - \tau)|R_{3g}(t, T, \tau) + R_{2g}(t, T, \tau)|^2, \tag{4.87}$$

where $I_{inh}(t) = e^{-\Delta_{in}^2 t^2/2}$ is the inhomogeneity factor, introduced in Section 4.3.2. If we denote the real and imaginary parts of the lineshape function $g_{ee}(t)$ by $P(t)$ and $Q(t)$ respectively, we can write

$$|R_{3g}(t, T, \tau) + R_{2g}(t, T, \tau)|^2 = e^{-2\{P(\tau)-P(T)+P(t)+P(\tau+T)+P(\tau+t)-P(\tau+T+t)\}} \times \cos^2[Q(T) + Q(t) - Q(T + t)]. \tag{4.88}$$

Because the signal is expected to decay quickly with t, an expansion of the exponent in t, up to the second order, could yield a reasonable approximation. We will follow the derivation of Cho *et al.* (1996) to reveal striking insights into the relation between the photon-echo peakshift and the energy gap correlation function. To this end we expand the $g(t)$ function as $g(a + t) \approx g(a) + \dot{g}(a)t + \ddot{g}(a)t^2/2$, where the superscript dot denotes a time derivative. The definitions in Section 4.3.1 lead to

$$P(0) = 0, \quad \dot{P}(0) = 0, \quad \ddot{P}(t) = \mathrm{Re}C_{ee}(t) \tag{4.89}$$

and

$$Q(0) = 0, \quad \dot{Q}(0) = 0, \quad \dot{Q}(t) = \mathrm{Im}\int_0^t d\tau C_{ee}(\tau). \tag{4.90}$$

The cosine part of Equation (4.88) can be expressed in a second cumulant expansion as $\cos^2[Q(T) + Q(t) - Q(T + t)] \approx \exp\{-(\dot{Q}(T))^2 t\}$, and the exponent of the exponential part could also be expanded to the second order in t. Equation (4.85) will thus become

$$S_{PE}(T, \tau) \approx \int_0^\infty \mathrm{d}t \; e^{B(T,\tau)t-A(T,\tau)t^2-2P(\tau)-\Delta_{in}^2\tau^2}, \tag{4.91}$$

where

$$A(T,\tau) = \Delta_{in}^2 + (\dot{Q}(T))^2 + \ddot{P}(0) - \ddot{P}(T) - \ddot{P}(\tau + T), \tag{4.92}$$

$$B(T,\tau) = 2\Delta_{in}^2\tau - 2[\dot{P}(T) - \dot{P}(\tau + T)]. \tag{4.93}$$

Integration of Equation (4.91) gives

$$S_{PE}(\tau,T) \approx \frac{\sqrt{\pi}}{2\sqrt{A(T,\tau)}} \exp\left\{-2P(\tau) - \Delta_{in}^2\tau^2 + \frac{B(T,\tau)^2}{4A(T,\tau)}\right\} \times \left[1 + \mathrm{erf}\left(\frac{B(\tau,T)}{2\sqrt{A(T,\tau)}}\right)\right]. \tag{4.94}$$

Equation (4.94) represents an approximate result for the photon-echo signal, and it can be used to calculate the photon-echo peakshift.

If the inhomogeneous broadening is much larger than the homogeneous broadening, i.e. $\ddot{P}(0) \ll \Delta_{in}^2$, the integrated photon echo is approximately given by $S_{PE}(T,\tau) \approx \exp\{-2P(\tau)\}$ for $\tau > \Delta_{in}^{-1}$. This can be verified by noticing that in this limit, $B \approx 2\Delta_{in}^2\tau$ and $A \approx \Delta_{in}^2$, and the error function $\mathrm{erf}(x) \approx 1$ for $x > 1$. The integrated photon-echo signal does not depend on T here, and the time-dependent photon-echo signal has its maximum at $t = \tau$. The peakshift is given by the condition $\dot{P}(\tau)|_{\tau=\tau^*(T)} = 0$, which leads to $\tau^*(T) = 0$ at $\tau > \Delta_{in}^{-1}$.

However, if the inhomogeneous and homogeneous line widths Δ_{in}^2 and $\ddot{P}(0)$ are comparable, one can apply another expansion of the signal, this time in τ. We expand all quantities in the second order of τ, which leads to $B \approx 2[\Delta_{in}^2 + \ddot{P}(T)]\tau$, $A \approx \ddot{P}(0) + \Delta_{in} + (\dot{Q}(T))^2$, $P(\tau) \approx \ddot{P}(0)\tau^2/2$, and then expand Equation (4.94) to the second order in τ. Taking the derivative of such an expanded integrated signal S_{PE} according to τ and solving the resulting linear equation leads to the following expression for the peakshift:

$$\tau^*(T) = \frac{1}{\sqrt{\pi}} \frac{[\Delta_{in}^2 + \ddot{P}(T)]\sqrt{\ddot{P}(0) + \Delta_{in}^2 + (\dot{Q}(T))^2}}{\ddot{P}(0)[\ddot{P}(0) + 2\Delta_{in}^2 + (\dot{Q}(T))^2] + \Delta_{in}^2(\dot{Q}(T))^2}. \tag{4.95}$$

In the case where inhomogeneous broadening can be completely ignored (e.g. in liquids), the peakshift expression simplifies dramatically to

$$\tau^*(T) = \frac{1}{\sqrt{\pi}} \frac{\ddot{P}(T)}{\ddot{P}(0)\sqrt{\ddot{P}(0) + (\dot{Q}(T))^2}}. \tag{4.96}$$

Often $\dot{Q}(T)$ is small compared with $\ddot{P}(0)$ and it can be neglected, and thus the peakshift directly reveals the real part of the energy gap correlation function $\ddot{P}(t) = ReC(t)$. The long time peakshift is correspondingly equal to zero, $\tau^*(T \to \infty) = 0$.

Most interestingly, if we take the limit of long population times $T \to \infty$ in Equation (4.95), i.e. for Δ_{in}^2 comparable to $\ddot{P}(0)$, the long time peakshift remains non-zero. We have $\dot{Q}(T \to \infty) \approx \lambda$, $\ddot{P}(T \to \infty) \approx 0$, where λ is the reorganization energy (Mukamel, 1995) and thus,

$$\tau^*(T \to \infty) = \frac{1}{\sqrt{\pi}} \frac{\Delta_{in}^2 \sqrt{\ddot{P}(0) + \Delta_{in}^2 + \lambda^2}}{\ddot{P}(0)[\ddot{P}(0) + 2\Delta_{in}^2 + \lambda^2] + \Delta_{in}^2 \lambda^2}. \tag{4.97}$$

Consequently, the non-zero long time value of the peakshift is evidence of finite inhomogeneity. An interesting biological application of the peakshift measurement is reported in Zimmermann *et al.* (2006). A chromophore molecule was injected into a mouse, and the evolution of the photon-echo peakshift of the chromophore captured by the mouse antibody was monitored as the antibody matured (on a timescale of days). Over time the antibody became more specifically designed for binding the chromophore, which resulted in a decrease in the inhomogeneity of the chromophore's environment in the trapping site. The chromophore fitted better into the trapping site of a matured antibody. This was observed as a decrease in the long time limit of the peakshift.

According to Equations (4.95) and (4.96), peakshift can be used to estimate the bath correlation function, i.e. it can be used to investigate fluctuations of the energy gap. If a complex of two or more molecules is measured, several molecules may contribute to a single electronic transition, and the total energy gap correlation function is built from the contributions of individual molecules. For the case of two identical molecules with uncorrelated energy gap fluctuations, a certain combination of one-colour peakshifts (measured with all three pulses of the same frequency) and two-colour peakshifts (measured with the first two pulses of one frequency and the third pulse of another frequency) allows us to estimate excitonic mixing and thus the value of excitonic coupling between the two molecules (Yang and Fleming, 1999). For two different molecules, one can even estimate the difference between the correlation functions on the two molecules (Mančal and Fleming, 2004).

4.4.2 Two-dimensional spectroscopy: resonance coupling, population transfer and coherence dynamics

The signal recorded in a photon-echo peakshift measurement is by itself already two-dimensional. One special feature of the two-dimensional signal, the position of its maximum along the τ axis as a function of the population time T, is used to extract interesting information from this two-dimensional data set. Now we introduce yet another multidimensional spectroscopic method (Jonas, 2003) that has yielded interesting insights into the properties of molecular systems on femtosecond timescales, and whose potential has certainly not yet been fully exploited.

We start by comparing the general structure of the first-order and the third-order response. We compare the first-order response function of Equation (4.60) which we write as

$$J(t) = i\mathrm{tr}\{\hat{\mu}^{(ge)}\mathcal{J}^{(eggg)}(t)W^{(gg)}\}e^{-i\Omega t}, \tag{4.98}$$

where $\mathcal{J}^{(eggg)}(t) = \tilde{\mathcal{U}}^{(egeg)}(t)\mathcal{V}^{(eggg)}$, with some representative R function, e.g. R_{2g} of Equation (4.36). The R function can be written as

$$R_{2g}(t_3, t_2, t_1) = -i\mathrm{tr}\left\{\hat{\mu}^{(ge)}\mathcal{J}^{(egee)}(t_3)\mathcal{U}^{(eeee)}(t_2)\hat{\mu}^{(eg)}\mathcal{J}^{(gegg)}(t_1)W^{(gg)}\right\}e^{-i\Omega(t_3-t_1)}. \tag{4.99}$$

The comparison suggests an analogy between the absorption spectrum given by the Fourier transform of the response function $J(t)$ and a possible non-linear spectrum given by Fourier transforms along the times t_3 and t_1. In this analogy, the frequency ω_1 dependence obtained by the Fourier transform of the third-order response in time t_1 would be the same as the frequency dependence of an absorption spectrum. The frequency ω_3 dependence obtained by the Fourier transform in time t_3 would be the same as that of an 'absorption' of a system out of equilibrium, whose statistical operator is initially $\hat{W}^{(ee)} = \mathcal{U}^{(eeee)}(t_2)\hat{\mu}^{(eg)}\mathcal{J}^{(gegg)}(t_1)\hat{W}^{(gg)}$. In this particular case, $\hat{W}^{(ee)}$ corresponds to the system in the excited state, and thus the non-equilibrium 'absorption' in fact also includes stimulated emission. In between the two generalized absorption spectra the evolution superoperator $\mathcal{U}^{(eeee)}(t_2)$ resides, and the spectrum evolves with a delay t_2. If we were able to record the full time dependence of the third-order response in t_1, t_2 and t_3 times, we could construct two-dimensional plots correlating the absorption and the non-equilibrium absorption–emission spectra at any given delay t_2. The t_2 evolution of the spectra may involve features that reveal relaxation of energy among several excited states. From Section 4.2.4, we know that for $t_1 > 0$, i.e. when the pulse $\boldsymbol{k}_1$ precedes the pulse $\boldsymbol{k}_2$, only the rephasing pathways contribute to the signal in direction $-\boldsymbol{k}_1 + \boldsymbol{k}_2 + \boldsymbol{k}_3$ (here we neglect pulse overlap effects). Thus we can define the *rephasing* 2D spectrum as

$$\Xi_{\mathrm{R}}(\omega_3, t_2, \omega_1) = \int_0^\infty \mathrm{d}t_3 \int_0^\infty \mathrm{d}t_1 \tilde{S}^{(3)}_{\mathrm{R}}(t_3, t_2, t_1)e^{-i\Omega(t_3-t_1)}e^{i\omega_3 t_3 - i\omega_1 t_1}. \tag{4.100}$$

Alternatively, we could also measure the *non-rephasing* signal and construct a similar quantity as above,

$$\Xi_{\mathrm{NR}}(\omega_3, t_2, \omega_1) = \int_0^\infty \mathrm{d}t_3 \int_0^\infty \mathrm{d}t_1 \tilde{S}^{(3)}_{\mathrm{NR}}(t_3, t_2, t_1)e^{-i\Omega(t_3+t_1)}e^{i\omega_3 t_3 + i\omega_1 t_1}. \tag{4.101}$$

The limits of the integration are set naturally, because for $t_3 < 0$ and $t_1 < 0$, the response is zero. The signs in the Fourier transforms are chosen to reflect the phase factors of the response. The quantities defined in Equations (4.100) and (4.101) can be directly measured experimentally. The heterodyne detection scheme of Section 4.1.4 enables us to measure the third-order signal electric field, which is in the impulsive limit directly proportional to $S^{(3)}$.

What the actual relation is between absorption spectra and the 2D spectra defined above can be found by considering a simple case of a two-level system. For long population times t_2 the dependence of the $g(t)$ function on t_2 becomes linear. Both contributing rephasing response functions then have a form,

$$R_{2g}(t_3, t_2, t_1) = R_{3g}(t_3, t_2, t_1) \approx e^{-g(t_3)-g^*(t_1)-i\Omega(t_3-t_1)}. \tag{4.102}$$

The Fourier transform, Equation (4.100) then leads to

$$\Xi_{\mathrm{R}}(\omega_3, t_2, \omega_1) \approx G(\omega_3 - \Omega)G^*(\omega_1 - \Omega), \tag{4.103}$$

where we define,

$$G(\omega) = \int_0^\infty \mathrm{d}t\; e^{-g(t)-i\omega t} \approx \frac{\Gamma + i\omega}{\Gamma^2 + \omega^2}. \tag{4.104}$$

The G function is related to the susceptibility $\chi^{(1)}$ and it determines the linear absorption spectrum $\alpha(\omega) \approx \mathrm{Re}\; G(\omega - \Omega)$. In Equation (4.104) we use the form $g(t) = \Gamma t$, with some real dephasing rate Γ corresponding to the so-called homogeneous limit of $g(t)$ (Mukamel, 1995). We can see that the rephasing 2D spectrum is not strictly proportional to the absorption spectrum, but it is a mixture of the absorptive and dispersive (related to Im $G(\omega - \Omega)$) contributions. To isolate the absorptive contributions, we have to investigate the non-rephasing spectrum, too. In the same limit as above we notice that

$$\Xi_{\mathrm{NR}}(\omega_3, t_2, \omega_1) \approx G(\omega_3 - \Omega)G(\omega_1 - \Omega), \tag{4.105}$$

and thus,

$$\begin{aligned} \Xi(\omega_3, t_2, \omega_1) &= \Xi_{\mathrm{R}}(\omega_3, t_2, \omega_1) + \Xi_{\mathrm{NR}}(\omega_3, t_2, \omega_1) \\ &\approx G(\omega_3 - \Omega)\mathrm{Re}\; G(\omega_1 - \Omega). \end{aligned} \tag{4.106}$$

This means that the real part of the sum 2D spectrum,

$$\mathrm{Re}\,\Xi(\omega_3, t_2, \omega_1) \approx \mathrm{Re}\; G(\omega_3 - \Omega)\mathrm{Re}\; G(\omega_1 - \Omega) \tag{4.107}$$

corresponds to a product of two absorption spectra, i.e. to the desired correlation plot of the absorption and absorption–emission frequencies. With a simple

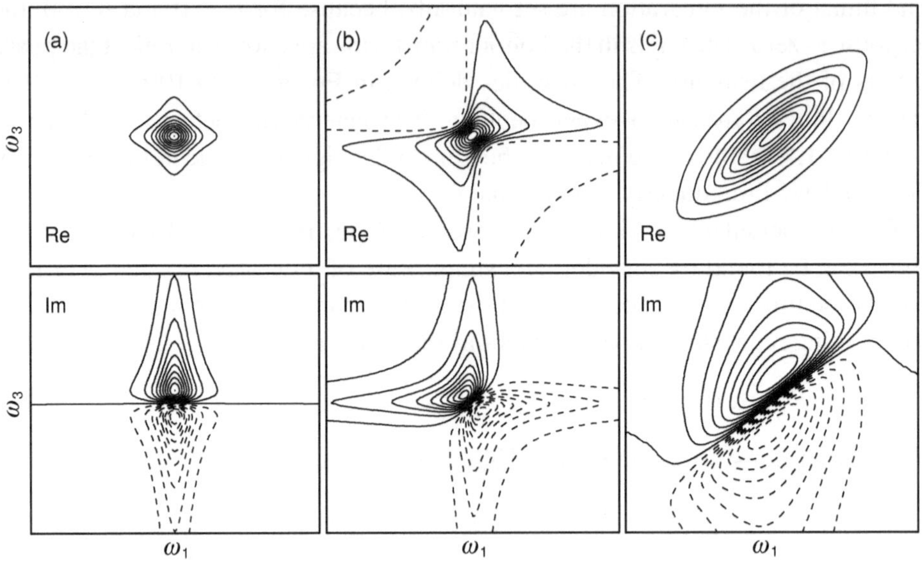

Figure 4.5 Basic homogeneous and inhomogeneous lineshapes of 2D spectroscopy. (a) Shows real (upper) and imaginary (lower) parts of the homogeneous 2D spectrum of a two-level system. (b) Shows a rephasing 2D spectrum, the non-rephasing one can be obtained by mirroring the picture with respect to the axis $\omega_1 = \Omega$. (c) Represents the 2D spectrum with a Gaussian distribution of the transition frequencies. Full (dashed) contours denote positive (negative) values of the spectral amplitude.

homogeneous form of the $g(t)$ function, we can construct the expected 2D spectrum of a simple two-level system in the homogeneous limit (see Figure 4.5a). For more on 2D lineshapes, see e.g. Tokmakoff (2000). Interpretation of many 2D spectra can be based on this simple shape. Firstly, the rephasing and non-rephasing spectra have certain characteristic orientations (Figure 4.5b), and both their real and imaginary parts contain positive and negative contributions. The real part of a total homogeneous 2D spectrum is however positive. The contribution of the R_2 pathway, Equation (4.99), which contains evolution in the excited state during the waiting time t_2, could be readily interpreted as a stimulated emission (SE). Combined with the corresponding non-rephasing pathway, this would lead to a decrease in absorption, if they were measured in a pump–probe ($\mathbf{k}_1 = \mathbf{k}_2$) configuration. This defines the meaning of the positive features in the (total!) 2D spectrum. The R_3 pathway contains propagation in the ground state, and it has the same sign as R_2. It cannot therefore represent absorption. Rather, it has to stand for ground state bleaching (GSB). The non-rephasing counterparts of this signal can be classified in the very same way. For higher-lying excited states, pathways with negative sign containing propagation in the excited state during waiting time t_2, contribute to both the rephasing and non-rephasing signals. Their combined negative signal is the one corresponding to the so-called excited state absorption (ESA).

So far, we have considered all molecules in the sample to be identical. Now, let us introduce inhomogeneity in the form of energetic disorder. We will see that with respect to disorder there is more to a 2D spectrum than just a product of absorption spectra. The response function formalism can handle simple disorder by including it into the lineshape function $g(t)$, as we have seen in Section 4.3.2. The product expressions, Equation (4.103) and Equation (4.105), are not valid any more, because the response

$$R_{2g}(t_3, t_2, t_1) = R_{3g}(t_3, t_2, t_1) \approx e^{-g(t_3)-g^*(t_1)-\Delta(t_3-t_1)^2-i\Omega(t_3-t_1)} \tag{4.108}$$

cannot be split into a product of t_3 and t_1 dependent functions. To perform the double Fourier transforms, Equations (4.100) and (4.101), analytically is possible, but rather tedious. The result of the analytical calculation is depicted in Figure 4.5c. To guess this 2D lineshape, we could have employed the fact that 2D spectra of non-interacting sub-ensembles are additive. The inhomogeneity can thus be represented as an averaging over 2D spectra of species with different transition frequencies Ω. Because all such spectra differ only by their position at the diagonal of the spectrum, the combined lineshape becomes elongated along the diagonal (see Figure 4.5). Anti-diagonal line width is not affected by the disorder, and consequently, both the *homogeneous* and the *inhomogeneous line widths* can, in principle, be estimated from a single measurement of the 2D spectrum.

An important effect which can be monitored by 2D (and also by pump–probe) spectroscopy is the excitation energy transfer (EET). The pathways R_2 and R_1 that contain contributions from excited state propagation (the SE and ESA contributions) will display changes due to excited state population transfer; while ground state pathways R_3 and R_4 (the GSB contributions) will not. As the contributions of the SE and GSB to the positive peaks are of a similar amplitude, population transfer to the neighbouring state exhibits itself in a 2D spectrum as a transfer of one half of the peak amplitude to a different spectral position. This results in a cross-peak between two frequency regions, as in Figure 4.6c. In the same way, the ESA contributions can appear in a new spectral position, if the acceptor molecule possesses a higher-lying excited state.

Population transfer is not the only way a cross-peak can occur in a 2D spectrum. In the dimer from Figure 4.6a, one can imagine e.g. an R_3 pathway starting with a transition from the ground state of the monomer A to its excited state, followed by its de-excitation by the second pulse, and further followed by an excitation of monomer B by the third pulse. Such an R_3 pathway contributes a cross-peak in the same spectral position as the one previously identified to be due to population transfer. With contributions like this, the question is, why should we actually

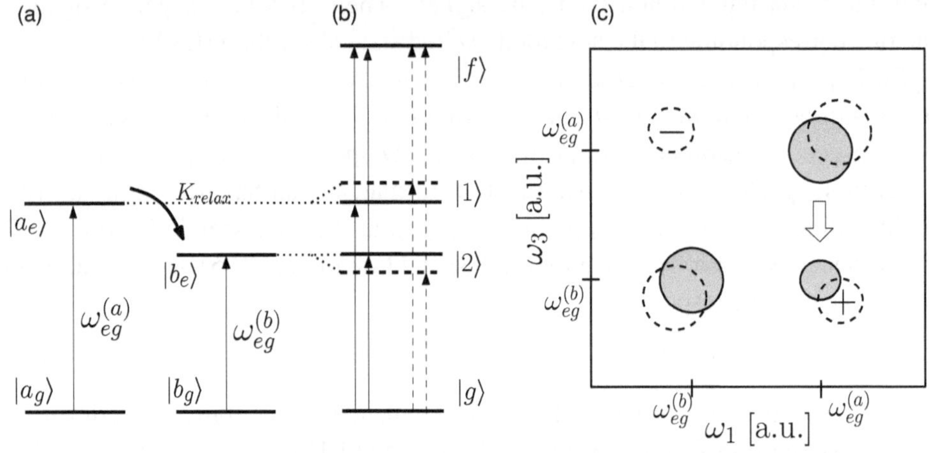

Figure 4.6 Schemes of energy levels and 2D spectra of a weakly coupled dimer. (a) Represents the dimer by the states of its monomers, (b) shows the same by collective states $|g\rangle = |a_g\rangle|b_g\rangle$, $|1\rangle = |a_e\rangle|b_g\rangle$, $|2\rangle = |a_g\rangle|b_e\rangle$ and $|f\rangle = |a_e\rangle|b_e\rangle$. (c) Demonstrates the features of the corresponding 2D spectra. The full circles represent diagonal peaks of a weakly coupled dimer including the energy relaxation peak. Dashed circles represent the peaks of a coupled dimer.

expect two uncoupled two-level molecules to yield a cross-peak-free spectrum at all? (We have stated before that 2D spectra of uncoupled transitions are additive.) The answer lies in counting correctly all pathways expected in the dimer. This can be done with the help of Figures 4.6a and 4.6b. The latter figure represents the dimer in terms of collective states, introduced in Section 4.2.1. The two schemes of levels are equivalent for an uncoupled dimer. Let us now count the rephasing pathways, Equation (4.41), that can contribute to the cross-peak, i.e. those that have different first and third interval frequencies. These are summarized in Figure 4.7, where one can immediately see that there are four positive (R_{2g} and R_{3g}) and four negative (R_{1f}) contributions to the signal. Although the doubly excited state $|f\rangle$ is reached in the R_{1f} part of the rephasing response, all transitions can be viewed as normal ground state to excited state transitions on the monomers. For example, $|1\rangle \rightarrow |f\rangle$ is the $|b_g\rangle \rightarrow |b_e\rangle$ transition, while monomer a remains in its excited state $|a_e\rangle$. In fact, for each positive contribution to the signal, we can find a corresponding negative one, and all cross-peaks cancel. For complexes of uncoupled monomers, no cross-peak can appear. In other words, the cross-peaks can appear only in complexes of coupled chromophores.

As we have seen, cross-peaks are of two kinds. Firstly, they can result from population transition. Such cross-peaks have to build up gradually as the population transfer proceeds over time t_2. Secondly, cross-peaks can appear instantaneously when the cancelling between different pathways is not perfect. The resonance

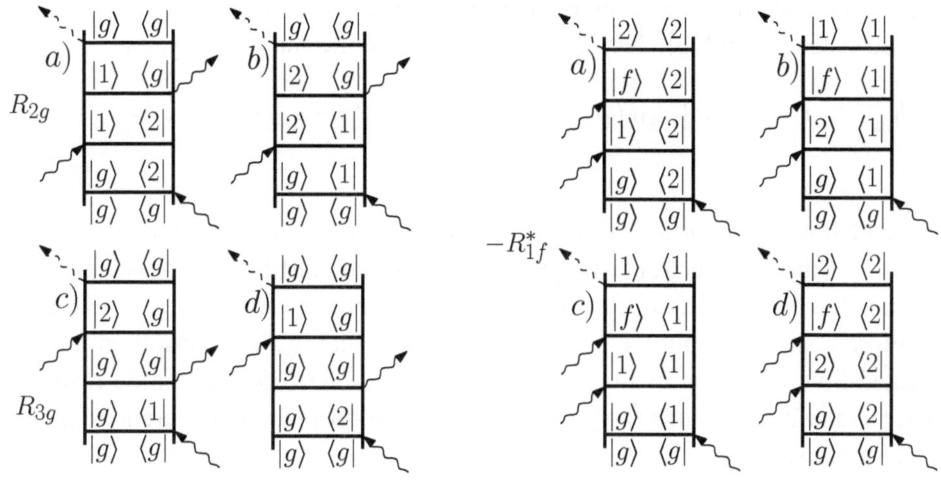

Figure 4.7 Rephasing Liouville pathways for a cross-peak in a dimer 2D spectrum. The pathways denoted by the same lower case letter cancel exactly for an uncoupled dimer. Using the definitions from Figure 4.6, the R^*_{1f} pathways can be shown to contain the same transitions as those of the R_{2g} and R_{3g} type.

coupling between molecules in a complex introduces, among other things, transition dipole moment redistribution among the collective transitions (van Amerongen *et al.*, 2000), and a perfect cancellation of the cross-peaks thus becomes impossible. The presence of a cross-peak at $t_2 = 0$ reveals resonance coupling. The effect of the resonance coupling can therefore be understood as lifting a cancellation due to a certain special relation between the Liouville pathways. This is not exclusive to optical spectroscopy. In the IR version of the same experiment, which is performed on molecular vibrations, no signal at all would appear due to a similar cancellation effect, if the molecular vibrations were perfect harmonic oscillators (Hamm and Zanni, 2011). All detectable signal is a consequence of anharmonicities. The main message of the present analysis is that one must not neglect higher-lying excited states. In the situation where one would expect it the least, i.e. when no or very weak coupling is present between individual chromophores in a complex, the omission of two-exciton states in calculating a 2D spectrum leads to a forest of unphysical cross-peaks.

Let us now assume that a non-zero coupling is present between the energy levels in Figure 4.6a. Only Figure 4.6b is then valid for assigning the Liouville pathways of Figure 4.7. We can see that the cross-peaks get half of their contributions from the pathways that contain evolution of off-diagonal density matrix elements in the one-exciton band. These elements are the so-called one-exciton coherences. A non-zero coherence, $\rho_{ab}(t) = \langle a|\mathrm{tr}_{\mathrm{bath}}\{\hat{W}(t)\}|b\rangle$, means that the corresponding state of the system can be written (at least partially) as a linear combination of the excited states $|a\rangle$ and $|b\rangle$. The corresponding Liouville sub-pathway, e.g.

$$R_{2g}^{(12)}(t_3, t_2, t_1) = \mathrm{tr}\{\mu_{g1}\mathcal{U}_{1g1g}(t_3)\mathcal{V}_{1g12}\mathcal{U}_{1212}(t_2)\mathcal{V}_{12g2}\mathcal{U}_{g2g2}(t_1)\mathcal{V}_{g2gg}\hat{W}_{gg}^{(eq)}\}, \tag{4.109}$$

has a characteristic phase factor from the evolution superoperator $\mathcal{U}_{1212}(t_2) \approx e^{-i\omega_{12}t_2}$. In the 2D spectrum, the phase factor is acquired by all similar response functions corresponding to the cross-peaks. Part of the cross-peak contributions therefore changes amplitude and shape with the delay t_2. This results in a complicated time evolution of the cross-peaks during time t_2 which can be resolved experimentally, and which can yield direct information about the coherence dephasing processes.

The 2D spectrum thus reveals homogeneous and inhomogeneous lineshapes, resonance coupling by the presence of cross-peaks, relaxation processes between electronic levels by a build-up of cross-peaks in time t_2, the presence of coherence between pairs of levels by oscillation of the cross-peaks and many other detailed pieces of information. All this in an ideal case of ultra-broadband excitation and well-separated peaks. In real-world measurements, many overlapping 2D features of interest may be congested into a single spectral region, and it becomes difficult to separate them from each other. Adding new degrees of freedom within the experiment, such as varying the polarization of the incoming pulses, may help to distinguish different contributions. The method is based on one aspect of the response functions that we have so far omitted from our discussion, namely, on the tensor nature of the response. Both the three electric fields entering the sample, and the polarization generated by them are vectors. The four dipole moment operators entering each third-order response have directions, and the response is thus a rank four tensor. In an isotropic sample, we have to average over all possible orientations of the molecular dipoles with respect to polarizations of the entering fields. By varying them we can enhance or suppress the contributions of certain Liouville pathways (Hochstrasser, 2001; Dreyer *et al.*, 2003; Read *et al.*, 2007) and correspondingly of some spectral features.

A further increase in the information content of non-linear spectroscopy, or simply a better resolution of the currently available information, might be achieved by going into higher dimensions. One can e.g. devise new pulse sequences providing more pulse delays. Three-dimensional spectra can thus be constructed, yielding more detailed information and requiring higher-order perturbation theory (Hamm, 2006; Fidler *et al.*, 2010). The utility of higher-dimensional spectroscopic techniques will greatly depend on the development of suitable representations of

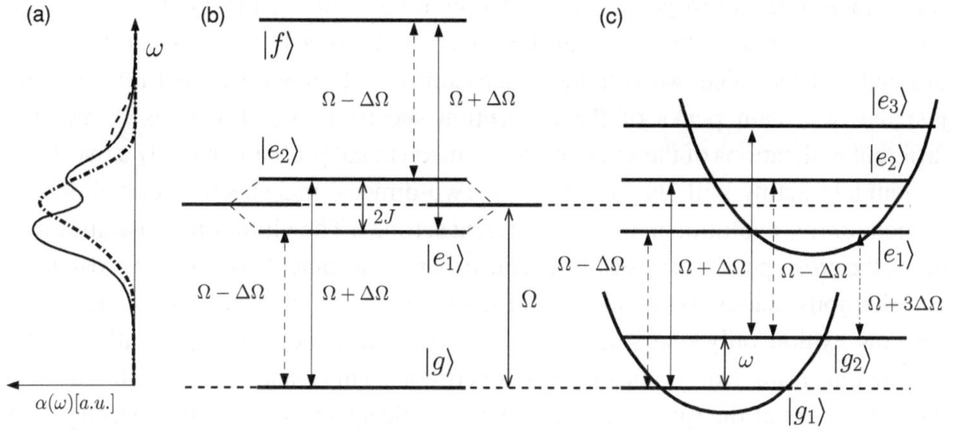

Figure 4.8 Electronic homodimer (b), and a two-level system with pronounced vibrational features (c). The absorption spectra of the two systems (a) are almost identical for a suitably chosen structure of the dimer. The mean transition frequency into the excited state is chosen the same, and the distance between the absorption maxima is $2J = 2\Delta\Omega$.

multidimensional data, which will allow us to overcome intrinsic human limitations to our (3 + 1)-dimensional world.

4.4.3 Electronic versus vibrational coherences

One of the most interesting early results of electronic 2D spectroscopy on photosynthetic systems was the experimental detection of long-lived coherent oscillations. These oscillations were interpreted as signatures of electronic coherence (Engel *et al.*, 2007). The consequences of its presence in a biologically relevant process of energy transfer and its relatively long lifetime became a matter of intense theoretical and experimental research. Oscillation patterns in non-linear spectra could, however, also originate from the presence of molecular vibrations of a suitable frequency.

A few simple electronic level systems with well-defined vibrations present an important case study of the general phenomena of *electron–phonon interaction* in molecular systems. Figure 4.8 presents a level structure of two systems, a molecular homo-dimer with resonance coupling J and a two-level molecule with one dominant vibrational mode. The frequency of the vibrational mode is equal to $2\Delta\Omega$, which equals the splitting of the dimer excited state energies $2J$. Absorption spectra of both systems can look very similar if the transition dipole moments from the ground states to the excited state manifold are the same. There may be some trace of the third peak in the absorption spectrum of the oscillator, but in a more complicated system it might be mixed with other transitions and therefore difficult to detect. The two systems, when excited by light with a spectrum covering just the two main peaks of the absorption spectrum (see Figure 4.8), are two different realizations of the same quantum-mechanical problem, namely a two-level system interacting with its environment. Two-dimensional spectroscopy allows us to monitor this evolution in the two different settings. The vibrational case allows us to see the cross-peaks fairly easily, because there is no cancellation of the cross-peak contributions thanks to the absence of ESA. On the other hand, purely electronic systems exhibit only one state in the ground-state band, and the oscillating 2D feature thus always originates in the excited state. This leads to clear differences in the 2D spectra of the two systems. The electronic spectrum contains a strong ESA feature with position depending on the structure of the dimer. The vibrational case should contain only positive contributions with well expressed cross-peaks. The cross-peaks evolve in time and may become negative during this time evolution. However, negative cross-peaks in a vibrational case do not automatically mean ESA contribution, as there is no ESA in the model in Figure 4.8c.

The presence of vibrational modes achieves a distinct effect in the electronic 2D spectrum, even if the vibrational frequency is so small that the cross-peaks completely overlap with the homogeneous lineshape of the electronic transition (Nemeth *et al.*, 2008). An electronic two-level system can be studied theoretically within the response function formalism, Equations (4.82) and (4.83). A Brownian oscillator model (Mukamel, 1995) of both the solvent and the molecular oscillator provides a good theoretical base for developing the suitable two-component $g(t)$, describing the solvent and vibrational mode interaction with the electronic transitions. When all transient (t_2 dependent) effects of the solvent reorganization disappear, long living oscillations due to the vibrational motion modulate the real and imaginary part of the 2D spectrum. The response function theory leads to a good quantitative agreement with the measured spectra (Nemeth *et al.*, 2008), and it enables us to make predictions of lineshapes and their time evolution, depending on the strength of the electron–phonon interaction.

In complexes of coupled molecules with transitions possibly modulated by vibrational modes, it might be difficult to decide what is the origin of the oscillations in a 2D spectrum. Separating the 2D spectra into rephasing and non-rephasing parts helps to achieve the classification. Among the rephasing Liouville pathways of a purely electronic system which yield a signal on the diagonal of the 2D spectrum, we cannot find any that would oscillate during the interval t_2. In the rephasing part of the purely electronic 2D spectrum, only the cross-peaks should oscillate. Analysis of the non-rephasing part of the response yields, however, several pathways (R_{2f} and R_{4g} type) that involve t_2-oscillations contributing to the diagonal peaks. Thus a characteristic feature of the 2D spectra of purely electronic coherence is that diagonal peaks oscillate only in the non-rephasing part of the 2D spectrum (Cheng and Fleming, 2009). The rephasing vibrational spectrum, on the other hand, contains a R_{3g} Liouville pathway with a ground state vibrational coherence $|g_2\rangle\langle g_1|$, which oscillates with the vibrational frequency and contributes to the peak at $\omega_1 = \Omega + \Delta\Omega$ ($|e_1\rangle \rightarrow |g_2\rangle$ transition) and $\omega_3 = \Omega + \Delta\Omega$ ($|e_3\rangle \rightarrow |g_2\rangle$ transition). Thus unlike the pure electronic case, vibrational coherence leads to oscillations of the diagonal peaks even in the rephasing 2D spectrum. Also, the additional ground state level $|g_2\rangle$ contributes a peak into the 2D spectrum which cannot appear in a purely electronic system, and this peak provides clear evidence of the presence of a vibration DOF (Mančal *et al.*, 2012).

The most complicated situation occurs when vibrational levels participate in the resonance interaction between chromophores. The resulting eigenstates of the system Hamiltonian have a mixed exciton–vibrational character, and they are often referred to as vibrons, vibrational excitons. The coherences between vibronic states can have a comparatively long lifetime due to their partially vibrational character, and the state can have increased transition dipole moments due to their partially electronic character (Christensson *et al.*, 2012). These vibronic states might hold a key to the long-lived coherences observed in 2D spectra of some photosynthetic systems, while at the same time rendering attempts to classify the origin of coherence into either vibrational or electronic classes meaningless.

4.5 Conclusions

We started by stating that the molecular systems investigated in this chapter have to be treated quantum mechanically, while light can be treated classically. We have based the response theory on this statement, and the success of experimental methods discussed in this chapter confirms its validity. Studying molecular systems with a method so well described by semi-classical theory has one more important advantage. As the light enters the Hamiltonian as a classical parameter, all quantum effects observed in the experiment originate, undeniably, in the molecular system itself. Non-linear spectroscopy therefore provides a faithful picture of the nature of microscopic processes initiated by photo-excitation.

The greatest advantage of the response theory is its close relationship to the density matrix description of molecular systems. Often, using response functions, specific spectral features can be related to the properties of density matrix elements, the evolution of which can therefore be revealed by analysing non-linear experiments. An opposite approach, when response functions are constructed from the elements of the reduced density matrix calculated from some equations of motion, also often provides a viable way of understanding photo-induced molecular processes. In principle, however, one has to be careful not to overestimate this close relationship. For example, the reduced density matrix does not carry information about the part of the system that we consider as environment, and as a result it does not correctly account for evolution of the environmental DOF between the interactions of the system with light (Mančal and Šanda, 2012). This does not seem to be a problem for recently proposed propagation schemes based on the Kubo–Tanimura hierarchy (Ishizaki and Fleming, 2009b). The choice of model and especially the formal splitting between the system and its environment plays a crucial role in successful calculation of the response functions from the density matrix evolution.

One must also not forget that non-linear spectroscopies of certain order provide insight only into a specific term of a perturbation series. One often describes response functions as a sequence of excitation processes where 'coherence states' and 'population states' are excited and de-excited. Such 'perturbative' language is very useful in recording the contributions to the non-linear spectra, but does not describe actual state of the system. Only the complete perturbation series must have the properties required for the proper density matrix, and it is not surprising that the 'coherence states' do not satisfy the requirements.

In this chapter, we have followed the main ideas of non-linear response theory to derive relations between the external stimulus and non-linear signals. We have demonstrated how to translate the general theory into a description of particular experimental methods, the photon-echo peakshift and 2D coherent spectroscopy. Together with some more extended accounts on the theory of non-linear spectroscopy available in the literature (Mukamel, 1995; Cho, 2009; Hamm and Zanni, 2011), the reader should now be able to apply the formalism discussed here to solve their own problems, and to follow the details in some of the following chapters.

Part II

Quantum effects in bacterial photosynthetic energy transfer

5

Structure, function, and quantum dynamics of pigment–protein complexes

IOAN KOSZTIN AND KLAUS SCHULTEN

5.1 Introduction

Photosynthesis is fundamental to life on Earth as it establishes access to the main energy source of the biosphere, sunlight (Blankenship, 2002). Photosynthesis is based on the interaction between living matter and the sun's radiation field, mainly visible light. This interaction involves the electrons of biological macromolecules and, accordingly, the process of light absorption is governed by quantum physics. During the course of biological evolution, photosynthetic lifeforms learned to exploit quantum physics in ingenious ways, in particular, under the circumstances of physiological temperature. A description of quantum phenomena under the influence of strong thermal effects as arise under these circumstances is challenging. Indeed, the quantum biology of photosynthesis is an active and fascinating research area.

Photosynthesis, in general, is understood to encompass the various processes in living cells by which lifeforms utilize sunlight to drive chemical synthesis. This involves primary processes of light-harvesting, transformation of electronic excitation energy into a membrane potential, as well as the splitting of water into oxygen, abstracting electrons that are added to molecules of nicotinamide adenine dinucleotide phosphate ($NADPH^+$) at a high redox potential. The membrane potential drives the synthesis of adenosine triphosphate (ATP) which is used to fuel many processes in living cells. In plant photosynthesis $NADPH^+$ and ATP are needed for the synthesis of sugar and starch, the most widely known products of photosynthesis. Because of its fundamental importance in cellular energetics, photosynthesis has been the subject of great evolutionary pressure such that, amidst a deep overall similarity, many variants have developed in the competition for habitats and efficiency.

Quantum Effects in Biology, ed. Masoud Mohseni, Yasser Omar, Gregory S. Engel and Martin B. Plenio. Published by Cambridge University Press.

Purple bacteria carry out only a highly rudimentary variant of photosynthesis, one that neither splits water nor produces $NADPH^+$, but leads to synthesis of ATP through provision of a membrane potential (Hu *et al.*, 2002). The photosynthesis variant of purple bacteria involves, nevertheless, the key molecular apparatus and primary processes present in all photosynthesis found on Earth. In particular, the apparatus realizes the main facets of light-harvesting as also found, for example, in plants.

Light-harvesting in purple bacteria and other photosynthetic lifeforms funnels electronic excitation energy into a molecular apparatus that converts it first into electron transfer across a cell membrane in a protein complex called the photosynthetic reaction centre (RC) (Hu *et al.*, 2002; Strümpfer *et al.*, 2011). The light-harvesting apparatus must absorb photons at a high enough rate to feed electronic excitation energy into the reaction centres, which requires hundreds to thousands of light absorbing molecules. Because of the size of the pool of absorbers, these molecules actually need to fulfil two functions, namely, light absorption as well as excitation transfer to the RCs (Sener *et al.*, 2011). This is obvious if one considers the fact that the many absorbers are necessarily spread over a wide area and, if a molecule distant from an RC absorbs light, the molecules between the primary absorber and the RC need to conduct the electronic excitation energy. To achieve an optimal structure for the dual role, the light absorbing molecules are organized and held into place by a scaffold of protein molecules, the light-harvesting complexes. Typically, about a hundred light-harvesting complexes hold 3000 light absorbing molecules.

The overall structure and function of the light-harvesting and RC system of purple bacteria have been covered in several recent reviews (Hu *et al.*, 1998, 2002; Sener *et al.*, 2011; Strümpfer *et al.*, 2011). The present chapter focuses only on the quantum biology of initial light absorption in the light-harvesting system and, in addition, but more briefly, on the initiation of electron transfer in the RC. The light absorbing molecules are chlorophylls and carotenoids. Carotenoids absorb light in the middle of the visible spectrum, i.e. at around 500 nm (Damjanovic *et al.*, 1999). Light excitation reaches an optically strongly allowed electronic excitation and is transferred quickly, namely within about 100 fs, to chlorophyll excitations that absorb at lower energy like themselves, namely in the range 800–900 nm (Ritz *et al.*, 2000a). The quantum physics of light absorption and excitation transfer by carotenoids, including the additional role of carotenoids in quenching chlorophylls in their triplet state, which can be harmful to the organism otherwise, is remarkable and has been described at length (Damjanovic *et al.*, 1999).

Electron transfer in the RC is initiated by a pair of electronically excited chlorophyll molecules. This pair is actually realizing the first energy conversion step in photosynthesis, as it changes light excitation energy, received indirectly from a large pool of chlorophylls and carotenoids, into a membrane potential generated through the electron transfer (Berg *et al.*, 2002).

In this chapter we want to discuss first light absorption by chlorophylls. This process poses two conceptual challenges to quantum biologists. Firstly, the chlorophylls develop their broad absorption characteristic, due to coupling between vibrational and electronic degrees of freedom, only when close to physiological temperature. As a broad absorption, capable of utilizing the continuous solar spectrum, is key to biological function, the role of temperature for chlorophyll spectral shape is of great relevance from a biological perspective.

The spectra of chlorophylls in bacterial photosynthesis are also shaped by chlorophyll–chlorophyll interactions. Indeed, the light-harvesting proteins of the purple bacteria contain rings of closely (about 10 Å centre–centre distance) spaced chlorophylls that form excitons after light absorption. The excitonic interaction affects the light absorption characteristics of chlorophyll rings present in the light-harvesting proteins (Hu and Schulten, 1997; Hu *et al.*, 1998; Damjanovic *et al.*, 2002).

In the following, we will introduce firstly the light-harvesting proteins of purple bacteria. Secondly, we will describe the optical transitions in the chlorophyll–protein systems under the condition of physiological temperature. Lastly, the description is extended briefly to RC electron transfer.

5.2 Light-harvesting complexes from purple bacteria: structure, function and quantum dynamics

The light-harvesting apparatus in purple bacteria is organized in a special cellular membrane formed through invagination of the so-called inner bacterial membrane, which actually surrounds the bacterial cell, into the cell interior. The invaginations come about either in the form of hundreds of spherical bulbs, roughly 60 nm in diameter, or in the form of lamellar sheets, both filling much of the bacterial cell. Figure 5.1a shows the spherical bulb as found in *Rhodobacter (Rba.) spheroides* (Sener *et al.*, 2010); lipids are not shown. The sphere is formed by a lipid bilayer that consists of about 200 trans-membrane proteins. These proteins, altogether, hold about 1000 carotenoids and 3000 bacteriochlorophylls (BChls). The arrangement of the BChls, without their protein scaffolds, is shown in Figure 5.1b, where one can recognize a hierarchical arrangement: four (blue) BChls forming a small cluster in each of the RCs, 56 BChls (red) form a figure eight line in a dimer of light-harvesting complexes 1 (LH1), and 27 BChls (green) form a ring (18 BChls) and a 'crown' (nine BChls) in each of the light-harvesting complexes 2 (LH2).

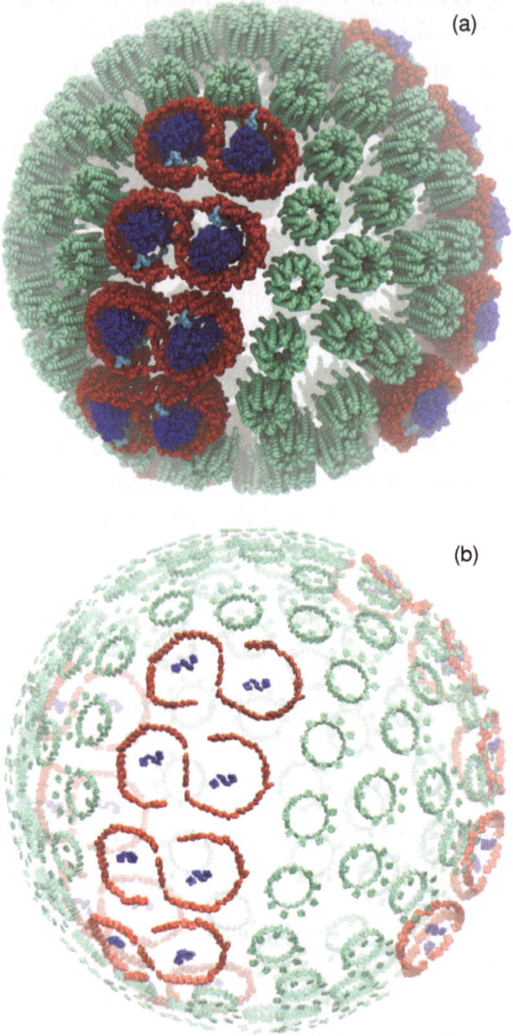

Figure 5.1 Supramolecular architecture of the chromatophore. (a) The chromatophore, a membrane vesicle of approximately 60 nm radius, consists of about 200 membrane proteins, shown in three different colours (blue, red and green), as there are three different proteins, reaction centres (RC, blue), light-harvesting complexes 1 (LH1, red), and light-harvesting complexes 2 (LH2, green). BChls and carotenoids in the chromatophore are held by LH1, LH2 and RC. (b) Here the 3000 BChls are shown solely, without protein and carotenoids. One can recognize an arrangement of LH1 and LH2 BChls in ring-like clusters. The LH1s in the particular chromatophore shown are dimerized into complexes $(LH1–RC)_2$ with their LH1 BChls forming a figure eight (red) and RC BCH1s forming a tetramer (blue). The BChls of LH2 form a crown-like ring (green).

In the light-harvesting apparatus, conventionally called a chromatophore, the RC BChls are the ultimate recipients of the light energy absorbed. The RC BChls either absorb light directly or receive excitation energy from the LH1 BChls. The LH1 BChls, being most proximate to the RC BChls, furnish excitation energy to the RC BChls after they either absorb sunlight directly, or receive excitation energy from neighbouring LH1 BChls or LH2 BChls. LH2 BChls are the main entry route for light energy into the chromatophore, as they contribute most to the BChl pool, outweighing RC and LH1 BChls by a factor of about four, the exact value depending on light conditions in the growth medium (Sener *et al.*, 2010).

The chromatophore, as shown in Figure 5.1, is an exquisite quantum-biological device that rivals in its simplicity, efficiency and functionality man-made solar energy devices. The design reached in biological evolution packs a maximum number of BChls into the available membrane space, while leaving space in the membrane for key chemical reactions that require diffusion of molecules, in particular, of quinones. The protein system is extremely modular and achieves its assembly into a spherical shape and proper stoichiometry of RC, LH1, LH2 in a self-organized process. The system is extremely stable functionally against unavoidable radiation damage of its components and, because of its modular form, repair or reuse of components is straightforward.

While all these aspects of the biological function of chromatophores deserve in-depth study, we will focus in this chapter mainly on the light absorption characteristics of the chromatophore, the very first step of light-harvesting. In the following, we will briefly describe the BChl systems of RC, LH1, and LH2.

5.2.1 The photosynthetic reaction centre

The photosynthetic reaction centre is depicted in Figure 5.2a, while Figure 5.2b shows how the RC is fitted into the inner part of a ring-shaped LH1 protein. The non-proteinaceous prosthetic groups in the RC are the main carriers of function: four BChls, two bacteriopheophytins (these are BChls without a central Mg ion), a permanently bound quinone, a temporarily bound quinone and an iron atom. The prosthetic groups are involved in accepting, or directly absorbing, electronic excitation energy and utilizing it for electron transfer to a temporarily bound quinone, where the electron, through recruitment of a proton, becomes bound in the form of a hydrogen atom.

Of the four BChls in the RC, two BChls lying closest to an approximate, two-fold symmetry axis (oriented vertically in Figure 5.2a), form a so-called 'special pair' (SP) that is the main acceptor of light excitation into the RC, although all four BChls are actually involved in light absorption or electronic coupling to the LH1 BChl system via fluorescent resonant energy transfer (Strümpfer and Schulten, 2012).

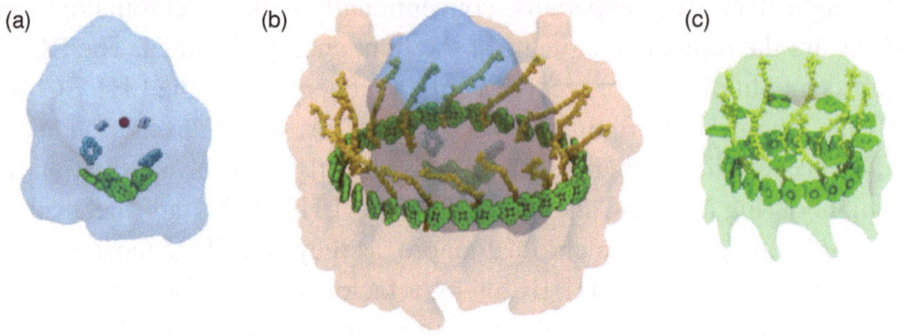

Figure 5.2 Pigments and cofactors of proteins found in the chromatophores of purple bacteria and acting as a scaffold for BChls, carotenoids and a few other prosthetic groups. (a) Reaction centre (RC) with the protein shown in transparent blue. Four molecules of BChl (green), two molecules of bacteriopheophytin (cyan), two molecules of quinones (grey) and one iron atom (red) in the RC interior are also shown; two carotenoids are not depicted. Of the four molecules of chlorophyll shown, the two central ones form the so-called special pair of chlorophylls that, after it receives electronic excitation through the LH1 chlorophylls, initiates electron transfer going first to one of the bacteriopheophytins and then, in further steps, to one of the quinones; since the RC is located in a membrane, the special pair-to-quinone electron transfer generates a membrane potential. (b) LH1–RC complex with the proteins shown in transparent red and blue, respectively. One can recognize a ring of 32 BChls (green) and carotenoids (yellow). (c) LH2 with the protein shown in transparent green; interior BChls are shown in green, carotenoids in yellow.

A subject of intense discussion regarding the RC prosthetic groups is that, in spite of near two-fold symmetry, the electron transfer from the excited SP state proceeds along just one of the two branches of the symmetric system. This fact is intriguing and brings up the question of why the RC exhibits a two-fold symmetry, i.e. two potential electron transfer branches, when only one is utilized for electron transfer. A likely answer is that the two-fold symmetry is used, in fact, solely to enhance the RC absorption characteristic and electronic coupling to the LH1 BChl system through so-called superradiance, as discussed in (Strümpfer and Schulten, 2011). Superradiance, a quantum coherence effect that speeds up excitation transfer between RC BChls and LH1 BChls, involves in this case a redistribution of absorption and emission strength (transition dipole moments) from all BChls to key low energy exciton states (Strümpfer *et al.*, 2012; Strümpfer and Schulten, 2012).

5.2.2 The light-harvesting complex LH1

The light-harvesting complex LH1 is depicted in Figure 5.2b along with its associated RC. LH1 contains 32 BChls and 16 carotenoids. The carotenoids absorb light at 600 nm and transfer the excitation absorbed within about 100 fs, i.e. extremely fast, to the BChls (Strümpfer *et al.*, 2012; Strümpfer and Schulten, 2012).

The BChls are arranged in a ring and are strongly interacting. The ring in its equilibrium structure exhibits a 16-fold symmetry, each symmetry unit containing two oppositely oriented BChls beside one carotenoid. The centre (Mg)–centre (Mg) distance between neighbouring BChls measures about 10 Å; these BChls exhibit strong excitonic coupling of about 300 cm^{-1}; non-nearest neighbour BChls are coupled in leading order through induced dipole–induced dipole interaction whose strength is described by the formula,

$$V_{nm} = \frac{1}{4\pi\varepsilon_0\varepsilon_r}\left[\frac{\mathbf{d}_n\mathbf{d}_m}{r_{nm}^3} - 3\frac{(\mathbf{d}_n\cdot\mathbf{r}_{nm})(\mathbf{d}_m\cdot\mathbf{r}_{nm})}{r_{nm}^5}\right], \quad (5.1)$$

where ε_r is the relative dielectric permittivity of the medium, $\mathbf{d}_n$ is the transition dipole moment of the nth BChl and r_{nm} is the centre–centre distance between BChls n and m.

The BChls are excited by light into either their so-called Q_x state absorbing at around 600 nm or their Q_y state, absorbing at around 800 nm; due to sub-picosecond internal conversion the lower energy Q_y is quickly the only excited state populated. The excitonic coupling just mentioned leads to delocalization of Q_y excitations across the BChl ring. The resulting excitations, termed excitons, form a band of excited states. The states around 875 nm carry the main oscillator strength and form the strongest absorption band.

The role of the LH1 protein is to furnish the scaffold into which BChls and carotenoids are inserted, assuring effective interaction that channels, within a picosecond, all light absorbed into the lower energy part of the exciton band formed by the Q_y excitations. The states in this exciton band are coupled through their transition dipole moments to the BChl exciton states of the RC, such that within about 20–30 picoseconds the excited LH1 BChl ring transfers its energy to the RC BChls (Strümpfer and Schulten, 2012). The excited RC BChls utilize their energy for electron transfer as described above, or send the energy back to the LH1 ring. The latter, seemingly wasteful, process is actually essential for spreading energy loss and heat dissipation under intense light condition equally over the chromatophore system, rather than absorbing all energy into the RC where heat and other radiation damage could be destructive (Sener *et al.*, 2007, 2010).

Altogether, the function of the LH1 system is to enlarge the cross-section for light absorption beyond what the four BChls in the RC can achieve by themselves. The RC is a device of about 14 000 atoms that would be too costly if it were served only through four BChls and one carotenoid absorbing the sun's energy; the LH1 system adds about 33 000 atoms, i.e. triples the mass of the system, but increases the BChl and carotenoid counts by factors of nine and seventeen, respectively;

clearly, the increase in protein mass is outweighed by the increase in absorption cross-section.

5.2.3 The light-harvesting complex LH2

In fact, even the addition of an LH1 does not tax the operating speed of a single RC and purple bacteria add, therefore, further protein mass and BChl–carotenoid light absorption strength, namely in the form of LH2s (Strümpfer *et al.*, 2012). These proteins are smaller versions of LH1 and come in two similar sizes, one which adds about 18 000 atoms with 27 BChls and nine carotenoids and one which adds about 16 000 atoms with 24 BChls and eight carotenoids.

Either LH2 forms, like LH1, a ring, but with about half the diameter. The symmetry units of the LH2s are very similar to the building blocks of LH1, except that they carry three BChls and one carotenoid in the LH2 case, as opposed to two BChls and one carotenoid in the LH1 case. The overall LH2 proteins exhibit a nine-fold or an eight-fold symmetry axis for the larger or smaller versions, respectively. The structures of the two sizes of LH2 have been determined crystallographically, namely, for LH2 of *Rhodopseudomonas (Rps.) acidophila* (nine-fold symmetric) (McDermott *et al.*, 1995; Papiz *et al.*, 2003) and of *Rhodospirillum molischianum* (eight-fold symmetric) as presented in (Koepke *et al.*, 1996). The third BChl in the LH2 building blocks is oriented parallel to the ring plane. The structure of an LH2 is shown in Figure 5.2c, where one can recognize that LH2 forms two BChl rings, one similar (but smaller) to the one in LH1 and one with wider spaced (Mg–Mg distance of about 20 Å) BChls that adopt a perpendicular orientation.

The LH2s, as described, are the most prevalent light-harvesting proteins in purple bacteria, outnumbering LH1s and RCs by a factor of about ten; the exact ratio depending on light intensity in the bacteriums' habitat or growth medium (Sener *et al.*, 2010). The LH2s are the smallest light-harvesting proteins and structurally better characterized than LH1s, as the latter have been described structurally only through electron microscopy images below atomic resolution, combined with homology modelling based on LH2 crystallographic structures (Hu and Schulten, 1998). The LH1s also come in 1:1 complexes with RCs. The better characterization and easier handling of LH2s have made them the favourites of spectroscopists who have achieved characterization of the complexes' optical properties down to the single LH2 level.

The remainder of this chapter will, hence, focus on LH2 spectra at low and physiological temperatures. We will also discuss, but more briefly, the characteristics of electron transfer under low and physiological temperature conditions.

5.3 Optical transitions in pigment–protein complexes

Membrane bound pigment–protein complexes (PPCs) fulfil crucial biochemical functions in photosynthetic organisms, e.g. conversion of absorbed light energy into electronic excitations and facilitation of primary electron charge separations that lead to electrochemical potential gradients. As the number of available high resolution crystal structures of a variety of PPCs continues to increase, it becomes more crucial than ever to develop new efficient and accurate computational methods to describe and predict their quantum-biophysical properties, using as input only structural information. The theoretical description of PPCs at physiological temperature in their native environment requires one to determine and characterize the quantum states of the interacting pigment molecules in the presence of thermal fluctuations.

While useful for interpreting experimental data, standard theoretical approaches for calculating electronic transfer rates and optical spectra of PPCs, which are based on empirical stochastic models containing several fitting parameters, are neither capable of predicting experimental outcomes, nor very helpful in guiding and designing new experiments. Here we present a general approach, which combines molecular dynamics (MD) simulations, quantum chemistry (QC) calculations and quantum many-body theory, for predicting and characterizing charge transfer, and spectral and optical properties (e.g. linear absorption and circular dichroism spectra) of PPCs (Damjanovic *et al.*, 2002; Janosi *et al.*, 2006; Kosztin and Schulten, 2008). The method requires only atomic-level crystal structure information. Firstly, the conformational dynamics of the PPC embedded into a fully solvated lipid bilayer is followed by means of classical MD simulations. Next, the lowest energy quantum states of the pigment molecules are determined along the MD trajectory by means of QC calculations. Finally, the transfer rate and/or optical spectra of the PPC are determined in terms of a lineshape function which, within the cumulant approximation, can be calculated from the results of the QC calculations. The features and viability of the method are exemplified by calculating the linear optical (OD) and circular dichroism(CD) spectra of the light-harvesting complex LH2 from *Rs. molischianum* (Koepke *et al.*, 1996) and the electron transfer rates in the photosynthetic reaction centre from *Rb. sphaeroides* (Deisenhofer *et al.*, 1985).

In order to calculate the OD spectrum of a PPC, one assumes that the electronic properties of individual pigment molecules can be described in terms of a two-level system, formed by the ground state and the lowest excited singlet state (in the case of BChls, the Q_y state) involved in the optical absorption process. When the interaction between pigments can be neglected (e.g. when the spatial separation between them is sufficiently large, as in the case of B800 BChls in LH2), one denotes these two states for pigment $n = 1, \ldots, N$, as $|0_n\rangle$ and $|1_n\rangle$, respectively. In the presence of a fluctuating environment composed of the protein matrix, lipid membrane and solvent molecules, these two levels transform into energy bands $|0_n;\lambda_{0n}\rangle = |0_n\rangle|\lambda_{0n}\rangle$ and $|1_n;\lambda_{1n}\rangle = |1_n\rangle|\lambda_{1n}\rangle$, characterized by the quantum numbers λ_{0n} and λ_{1n}, which specify the state of the nth pigment on the ground- and excited-state potential energy surface, respectively. Because exact calculation of the eigenstates $|0_n;\lambda_{0n}\rangle$, $|1_n;\lambda_{1n}\rangle$ and of the corresponding energy eigenvalues $\mathcal{E}_{0_n,\lambda_{0n}}$, $\mathcal{E}_{1_n,\lambda_{1n}}$ is not practical, an approximate solution is usually sought by identifying the quantum numbers λ_{0n} and λ_{1n} with the vibronic states of the PPC, which can be modelled within the harmonic approximation as a phonon heat bath. An alternative, even better approach is to follow the dynamics of the nuclear degrees of freedom of the PPC by all-atom MD simulations, and determination of the energy gap time series $\Delta E_n(t) = \mathcal{E}_{1n}(t) - \mathcal{E}_{0n}(t)$ at each MD timestep by means of QC calculations. The main assumption of this approach is that the obtained energy gap time series, $\Delta E_n(t)$, can be used to calculate approximately equilibrium quantities (e.g. energy gap density of states and time autocorrelation functions) of the original system without a knowledge of the exact energy gap spectrum $\Delta\mathcal{E}_{1_n,\lambda_{1n},\lambda_{0n}} = \mathcal{E}_{1_n,\lambda_{1n}} - \mathcal{E}_{0_n,\lambda_{0n}}$.

The Hamiltonian of the PPC in the absence of the excitonic coupling between the pigment molecules can be written as $\mathcal{H} = H_0 + H$, where

$$H_0 = \sum_{n,\lambda_{0n}} |0_n;\lambda_{0n}\rangle \mathcal{E}_{0_n,\lambda_{0n}} \langle 0_n;\lambda_{0n}|, \tag{5.2}$$

and

$$H = \sum_n H_n = \sum_{n,\lambda_{1n}} |1_n;\lambda_{1n}\rangle \mathcal{E}_{1_n,\lambda_{1n}} \langle 1_n;\lambda_{1n}|. \tag{5.3}$$

The electric dipole moment operator through which the incident light field couples to the nth pigment molecule is given by

$$\hat{\boldsymbol{\mu}}_n = \sum_{\lambda_{1n},\lambda_{0n}} \mathbf{d}_{n,\lambda_{1n},\lambda_{0n}} |1_n;\lambda_{1n}\rangle\langle 0_n;\lambda_{0n}|, \tag{5.4}$$

where the transition dipole moment (TDM) matrix element $\mathbf{d}_{n,\lambda_n,\lambda_0}$ in the Condon approximation (May and Kühn, 2011) is

$$\mathbf{d}_{n,\lambda_n,\lambda_0} \approx \mathbf{d}_n \langle \lambda_{1n}|\lambda_{0n}\rangle. \tag{5.5}$$

Here $\mathbf{d}_n = \langle 1_n|\hat{\boldsymbol{\mu}}_n|0_n\rangle$ is the real TDM vector whose time series can be determined from the same combined MD/QC calculations as $\Delta E_n(t)$. Note that while $\langle 1_n|0_n\rangle = 0$, in general the Franck–Condon factors $\langle \lambda_{1_n}|\lambda_{0_n}\rangle$ are finite (May and Kühn, 2011).

In the case of N excitonically coupled pigment molecules (e.g. the B850 BChls in LH2), formally Equations (5.2)–(5.5) remain valid provided that the site index n is replaced with the excitonic index J. Because the QC calculation of the energies

$\mathcal{E}_{J,\lambda_J}$ and TDMs $\mathbf{d}_J$ of the excitonic states $|J;\lambda_J\rangle$, $J = 1, \ldots, N$, of a quantum system formed by N coupled pigments is still prohibitively expensive, a practical solution is to employ an *effective Hamiltonian* for determining the time series $\Delta E_J(t) = \mathcal{E}_J(t) - \mathcal{E}_0(t)$ and $\mathbf{d}_J(t)$ from $\Delta E_n(t)$ and $\mathbf{d}_n(t)$ of the individual pigments. Assuming a point dipole–dipole coupling between the pigment molecules, given by Equation (5.1), the eigenvalue equation one needs to solve at every MD time step is

$$\sum_m [(\Delta E_n \delta_{nm} + V_{nm}) - \Delta E_J \delta_{nm}] c_m^{(J)} = 0. \tag{5.6}$$

In term of the coefficients $c_n^{(J)} = \langle J|1_n\rangle$ the excitonic TDMs are

$$\mathbf{d}_J = \sum_n \langle J|1_n\rangle\, \mathbf{d}_n. \tag{5.7}$$

5.3.1 Linear absorption and lineshape function

According to standard linear response theory, the OD spectrum $I(\omega)$ of a PPC is proportional to the dipole–dipole correlation function (Mukamel, 1995; May and Kühn, 2011),

$$I(\omega) \propto \omega \sum_{n,m} \mathrm{Re}\left[\int_0^\infty dt e^{i\omega t} \left\langle \hat{\mu}_{m,i}^\dagger(0) \hat{\mu}_{n,i}(t) \right\rangle\right], \tag{5.8}$$

where $\hat{\mu}_{n,i}(t) = e^{-iHt}\hat{\mu}_{n,i}(0)e^{iH_0 t}$ is the $i \in \{x, y, z\}$ component of the time-dependent electric dipole operator, and $\langle \ldots \rangle = Z_0^{-1}\mathrm{Tr}\{\exp(-\beta H_0)\ldots\}$ with $\beta = 1/k_B T$ being the usual temperature factor and Z_0 the corresponding partition function. To simplify notation we use units in which $\hbar = 1$, and apply the convention of implicit summation over repeated indices.

It can be shown (Janosi *et al.*, 2006) that for an aggregate of non-interacting pigments in their native environment the OD spectrum given by Equation (5.8) can be expressed in terms of the *lineshape function*,

$$A_n(\omega) \equiv \mathrm{Re} \int_0^\infty dt e^{i\omega t} \left\langle e^{iH_0 t} e^{-iH_n t} \right\rangle, \tag{5.9}$$

namely as

$$I(\omega) \propto \omega \sum_n d_n^2 A_n(\omega). \tag{5.10}$$

In a similar fashion, for N excitonically coupled pigment molecules, which can be regarded as a system of N non-interacting excitons, one obtains for the OD

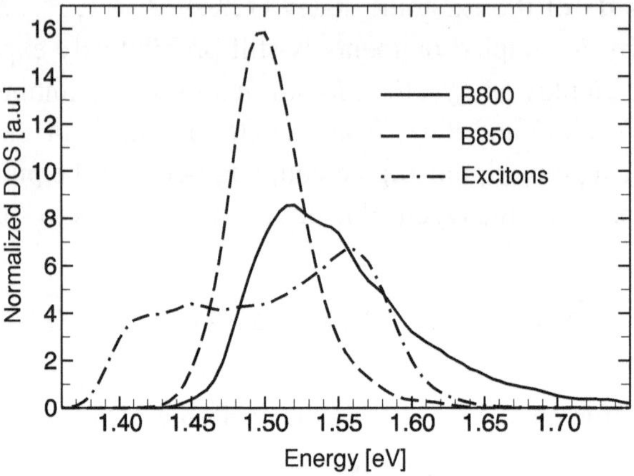

Figure 5.3 Normalized density of states, $\mathcal{N}(\omega)$, for individual B800 BChls (solid line) and B850 BChls (dashed line), and B850 excitons (dashed-dotted line) in LH2 from *Rs. molischianum* computed as binned histograms of the corresponding Q_y excitation energy time series obtained from combined MD/QC simulations.

spectrum (Janosi *et al.*, 2006),

$$I(\omega) \propto \omega \sum_J d_J^2 A_J(\omega), \tag{5.11}$$

where the lineshape function is

$$A_J(\omega) = \text{Re} \int_0^\infty dt e^{i\omega t} \left\langle e^{iH_0 t} e^{-iH_J t} \right\rangle. \tag{5.12}$$

In general, the Hamiltonians H_0 and H_n do not commute and thus calculation of the quantum time correlation function in Equation (5.9) is non trivial. Otherwise, one would have $\langle e^{iH_0 t} e^{-iH_n t} \rangle \approx \langle \exp(-i\Delta H_n t) \rangle$, with $\Delta H_n = H_n - H_0$, and the lineshape function could be expressed in terms of the energy gap density of states (DOS) as

$$A_n(\omega) \approx \pi \mathcal{N}(\omega), \tag{5.13}$$

$$\mathcal{N}(\omega) \equiv \langle \delta(\omega - \Delta H_n) \rangle \approx \langle \delta(\omega - \Delta E_n(t)) \rangle, \tag{5.14}$$

where the DOS $\mathcal{N}(\omega)$ is approximated by the binned histogram of the energy gap fluctuations $\Delta E_n(t)$ obtained from combined MD/QC calculations (Mercer *et al.*, 1999; Damjanovic *et al.*, 2002; Janosi *et al.*, 2006).

$\mathcal{N}(\omega)$ calculated for the individual B800 (solid lines) and B850 (dashed lines) BChls are shown in Figure 5.3. The striking difference between these two DOSs is due to the fact that in LH2 from *Rs. molischianum*, the B800s and B850s are surrounded by mainly polar and non-polar residues, respectively. $\mathcal{N}_{B850}(\omega)$ is only slightly red-shifted to 1.502 eV (825 nm) and has essentially the same shape, with

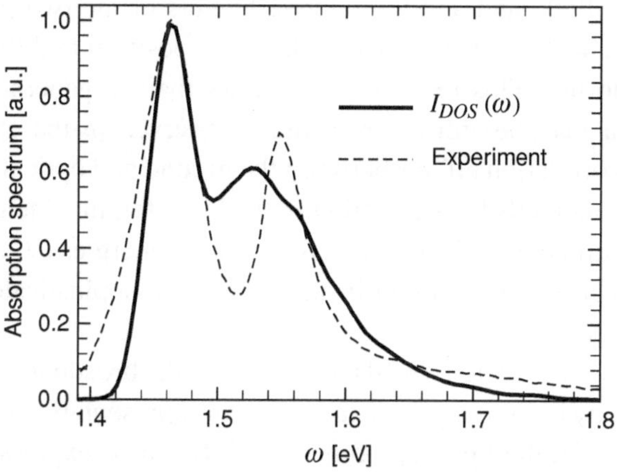

Figure 5.4 Absorption spectrum $I_{DOS}(\omega)$ of LH2 for *Rs. molischianum*. The spectrum was calculated as a combined DOS of B800 BChls and B850 weighted by the corresponding dipole strengths (solid line). $I_{DOS}(\omega)$ was blue-shifted by 20 meV in order to overlay its B850 peak with the corresponding one in the experimental OD spectrum (Zhang *et al.*, 2000) (dashed line).

full width at half maximum FWHM $\approx$ 53 meV, as the DOS obtained in similar MD/QC calculations by (Mercer *et al.*, 1999) for BChl solvated in methanol, also at room temperature. This observation indicates that the thermal motion of the nuclei in individual BChls leads to Q_y energy gap fluctuations that are insensitive to the actual nature of the *non-polar* environment (Janosi *et al.*, 2006). By contrast, the *polar* environment has a profound effect on $\mathcal{N}_{B800}(\omega)$, making it broad and asymmetric, characterized by: (i) a blue-shifted peak at 1.528 eV (811 nm); (ii) a considerably increased mean energy gap $\langle \Delta E_{B800} \rangle$ of 1.556 eV (797 nm), which matches rather well the experimental value of 800 nm; and (iii) a large FWHM of about 100 meV. The DOS of the excitonic energies, computed as a binned histogram of $\Delta E_J(t_\ell)$, $J = 1, \ldots, 16$ (calculated by solving for each MD snapshot, within the point-dipole approximation, the eigenvalue Equation (5.6)), are also shown in Figure 5.3 (dashed-dotted line) (Janosi *et al.*, 2006).

According to Equations (5.10) and (5.13)–(5.14), a rough estimate of the OD spectrum of the B800 BChls and B850 excitons is given by the corresponding TDM strength-weighted DOS,

$$I_{DOS}(\omega) \propto \omega \left[\sum_J d_J^2 \langle \delta(\omega - \Delta E_J) \rangle + \sum_{B800} d_{B800}^2 \langle \delta(\omega - \Delta E_{B800}) \rangle \right], \quad (5.15)$$

where the $B800$ index means summation over all B800 BChls. The calculated $I_{DOS}(\omega)$ (blue-shifted by 20 meV) is shown in Figure 5.4 (solid line) along with the experimental OD spectrum (Zhang *et al.*, 2000; Ihalainen *et al.*, 2001) (dashed line). While the B850 band and the relative heights of the two peaks in $I_{DOS}(\omega)$ match rather well with the experimental data, the position and the broadening of the B800 peak do not. This result clearly shows that, in general, peak positions in optical spectra may be shifted from those observed in the excitation energy spectrum due to correlation effects between the ground and optically active excited states. Therefore, methods for simulating optical spectra, in which the positions of the peaks are identified with the computed excitation energies (stick spectrum), are not entirely correct and one should instead use more sophisticated methods that include quantum correlation effects.

In general, Equations (5.13)–(5.14) overestimate the broadening of the lineshape function. Indeed, by employing the exact spectral representation of the correlation function $\langle e^{-iH_0t}e^{iHt}\rangle$, the lineshape function (5.9) can be expressed as a Franck–Condon-weighted and thermally averaged density of state (May and Kühn, 2011), i.e. as

$$A_n(\omega) = 2\pi \sum_{\lambda_{0n},\lambda_{1n}} \rho_{\lambda_{0n}} |\langle\lambda_{0n}|\lambda_{1n}\rangle|^2 \delta(\omega - \Delta\mathcal{E}_{n,\lambda_{1n},\lambda_{0n}}), \tag{5.16}$$

where $\rho_{\lambda_{0n}} = Z_0^{-1}\exp(-\beta\mathcal{E}_{0_n,\lambda_{0n}})$ is the statistical matrix of the electronic ground state. Indeed, Equations (5.13)–(5.14) follow directly from Equation (5.16) if one sets the Franck–Condon factors $\langle\lambda_{0n}|\lambda_{1n}\rangle$ equal to unity.

A systematic way of calculating the correlation function in Equation (5.9) is to employ the cumulant expansion method (Mahan, 1990; Mukamel, 1995; May and Kühn, 2011). Within the second-order cumulant approximation (Mukamel, 1995) one has,

$$\langle e^{iH_0t}e^{-iH_nt}\rangle \approx \exp\left[-i\langle\Delta H_n\rangle t - \int_0^t d\tau(t-\tau)\mathcal{C}_n(\tau)\right], \tag{5.17}$$

where $\Delta H_n(t) = e^{iH_0t}\Delta H_n e^{-iH_0t}$, $\mathcal{C}_n(t) = \langle\delta H_n(t)\delta H_n(0)\rangle$, and $\delta H_n(t) = \Delta H_n(t) - \langle\Delta H_n\rangle$. The quantum statistical averages in Equation (5.17) can be approximated by the corresponding classical ones, involving the energy gap time series $\Delta E_n(t)$, as follows:

$$\langle\Delta H_n\rangle \approx \langle\Delta E_n(t)\rangle \equiv \omega_n, \tag{5.18}$$

$$\mathrm{Re}[\mathcal{C}_n(t)] \approx C_n(t) \equiv \langle\delta E_n(t)\delta E_n(0)\rangle. \tag{5.19}$$

Note that while Equation (5.19) is widely used (Schulten and Tesch, 1991; Makri, 1999; Mercer *et al.*, 1999), other approximation schemes have also been applied and tested (Egorov *et al.*, 1999).

By employing the *fluctuation dissipation theorem* (Mukamel, 1995), the imaginary part of the quantum correlation function $\mathcal{C}_n(t)$ can also be expressed in terms of the energy gap correlation function $C_n(t)$. One finds (Mukamel, 1995;

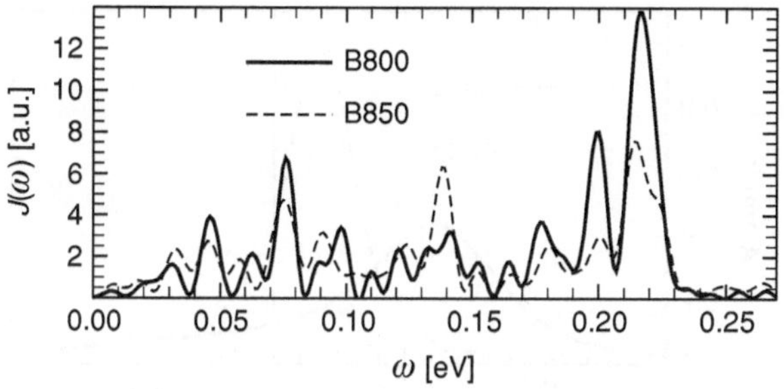

Figure 5.5 Spectral density function $J(w)$ for B800 (solid line) and B850 (dashed line) obtained according to Equation (5.21).

Janosi *et al.*, 2006),

$$\mathrm{Im}[C_n(t)] = \int_0^\infty \frac{d\omega}{\pi} J_n(\omega) \sin \omega t, \tag{5.20}$$

where the real *spectral density* is given by

$$J_n(\omega) = 2 \tanh(\beta\omega/2) \int_0^\infty dt\, C_n(t) \cos \omega t. \tag{5.21}$$

The spectral densities $J(\omega)$ calculated for the B800 and B850 BChls are shown in Figure 5.5. The prominent peaks (especially for B800) at about $\omega_p = 0.22$ eV, are due to the fast initial decay of $C(t)$, which is most likely caused by a strong coupling of the BChls to an intra-molecular C=O vibronic mode (Mercer *et al.*, 1999; Damjanovic *et al.*, 2002). The complex structure of the spectral functions indicate that all inter- and intra-molecular vibronic modes with frequency below ω_p will contribute to the lineshape function. Hence, attempts to use simplified model spectral functions appear to be unrealistic, even if these may lead to absorption spectra that match the experimental results.

Finally, the lineshape function within the second cumulant approximation is

$$A_n(\omega) \equiv \overline{A}_n(\omega - \omega_n) = \int_0^\infty dt\, e^{-\phi_n(t)} \cos[(\omega - \omega_n)t + \varphi_n(t)], \tag{5.22}$$

where the broadening and frequency shift functions are given by

$$\phi_n(t) = \int_0^t d\tau\, (t - \tau) C_n(\tau), \tag{5.23}$$

and

$$\varphi_n(t) = \int_0^\infty \frac{d\omega}{\pi} J_n(\omega) \frac{\omega t - \sin \omega t}{\omega^2}. \tag{5.24}$$

The lineshape functions of individual B800 and B850, calculated from Equation (5.22), are plotted in Figure 5.6 (Janosi *et al.*, 2006). The origin of the

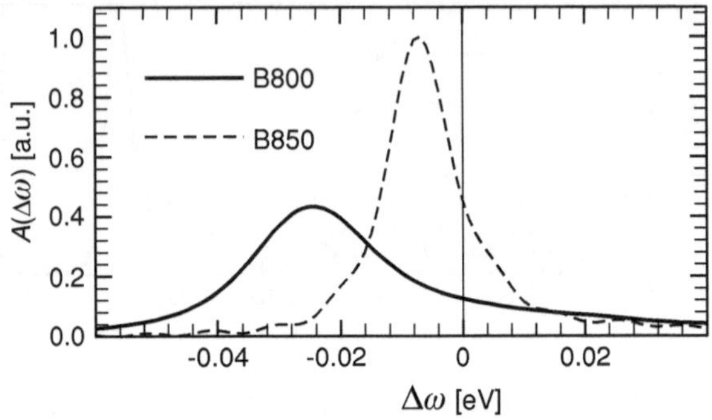

Figure 5.6 Lineshape functions $\overline{A}_{B800}(\Delta\omega)$ (dashed line) and $\overline{A}_{B850}(\Delta\omega)$ (solid line).

frequency axis corresponds to the mean energy gaps ω_{B800} and ω_{B850}, respectively. The highly polarized surrounding of the B800 BChls in *Rs. molischianum* renders $A_{B800}(\omega)$ twice as broad (FWHM $\approx$ 26 meV) as $A_{B850}(\omega)$ (FWHM $\approx$ 13 meV). Also, the red-shift of the peak of the former ($\Delta\omega \approx$ 25 meV) is more than three times larger than that of the latter ($\Delta\omega \approx$ 7 meV).

Although the 1 ps long energy gap time series provides a proper estimate of the B800 and B850 lineshape functions, the same data is insufficient to determine with reasonable accuracy the individual excitonic lineshape functions $A_J(\omega)$.

By neglecting the effect of *exchange narrowing* (Somsen *et al.*, 1996; van Amerongen *et al.*, 2000), one can approximate with reasonable accuracy the individual excitonic lineshape functions $A_J(\omega) \approx A_{B850}(\omega)$ and, similarly to Equation (5.15), the OD spectrum of the LH2 BChls can be calculated as (Janosi *et al.*, 2006)

$$I(\omega) \propto \omega \left[\sum_J d_J^2 \overline{A}_{B850}(\omega - \omega_J) + 8 d_{B800}^2 \overline{A}_{B800}(\omega - \omega_{B800}) \right], \tag{5.25}$$

where $\omega_J = \langle \Delta E_J \rangle$. As shown in Figure 5.7, $I(\omega)$ (subject to an overall blue-shift of 20 meV) matches remarkably well with the experimental OD spectrum, especially if one takes into account that it was obtained from the sole knowledge of the high resolution crystal structure of LH2 from *Rs. molischianum* (Janosi *et al.*, 2006). The reason why both the B800 and B850 peaks of $I(\omega)$ are somewhat narrower than the experimental ones is most likely the fact that, in the above calculations, the effect of static disorder was ignored. Indeed, the calculations were based on a single LH2 ring, while the experimental data is averaged over a large number of such rings. While computationally expensive, in principle, the effect of static disorder can be taken into account by repeating the above calculations for different initial configurations of the LH2 ring, and then averaging the corresponding OD spectra.

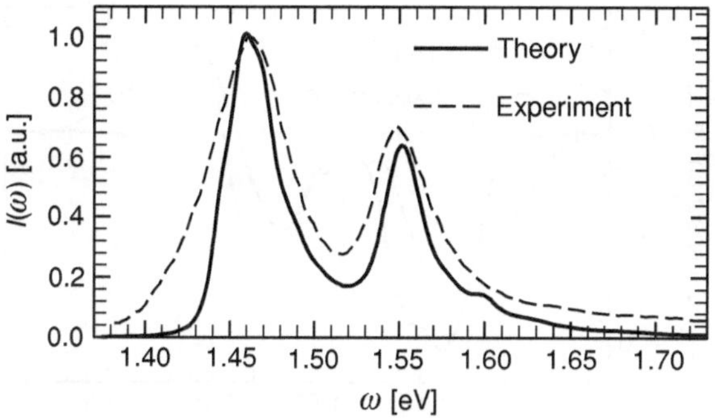

Figure 5.7 Absorption spectrum of the BChl aggregate in *Rs. molischianum* LH2. Shown is the comparison of the computed (solid line) and experimental (dashed line) absorption spectrum (in arbitrary units). The computed spectrum has been blue shifted by 20 meV for best match.

5.3.2 Circular dichroism

The CD spectrum is defined as $I_{CD}(\omega) = I_L(\omega) - I_R(\omega)$, where $I_{L/R}(\omega)$ is the OD spectrum for left/right circularly polarized light. Unlike the case of the OD spectrum, calculation of $I_{CD}(\omega)$, even within the leading order approximation, requires taking into account the spatial variation of the light field across the PPC as well as the excitonic coupling between the pigment molecules, regardless of how small this may be (Somsen *et al.*, 1996). The sensitivity of the CD spectrum to geometrical and local details of the PPC makes it a quantity difficult to predict by theoretical modelling. The CD spectrum can be calculated as (van Amerongen *et al.*, 2000)

$$I_{CD}(\omega) \propto \omega \sum_J R_J A_J(\omega), \tag{5.26}$$

where

$$R_J = \frac{\pi}{\lambda} \sum_{n,m} \langle J|1_n\rangle [\mathbf{r}_n \cdot (\mathbf{d}_n \times \mathbf{d}_m)] \langle 1_m|J\rangle \tag{5.27}$$

is the *rotational strength* of the excitonic state J. The rotational strength plays the same role for the CD spectrum as the TDM strength for the OD spectrum. According to Equation (5.27), in the absence of the excitonic coupling $\langle 1_m|J\rangle = \delta_{mJ}$ and $R_J = 0$ holds. Consequently, as already mentioned above, the CD spectrum vanishes in the case of vanishing excitonic coupling. R_J accounts for the coupling between the TDM of the excitonic state J and the orbital magnetic moment of the other excitons. The coupling to the local magnetic moment (Cotton effect) is assumed to be small and is usually neglected (Somsen *et al.*, 1996; van Amerongen *et al.*, 2000).

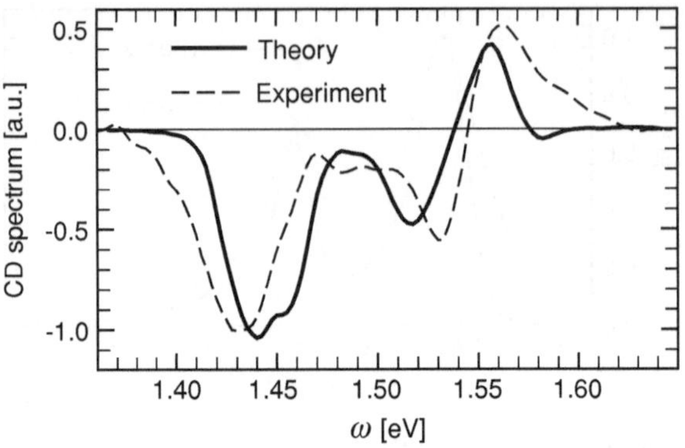

Figure 5.8 Comparison between the computed (solid line) and experimental CD spectrum of the BChl aggregate in *Rs. molischianum* LH2.

Employing the same MD/QC data used to determine the OD spectrum, the CD spectrum of the BChls in LH2 from *Rs. molischianum* was calculated using Equation (5.26), where the summation index J runs over all B850 and B800 excitonic states and $A_J(\omega) = \overline{A}_\alpha(\omega - \omega_J)$, with $\alpha \in \{B850, B800\}$ (Janosi *et al.*, 2006). The obtained $I_{CD}(\omega)$, shown in Figure 5.8 (solid line), matches surprisingly well with the experimental spectrum (dashed line) (Ihalainen *et al.*, 2001).

5.4 Electron transfer in pigment–protein complexes

Other quantum processes of great interest in PPCs involve electrons switching between two states. Two examples are: (1) electron transfer processes when an electron moves in the PPC from an orbital on the donor moiety D to an orbital on the acceptor moiety A; and (2) bond formation/breaking processes in an enzyme when electrons shift between a non-bonding state and a bonding state. Here we focus only on the electron transfer processes in a PPC. Our goal is to present a general approach for calculating the rate using only high resolution crystal structure information and MD data.

Quite generally, the electron transfer process can be described as a chemical reaction, $AD \rightarrow A^+D^-$, where $S_1 \equiv AD$ and $S_2 \equiv A^+D^-$ are the reactant and product states, respectively. The energy of the two states, $E_1(t)$ and $E_2(t)$, changes in time due to (i) motions along a reaction coordinate (that describes the electron transfer process) and (ii) thermal fluctuations of the remaining degrees of freedom of the PPC. Often the interaction energies that couple the two electronic states involved in the reaction are small compared to the temporal variations of $E_1(t)$ and $E_2(t)$. In this rather typical case, the actual reaction process is confined to moments when the two electronic states become energetically degenerate $[E_1(t) = E_2(t)]$.

In the case of membrane bound PPCs, such *curve crossing* processes are strongly dependent on the thermal fluctuations of the entire system, including the protein matrix, the solvent and the lipid membrane. In a quantum-mechanical description one defines the Hamiltonians H_1 and H_2 which describe the collective motion of the system in the reactant (initial) and product (final) electronic states of the PPC. The weak coupling between the two states can be described by a tunnelling matrix element V.

Similarly to the case of optical absorption spectra, the effect of dynamic disorder on the electron transfer processes in PPCs can also be determined by employing the combined MD/QC method described in the previous section. Once the A and D moieties have been identified, the time series of the electronic ground state energies $E_{1,2}(t_i)$, $i = 0, 1, \ldots$, of the two redox states can be determined by QC calculation for each snapshot t_i along the MD trajectory.

5.4.1 Electron transfer rate

Assuming that the tunnelling matrix element V does not change significantly due to the thermal motion of the protein matrix, within the lowest order of perturbation theory in V, the electron transfer rate k_{ET} in a PPC can be expressed as (May and Kühn, 2011)

$$k_{ET} = |V|^2 \int_{-\infty}^{\infty} dt \left\langle e^{iH_1t} e^{-iH_2t} \right\rangle. \tag{5.28}$$

Similarly to Equation (5.17), by employing the cumulant approximation one obtains

$$k_{ET} \approx |V|^2 \int_{-\infty}^{\infty} dt \exp\left[-i\langle \Delta H \rangle t - \int_0^t d\tau (t-\tau) C(\tau) \right], \tag{5.29}$$

where $\Delta H \equiv H_2 - H_1$, $\Delta H(t) = e^{iH_1t} \Delta H e^{-iH_1t}$, $C(t) = \langle \delta H(t) \delta H(0) \rangle$, and $\delta H(t) = \Delta H(t) - \langle \Delta H \rangle$. By following the same methodology as in the derivation of the OD spectrum of a PPC [Equations (5.18), (5.19) and (5.22)], the rate can be brought to the form

$$k_{ET} = 2|V|^2 \int_0^{\infty} dt\, e^{-\phi(t)} \cos[\epsilon t - \varphi(t)], \tag{5.30}$$

where $\epsilon = \langle \Delta H \rangle \approx \langle \Delta E(t) \rangle$ is the mean energy gap,

$$\phi(t) = \int_0^t dt' (t - t') C(t), \tag{5.31}$$

with $C(t) = \mathrm{Re}[\mathcal{C}(t)] \approx \langle \delta E(t) \delta E(0) \rangle$, and

$$\varphi(t) = \int_0^\infty \frac{d\omega}{\pi} J(\omega) \frac{\omega t - \sin \omega t}{\omega^2}, \tag{5.32}$$

where the spectral function has the usual form,

$$J(\omega) = 2 \tanh(\beta\omega/2) \int_0^\infty dt\, C(t) \cos \omega t. \tag{5.33}$$

Thus, the calculation of k_{ET} and $J(\omega)$ requires only a knowledge of the energy gap time series $\Delta E(t)$ and not those of the individual energies $E_{1,2}(t)$. This simple observation is important because most QC methods do not allow the accurate determination of individual energy levels, but they can provide accurate energy differences.

One can show that in the high temperature limit (i.e. $\beta\omega \ll 1$), the cumulant approximation of the electron transfer rate [Equations (5.30)–(5.33)] approaches the classical Marcus theory (Kosztin and Schulten, 2008). Indeed, assuming that $C(t) \equiv C_M(t) = \Delta^2 \exp(-t/\tau)$, where $\Delta^2 = \langle \delta E^2 \rangle = \langle \Delta E^2 \rangle - \langle \Delta E \rangle^2$ is the variance of the energy gap fluctuations and τ is the corresponding relaxation time, the integrals in Equations (5.31) and (5.33) can be performed exactly with the results,

$$\phi_M(t) = \Delta^2 \tau [t - \tau(1 - e^{-t/\tau})] \approx \frac{\Delta^2}{2} t, \qquad \text{for} \quad t \ll \tau, \tag{5.34}$$

and

$$J_M(\omega) = 2 \tanh(\beta\omega/2) \frac{\beta\Delta^2}{1 + (\omega\tau)^2} \approx \frac{\beta\Delta^2 \omega\tau}{1 + (\omega\tau)^2}. \tag{5.35}$$

Note that the maximum of the spectral function $J_M(\omega)$ corresponds to the energy gap,

$$\epsilon_M = \frac{\beta\Delta^2}{2}. \tag{5.36}$$

Within the same range of approximations the phase factor (5.32) becomes

$$\varphi_M(t) \approx t \int_0^\infty \frac{d\omega}{\pi} \frac{J_M(\omega)}{\omega} = \epsilon_M t. \tag{5.37}$$

Inserting Equations (5.37) and (5.34) into (5.30) and performing the Gaussian integral, one obtains the well-known Marcus formula (Marcus, 1956b, a; May and Kühn, 2011),

$$k_M = 2\sqrt{2\pi} \frac{|V|^2}{\Delta} \exp\left[-\frac{(\epsilon - \epsilon_M)^2}{2\Delta^2} \right]. \tag{5.38}$$

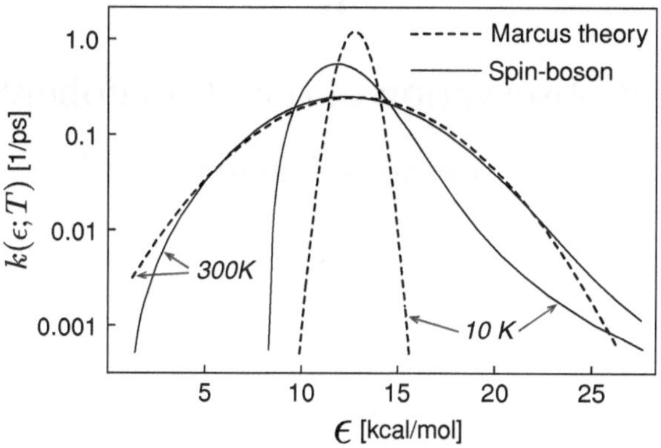

Figure 5.9 Comparison of electron transfer rates $k(\epsilon; T)$ shown as a function of ϵ evaluated in the framework of the spin–boson model (solid lines) and by Marcus theory (dashed lines) at temperatures 10 K and 300 K.

Because, in general, the spectral function (5.33) has a complex structure (Damjanovic *et al.*, 2002; Janosi *et al.*, 2006), the simple exponential approximation of the correlation function $C(t)$ may not be justified, so that differences between the electron transfer rates calculated with the simple Marcus formula (5.38) and with the cumulant approximation (5.30) may be expected even at high temperatures.

A detailed study of electron transfer rates k_{ET} in the photosynthetic reaction centre of *Rhodopseudomonas viridis* by employing the *spin–boson model* (which yields, formally, the same results as the cumulant approach with $J(\omega)$ representing the phonon spectral function) was reported in (Xu and Schulten, 1992, 1994). The model parameters Δ and τ were determined by means of all atom MD simulations. Because of large errors in calculating the mean redox energy gap ϵ, the authors used this as a fitting parameter. The calculated $k_{ET}(\epsilon; T)$ for temperatures $T = 10$ K and $T = 300$ K are shown in Figure 5.9, and are compared with the corresponding results predicted by Marcus theory (Marcus, 1956b, a). As expected, at high (physiological) temperature the rate evaluated from Marcus theory in a wide range of ϵ values agrees well with the rate evaluated from the spin–boson model at $T = 300$ K. However, Marcus theory and spin–boson model differ significantly at $T = 10$ K. At such a low temperature the rate as a function of ϵ for the spin–boson model is asymmetrical. This result agrees with observations reported in Gunn and Dawson (1989), which show a distinct asymmetry with respect to ϵ_M at low temperatures. Such an asymmetry is not predicted by the models of Marcus and Hopfield (Hopfield, 1974; Marcus and Sutin, 1985; Sumi and Marcus, 1986).

6

Direct observation of quantum coherence

GREGORY S. ENGEL

6.1 Detecting quantum coherence

Observing quantum coherence is a tricky business. Even before we fire up the femtosecond laser systems and start our search, we need to ask ourselves some basic questions: What is coherence? How does coherence manifest in measurements? What methods can we use to detect and measure coherence? Only then, can we begin to interpret data to search for coherence and to try to make sense of the observed signals. In this section, I will discuss detecting quantum coherence both in theory and in practice. The goal of this section is to present a unified view of experimental approaches in broad strokes and to provide useful references to guide further exploration.

6.1.1 What is 'quantum coherence'?

Firstly, the precise meaning of 'quantum coherence' must be pinned down. I define quantum coherences to be off-diagonal elements of the density matrix representing the ensemble. (I am explicitly setting aside questions of 'quantum' versus 'classical' coherences; let us simply assume that the system in question is best described quantum mechanically and presume that the coherences in such a system are also quantum in nature. The simple fact is that we don't yet have the tools to answer this debate definitively.) This definition, however, is still insufficient. The density matrix and the magnitudes of its off-diagonal elements in particular depend on the basis set used to write down the matrix. For the purposes of this chapter, I will consider the density matrix only in the Hamiltonian eigenbasis.

Constructing the definition of quantum coherence using the Hamiltonian eigenbasis creates enormous experimental and interpretive simplifications. Firstly, this

Quantum Effects in Biology, ed. Masoud Mohseni, Yasser Omar, Gregory S. Engel and Martin B. Plenio. Published by Cambridge University Press.

definition is very well suited to spectroscopic methods that measure the energy of transitions. Because the Hamiltonian operator corresponds with energy observations, the population elements along the main diagonal of the density matrix represent the probability of finding the system in a given energy level. Secondly, the Hamiltonian governs temporal evolution in quantum mechanics. Therefore, unitary dynamics of both the populations (diagonal) and coherences (off-diagonal) density matrix elements in the Hamiltonian eigenbasis are also quite simple. Populations do nothing. Coherences evolve phase based on the energy difference between the two associated Hamiltonian eigenstates. The equation of motion for the density matrix in any basis set is

$$\frac{\partial \rho}{\partial t} = -\frac{i}{\hbar}[\hat{H}, \rho]. \tag{6.1}$$

When expressed in the Hamiltonian eigenbasis, this equation simplifies to

$$\frac{\partial \rho_{ij}}{\partial t} = -\frac{i}{\hbar}(\epsilon_i - \epsilon_j)\rho_{ij}, \tag{6.2}$$

where ϵ_i is the energy of the ith eigenstate. Therefore, populations ($i = j$) remain constant while coherences ($i \neq j$) evolve phase with time according to the relation

$$\rho_{ij}(t) = e^{\frac{-i(\epsilon_i - \epsilon_j)t}{\hbar}} \rho_{ij}(0) \quad \text{for} \quad \mathrm{i} \neq \mathrm{j}. \tag{6.3}$$

This bright line rule showing that populations of Hamiltonian eigenstates remain constant while coherences oscillate appears to point to a clear strategy for observing quantum coherence. Unfortunately, it's not quite that easy.

6.1.2 How does quantum manifest in observations?

We have now shown that, under unitary evolution, populations of Hamiltonian eigenstates remain constant while coherences oscillate. For the moment, we will even set aside the fact that temporal evolution need not be unitary when describing a subsystem in a reduced dimensional description. (That is, if we ignore states of the system such as a bath, energy can flow into those states giving rise to dissipative dynamics.) Instead, let us focus on the oscillatory beating signals as a marker of quantum coherence in the Hamiltonian eigenbasis. Such oscillatory beating signals do not generally arise from coherence in other basis sets. For example, delocalization, which is coherence in the site basis, can exist without creating any observable oscillatory signals.

It would appear that we now have a simple rule identifying quantum: coherences oscillate. However, we need to actually observe this oscillation, and for that we need to make observations of the system. Within the density matrix formalism, making an observation, A, corresponding to operator, $\hat{A}$, is simple:

$$\langle A \rangle = \mathrm{Tr}(\hat{A}\rho). \tag{6.4}$$

So far, we have only used the Hamiltonian operator in our discussion. If we try a measurement of the energy of any system evolving under unitary dynamics, we find that we are blind to coherences. (This result should not come as a surprise– unitary dynamics conserve energy so no oscillations are possible!) We can show this fact explicitly by simply recalling that the Hamiltonian matrix is diagonal when expressed in its own eigenbasis,

$$\langle E \rangle = \mathrm{Tr}(\hat{H}\rho) \tag{6.5}$$

$$= \sum_i \epsilon_i \rho_{ii}. \tag{6.6}$$

Recall that populations are constant under unitary dynamics. The equation above shows that the observed energy is independent of the coherences (and any phase associated with them). Thus, measurements of energy cannot probe quantum coherences. Clearly, we need a new measurement.

In selecting a new approach to observing quantum coherence, we must select a measurement that does not commute with the Hamiltonian. Operators that do not commute do not (in general) share eigenfunctions, and we therefore ensure that the operator is not diagonal in the Hamiltonian eigenbasis. Spectroscopically, the most convenient operator to select is the dipole operator, although we could select any operator.[1]

We know that the dipole operator, which governs the interaction between light and matter, does not commute with the Hamiltonian. Firstly, we have seen detailed perturbative expansions of this interaction in Chapter 2. Secondly and more intuitively, we know that molecular systems absorb light and change energetic states. In the most simple form, the light of a proper colour (through the dipole operator) couples the ground and excited states leading to some probability of being found in the excited state after the interaction. Because this operator does not commute with the Hamiltonian, it must contain off-diagonal elements when written in the Hamiltonian eigenbasis. Thus, from Equation 6.4 we can see that these off-diagonal elements will allow coherences to contribute to the observable quantity.

In a spectroscopic measurement, the manifestation of the observable associated with the dipole operator is amplitude (and therefore intensity). That is, when we set out to measure quantum coherence, we will look for periodic oscillations in the amplitude of the signal. These amplitude fluctuations are called 'quantum beats.'

[1] Commonly, theory papers work in the site basis to simplify interpretation of calculations. Although no simple physical operator coincides with the site basis per se, this approach is very intellectually satisfying, because it conveniently represents spatial locations.

Quantum beating is a ubiquitous feature of quantum coherence. In the practical example above, we examined the dipole operator, but the same beating would occur with any operator that does not commute with the Hamiltonian. Thus, we can think of this beating as probability washing back and forth across sites, or equally well as oscillations of the electrodynamic polarization via the dipole operator. In reality, it is all of these things.

6.2 Observation of quantum coherence using 2D electronic spectroscopy

On one hand, observations of quantum coherence are inherent to all spectroscopic measurements. That is, absorption involves a momentary coherence between the ground and excited states induced by the light. This coherence generates the oscillating field that we observe. (For example, the signal field oscillates out of phase with the driving field in linear absorption spectroscopy, creating the destructive interference that we see as absorption.) We are looking for something more, however. In particular, we want to know the fate and role of coherence after excitation – especially if such coherences persist on the timescale of chemical dynamics. To understand this phenomenon and whether it can matter, we have to create coherences and then probe the fate of the coherences among excited states.

The first hint of coherence among excited states in a photosynthetic light-harvesting complex came in 1997 when Savikhin, Buck and Struve saw unusual oscillations in their pump–probe data (Savikhin *et al.*, 1997). Oscillations in pump–probe data were commonplace by this time, but interestingly, these oscillations appeared in anisotropy measurements indicating that a simple vibrational wave packet was not the cause. The authors created a model wherein the seven accessible electronic states of the system were grouped into two bins and argued that electronic coherence among the binned states was the cause. This insightful model was later verified with 2D electronic spectroscopy (Engel *et al.*, 2007).

The first direct observations of quantum coherence biological systems occurred in 2007, when the same Fenna–Matthews–Olson complex (Fenna and Matthews, 1975; Camara-Artigas *et al.*, 2003; Tronrud *et al.*, 2009) was interrogated with 2D spectroscopy (Engel *et al.*, 2007). Before I discuss this measurement in detail, I would like to introduce the excitonic energy transfer system that has provided the platform for all of this early work: the Fenna–Matthews–Olson complex.

6.2.1 Biology of the Fenna–Matthews–Olson photosynthetic complex

The Fenna–Matthews–Olson complex (FMO) comes from green sulphur bacteria, which live in extremely low light environments such as under microbial mats or in the deep ocean. As such, they have evolved large, extremely broadband antennae that operate at near perfect quantum efficiency. (Note that thermodynamic efficiency is nowhere near unity. This lack of thermodynamic efficiency arises from a series of 'downhill' energy transfer steps. These steps make the process fast and irreversible at the cost of thermodynamic efficiency.) Thus, this archaea provides a fabulous model organism for studying strategies for light-harvesting and energy transfer. Effectively, we are exploiting 2.4 billion years of evolutionary research and development.

The Fenna–Matthews–Olson complex provides an ideal model system for photosynthetic light-harvesting because:

- The X-ray structure is known to within 2.2 Å;
- The complex is (sparingly) water soluble;
- The complex is highly asymmetric, yielding not optically dark states;
- With only seven strongly coupled bacteriochlorophyll pigments, each monomer is computationally tractable;[2] and
- The three monomeric units in the trimer are extremely weakly coupled.

The crystal structure solved by Tronrud *et al.* is shown in Figure 6.1 (Tronrud *et al.*, 2009). This protein acts as a 'spacer' that separates the chlorosome antenna (a gigantic almost crystalline rod-like structure containing 250 000 bacteriochlorophylls) from the reaction centre to permit reductant to diffuse to the reaction centre and refill the hole after charge separation. Effectively, the Fenna–Matthews–Olson complex serves as an excitonic wire linking the chlorosome to the reaction centre, as shown in Figure 6.2. The complex itself is trimeric, but the monomeric units are very weakly coupled. Recently, an additional eighth bacteriochlorophyll molecule has been discovered by Tronrud *et al.* to sit between the baseplate and the rest of the FMO complex (Tronrud *et al.*, 2009).

Thinking spectroscopically, we are blind to everything but the optically active modes. The FMO complex contains seven bacteriochlorophyll molecules each with an electronic transition near 800 nm (formally a Q_y transition). The seven states couple to one another electrostatically (typically estimated with dipole–dipole interactions). Thus, we can construct a 7×7 Hamiltonian, as has been done by Vulto *et al.*, Renger *et al.* and Hayes and Engel (Vulto *et al.*, 1999; Renger *et al.*, 2001; Hayes and Engel, 2011). After diagonalizing the matrix, we get seven delocalized, excitonic states. It is these states that we will probe.

[2] Recently, an eighth bacteriochlorophyll pigment has been discovered; this new pigment is weakly bound to the complex and is not present in all isolated samples. The new bacteriochlorophyll is sited between FMO and the chlorosome. It has not yet been observed spectroscopically.

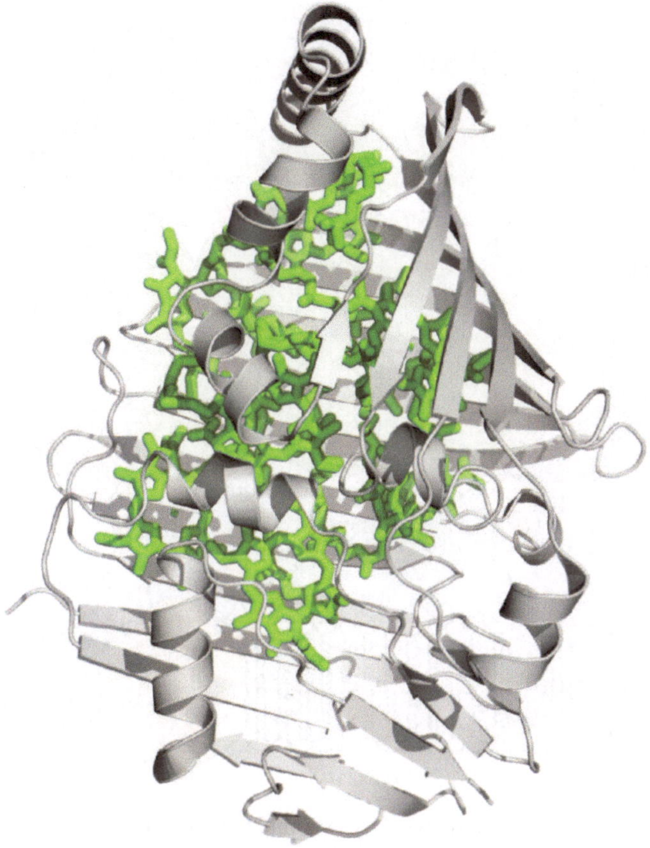

Figure 6.1 The Fenna–Matthews–Olson complex consists of a protein backbone (grey) containing a hydrophobic pocket that holds seven strongly coupled bacteriochlorophyll molecules (green).

6.2.2 Detecting quantum coherence among excited states

With the sample of the FMO complex at the ready, we can now think about how to detect quantum coherence signals. To make the signal detectable, we seek to create large, prescribed coherences across the entire ensemble using our laser. Ultra fast pulses have sufficient bandwidth to span all the excitonic states within the bacteriochlorophyll Q_y band.[3] Further, pulses less than 40 fs permit detailed observations of coherent quantum beating among these states. From the energy gaps in the Hamiltonian, we find that electronic coherences will exhibit oscillatory

[3] For details of the electronic transitions of chlorin pigments and their nomenclature, the reader is directed to (Blankenship, 2002).

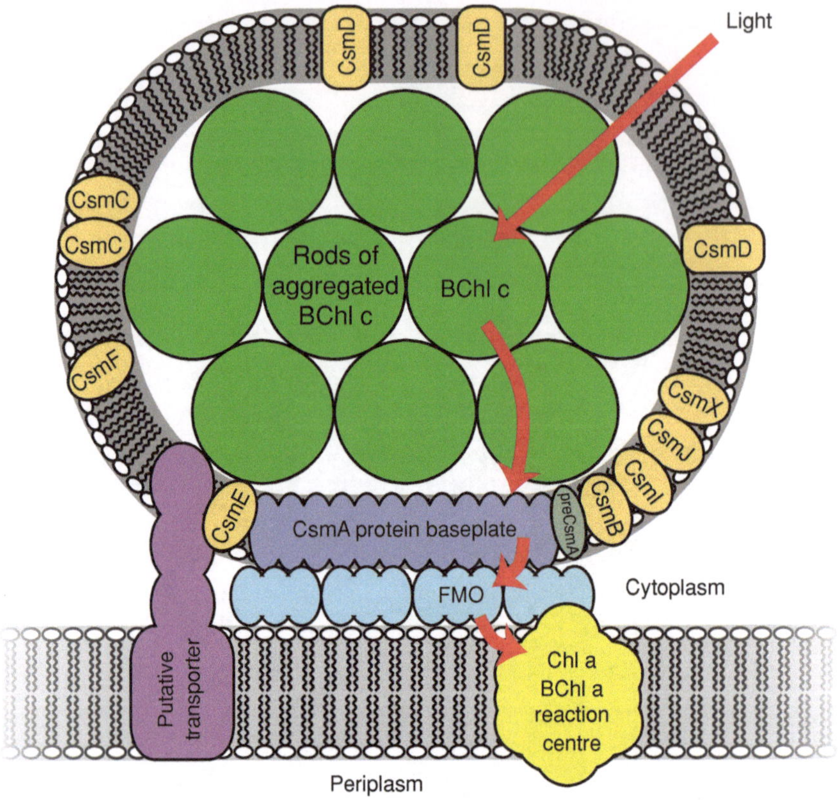

Figure 6.2 The Fenna–Matthews–Olson complex serves as an excitonic wire linking the chlorosome to the reaction centre.

periods ranging from 80–500 fs (corresponding with the energy differences of 70 cm^{-1} to 400 cm^{-1}).

Clearly, we require an ultra fast spectroscopy, but the problem is harder than that. Proteins are not terribly ordered objects, with every protein folded slightly differently. That is, we also need a spectroscopy that can help us eliminate inhomogeneity from our signal. Photon-echo spectroscopy (the same strategy as a spin-echo) does this for us. Now, all that is left is getting sufficient spectral resolution to see the different states. However, spectral resolution has a natural relationship with temporal resolution through the Fourier transform. Just like the uncertainty principle, these conjugate variables are not generally compatible. However, 2D electronic spectroscopy, while not escaping the Fourier limit, naturally allows for an optimal combination of the temporal and spectral resolutions because the spectral resolution is effectively determined by the molecular response.

Thus, we implement 2D electronic spectroscopy, described in Chapter 4. Experimental implementation of this spectroscopy has been described in detail elsewhere (Brixner *et al.*, 2004; Cowan *et al.*, 2004). Briefly, a sequence of three pump pulses interacts with the sample and causes stimulation of a third-order coherent response, which is then frequency and phase resolved using frequency domain heterodyne detection. The resulting interferogram provides both the magnitude and phase of the signal. During the time period between the first two pulses, the response evolves as a coherence between the ground and resonant excited states, defined as a one-quantum coherence, and the initial time delay is therefore referred to as the coherence time (τ). The second interaction promotes the evolution to populations of the excited states, populations of the ground state and coherences between excited states, zero-quantum coherences. The time delay between the second and third pulses is referred to as the waiting time (T). The final pulse stimulates the sample into a radiative coherence which emits the third-order signal. The delay between the third pulse and the emitted signal is referred to as the rephasing time (t). Two-dimensional spectra at fixed waiting times are then generated by taking a two-dimensional Fourier transform of the properly apodized signal over the τ and t dimensions. The relative pulse ordering of the first two beams can be varied to access additional non-rephasing pathways, which provide complementary information. The total signal can thus be separated into rephasing and non-rephasing contributions.

For multi-chromophoric systems, however, the spectroscopic description above is incomplete. Because the ultra fast pulses have broad bandwidth that will generally span multiple transitions. Thus, after the initial excitation of a coherence between the ground and excited state, the second pulse may or may not interact with the same excited state. (In general, all such interaction necessarily happens, and our focus on specific terms in the perturbative sum is a bit disingenuous; nevertheless, this approach provides a very convenient physical and intuitive description of the process.) If the second pulse interacts with a different state, we are left in a coherence between excited states rather than a population. This coherence evolves phase at the energy difference between the excitonic excited states. The period of this phase is slow enough to be observed in the waiting time dimension of the 2D spectra. A beating pathway is shown diagrammatically in Figure 6.3.

6.3 Identifying and characterizing quantum coherence signals

The quantum beating signals that indicate the presence of quantum coherence are visible during the waiting time delay T. The signals manifest as oscillations in both the absolute amplitude of the response and in the real portion of the response.

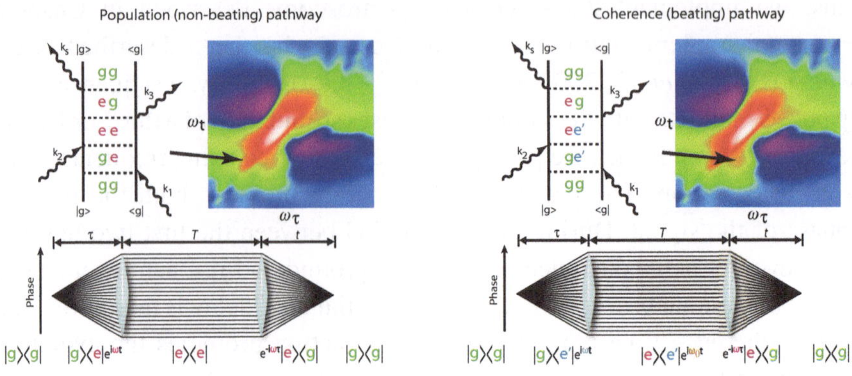

Figure 6.3 The left panel shows a constant (population) pathway that does not beat during the waiting time, *T*. The pathway is expressed both as a Feynman diagram and also as a lens diagram showing how phase spreads and rephases. The time evolution is shown below the lens diagram in the form of a time-dependent factor in front of the density matrix element. The right panel shows a pathway that beats during the waiting time because the pathway involves a coherence between excited states.

Interestingly, the signal appears in the absolute magnitude because it interferes with a constant, non-oscillating signal. Firstly, there is constructive interference, then destructive interference, and so on. Two pathways that appear in the same place in the spectrum and interfere in this way are shown in Figure 6.3.

Prior to the data published in 2007 (Engel *et al.*, 2007) shown in Figure 6.4, electronic coherence had never been observed in 2D spectra. The beating signals observed were reproducible in both phase and frequency, indicating that the signal was of molecular origin. (Nevertheless, over the course of about six months, control experiments were run and double-checked to ensure that simple explanations such as cryostat expansion and contraction upon filling, yaw in translation stages, laser power fluctuations, etc. were not the cause of the beating.) The signals observed were fully consistent with the expected beating pattern, predicted by looking at excitonic energy differences and weighting the intensity with the appropriate excitonic transition dipole moments. Because this first signal came as a surprise in an experiment expected to show beating for only 70 fs based on theoretical predictions (Pisliakov *et al.*, 2006), the sampling times were not uniform. (The sampling started with 10 fs steps, moving to 15 fs steps, then 20 fs steps until finally sheer panic caused 30 fs step sizes as the beating persisted.) Such non-uniform steps required adaptation of the non-uniform fast Fourier transform (NFFT) (Potts and Kunis, 2007). This algorithm identifies a large family of solutions consistent with the data. One such solution is show in Figure 6.4. Much later, a more thorough analysis on a more complete data set would confirm this assignment (Hayes and Engel, 2011).

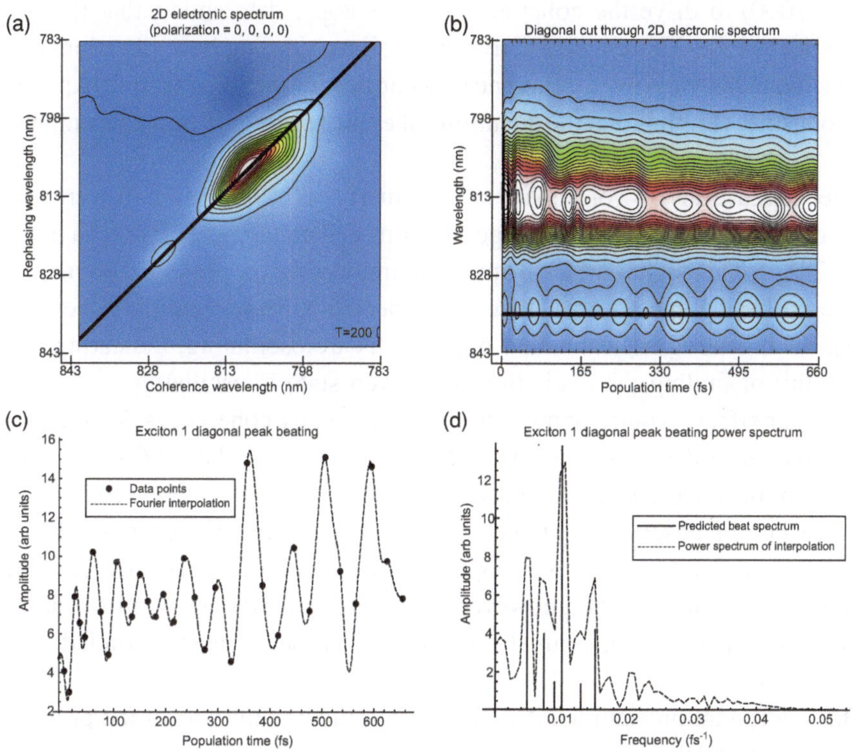

Figure 6.4 The first data to show electronic coherence between single states comes from FMO. The non-rephasing portion of the data beats along the main diagonal (a). The time course of this beating is shown in (b). The beating pattern of the lowest exciton peak (c) shows frequencies that correspond well to the predicted beat frequencies and intensities (d).

The precise origin of the beating pattern itself was only explained later by carefully dissecting the Feynman diagrams associated with quantum coherence (Cheng *et al.*, 2007). This work led directly to a detailed understanding of the lineshape of the beating signals as well.

The quantum beating in FMO can be used to interrogate many new aspects of the Hamiltonian that govern energy transfer and relaxation. For example, the beating signals arise from energetic differences among states and therefore the beating spectrum can be used to understand the Hamiltonian. Fourier transforming along the waiting time axis shows many new features within the still congested 2D spectrum (Hayes and Engel, 2011). Additionally, the beating signals are remarkably robust (Hayes *et al.*, 2011). Attempts to scramble vibrational modes or to shift resonances with isotopic substitution miserably failed to affect the beating signals. This failure significantly constrains microscopic models of the role of the protein bath, because the most simple models invoked a single finely tuned mode (out of over 100 000) to drive the coherence – the isotopic data shows that this simply cannot be the case. There is still active discussion regarding how vibrational modes may mediate this long-lived coherence. Regardless of the microscopic explanation, such robustness implies that the quantum coherence presents a tractable engineering target.

After suggestions by Voronine *et al.* (Palmieri *et al.*, 2009), signals were isolated that show direct coupling between populations and coherences (Panitchayangkoon *et al.*, 2011). This signal arises from quantum beating on peaks in the main diagonal in rephasing spectra that were not expected to show beating (Cheng *et al.*, 2007). Such beating indicates that populations are oscillating, indicating that the probability of finding the excitation in a given state is oscillating in time. Such non-secular effects require populations to be driven by coherences. When an oscillatory coherence drives a population, the population oscillates 90 degrees out of phase from the coherence, which is exactly the signal that is observed.

However, this observation creates many new questions. For example, FMO is a non-degenerate system such that population oscillation implies energetic oscillations. It is important to recall, however, that energy oscillations within a subsystem are not forbidden – energy can still be conserved, but it must be traded with the environment.

This interpretation significantly changes our view of the role of the protein bath again. The protein is strongly coupled to the bacteriochlorophylls and actively trades energy with the system. In some sense, this observation is a relief! Scientists had long wondered how chlorophyll could sit in a soft, polarizable environment yet not dissipate energy. Coherence solves this problem. The coupling exists, but prior measurements (hole burning, transfer efficiency, Stokes shifts) were blind to the coupling, because interpretation of that data implies incoherent, dissipative coupling. In addition, this coupling between coherences and populations explains not only why populations oscillate, but it also explains why coherences persist for so long. The long-lived coherence borrows lifetime from the populations to which it couples.

Finally, analysis of the dephasing rates of coherences provides yet another handle to analyse quantum beating. Early on, Hayes *et al.* recognized that different coherences dephase with rather different rates (Hayes *et al.*, 2010). The pattern does not match any simple scheme such as higher frequency beats dephasing faster, nor are resonances evident with any particular bath modes. As more complete data sets emerged, a more formal analysis approach was introduced that exploits a modified Fourier transform (z-transform) which involves complex frequencies – essentially isolating both dephasing rates and beat frequencies at the same time (Caram and Engel, 2011; Caram *et al.*, 2012).

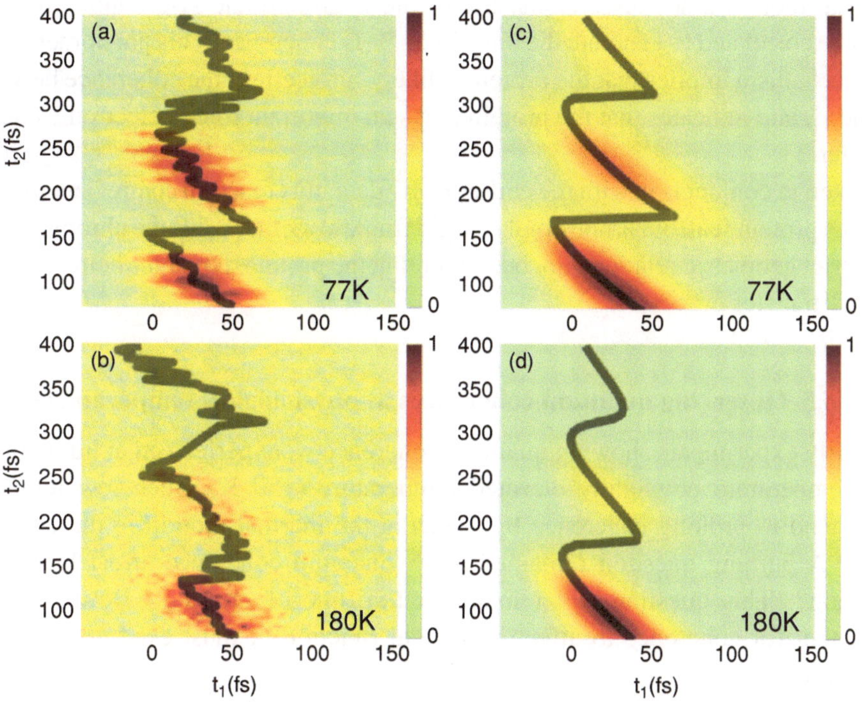

Figure 6.5 Lee *et al.* (2007) captured the first data to show long-lived quantum coherence in the bacterial reaction centre, proving that quantum beating signals were not unique to FMO. The data also showed that the coherence was of unusual character such that excited states maintained phase relative to one another, but not with the ground state.

6.4 Quantum coherence in reaction centres using two colour electronic coherence photon echo spectroscopy

Almost simultaneously, a new spectroscopic technique was developed by Lee, Cheng and Fleming that was exquisitely sensitive to electronic coherences, Two Colour Electronic Coherence Photon Echo Spectroscopy (2CECPES) (Lee *et al.*, 2007). This new variant of photon-echo peakshift spectroscopy has many advantages over 2D spectroscopy, such as improved sensitivity, but 2CECPES lacks frequency resolution within the pulse envelope.

Using this new approach, Lee, Cheng and Fleming simultaneously measured dephasing of the coherences within the reaction centre between the B exciton (accessory bacteriochlorophyll) and the ground state, between the H exciton (bacteriopheophytin) and the ground state and between B and H. The data showed that the B–H coherence lasted more than an order of magnitude longer than either the B–ground or H–ground coherence (Figure 6.5). Lee *et al.* then intuited that the spectral motion that causes dephasing of the B-ground and H-ground coherence must be correlated to generate the long-lived B–H coherence. This phenomenological mechanism implies that the protein is likely 'protecting' the coherence between B and H, and indicates that the protein plays an important role in electronic energy transfer.

Taken in conjunction with the earlier FMO data, this reaction centre data implies that the protein scaffold is not simply a rigid framework that holds the chromophores in a strict geometric relationship, but rather that the protein dynamics help to govern energy transfer processes.

6.5 Observing quantum coherences at physiological temperatures

All early experiments showing quantum coherence were performed at liquid nitrogen temperature or slightly elevated temperature (150 K). Such low temperatures tend to prolong coherence by reducing the thermal motion in the environment. A burning question remained: does quantum coherence affect biological function? (This question still remains at the time of printing.) However, it is clear that for coherence to affect biological function, it must exist at biological temperature.

The first reported results of quantum coherence at high temperature came from Scholes and co-workers showing quantum coherence among phycobiliproteins from cryptophyte algae using two time anisotropy spectroscopy (Collini *et al.*, 2010). Shortly thereafter, quantum beating was reported in FMO at physiological temperature as well, as shown in Figure 6.6 (Panitchayangkoon *et al.*, 2010). Both of these experiments require significantly improved signal-to-noise over earlier measurements. However, the heart of the experiment is identical to earlier observations of quantum coherence.

In each case, the observed dephasing rates match very well with theoretical predictions of the broad optimum transport regime predicted by Plenio and Huelga, and separately by Aspuru and co-workers (Plenio and Huelga, 2008; Rebentrost *et al.*, 2009a). Also, in each case, the coherence persists on the same timescale as energy transfer. Even in light-harvesting complex 2 (LH2) from purple bacteria, the coherence appears to persist at room temperature on the same timescale as energy transfer (although inhomogeneity in the ensemble frustrates direct measurements) (Fidler *et al.*, 2012).

These data stop short of proving biological function, but they unequivocally establish the presence of long-lived coherence after excitation at physiological temperatures. Furthermore, the agreement with phenomenological models indicates that our understanding of this coherence is reasonably good. However, our microscopic understanding of the effect remains incomplete.

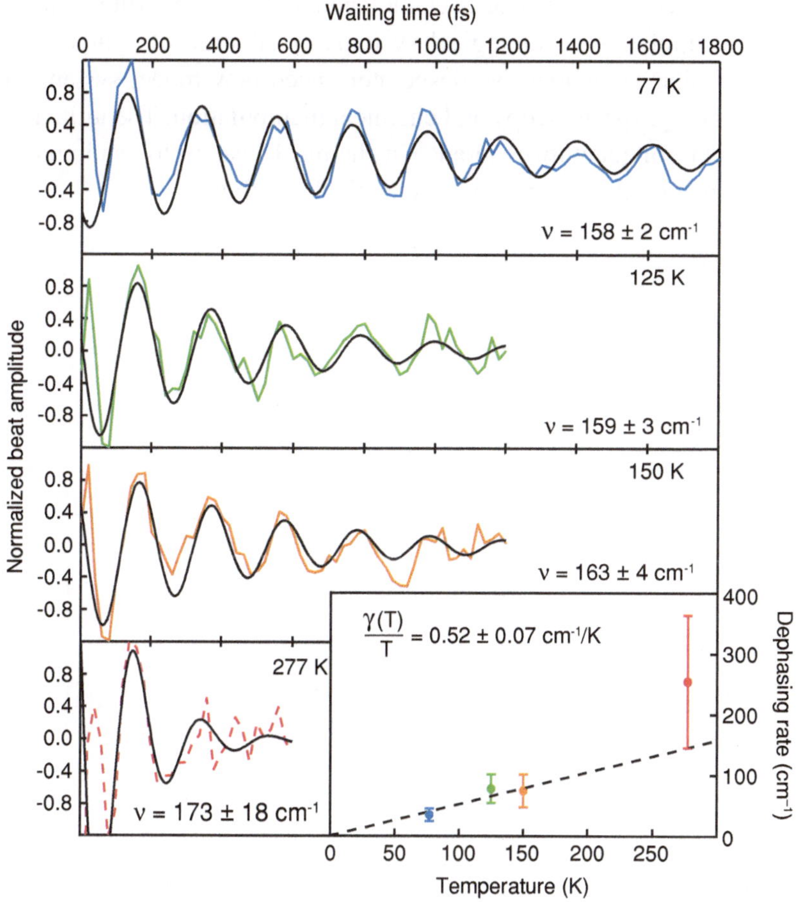

Figure 6.6 Quantum beating signals in photosynthetic complexes persist to room temperature. The signals agree in phase and frequency with low temperature measurements (Panitchayangkoon *et al.*, 2010).

6.6 Outlook for future measurements of coherence

The experiments conducted so far create at least as many questions as they answer. There are many frontiers for new measurements of quantum coherence and quantum effects in photosynthesis. Single-molecule experiments pose one of the most interesting and difficult challenges. Non-linear single-molecule measurements prior to bleaching or photodamage would answer many questions about inhomogeneity and provide clear new insights into the microscopic basis for long-lived coherence. Another equally important measurement involves inter-complex energy transfer. Studies on isolated proteins illuminate energy transfer within complexes, but how transfer among complexes may behave is still entirely speculative.

Finally, we need more control over the chemical systems. Either in the form of genetic manipulation of biological systems or creation of synthetic systems that exhibit similar behaviour, we desperately need new model systems that are amenable to theory, spectroscopy and chemical manipulation. The ultimate goal is to create and control quantum coherence in chemical systems to manipulate energy transfer, information processing and/or chemical reactivity.

7

Environment-assisted quantum transport

MASOUD MOHSENI, ALÁN ASPURU-GUZIK, PATRICK REBENTROST, ALIREZA SHABANI, SETH LLOYD, SUSANA F. HUELGA AND MARTIN B. PLENIO

Transport phenomena at the nanoscale exhibit both quantum (coherent) and classical (noisy) behaviour. Coherent and incoherent transfer are normally viewed as limiting cases of a certain underlying dynamics. However, there exist parameter regimes where an intricate interplay between environmental noise and quantum coherence emerges, and whose net effect is an increase in the efficiency of the transport process. In this chapter we illustrate this phenomenon in the context of excitation transport across quantum networks. These are model systems for the description of energy transfer within molecular complexes and, in particular, photosynthetic pigment–protein molecules, a type of biologically relevant structures whose dynamics has been recently shown to exhibit quantum coherent features. We show that nearly perfect transport efficiency is achieved in a regime that utilizes both coherent and noisy features, and argue that Nature may have chosen this intermediate regime to operate optimally.

7.1 Introduction

The dynamical behaviour of a quantum system can be substantially affected by interaction with a fluctuating environment and one might initially be led to expect a negative effect on quantum transport involving coherent hopping of a (quasi-) particle between localized sites. In this section, however, we demonstrate that quantum transport efficiency can be enhanced by a dynamical interplay of the quantum dynamics imposed by the system Hamiltonian with the pure dephasing induced by a fluctuating environment. Within the context of this book, one of the most relevant transport processes concerns energy transfer across molecular systems

Quantum Effects in Biology, ed. Masoud Mohseni, Yasser Omar, Gregory S. Engel and Martin B. Plenio. Published by Cambridge University Press.

(May and Kühn, 2011), whose quantum nature has been recently observed experimentally, for example in chromophoric light-harvesting complexes (Engel *et al.*, 2007; Lee *et al.*, 2007). Many aspects of the role of the environment in such chromophoric systems (Grover, 1971; Yang and Fleming, 2002; Gilmore and McKenzie, 2008) and model geometries (Gaab and Bardeen, 2004a) have been widely studied. As we will explain here, in a large variety of quantum networks and under appropriate conditions, interaction with an environment can result in an increased transport efficiency.

Indeed, in model systems of chromophoric complexes, it was suggested that quantum transport, assisted by environmental noise, relaxation and dephasing, explains the observed high energy transfer efficiency (Mohseni *et al.*, 2008; Plenio and Huelga, 2008). This approach was also used to quantify the percentage contributions of quantum coherence and environment-induced relaxation to the overall efficiency (Rebentrost *et al.*, 2009a) and the basic building blocks governing the dynamics of noisy quantum network dynamics were discussed in some detail in (Caruso *et al.*, 2009; Chin *et al.*, 2010b) and applied to the description of exciton energy transfer across the Fenna–Matthews–Olson (FMO) pigment–protein complex, a prototype for larger photosynthetic energy transfer systems. While a fully quantitative treatment of the system environment interaction is challenging and requires the development of sophisticated and numerically intensive methods (see Chapter 2 for further details), some first insights into the phenomenon, that is variously termed environment-assisted quantum transport (ENAQT) or noise-assisted transport (NAT), may be gained from simplified noise models based on the Haken–Strobl model, which is used to describe a classical Markovian bath in the high temperature limit (Haken and Strobl, 1973; Leegwater, 1996; Gaab and Bardeen, 2004a).

In the following, we present a simple set of equations of motion that capture the essential dynamical features required to exhibit ENAQT or NAT, and identify the fundamental processes involved in assisting excitation transfer by dephasing noise. Then we proceed to consider this phenomenon analytically for a two-site system and numerically for the FMO complex. In the final two sections, we employ an efficient technique to estimate energy transfer efficiency of complex excitonic systems, within more realistic interactions with vibrational and radiative environments beyond Markovian and perturbative assumptions. We show that the FMO energy transfer efficiency is optimum and robust with respect to important environmental parameters including reorganization energy λ, bath frequency cut-off γ and temperature T. We identify the ratio of $kT\lambda/\hbar\gamma g$ as a single key parameter governing quantum transport efficiency, where g is the average excitonic energy gap.

7.2 Master equations for quantum transport

The time evolution of a system of N chromophores in the presence of a single excitation can be modelled by a quantum network of N sites subject to a tight-binding Hamiltonian of the form (May and Kühn, 2011),

$$H_S = \sum_{m=1}^{N} \varepsilon_m |m\rangle\langle m| + \sum_{n<m}^{N} V_{mn}(|m\rangle\langle n| + |n\rangle\langle m|). \tag{7.1}$$

The states $|m\rangle$ denote the excitation being at site m while the other sites are in the ground state. The site energies and interactions are given by ε_m and V_{mn}, respectively. The site energies are due to different local environments of otherwise identical molecules. For chromophores, the coupling is mediated by the Coulomb interaction (Förster coupling) or electron exchange (Dexter coupling). Under low light conditions such as those found in the natural habitat of green sulphur bacteria, or in recent experiments, it is appropriate to study the dynamics in the *single exciton manifold*, spanned by the states $|m\rangle$.

A multi-chromophoric system interacts with the surrounding environment, such as a solvent or protein, which is usually a macroscopic system with many degrees of freedom. This coupling leads to irreversible dynamics characterized by dephasing and relaxation. Here, we employ the Haken–Strobl pure-dephasing model, which assumes classical Gauss–Markovian fluctuations (Haken and Strobl, 1973) (for more sophisticated, but also numerically more intensive methods, see Chapter 2 for further details). We consider only diagonal fluctuations which are typically larger than fluctuations of the inter-molecular couplings (Adolphs and Renger, 2006; Cho *et al.*, 2005). Additionally, we assume that fluctuations at different sites are uncorrelated, an assumption that is supported by recent numerical studies (Olbrich *et al.*, 2011) and that all chromophores experience the same coupling to their respective environments. With these assumptions, one obtains the Haken–Strobl equation for the density operator in the Schrödinger picture as (Haken and Strobl, 1973),

$$\dot{\rho}(t) = -\frac{i}{\hbar}[H_S, \rho(t)] + L_\phi \rho(t), \tag{7.2}$$

where the pure dephasing Lindblad operator is given by,

$$L_\phi \rho(t) = \gamma \sum_m \left[A_m \rho(t) A_m^\dagger - \frac{1}{2} A_m A_m^\dagger \rho(t) - \frac{1}{2}\rho(t) A_m A_m^\dagger \right], \tag{7.3}$$

with the generators $A_m = |m\rangle\langle m|$, and a pure dephasing rate is given by γ. The equation of motion presented here leads to exponential decay of all coherences in the density operator and, in the absence of a sink site, which models the final irreversible transfer, to an equalization of all populations in the long-time limit and hence corresponds to the high-temperature limit. There are several possible ways to measure or quantify the success rate of an energy transfer process. Given that the primary aim of the excitation energy transfer is to deliver energy to a particular site with the highest possible efficiency, it appears to be natural to use the energy transfer efficiency. To account for exciton recombination and exciton trapping, we augment the Hamiltonian (7.1) with anti-Hermitian parts (Mohseni *et al.*, 2008; Plenio and Huelga, 2008; Mukamel, 1995), $H_S - i\hbar H_{\text{recomb}} - i\hbar H_{\text{trap}}$, with $H_{\text{recomb}} = \Gamma \sum_m |m\rangle\langle m|$ and $H_{\text{trap}} = \sum_m \kappa_m |m\rangle\langle m|$. The exciton recombines with a rate Γ at every site (in general, this exciton's lifetime is of the order of $\sim$1 ns) and is trapped with a rate κ_m at certain sites. The probability that the exciton is successfully captured at a site m within the time interval $[t, t+dt]$ is given by $2\kappa_m \langle m|\rho(t)|m\rangle dt$. Thus, the efficiency $\eta(t)$ for transfer within the time interval $[0, t]$ can be defined as the integrated probability of trapping at multiple sites and is given by

$$\eta(t) = 2 \sum_m \kappa_m \int_0^t ds \, \langle m|\rho(s)|m\rangle. \tag{7.4}$$

7.3 Quantum transport in a two-chromophore system

A particularly simple and illustrative example is given by quantum transport in a system of two sites without trapping and recombination: a particle hops from sites 1 to 2 with a significant energy mismatch between 1 and 2 (see Plenio and Huelga (2008) for a treatment of the three-site case). With $|1\rangle$ and $|2\rangle$, the states where the exciton is localized at site 1 and 2, respectively, the Hamiltonian for such a system can be written $H = \frac{1}{2}\varepsilon(|1\rangle\langle 1| - |2\rangle\langle 2|) + \frac{1}{2}V(|1\rangle\langle 2| + |2\rangle\langle 1|)$, where ε is the energy mismatch between 1 and 2, and V is the strength of the hopping term. The coherent evolution of the system, starting from site 1, is simply a rotation about an axis displaced by angle $\theta = \sin^{-1}(V/\hbar\Omega)$ from the z-axis in the x–z plane. The maximum probability of finding the system at site 2 is $\sin^2 2\theta$, and the average probability of finding it there is $\sin^2\theta$. If the energy mismatch is sufficiently large, substantial hopping does not occur and the system remains localized at site 1.

In the presence of decoherence, the system obeys the Bloch equation. Pure dephasing corresponds to a Lindblad operator, $\sqrt{\gamma}(|1\rangle\langle 1| - |2\rangle\langle 2|)$, where $\gamma = 1/T_\phi$. T_ϕ is the pure dephasing time. The conventional Bloch analysis now holds. The system, instead of remaining localized at site 1, gradually diffuses, ultimately becoming a uniform mixture of $|1\rangle$ and $|2\rangle$. In the equilibrium state the system has a 50% chance of being found at site 2. The diffusion process can be thought of as a random walk on the Bloch sphere, with step length θ and time per step, γ^{-1}. Accordingly, the system must perform $\approx (\pi/\theta)^2$ steps and the diffusion time is $\tau_{\text{diff}} \approx (\pi/\theta)^2\gamma^{-1}$ to reach a steady state. For a system with more than two sites, the transport will be more complicated. Nevertheless, we still expect the transport rate to increase in direct proportion with the inverse of the individual site decoherence time. This is indeed true if the decoherence time does not substantially exceed the timescales defined by the transport terms in the Hamiltonian, and the energy mismatch from site to site. This fact supports the second prediction of environment-assisted quantum transport.

In the case of rapid dephasing, $\gamma > \Omega$, the angle ϕ that the system precesses before being decohered is $\approx \Omega/\gamma$. The probability of remaining in site 1 becomes $\cos^2\phi \approx 1 - (\Omega/\gamma)^2$, and the system enters the Quantum Zeno regime (Misra and Sudarshan, 1977). The system essentially performs a biased random walk with step size ϕ and an average time per step of $(\gamma/\Omega)^2\gamma^{-1} = \gamma/\Omega^2$. In time t, the system diffuses by an angle $\Omega\sqrt{t/\gamma}\cdot\Omega/\gamma = \Omega^2 t^{1/2}\gamma^{-3/2}$. In the case where the system has more than two states, we still expect this analysis to hold, taking γ to be the dephasing rate and Ω to be an average eigenfrequency. This supports our third prediction: as the dephasing rate grows larger than the Hamiltonian energy scale, the transport rate is suppressed by a polynomial in the dephasing rate. The system will obviously converge to the same statistical mixture as mentioned above, albeit on a very long timescale.

We now proceed to a quantitative discussion of the ENAQT effect for a two-level system in terms of efficiency. In this case, the efficiency can be obtained exactly as a function of the parameters γ, κ, ε and V. Firstly, we use the identity $\eta = -\frac{2}{\hbar}\sum_m \kappa_m \langle m|\mathcal{L}^{-1}\rho(0)|m\rangle$, which is obtained for Lindblad master equations as Equation (7.2). It involves the inverse of the superoperator $\mathcal{L}$, defined by $\mathcal{L}\rho(t) = -i/\hbar[H,\rho(t)] + L_\phi\rho(t) - \{H_{\text{trap}},\rho(t)\} - \{H_{\text{trap}},\rho(t)\}$, which can be readily computed. Here, we assume that the initial state is $\rho(0) = |m\rangle\langle m|$, $\kappa_1 = 0$, and $\kappa \equiv \kappa_2 \neq 0$. Then the efficiency for a two-chromophore system is

$$\eta_{\text{TLS}} = \frac{V^2\kappa(\gamma + 2\Gamma + \kappa)}{V^2(2\Gamma+\kappa)(\gamma+2\Gamma+\kappa) + \Gamma(\Gamma+\kappa)(\varepsilon^2 + \hbar^2(\gamma+2\Gamma+\kappa)^2)}. \tag{7.5}$$

Consistently, one finds perfect entrapment, $\eta_{\text{TLS}} = 1$, for $\Gamma = 0$ (note however, that is no longer the case in more complex geometries, when destructive interference can also become important (Caruso *et al.*, 2009)). From this analytical formula we can derive the optimal dephasing rate, $\gamma_{\text{opt}} > 0$ for observing ENAQT. This exists when $|\varepsilon|/\hbar > \kappa + 2\Gamma$ (Plenio and Huelga, 2008) and turns out to be

$$\gamma_{\text{opt}} = |\varepsilon|/\hbar - \kappa - 2\Gamma. \tag{7.6}$$

Thus, the optimal dephasing rate together with the level broadening induced by trapping and recombination exactly matches the energy level difference. The corresponding optimal efficiency is $\eta_{\mathrm{TLS,opt}} = V^2\kappa/(V^2(2\Gamma+\kappa)+2\hbar\Gamma|\varepsilon|(\Gamma+\kappa))$, which gives the maximal efficiency that can be obtained by tuning the dephasing rate given the other parameters. At large dephasing rates the efficiency is suppressed as the inverse of the dephasing rate, $\eta_{\mathrm{TLS},\infty} \approx V^2\kappa/\hbar^2\Gamma(\Gamma+\kappa)\cdot 1/\gamma$. These general predictions of ENAQT are also observed in numerical simulations of the FMO complex, as we show in a coming section.

7.4 The principles of noise-assisted quantum transport

Prior to discussing the numerical simulation of a model system exemplifying an actual chromophoric system, we will elucidate the basic building blocks underpinning the noise-assisted dynamics discussed above. In subsequent subsections we will demonstrate using explicit examples the existence of noise-assisted transport and will argue that both noise and coherence are important for the evolution of quantum networks in the presence of an environment. But, how does the interplay between these processes occur? Here we would like to precede these examples with a brief summary of the basic dynamical principles for noise-assisted transport, as have been developed and explained, for example, in Plenio and Huelga (2008), Caruso *et al.* (2009) and Chin (2010b, 2012).

Bridging energy gaps and blocking paths

Multi-chromophoric systems consist of a number of sites whose energies will generally differ from site to site. In the previous subsection we have seen that if this energy difference is larger than the intersite hopping matrix element in the relevant Hamiltonian, then transitions will be strongly suppressed. Dephasing noise can help to overcome these energy gaps, as it will lead to line broadening thus leading to increased overlap between sites without the loss of excitations from the system, see Figure 7.1. Alternatively, one may view dephasing noise as arising from the random fluctuations of energy levels. Then, the fluctuating energy levels will occasionally come energetically close with a level separation smaller than the coupling strength, thus allowing enhanced excitation energy transfer between the sites. A moderate amount of fluctuations serves to enhance the transport while excessive amplitude noise on the site energies will make the probability smaller of sites to be energetically close. Hence we expect an optimal finite noise strength that maximizes transport between two sites. This view is corroborated by the analysis in Section 7.3. However, the application of excessive noise and the concommitant

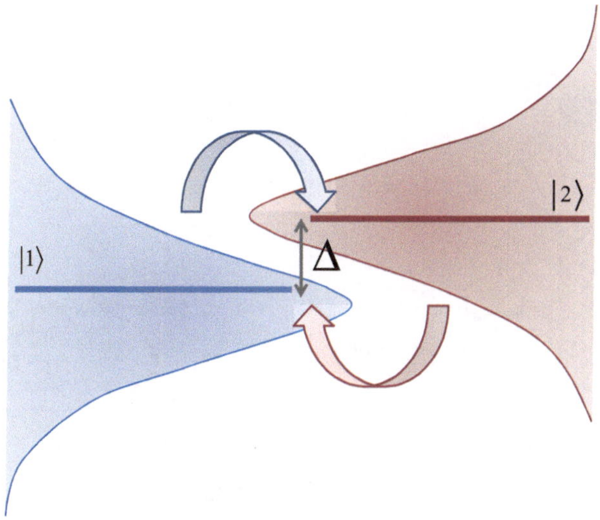

Figure 7.1 Local dephasing leads to random fluctuations of site energies and line-broadening of energy levels. As a result, overlap between sites is increased and excitation transport is facilitated by this type of noise. Alternatively, viewing these fluctuations dynamically, one obtains that the energy gap Δ between levels varies in time. A non-linear dependence of the transfer rate on the energy gap may therefore lead to enhancement of the average transfer rate in the presence of dephasing noise.

reduction of effective transition rate between sites can also be an advantage as it may effectively block unfavourable transfer paths from being followed.

Constructive and destructive interference

As we have seen in the previous subsection, the bridging of energy gaps may already occur in linear chains as short as two sites. A general quantum network, however, will possess many different paths between two sites of the network and hence it may exhibit a wealth of constructive and destructive interference effects. The essential nature of this type of effect may be seen in a simple network where the coherent interaction is described by a Hamiltonian with an exchange coupling term, $H = \sum_{k=1}^{3} \hbar\omega_i |i\rangle\langle i| + \sum_{k=1}^{2} j_k(|k\rangle\langle k| + h.c),$[1] where $|i\rangle$ corresponds to an excitation in site i and where we assume $j_1 = j_2$. Furthermore, we assume that site 3 is dissipatively coupled to the reaction centre, see Figure 7.2. An excitation initially prepared in the antisymmetric state, $|\psi\rangle = (|1\rangle - |2\rangle)/\sqrt{2}$, now forms an eigenstate of this Hamiltonian which has zero overlap with site 3. Under natural conditions a multi-chromophoric system is not excited in such an antisymmetric state. Rather, as the FMO complex it will tend to receive a single excitation locally, for example

[1] $h.c$ = hermitian conjugate.

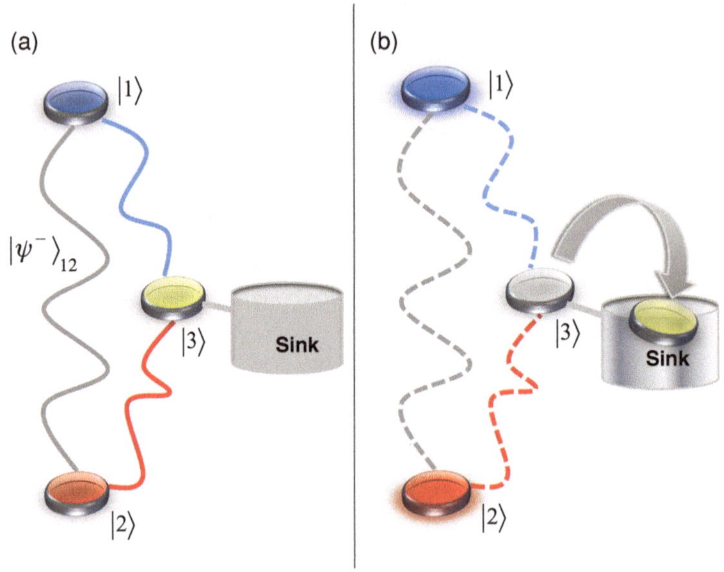

Figure 7.2 Inhibition of destructive interference. A three-site network provides the simplest scenario to illustrate NAT. Sites $|1\rangle$ and $|2\rangle$ are each coupled to a third site $|3\rangle$ via a coherent exchange interaction of the same strength, while site $|3\rangle$ is irreversibly connected to a sink (trapping site). In (a) the excitation is delocalized over sites $|1\rangle$ and $|2\rangle$ in a singlet-like state $|\psi^-\rangle$, so that there is an equal probability of finding the excitation at either site, but with a wavefunction that is antisymmetric with respect to the interchange of 1 and 2. Due to destructive interference of the tunnelling amplitudes, state $|\psi^-\rangle$ is not evolving and hence no excitation will ever reach the sink. In (b) pure dephasing causes the loss of phase coherence between sites $|1\rangle$ and $|2\rangle$ and the two tunnelling amplitudes to site $|3\rangle$ no longer cancel. The excitation is now eventually transferred to site $|3\rangle$ and transport is completed after a subsequent irreversible transfer to the trapping site.

on site 1. But, noting that $|1\rangle = [(|1\rangle - |2\rangle)/\sqrt{2} + (|1\rangle + |2\rangle)/\sqrt{2}]/\sqrt{2}$, we realize that through constructive interference the symmetric part of the initial state will propagate very rapidly into site 3, and from there into the reaction centre, while the antisymmetric part will not evolve at all. This trapping of population will be suppressed either via energetic disorder, which will release trapped population by inducing coherent oscillations between symmetric and anti-symmetric states, or by environmental dephasing noise which degrades interference effects and therefore destroys coherent trapping.

Splitting energy levels – the phonon antenna

The quantum coherent dynamics between energy levels can also be used to optimize the efficiency of energy transport in different ways. To this end, it should be noted

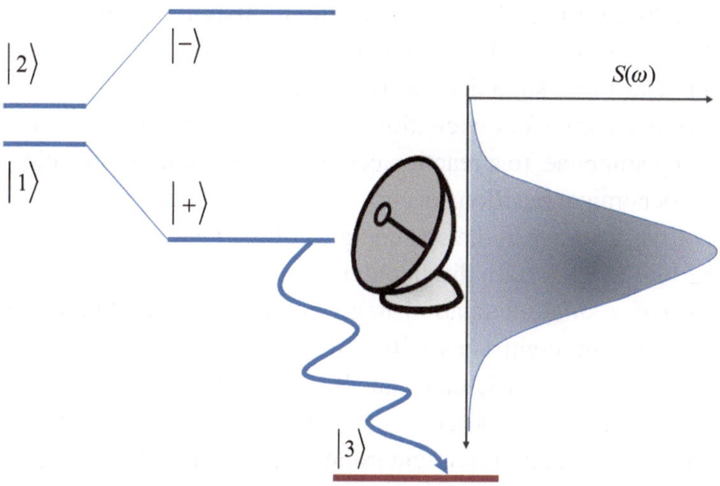

Figure 7.3 Phonon antenna. A typical situation encountered in energy transfer involves two closely spaced energy levels $|1\rangle$ and $|2\rangle$ that are separated from a third level $|3\rangle$ to which excitations should be delivered. A coherent interaction between the upper two energy levels leads to dressed levels $|\pm\rangle$ and an energy splitting which, if matched to the maximum of the environment spectral density $S(\omega)$, will optimize transport from the upper to the lower level. Hence the coherently split levels act as an antenna for environmental fluctuations.

that the energy levels of two sites that are coupled coherently will split, leading to new eigenstates of the global system, one of which is shifted upwards and one that is shifted downwards. In the dressed state basis, dephasing noise will induce transitions between these eigenstates (phase noise in the site basis becomes amplitude noise in the dressed basis) leading to energy transport towards the lower lying of the two energy levels, see Figure 7.3. The transition rate between these two states will depend on temperature, the matrix element between these two eigenstates *and*, crucially, the spectral density at the energy difference between the two eigenstates. Matching the energy level splitting to the maximum of the spectral density of the environmental fluctuations can thus Optimize energy transport. In this sense, we can argue that the two eigenstates of the coupled Hamiltonian harvest environmental noise to enhance excitation energy transport through the formation of a 'phonon antenna' (Chin *et al.*, 2012; del Rey *et al.*, 2013).

7.5 Quantum transport in the Fenna–Matthews–Olson protein complex

The trimeric Fenna–Matthews–Olson pigment–protein complex is found in green sulphur bacteria *Chlorobium tepidum* (Engel *et al.*, 2007; Cho *et al.*, 2005; Müh *et al.*, 2007) which live in stratified lakes and close to hydrothermal oceanic vents

(Blankenship, 2002). Recently, each of the three subunits was found to have eight chlorophyll molecules spatially arranged within several nm (Schmidt am Busch *et al.*, 2011). The three subunits can be treated independently from each other. The FMO complex transfers excitation energy from the chlorosomes, the main light-harvesting antennae, to a reaction centre where a charge separation event and subsequent biochemical reactions occur.

Because of the disorder in the site energies of the FMO complex we expect, as before in the two-level system discussed above, the presence of ENAQT or NAT. The dynamics of a single excitation is governed by a Hamiltonian of the form Equation (7.1) for the eight sites with a distribution of site energies and inter-site couplings, as given in (Schmidt am Busch *et al.*, 2011). The chromophoric Förster couplings are up to 100 cm^{-1}. The chlorophyll transition frequencies are shifted by the electrostatic protein environment, resulting in site-dependent electrochromic shifts of several 100 cm^{-1} (Adolphs and Renger, 2006; Müh *et al.*, 2007; Schmidt am Busch *et al.*, 2011). Fluctuations of the protein in the solvent lead to fluctuations of the transition frequency of the chlorophyll molecules and therefore to loss of excitonic phase coherence. We use the master Equation (7.2) with a site-independent dephasing rate γ according to the Haken–Strobl model. This model is expected to deliver insight into the high-temperature dynamics of an excitation in the FMO complex and, under the given assumptions, does not place any restrictions on the dephasing rate γ.

The initial state for our simulation of the system is localized at the eighth chromophore, which is believed to be physically close to the chlorosome antenna complex (Schmidt am Busch *et al.*, 2011). In the FMO complex, chromophore 3 is in the vicinity of the reaction centre (Li *et al.*, 1997; Adolphs and Renger, 2006; Müh *et al.*, 2007). Thus, one can assume that chlorophyll 3 is the main excitation donor to the reaction centre (Mohseni *et al.*, 2008; Plenio and Huelga, 2008). The precise transfer rate to the reaction centre is not fully characterized. Yet, based on typical transfer rates in chromophoric complexes with similar inter-molecular distances, we estimate it to be $\kappa_3 = 1\ ps^{-1}$ (Mohseni *et al.*, 2008). Thus, the efficiency of energy transfer according to Equation (7.4) becomes $\eta = 2\kappa_3 \int_0^\infty dt \langle 3|\rho(t)|3\rangle$.

In Figure 7.4 the efficiency of transfer and the transfer time is given as a function of the dephasing rate γ. At low dephasing, purely quantum-mechanical evolution leads to an efficiency of below 20%. With increasing dephasing the efficiency increases considerably, up to almost 100%, where it remains approximately constant for a range of γ of one order of magnitude. For stronger dephasing the efficiency is slowly suppressed again, delocalization is destroyed, and the overlap with the target site vanishes. The transfer time is around 300 ps in the fully quantum limit and improves significantly to 7 ps in the intermediate ENAQT regime. For large dephasing, the transfer slows down again entering a Quantum Zeno regime

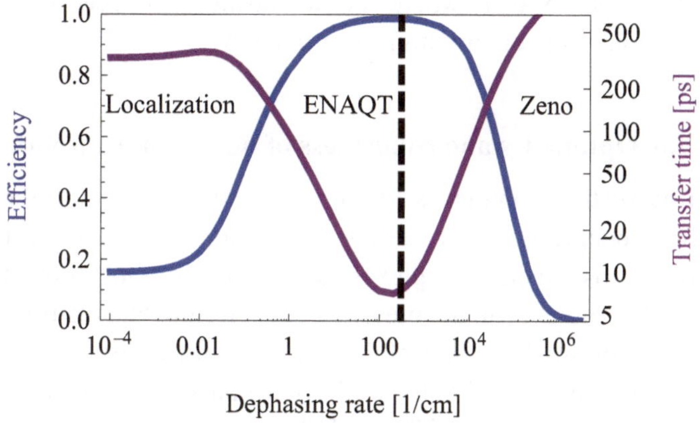

Figure 7.4 Environment-assisted quantum transport. Efficiency (blue) and transfer time (purple) as a function of the pure-dephasing rate is demonstrated for the eight chromophore Fenna–Matthew–Olsen sub-complex. A clear picture of the three dephasing regimes is obtained: from left to right, the fully quantum regime which is dominated by localization induced by the intrinsic static disorder of the pigment energies; the ENAQT regime, where unitary evolution and dephasing collaborate with the result of increased efficiency; finally, the Quantum Zeno regime, where strong dephasing suppresses quantum transport. The estimated dephasing rate at room temperature for the FMO spectral density is drawn as a dashed line.

(Misra and Sudarshan, 1977) more than 500 ps, the same order of magnitude as the excitation lifetime; the exciton is more likely to recombine than to be trapped.

One can estimate the dephasing rate as a function of temperature by employing a standard system–reservoir model (Breuer and Petruccione, 2002). In this context, the spectral density is given by $J(\omega) = \sum_i \omega_i^2 \lambda_i^2 \delta(\omega - \omega_i)$, where ω_i are frequencies of the harmonic-oscillator bath modes and λ_i are dimensionless couplings to the respective modes. In the continuum limit, we assume an Ohmic spectral density with cut-off, $J(\omega) = \frac{E_{\mathrm{R}}}{\hbar\omega_c} \omega \exp(-\omega/\omega_c)$. For the FMO complex, the reorganization energy is found to be $E_{\mathrm{R}} = 35 \text{ cm}^{-1}$ (Cho *et al.*, 2005) and the cut-off $\omega_c = 150 \text{ cm}^{-1}$, inferred from Fig. 2 in (Adolphs and Renger, 2006). In the Markovian regime, the dephasing rate γ is given as the zero-frequency limit of the Fourier transform of the bath correlator (Breuer and Petruccione, 2002). As a result, γ is found to be proportional to the temperature and the derivative of the spectral density at vanishing frequency, $\gamma(T) = 2\pi \frac{kT}{\hbar} \frac{\partial J(\omega)}{\partial \omega}|_{\omega=0}$. For the above spectral density, the rate turns out to be $\gamma(T) = 2\pi \frac{kT}{\hbar} \frac{E_{\mathrm{R}}}{\hbar\omega_c}$. This gives a rough estimate for the dephasing rate at room temperature of around 300 cm^{-1}, which is indicated in Figure 7.4. An experimental measurement of the dephasing rate delivered a value of $250 \pm 100 \text{ cm}^{-1}$ at $T = 277$ K (Panitchayangkoon *et al.*, 2010). Hence, the natural operating point of the FMO complex is estimated to be well within the regime of ENAQT, where the dephasing introduced by a fluctuating environment enhances the energy transfer efficiency.

7.6 Optimality and robustness of quantum transport

Natural light-harvesting complexes live in a variety of totally different ambient conditions, from deep regions of the Pacific Ocean to the hot springs of Tibet. Some of these photosynthetic complexes can manage to transfer excitonic energy with almost 100% efficiency in the presence of environmental disturbances. Hence in addition to the optimality, the stability of quantum transport process in such complexes is surprising. In previous sections we have introduced the notion and underlying mechanisms of ENAQT or NAT using simple dynamical equations of a pure-dephasing model. The validity of these types of equations is limited to the regime of Markovian and high temperature with a classical white noise bath. In order to explore the optimality and robustness of natural or artificial light-harvesting complexes, we need to compute the energy transfer efficiency landscape as a function of the independent degrees of freedom of the pigment–protein complex over a wide range of values. Such analysis necessitates employing a numerically tractable, while reliable, master equation to calculate transport efficiency beyond Markovian and perturbative regimes.

7.6.1 Efficient simulation of quantum transport beyond Markovian and perturbative limits

In Chapter 2, we discussed that the TC2 master equation can be applied beyond the weak system–bath coupling regime. Here we employ TC2 for simulating quantum transport phenomena. For a system–bath Hamiltonian $H_{SB} = H_S + H_{int} + H_B$, interaction term $H_{int} = \sum_j S_j B_j$ and bath correlation function $C_j(t-t') = \langle \tilde{B}_j(t)\tilde{B}_j(t')\rangle$, TC2 has the following convolutional form:

$$\frac{\partial}{\partial t}\rho(t) = \mathcal{L}_S\rho(t) - \sum_j \left[S_j, \frac{1}{\hbar^2}\int_0^t C_j(t-t')e^{-iH_S(t-t')/\hbar} S_j \rho(t') e^{iH_S(t-t')/\hbar} dt' - h.c. \right], \tag{7.7}$$

where $\mathcal{L}_S\rho(t) = -i/\hbar[H_S, \rho(t)]$. In the context of exciton transport, we need to add an extra non-unitary superoperator to incorporate exciton recombination and trapping processes, $\mathcal{L}_S + \mathcal{L}_{recomb} + \mathcal{L}_{trap}$ with $\mathcal{L}_{recomb} = -\Gamma \sum_m |m\rangle\langle m|$ and $\mathcal{L}_{trap} = -\sum_m \kappa_m |m\rangle\langle m|$. We can estimate the energy transfer efficiency (7.4) with

TC2 modelling without explicitly solving for the time evolution of $\rho(t)$,

$$\eta = 2\sum_m \kappa_m \langle m|\tilde{\rho}(0)|m\rangle, \tag{7.8}$$

where $\tilde{\rho}(s)$ is the Laplace transform of $\rho(t)$. The master equation (7.7) describes an approximation to the exact dynamics of an open quantum system in the presence of a Gaussian bath (see Chapter 2). This leads to an inevitable error in quantifying the efficiency η. In Shabani *et al.* (2012), using a combination of phenomenological and analytical approaches, we estimate this error for weak and intermediate system–bath couplings and bath memory timescales of the FMO complex, indicating the reliability of TC2 for estimating quantum transport efficiency in such regimes.

7.6.2 Optimality and robustness with respect to reorganization energy and cut-off frequency

The time-non-local master equation (7.7) was employed in Mohseni *et al.* (2014) to efficiently estimate the energy transfer efficiency *landscape* as a function of various independent system and environmental degrees of freedom over a wide range of values. This efficient simulation allows us to examine comprehensively all relevant regimes of the multiparameter space for finding possible high efficienly and robust neighbourhoods. Such studies could shed light on the maximum capabilities that can be achieved for optimal material design to engineer and characterize fault-tolerant artificial light-harvesting systems; this will be addressed in Chapter 14.

We first explore the variation of the FMO energy transfer efficiency versus reorganization energy and bath cut-off frequency using a Drude–Lorentzian spectral density,

$$\mathcal{C}_j(t) = \int_0^\infty d\omega \frac{2\hbar\lambda\gamma\omega}{\omega^2+\gamma^2}\left(\coth\left(\frac{\beta\hbar\omega}{2}\right)\cos(\omega t) - \sin(\omega t)\right). \tag{7.9}$$

The reorganization energy, λ, is proportional to the squared value of the system–bath couplings. The bath cut-off frequency is the inverse of the bath coherence timescale that captures the non-Markovian nature of the environment. That is, the non-Markovianity measure, defined as the information flow from the system to the phonon bath, by Breuer *et al.* (Breuer *et al.*, 2009), increases exponentially with decreasing bath cut-off frequency.

The optimality and robustness of ETE for the FMO protein complex at the experimentally estimated values of $\lambda = 35$ cm^{-1} and $\gamma = 50$–166 cm^{-1} are evident in Figure 7.5. An independent study on the optimality of ETE versus reorganization

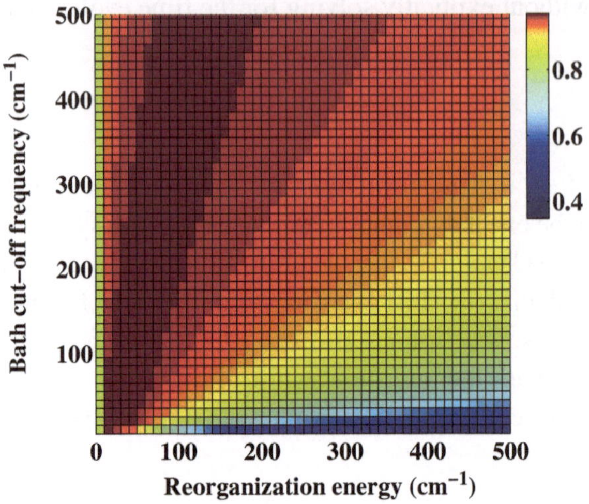

Figure 7.5 Top view of ETE landscape, indicating that the ratio of the reorganization energy over bath frequency cut-off can be considered as the parameter that governs the energy transfer efficiency at a fixed temperature. As we inspect this plot in an angular coordinate from the vertical axis, γ, toward the horizontal axis, λ, we can distinguish different regions of the ETE landscape that are separated by straight lines λ/γ. At a very small decoherence rate, the FMO complex experiences weak localization due to static disorder. As we increase this ratio, an optimal region of ETE emerges that is induced by an appropriate level of interplay between environmental fluctuations and coherent evolution. At higher levels of this parameter, ETE drops significantly due to strong localization induced by dynamical disorder. From Mohseni *et al.*, 2014.

energy has also been reported in Wu *et al.* (2010). It can be observed that the non-Markovianity of the bath can slightly increase ETE in the regimes of weak system–bath coupling. However, such slow bath behaviour can significantly decrease ETE when the system interacts strongly with it. The main question is how can one understand this phenomenon for all non-Markovian and Markovian regimes in the context of ENAQT described in previous sections? The landscape in Figure 7.5 shows a remarkable interplay of reorganization energies λ and bath frequency cut-off γ. The ETE takes values below unity if the FMO operates at the limit of very small λ and large γ. On the other limit, the FMO efficiency drops significantly when operating at large λ and very small γ. These two regimes can be understood as manifestations of weak- and strong-quantum localization, respectively (Mohseni *et al.*, 2014).

Careful examination of Figure 7.5 reveals three distinct regions of energy transfer efficiencies at room temperature that are governed by a single parameter proportional to λ/γ. For small and for large values of this parameter, the efficiency is low. The efficiency reaches its maximum for intermediate values of this parameter. In both strong and weak quantum localization limits the excitation will be spatially trapped in the regions typically far from the reaction centre and eventually dissipating to bath due to adversarial electron–hole recombination processes, which occur on a three orders of magnitude slower timescale. In the intermediate regime, the right amount of interplay of quantum coherence and environmental fluctuations can facilitate an optimal energy transport in a robust fashion by various physical mechanisms, including minimizing site energy mismatches, washing out potential destructive quantum interference effects and enhancing the energy funnelling by providing an appropriate vibrational energy sink.

We can understand these observations in terms of ENAQT by noting that the effective decoherence rate is given by λ/γ in the perturbative limit at a fixed temperature (Breuer and Petruccione, 2002). This ratio becomes smaller as we raise γ, therefore a stronger coupling λ is needed to guarantee the level of decoherence strength required for ENAQT. The overlap of a delocalized exciton wavefunction with the trap enables an almost complete, 98% quantum transport in the optimal range of λ/γ at the ambient temperature. As we increase the reorganization energy and bath coherence timescale, the ETE starts to drop. In this regime, the excitonic wavefunction again experiences localization as the environmental fluctuations change their role to exert a strong adversarial effect on the quantum transport, essentially as a source of strong dynamical disorder. The ETE landscape in Figure 7.5, clearly has level sets that exhibit a linear relationship with λ and γ. Indeed the ratio λ/γ, known as the Kubo number, is the parameter that governs Anderson localization transition in stochastic classical modelling of environmental interactions. (Krameri and MacKinnont, 1993; Castiglione, 2000; Goychuk and Hänggi, 2005). In the fixed high-temperature limit of ETE, illustrated in Figure 7.5, one can observe that λ/γ is a determining parameter for transport efficiency in the regions beyond the optimal ENAQT area. In the next section, we go beyond the Kubo number, by directly investigating the temperature-dependent energy transfer dynamics, leading to a general governing parameter as $\lambda T/\gamma$. For a more quantitative study of the degree of optimality and robustness of the energy transfer as functions of system–bath coupling strength and bath memory see Mohseni *et al.* (2014).

7.6.3 Optimality and robustness with respect to reorganization energy and temperature

As shown in the previous section, the parameter λ/γ governs the shape of the ETE landscape at a fixed high-temperature limit. However, in the perturbative limit, the decoherence rate can be expressed by $\lambda T/\gamma$ which captures the suboptimal ETE in the weak localization region. Now, we investigate if the ETE behaviour in all regimes can be globally captured by the parameter $\lambda T/\gamma$, for the given FMO Hamiltonian. Specifically, we need to verify whether one can predict the optimal

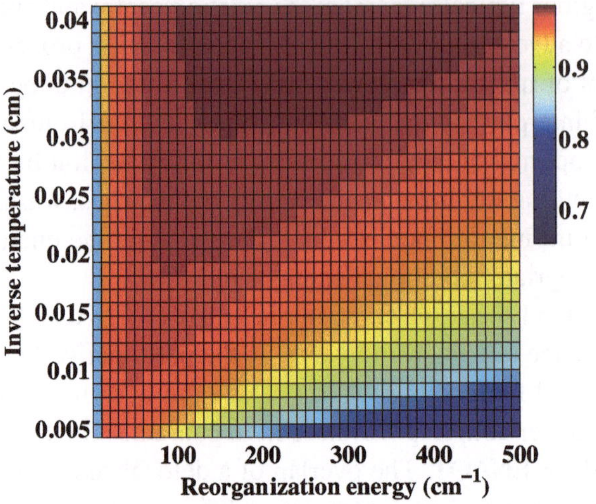

Figure 7.6 The ETE landscape as a function of inverse temperature and reorganization energy. In the high temperature regime ($\beta^{-1} > \gamma$), the three regions of weak localization, ENAQT and strong localization can be distinguished by the parameter $kT\lambda$, given the free Hamiltonian of FMO and $\gamma = 50$ cm^{-1}. These results, combined with Figure 7.5 suggest that the parameter $\lambda T/\gamma$ governs the shape of the overall FMO energy transfer landscape. From Mohseni *et al.*, 2014.

noise-assisted transport region as well as ETE suppression levels at the strong localization regions by a single parameter, $\lambda T/\gamma$. To examine the validity of this theory, we study ETE as a function of the reorganization energy and the inverse temperature for a fixed $\gamma = 50$ cm^{-1}, see Figure 7.6. Similarly to the plot of efficiency as a function of λ, γ, Figure 7.5, here the efficiency landscape is also divided by lines of approximately λT (with some deviations from linearity in the high λ and low temperature regime). It can be observed from the figure that for small λT weak localization is dominant. At the intermediate λT values environment-assisted energy transport occurs. As we move towards larger system–bath interactions and higher temperatures, strong dynamical disorder diminishes the coherence and the exciton migration can be fully described by an incoherent hopping process, since the wavefunction is essentially localized over spatial sites. At this regime, the effect of high temperature can be understood from the dynamics of BChls energy fluctuations which is described by the symmetrized correlation function $S(t) = \frac{1}{2}\langle\{\tilde{B}(t), \tilde{B}(0)\}\rangle_{ph}$ ($B_j = B$, for any BChl j). The function can be extracted experimentally by three-pulse photon-echo peakshift measurement. For a Drude–Lorentzian spectral density and fixed γ, the temperature T and the reorganization energy λ determine the amplitude of the site energy fluctuations. Our simulation demonstrates that a highly efficient energy transfer can be achieved at *moderate* site energy fluctuations away from both weak and strong localization limits.

Combining different regimes of three important environmental parameters, λ, γ, and T, the effective decoherence strength $kT\lambda/\gamma$ emerges as the parameter that governs the energy transfer efficiency landscape. For the FMO free Hamiltonian energy gaps, by increasing the single parameter $kT\lambda/\gamma$ from small to intermediate, and from intermediate to large values, one can describe the transition from weak localization to ENAQT, and from ENAQT to strong localization. More generally, when the effective decoherence rate $kT\lambda/\gamma$ is either much smaller or much larger than the typical energy splitting g between delocalized energy eigenstates, then transport is suppressed. Thus, in order to predict the general patterns of quantum transport in generic light-harvesting systems, we should compare the relative strength of $kT\lambda/\gamma$ to the average excitonic energy gap of the free Hamiltonian g.

7.7 Conclusion

Environment-assisted quantum transport is a fundamental effect which occurs in a wide variety of transport systems. Broadly speaking, ENAQT is similar in flavour to the phenomenon of stochastic resonance (Gammaitoni *et al.*, 1998): adding noise to a coherent system enhances a suitable figure of merit characterizing its performance. In the case of excitation transport across the type of molecular complexes we have focused on, that figure of merit is the transport efficiency in a given time from the chlorosome antenna to a reaction centre. While a purely coherent evolution would lead to inefficient transport as a result of the emergence of non-propagating quantum states, the presence of a fluctuating environment can assist the transport by means of a combination of effects whose building blocks can be traced back to very fundamental processes: the removal of destructive interference and/or the elimination of inefficient coherent transport paths and the exploitation of line broadening effects. Qualitative agreement with observed transport times and efficiencies can already be achieved with simple dynamical models using master equations of the Lindblad form. The maximum efficiency of ENAQT occurs when the decoherence rate is comparable to the energy scales of the coherent system, as defined by the energy mismatch between states and the hopping terms. By changing the energy mismatch and the hopping terms, the temperature at which the maximum transport efficiency occurs can be tuned. In the Fenna–Matthews–Olson protein complex within the pure dephasing model and with the spectral density as discussed above, this maximum occurs at approximately room temperature. However, accounting for efficient transport is only one part of the puzzle, and the accurate description of additional dynamical properties recently unveiled by 2D spectroscopy requires a more sophisticated approach to describe the system–environment interactions. Recent work has already paved the way towards a more complete understanding, and both non-Markovian (Rebentrost *et al.*, 2009b; Thorwart *et al.*, 2009; Rebentrost and Aspuru-Guzik, 2011; Shabani *et al.*, 2012; Mohseni *et al.*, 2013, 2014) and numerically exact techniques (Prior *et al.*, 2010) are being developed, as well as detailed quantum chemistry calculations (Shim *et al.*, 2012). Our simulation based on a TC2 master equation is an attempt in this direction. These analyses are expected to finally elucidate whether or not it is necessary for the environment to display quantum features and what system–environment interactions allow for long-lived delocalized excitations while simultaneously quickly destroy local (site) coherences. The proposed idea of delocalization as a facilitator for environmental sensing favours the idea of an environment supporting discrete features. Further experimental evidence would refute or corroborate this hypothesis but, importantly, the general concept of noise-assisted transport is clearly resilient to the specific details of the environmental mode structure, and is an essential ingredient to account for the observed high transfer efficiency in actual photosynthetic complexes.

Part III

Quantum effects in higher organisms and applications

8

Excitation energy transfer and energy conversion in photosynthesis

ELISABET ROMERO, VLADIMIR I. NOVODEREZHKIN
AND RIENK VAN GRONDELLE

Photosynthesis is the biological process by which the energy of the Sun is collected, converted and stored in chemical bonds needed to power life. Therefore, photosynthesis serves as the vital link between the light energy of the Sun and almost all living organisms on Earth. In this chapter we will focus on the first steps of photosynthesis: energy collection and conversion, i.e. light-harvesting and charge separation, highlighting the role of quantum effects on the ultrafast dynamics and quantum efficiency of these two remarkable processes. Both experimental and theoretical approaches will be described and combined.

8.1 Photosynthesis

In photosynthesis solar energy is absorbed by the light-harvesting antenna and transferred to the photosynthetic reaction centre (RC) within several tens of picoseconds. In the RC, the absorbed excitation energy is converted into electrochemical energy by means of an ultra fast charge separation. Photosynthetic purple bacteria employ a single reaction centre, in contrast, in photosynthesis of plants, algae and cyanobacteria, two reaction centres, Photosystem II (PSII) and Photosystem I (PSI), operate in series. PSII uses light to extract electrons from water (to produce oxygen), while PSI uses light to reduce $NADP^+$ to NADPH. The subsequent electron transfer from PSII to PSI is coupled to the build-up of a proton motive force (pmf) that is used to form ATP. NADPH and ATP are required in the Calvin–Benson cycle to produce a reduced sugar. In the following we will discuss photosynthetic charge separation and photosynthetic light-harvesting with an emphasis on the role of quantum effects.

Quantum Effects in Biology, ed. Masoud Mohseni, Yasser Omar, Gregory S. Engel and Martin B. Plenio. Published by Cambridge University Press.

8.2 Photosynthetic energy conversion: charge separation

Following elucidation of the structure of the bacterial RC (Deisenhofer *et al.*, 1984), a detailed picture of photosynthetic charge separation has emerged. During past decades the primary electron transfer processes of all of the different types of RCs, purple bacterial, Photosystem I (Ben-Shem *et al.*, 2003), Photosystem II (Ferreira *et al.*, 2004; Loll *et al.*, 2005; Umena *et al.*, 2011) have been extensively studied. In the following we will discuss charge separation in the bacterial RC and the Photosystem II RC in some detail.

In the bacterial RC (BRC), for which a structure with atomic resolution has been available since 1984 (Deisenhofer *et al.*, 1984), the bacteriochlorophyll (BChl) pigments are arranged in two branches labelled A and B (also called L and M) with the symmetry axis running through the centrally located 'special pair' (P) BChls (see Figure 8.1). Surprisingly, light-driven electron transfer occurs only along the A (or 'active') branch. In addition to the special pair P the BRC contains in each branch a 'monomeric' or assessory BChl (B_A, B_B), a bacteriopheophytin (BPhe) (H_A, H_B) and a quinone (Q_A, Q_B). Primary charge separation in BRCs occurs within $\approx$ 3 ps with the formation of $P^+H_A{}^-$ (Martin *et al.*, 1986). Currently, there is general agreement that B_A acts as an electron transfer intermediate, with electron transfer from the excited state of P (P*) to B_A occurring in 3 ps; almost three times slower than the electron transfer from B_A to H_A in about 1 ps (Holzapfel *et al.*, 1989, 1990; Arlt *et al.*, 1993).

In spite of the obvious symmetry of the BRC, the electron transfer essentially occurs only along the active or A-branch. Estimates of the electronic coupling strengths along both paths do indeed favour the active branch, but not in the experimentally observed 200 :1 ratio. Spectacular variations in the rate of electron transfer were observed upon variation of the identity of Tyr M210 (Nagarajan *et al.*, 1990; Beekman *et al.*, 1996), strategically positioned between the cofactors in the active branch. These could be explained within the framework of the Marcus theory for electron transfer, and consequently were largely ascribed to changes in the free energy of the initial radical pair as a result of the mutation.

In the Marcus model for electron transfer, the rate k_{ET} is given by (Marcus, 1993)

$$k_{ET} = \frac{2\pi}{\hbar}|V_{el}|^2(4\pi\lambda k_B T)^{-1/2}\exp\{-(\Delta G^0 + \lambda)^2/4k_B\lambda T\}, \tag{8.1}$$

where V_{el} is the electronic coupling matrix element, ΔG^0 is the free energy, λ is the reorganization energy, T is the temperature and k_B is the Boltzmann constant. Other types of mutants were also generated, for instance, mutants in which hydrogen bonds of the protein surrounding the 'special pair' P were modified produced dramatic changes in the redox potential of P, again without affecting the asymmetry of charge separation (Lin *et al.*, 1994). In a triple mutant with the

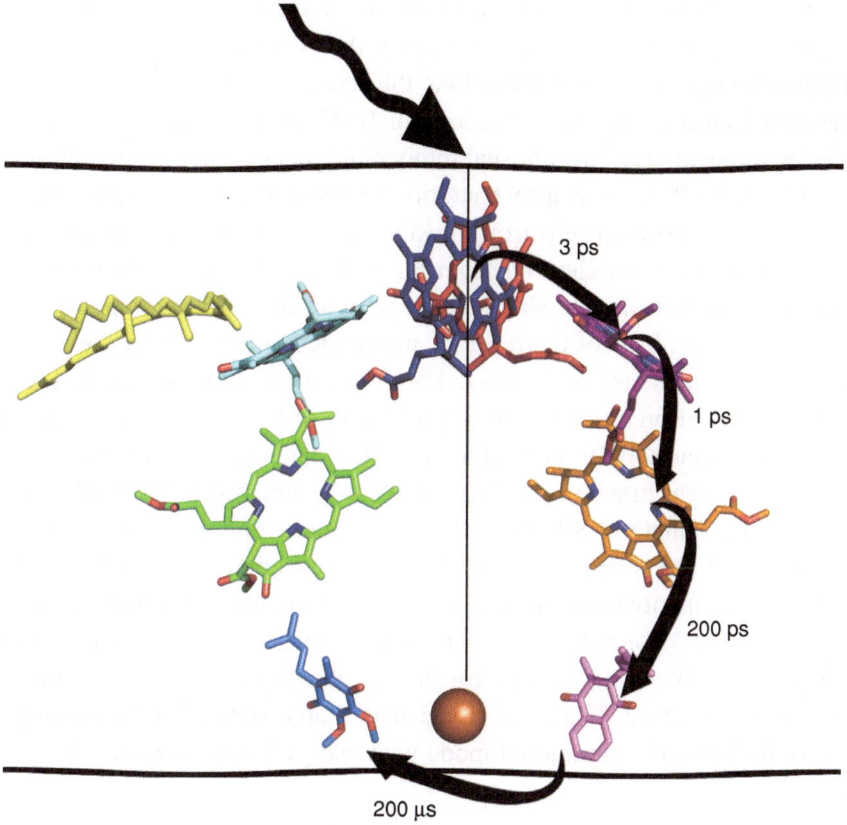

Figure 8.1 The photosynthetic reaction centre of the purple bacterium *Rhodopseudomonas viridis*. The reaction centre consists of three proteins that non-covalently bind four bacteriochlorophylls (P_A and P_B forming the 'special pair' in red and blue, respectively; B_A (purple), B_B (cyan)), two bacteriopheophytins (H_A (orange) and H_B (green)), two quinones (Q_A (light purple) and Q_B (light blue)), one Fe-atom (red), and one carotenoid (yellow). The arrows (black) show the sequence of electron transfer events following excitation of the 'special pair' (via direct light absorption or energy transfer from the antenna) leading to transmembrane charge separation. This was the first membrane protein for which a structure was solved and this gave the Nobel prize in Chemistry 1988 to Michel, Deisenhofer and Huber. X-ray structure adapted from (Deisenhofer *et al.*, 1995).

BPhe H_A replaced by a BChl, in which $P^+B_A^-$ was upshifted in free energy and $P^+B_B^-$ was stabilized, a measurable amount of $P^+H_B^-$ was obtained (Kirmaier *et al.*, 1999). This again suggests that the free energy of the charge separated states controls the asymmetry of the electron transfer, rather than the differences in electronic coupling. A third possibility is that the protein matrix plays an active role in driving the charge separation selectively along the A-branch. A hint that this is actually the case was obtained from Stark spectroscopy which revealed a substantially higher dielectric strength along the A-branch (Steffen *et al.*, 1994) clearly favouring charge separation along the active branch.

The initial electron transfer reactions in BRC show an increase in rate with decreasing temperature. This unusual temperature dependence is seen in the electron transfer from P^* to the nearby bacteriopheophytin (H_A), in the individual steps that underlie this reaction (electron transfer from P^* to B_A and from B_A to H_A) and in the transfer of an electron from H_A to Q_A (Fleming *et al.*, 1988). Equation (8.1) shows that the rate should increase with decreasing temperature if the potential energy surfaces of the reactant and product intersect near the zero-point level of the reactant state: $\Delta G^0 = -\lambda$. Early experiments by Devault and Chance, who observed electron transfer from a cytochrome to P^+ at cryogenic temperatures (Devault and Chance, 1966; Devault *et al.*, 1967) laid the basis for the concept of electron and nuclear tunnelling in biology (Marcus and Sutin, 1985; Moser *et al.*, 1992). Because Equation (8.1) neglects nuclear tunnelling it can overestimate the contribution of the exponential factor when T is small compared with the characteristic frequencies involved or when $\Delta G^0 < -\lambda$. A quantum-mechanical treatment based on the theory for radiationless transitions, that includes nuclear tunnelling, gave a qualitatively similar picture for the speeding up of charge separation with decreasing temperature. If it is assumed that charge separation is strongly coupled to some harmonic vibrational mode with energy $\hbar\omega$, then the rate is given by (Jortner, 1976),

$$k_{ET} = k_0\{\tanh(\hbar\omega/2k_BT)^{-1/2}\} \text{ with } k_0 = \frac{2\pi}{\hbar}|V_{el}|^2(2\pi\lambda\hbar\omega)^{-1/2}. \qquad (8.2)$$

The temperature dependence of the rate of electron transfer from P^* to H_A in *Rb. sphaeroides* RCs fits well with Equation (8.2), taking $\hbar\omega = 80\ cm^{-1}$ (Fleming *et al.*, 1988).

So far all of this is based on the assumption that vibrational relaxation occurs rapidly relative to electron transfer. Although relaxation phenomena in RCs are known to occur on timescales even longer than 10 ns, this assumption is probably valid for slow electron transfers like $P^+Q_A^-$ recombination. But what about the fast events?

8.2.1 In bacterial reaction centres charge separation is coupled to coherent nuclear motions

In the BRC the excited state of the special pair P is coupled to low-frequency vibrations (Vos *et al.*, 1993, 1994; Streltsov *et al.*, 1996), which represent collective nuclear motions of the pigments and their surroundings. The vibrational wavepacket, created upon excitation of P by a femtosecond laser pulse consists of modes in the 100–200 cm^{-1} range decaying on a few 100 fs timescale. The Fourier transform (FT) spectrum showed that the range <35 cm^{-1} also has significant vibrational strength. Remarkably, in native and pheophytin-modified *Rb. sphaeroides* R-26 RCs a wavepacket-like motion could be observed in the absorption band of the primary photoproduct $P^+B_A{}^-$ (Yakovlev *et al.*, 2000, 2002a, b). Thus, the FT spectra of the components corresponding to the stimulated emission (SE) of the reactant P* and the excited state absorption (ESA) of the product $P^+B_A{}^-$ contain two modes (30 and 130 cm^{-1}), with the low-frequency 30 cm^{-1} mode significantly increased in the photoproduct. The 30 cm^{-1} mode was ascribed to rotation of the H_2O molecule located between P and B_A (Yakovlev *et al.*, 2002a, b). Observation of the oscillatory components in the H_A-band (760 nm) suggested that the 30 cm^{-1} mode results in coherent $P^+H_A{}^-$ state formation.

To describe the observed spectral evolution, the electron-vibrational dynamics has been modelled using the density matrix equation with the Redfield superoperator in the basis of electron-vibrational states. The model includes two diabatic states, i.e. an excited state P* and a charge-transfer (CT) state $P^+B_A{}^-$. Strong coupling of these states with two collective nuclear modes is proposed (Novoderezhkin *et al.*, 2004). The mixing of diabatic states (with different displacements along each of the two nuclear coordinates) results in a complicated potential energy surface that determines the dynamics of the excited-state wavepacket.

This model resulted in a quantitative fit of the experimental kinetics of the SE near 900–950 nm and the ESA in the 1020 nm region of the pheophytin-modified *Rb. sphaeroides* R-26 RC (Novoderezhkin *et al.*, 2004). Thus non-equilibrated vibrational modes involved in electron transfer play an important role in photoproduct formation in the BRC. The configuration of the two vibrational coordinates involved plays an essential role in establishing the high efficiency of charge separation, both for coherent and non-coherent excitation. In particular, a strong coupling to the 130 cm^{-1} mode allows an effective electron transfer from the primary donor P* to the photoproduct $P^+B_A{}^-$ state, whereas strong coupling of the product state to the second 30 cm^{-1} mode (resulting in pronounced displacement of its potential surface along the *y*-coordinate, as shown in Figure 8.2), causes a motion of the $P^+B_A{}^-$ part of the wavepacket along the *y*-coordinate, i.e. away from the crossing point, thus stabilizing the charge-separated state.

8.2.2 *Alternative ultra fast pathways for charge separation in bacterial reaction centres*

Until the late 1990s, it was generally assumed that photosynthetic charge separation only occurs from the excited special pair P*, and that excitation of all other pigments results in energy transfer to P, followed by charge separation from P*

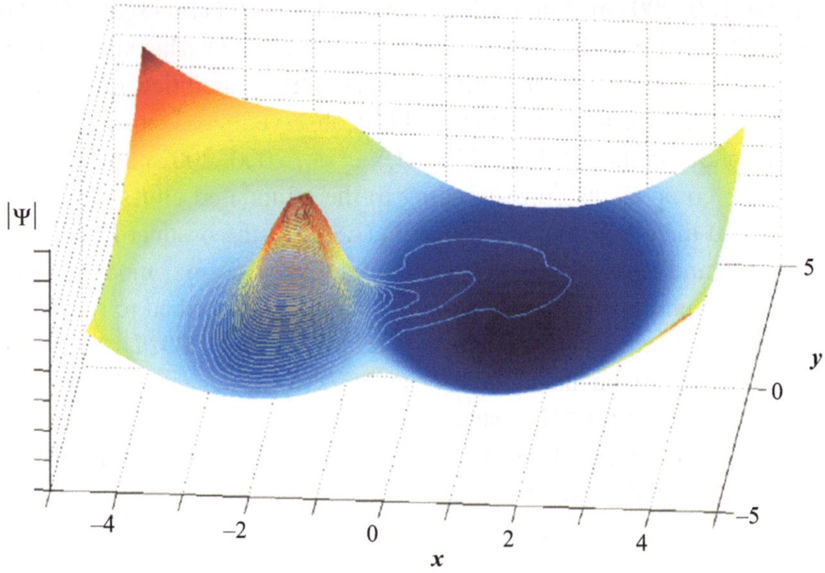

Figure 8.2 The lowest excited state potential energy surface of the BRC as determined by the mixing of the states P^*B_A (left minimum) and $P^+B_A^-$ (right minimum). Both states are assumed to be strongly coupled to two nuclear modes (130 cm^{-1} and 30 cm^{-1}) represented by the coordinates x and y. The 3D-contour shows the amplitude of the nuclear wavepacket 250 fs after impulsive excitation, when it reaches the crossing point between the two states. Part of it will flow into $P^+B_A^-$, part of it will bounce and try to reach the $P^+B_A^-$ state again. Reprinted with permission from (Novoderezhkin *et al.*, 2004). Copyright 2004 American Chemical Society.

(Jonas *et al.*, 1996). However, experiments on a mutant of the bacterial RC where the residue YM210 had been replaced by a Trp (YM210W) demonstrated the presence of another ultra fast pathway of charge separation (van Brederode *et al.*, 1997). In this mutant excitation of the primary donor P gives rise to very slow charge separation (40 ps at RT, 1 ns at 77 K). Most likely, in the YM210W mutant RC the first intermediate $P^+B_A^-$ is higher in energy than P* (it has indeed been speculated that the Tyr YM210 stabilizes B_A^- by its hydroxyl dipole (Alden *et al.*, 1996)). In contrast, excitation of the accessory BChl B_A in the YM210W RC led to a significant amount of $P^+B_A^-$ formation in less than 1 ps, without the involvement of P*. Then $P^+B_A^-$ decayed into $P^+H_A^-$ on a few ps timescale. The very same process was identified later for wild-type BRCs (van Brederode *et al.*, 1999).

8.2.3 Electronic coherence in bacterial reaction centres

Under broadband excitation it was possible to create electronic coherence between the two exciton states $(P^*)_+$ and $(P^*)_-$ (Arnett *et al.*, 1999). The coherent motion of the exciton wavepacket resulted in oscillations of the anisotropy with a frequency of 593 cm^{-1} (which is close to the exciton splitting) and damping constant of 35 fs. This indicates that electronic coherence may play a role in the early stages of charge separation in BRCs.

8.2.4 Charge separation in the photosystem II reaction centre

The reaction centre of Photosystem II (PSII RC) performs the initial charge separation reaction in oxygenic photosynthesis (Dekker and van Grondelle, 2000; Raszewski *et al.*, 2008). The thereby created oxidized state eventually leads to the oxidation of H_2O and the production of O_2 via the accumulation of four positive charges on the Mn_4CaO_5 cluster upon four subsequent illuminations. According to the X-ray structure (Ferreira *et al.*, 2004; Loll *et al.*, 2005; Umena *et al.*, 2011) the overall pigment organization in the PSII RC is similar to that of the BRC, with two additional chlorophylls bound to the periphery of the complex. Charge separation in the PSII RC occurs on a multitude of timescales ranging from sub-ps to hundreds of ps (Dekker and van Grondelle, 2000). In contrast with the BRC, visible-pump-mid-infrared probe experiments have shown initial formation of the radical pair $Chl_{D1}{}^{+}Pheo_{D1}{}^{-}$ in a significant fraction of the PSII RCs on a ps timescale (Groot *et al.*, 2005) (in the PSII RC the active branch is called D1 while the inactive is D2). In that work, the $P_{D1}{}^{+}$ formation was observed only after 5–6 ps, followed by radical pair relaxation. In this view, the observed multi-exponential kinetics of the charge separation process originates, in part, from the fact that, on average, only small energy differences exist between most of the excited and charge separated states of the PSII RC. A factor that should be taken into account in explaining these kinetics is the intrinsic disorder. Within the sample ensemble, many different protein configurations (realizations of the disorder) with different relative distance and orientation of the cofactors/protein residues are possible, due to slow protein motions (Novoderezhkin *et al.*, 2005c). This means that the sample ensemble contains a collection of energetically different reaction centres. Therefore, the small energy differences between the electronic states in the PSII RC, in combination with the intrinsic disorder, result in the observed highly multi-exponential kinetics.

The first attempt to explain the spectra and kinetics in the PSII RC was performed using the so-called 'multimer model', which was based on the idea that in the core of the PSII RC, in contrast with the BRC, all the chlorins are spectroscopically 'equal', giving rise to exciton states delocalized over 2–3 pigments (Durrant *et al.*, 1995). In a more realistic model, site energies were extracted from a simultaneous evolutionary-based fit of the linear spectra using the modified Redfield approach (Novoderezhkin *et al.*, 2005c; Raszewski *et al.*, 2005; Novoderezhkin *et al.*, 2007a; Raszewski *et al.*, 2008). The result of such a model is shown in Figure 8.3.

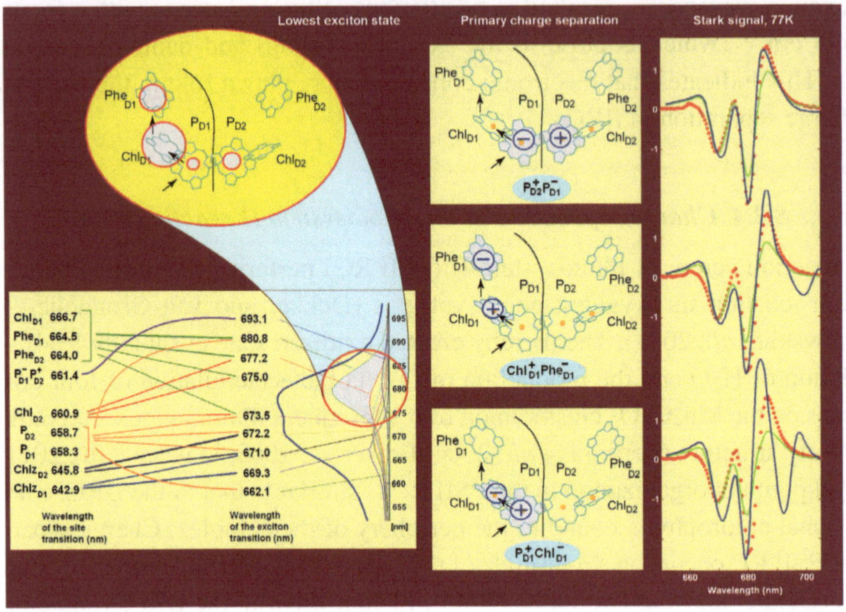

Figure 8.3 *Left frame*: Exciton structure of the PSII RC. Energies corresponding with the unperturbed site energies of eight pigments and the first CT intermediate $P_{D2}{}^{+}P_{D1}{}^{-}$ (lines indicate participation of the pigments in the exciton states), and the 77 K absorption spectra with individual exciton components. Inset on the top shows the averaged structure of the lowest exciton state, where the circles show the pigments that on average are coherently mixed in the lowest exciton state. The area under the circle is proportional to the population of the corresponding site. *Middle frames*: Possible pathways for primary charge separation in the PSII RC. Circles show localization of the electron and hole in the CT states (i.e. $P_{D2}{}^{+}P_{D1}{}^{-}$, $Chl_{D1}{}^{+}Phe_{D1}{}^{-}$ and $P_{D1}{}^{+}Chl_{D1}{}^{-}$) that can be coupled to the lowest exciton state. *Right frame*: Stark spectra calculated with the same CT states as shown in the middle frame. Red dots correspond with the experimental data, the Stark signal is calculated with coupling to CT (blue) and without coupling to CT state (green). Adapted from (Novoderezhkin *et al.*, 2007a), Copyright 2007, with permission from Elsevier.

The lowest state with an absorption band centred near 682 nm corresponds with the $P_{D2}{}^{+}P_{D1}{}^{-}$ charge-transfer (CT) state. This state has become weakly allowed, borrowing dipole strength due to mixing with the pure exciton states and due to reorganization has shifted significantly to the red. Thus, the introduction of a coupling between the excited and the CT state produces relatively small changes in the absorption-type spectra, but has a dramatic effect on the fluorescence (FL) profile. The FL spectrum is mostly determined by contributions from the two lowest states, i.e. a mixed exciton–CT state [$(P_{D2}P_{D1})^{*}$- $P_{D2}{}^{+}P_{D1}{}^{-}$ peaking at 682 nm] and a superradiant 'multimeric' exciton state (delocalized over the D1 branch and peaking at 680 nm), which means that the FL profile will be extremely sensitive to the precise energy of the CT state. On the other hand, the degree of mixing of the CT state with its huge static dipole with the exciton states is the factor that determines the shape and amplitude of the Stark spectrum (Figure 8.3, right frame).

The excited state from which the charge separation is initiated consists of a coherent superposition of four pigments, i.e. Chl_{D1}, Phe_{D1}, P_{D1}, and P_{D2} (Figure 8.3). Since there is a strong coherence between these four sites, some CT state coupled to any of these four pigments will be effectively mixed with the whole set of exciton states (the amount of mixing depending on the size of the energy gap between each of the excitonic states and a certain CT state). Thus, the primary electron transfer towards Phe_{D1} can, in principle, start from P_{D2}, Chl_{D1} or P_{D1} producing the first charge-separated configuration $P_{D2}{}^{+}P_{D1}{}^{-}$, $Chl_{D1}{}^{+}Phe_{D1}{}^{-}$ or $P_{D1}{}^{+}Chl_{D1}{}^{-}$, respectively.

The predominant population of Chl_{D1} is a strong argument for considering this pigment as the primary electron donor, producing the CT state $Chl_{D1}{}^{+}Phe_{D1}{}^{-}$. On the other hand, P_{D1} and P_{D2} are characterized by a bigger overlap of the electronic wavefunctions of the two pigments, thus creating a better coupling between the excited states of P_{D1} and P_{D2} and the $P_{D2}{}^{+}P_{D1}{}^{-}$ CT state. Clearly, these two factors compete. In this case, the relative participation of the pigments (P_{D1}, P_{D2} and Chl_{D1}) in the lowest exciton state as well as the energy gap between the lowest exciton state and the corresponding CT state will be strongly dependent on the specific realization of the disorder. For instance, in some realizations, i.e. in some of the RCs, the excitation can be strongly localized on Chl_{D1}; moreover, this localized state can be even lower in energy than the $P_{D2}{}^{+}P_{D1}{}^{-}$ CT state. Obviously, in such realizations, even in the presence of mixing between the excited states and the $P_{D2}{}^{+}P_{D1}{}^{-}$ intermediate, the charge separation will be initiated from Chl_{D1} with the formation of the $Chl_{D1}{}^{+}Phe_{D1}{}^{-}$ radical pair. On the other hand, in delocalized 'multimeric' realizations, the $P_{D2}{}^{+}P_{D1}{}^{-}$ state is much better connected with the whole excited-state manifold, and thus can play the role of the initial CT state.

In order to test experimentally the hypothesis formulated above, i.e. disorder-induced charge separation pathways, we have performed a detailed transient absorption investigation (Romero *et al.*, 2010). Because of the well-known spectral congestion in the PSII RC and similar timescales predicted for the proposed charge-separation pathways (where the first charge-separated state configurations are $P_{D2}{}^{+}P_{D1}{}^{-}$, $Chl_{D1}{}^{+}Phe_{D1}^{-}$ and/or $P_{D1}{}^{+}Chl_{D1}{}^{-}$), the transient absorption experiments were carried out at 77 K (to enhance spectral resolution and reduce back-reactions and uphill energy transfer) and various excitation conditions had to be used to allow photoselection of subpopulations with different proportions of the different pathways. Additionally, the changes in absorption due to excitation had to be probed over the whole visible range and during an extensive time range.

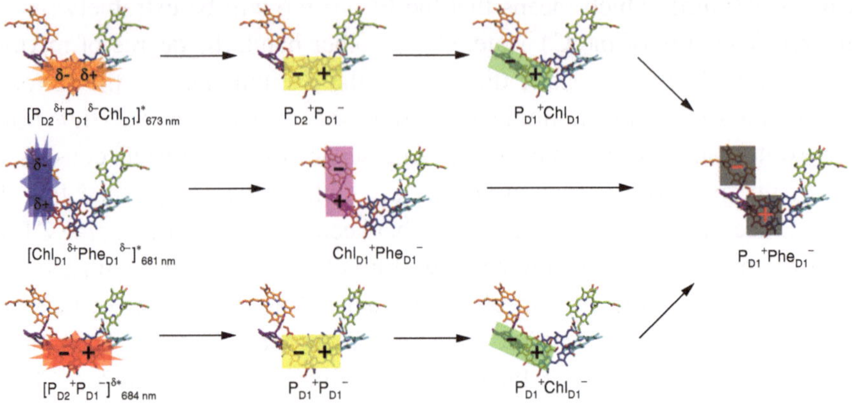

Figure 8.4 Two different pathways for charge separation in the PSII RC: excitation energy and charge distribution of the electronic states involved in charge separation. *Top and bottom*: Charge separation via the P_{D1} path. *Centre*: Charge separation via the Chl_{D1} path. The excited states are represented as stars, the radical pairs are represented as rectangles. X-ray structure adapted from (Umena *et al.*, 2011). Adapted with permission from (Romero *et al.*, 2010). Copyright 2010 American Chemical Society.

Combining all of the obtained results (13 experiments) and using global and target analysis according to a kinetic scheme (van Stokkum *et al.*, 2004), we have demonstrated that, in agreement with the theoretical model (Novoderezhkin *et al.*, 2007a), at least two different excited states, $(Chl_{D1}Phe_{D1})^*$ and $(P_{D1}P_{D2}Chl_{D1})^*$, give rise to two different pathways for ultra fast charge separation. These results indicate that, indeed, the disorder produced by the slow protein motions causes energetic differentiation among the reaction centre complexes, which opens up the possibility of different pathways for charge separation (Figure 8.4).

As can be seen in Figure 8.4, the charge separation process consists of two steps: (1) the energy conversion step, exciton $\rightarrow$ CT state, in which the sunlight excitation energy is converted into a charge-separated state; and (2) the separation of charge step, CT1 $\rightarrow$ CT2, in which the positive and negative charges must be physically separated to avoid energy losses by charge recombination.

The efficiency of the energy conversion step depends on the electronic characteristics of the exciton states. The efficiency is high if the electron density distribution in the exciton is similar to that in the CT state. To investigate the presence and properties of the CT states in the PSII RC, Stark spectroscopy is the most suitable technique (Boxer, 1996). However, due to spectral congestion, this technique has to be applied to site-directed mutant PSII samples in order to allow unambiguous assignment of the absorption spectral bands (Diner *et al.*, 2001). In that work, each of the site-directed mutants studied contains a single amino acid mutation, near one of the chlorins involved in charge separation (P_{D1}, P_{D2}, Chl_{D1}, Phe_{D1}), which induces an energy shift of its transition energy with respect to the wild-type (WT) energy. Comparison of the Stark spectra of WT with eight mutants (mutants studied close to: P_{D1} (3), P_{D2} (3), Chl_{D1} (1), Phe_{D1} (1)), unravels the configuration and transition energy of three electronic states which initiate charge separation with both exciton and CT character. These states are: two exciton states with CT character, $(P_{D2}{}^{\delta+}P_{D1}{}^{\delta-}Chl_{D1})^*_{673nm}$ and $(Chl_{D1}{}^{\delta+}Phe_{D1}{}^{\delta-})^*_{681nm}$; and one CT state with exciton character, $(P_{D2}{}^{+}P_{D1}{}^{-})^{\delta *}{}_{684nm}$ (Romero *et al.*, 2012). In this exciton–CT representation, the subscript indicate the approximate centre wavelength of the electronic transition and the $\delta+/\delta-$ and $\delta*$ indicate the CT and exciton character, respectively.

In addition, these results show that the pigment–protein interactions fine-tune the energy of the exciton and CT states, and hence the mixing between these states which ultimately controls the selection and efficiency of a specific charge separation pathway. In this view, the protein is not just a passive spectator of the electron transfer reactions, the protein is the active director of the action (Lin *et al.*, 2005; Wang *et al.*, 2007; Brecht *et al.*, 2009). Combining the conclusions obtained by transient absorption and Stark spectroscopy, we obtain the following charge separation pathways:

P_{D1} path:

$$(P_{D2}{}^{\delta+}P_{D1}{}^{\delta-}Chl_{D1})^*{}_{673nm} \rightarrow P_{D2}{}^{+}P_{D1}{}^{-} \rightarrow P_{D1}{}^{+}Chl_{D1}{}^{-} \rightarrow P_{D1}{}^{+}Phe_{D1}{}^{-}$$

Chl_{D1} path:

$$(Chl_{D1}{}^{\delta+}Phe_{D1}{}^{\delta-})^*{}_{681nm} \rightarrow Chl_{D1}{}^{+}Phe_{D1}{}^{-} \rightarrow P_{D1}{}^{+}Phe_{D1}{}^{-}$$

P_{D1} path:

$$(P_{D2}{}^{+}P_{D1}{}^{-})^{\delta *}{}_{684nm} \rightarrow P_{D2}{}^{+}P_{D1}{}^{-} \rightarrow P_{D1}{}^{+}Chl_{D1}{}^{-} \rightarrow P_{D1}{}^{+}Phe_{D1}{}^{-}$$

8.2.5 Quantum coherence and charge separation in the photosystem II reaction centre

Recently, this model of charge separation has been further studied using two-dimensional electronic spectroscopy (2DES) because of the suitability of this technique to the study of quantum effects in photosynthetic complexes (Ginsberg *et al.*, 2009; Read *et al.*, 2009; Schlau-Cohen *et al.*, 2012). In these experiments the sample is excited with a pair of temporally ultra short and spectrally broad excitation laser pulses, while the photon echo stimulated by a third laser pulse is measured. The 2D spectrum is obtained by a double Fourier transform of the measured signal.

The 2D electronic spectra of the PSII RC contains cross-peaks, reflecting couplings between exciton states (Brixner *et al.*, 2005), as well as oscillatory features in the spectral traces which reveal the dynamics of quantum coherences (Engel *et al.*, 2007; Cheng and Fleming, 2008; Collini *et al.*, 2010; Panitchayangkoon *et al.*, 2010). However, the presence of electronic and/or vibrational coherences as well as the coupling between them is under investigation. Our current model predicts that the degree of coherent mixing between exciton–CT states (determined by energetic disorder induced by slow conformational changes of the complex) is related to the high speed and efficiency of the charge separation process. Nevertheless, the role of quantum coherence in the charge separation process in the PSII RC has to be confirmed by exhaustive analysis and modelling of the 2DES experimental data.

8.3 Light-harvesting

In photosynthesis most of the pigments are organized in a light-harvesting antenna (LH), to collect solar photons and transfer the electronic excitations to the RC to drive charge separation. But, why is the LH needed? Photosynthesis must be able to operate at low light levels, such as those that generate less than one electronic excitation per Chl per second (normal sunlight). Yet the most important biochemical reactions associated with photosynthesis require several electron-transfer events. For example, water oxidation in the PSII RC requires the cumulative effect of four electronic excitations, all of which must occur within a certain time. Light-harvesting overcomes this problem by concentrating the available light energy and feeding the electronic excitations from hundreds of light-absorbing pigment molecules into a single RC. Another reason for having an LH is that this allows photosynthetic organisms to survive using fewer RCs. This is beneficial because RCs are 'expensive' – each one requires a large investment of resources from the organism. The LH also allows a broad range of the spectrum to be exploited for photosynthesis, because it may be composed of a variety of pigment proteins that contain different pigments (and which therefore absorb different colours of light) connected to one RC. Finally, in a multi-protein LH the flow of excitation energy can be regulated by modulating the quenching properties of one of the constituent antenna proteins; which provides a way of protecting plants from potentially harmful absorbed energy from excess sunlight (Holt *et al.*, 2004).

8.3.1 Excitation energy transfer and excitons

Figure 8.5 summarizes the excitation energy transfer for the purple bacterial photosynthetic membrane for which this process of energy transfer and capture by the BRC has been extensively studied (van Grondelle *et al.*, 1994; Hu *et al.*, 1998; Sundstrom *et al.*, 1999; van Grondelle and Novoderezhkin, 2006).

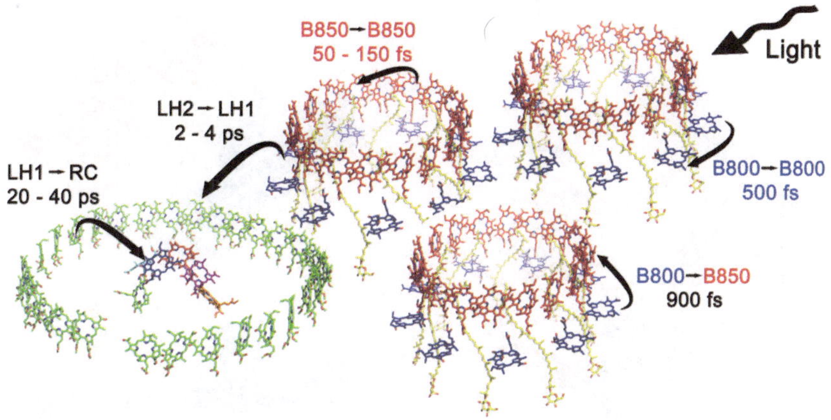

Figure 8.5 Bacterial light-harvesting. In the bacterial photosynthetic apparatus LH1–RC complexes are surrounded by LH2 complexes. Both in LH1 and LH2 the BChls are organized in ring-like structures (McDermott *et al.*, 1995; Roszak *et al.*, 2003; Codgell *et al.*, 2006a). The BRC is positioned in the centre of the LH1 ring. In the LH2 complex the BChls are shown in blue (the B800 ring, named due its absorption maximum) and red (B850 ring) while the carotenoids are shown in yellow. The excitations around 400–500 nm are mainly absorbed by the carotenoids and transferred to the BChls in less than 1 ps (Polivka and Sundstrom, 2004). The excitation absorbed by the B800 ring is transferred to the red B850 ring in about 1 ps (van der Laan *et al.*, 1990). Within a single LH2 or LH1 the excitations move around very rapidly, typically in a few hundred fs (Bradforth *et al.*, 1995; Visser *et al.*, 1996). The transfer between rings is slower, between 1–10 ps. Since the LH1 rings absorb more to the red than the LH2 ring (875 nm versus 850 nm) the excitations are concentrated around the BRC. In fact, the transfer of electronic excitations from the LH1 antenna to the BRC is the rate limiting step that occurs in about 40 ps (Visscher *et al.*, 1989). Once the BRC gets excited, a fast (3 ps) charge separation occurs.

In this case, the pigments BChl and carotenoid, bound to proteins, are organized as an energetic funnel. In the periphery we find the circular LH2 complex (McDermott *et al.*, 1995), absorbing light at 800 nm and 850 nm and surrounding the RC we have the LH1 complex (Roszak *et al.*, 2003), absorbing at around 870 nm, also the main transition of the RC is at about 870 nm. Carotenoids absorb in the visible and transfer the absorbed solar photons often in less than 1 ps (Polivka and Sundstrom, 2004) to a BChl close by. The overall probability for an excitation to be transferred to the RC may easily exceed 95%. The observed lifetime of an excitation in the bacterial antenna is about 50 ps (van Grondelle, 1985; Sundstrom *et al.*, 1986).

Plant antenna systems operate in a very similar manner, they employ Chl *a* and Chl *b* as major pigments, complemented by a variety of carotenoids. Chl *a* absorbs light at around 670–680 nm and Chl *b* at around 650 nm. Carotenoids,

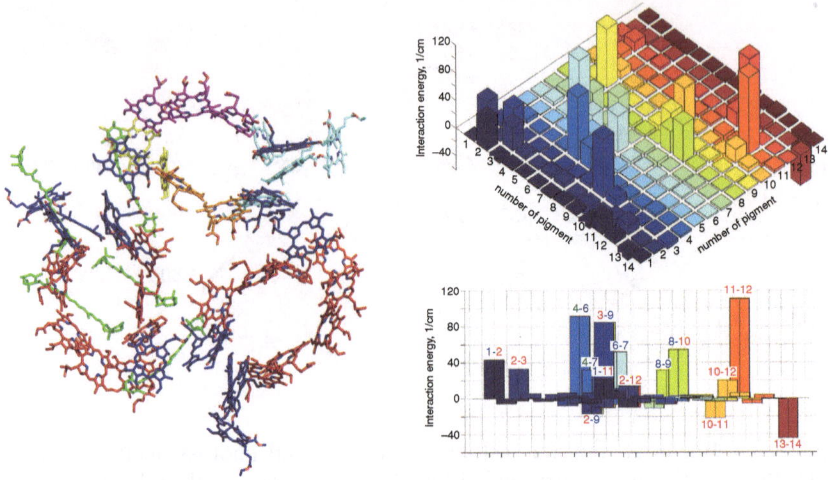

Figure 8.6 Plant light-harvesting complex II (LHCII). *Left:* Trimeric LHCII structure (view from the top of the thylakoid membrane) as determined by (Liu *et al.*, 2004). In the monomers shown at the bottom, Chls *a* are shown in red, Chls *b* are shown in blue and the carotenoids are shown in green. In the top monomer some Chls clusters are indicated: Chls *a* 610–612 (magenta), Chls *a* 602–603 (orange), Chls *a* 613–614 (yellow) and the Chls *a*/*b* 604–607 (cyan). *Right:* Interaction energies for the LHCII monomer with pigments 1–14 representing Chl601–Chl614. Pigments are numbered following the crystal structure. Note the strongly interacting clusters Chls *a* 610–612, Chls *a* 613–614, Chls *a* 602–603 and the Chl *a*/*b* cluster 604–607. Intermonomer interactions occur mainly between Chls *b* 601 and Chls *b* 608/609 from adjacent subunits. The bottom frame shows the side view of the bar plot shown in the top frame. Reprinted with permission from (Novoderezhkin *et al.*, 2005). Copyright 2005 American Chemical Society.

like lycopene or β-carotene absorb light in the 400–500 nm spectral region. Also, in plants, the pigments are organized in pigment–proteins, which occur highly organized in the thylakoid membrane. The structure of he LHCII pigment–protein is shown in Figure 8.6. There are two characteristic distinctions between plant and bacterial antennae: (1), there are no highly symmetric structures like LH1 and LH2 in plants; and (2), the number of pigments per unit volume can easily be two to three times higher than in the bacterial antennae, with the distance between neighbouring Chls less than 1 nm.

What is the mechanism of this amazingly efficient energy transfer process by which the excitations of hundreds of pigments are collected at a single site? In PSI of plants, the energy absorbed by about 200 chlorophylls is transported in a few tens of picoseconds to the PSI RC in the centre of the complex, to drive a transmembrane charge separation. In the RC-LH1 core complex, about 30 bacteriochlorophylls surround the RC and the absorbed energy is transferred to the RC in about 40 ps. In such densely packed structures the excited states of the pigments are coupled via dipole–dipole interactions. The strength of this coupling is one of the major factors that determine how fast the excitation moves through the assembly of coupled pigments.

If the dipole–dipole coupling is very weak, say much weaker than the coupling of the excited state to its environment (electron–phonon coupling), the excitation will be essentially localized before energy transfer can occur and the energy transfer will take place according to a hopping process, with a rate W_{DA}, given by the Förster equation,

$$W_{DA} = 8.8 \cdot 10^{17} \cdot \frac{k_R^D}{n^4} \cdot \frac{\kappa^2}{R^6} \int \frac{\epsilon_A(\tilde{\upsilon}) F_D(\tilde{\upsilon})}{\tilde{\upsilon}^4} d\tilde{\upsilon}, \tag{8.3}$$

where R is the centre-to-centre distance between donor and acceptor in nm and $\tilde{\upsilon}$ expressed in cm^{-1}, k_R^D is the radiative rate of the donor, n is the index of refraction, and κ is a factor describing the mutual orientation of the transition dipoles of donor and acceptor. As it is obvious from Equation (8.3), the rate depends critically on the overlap of the emission spectrum of the donor $F_D(\tilde{\upsilon})$ with the absorption spectrum of the acceptor $\epsilon_A(\tilde{\upsilon})$, essentially reflecting energy conservation during the transfer process. It is straightforward to calculate the Förster rate for pigments like Chl and BChl at a distance of 2 nm; W_{AD} can easily be of order 10^{11}–10^{12} s^{-1}.

However, in photosynthetic pigment–proteins the coupling between neighbouring pigments is of the order of several tens to several hundreds of wavenumbers, due to the close proximity of the pigments (< 1 nm). This is shown in Figure 8.6 by the interaction Hamiltonian of LHCII. In that case the excitation will not be localized, but becomes 'delocalized'. Such a delocalized excitation is called an exciton and the excited state wavefunction ψ_k of the exciton is given by the following expression:

$$\psi_k = \sum_n c_{kn} \varphi_n, \tag{8.4}$$

in which the φ_n represent the locally excited states. The ψ_k are delocalized and the participation of each locally excited state in ψ_k is given by the coefficient c_{nk}. As a result, the whole antenna is generally characterized by a complicated manifold of excited states, including collective electronic excitations (excitons) with a high degree of delocalization, in combination with more localized excitations due to the presence of weakly coupled pigments.

Modulation of the electronic transition energies by slow conformational motion of the protein matrix produces disorder of the site energies within a single complex (thus resulting in more localized exciton wavefunctions), as well as inhomogeneous broadening of the electronic transitions due to ensemble averaging. Evolution of the antenna complex through a number of conformational sub-states can be monitored directly using single-molecule techniques (Bopp *et al.*, 1999; van Oijen *et al.*, 1999; Rutkauskas *et al.*, 2004; Novoderezhkin *et al.*, 2006; Rutkauskas *et al.*, 2006; Kruger *et al.*, 2010). Coupling of excited states to fast nuclear motions (intra- and interpigment vibrations, phonons) results in homogeneous broadening of the electronic transition spectra and their red shift due to reorganization effects, together with a further decrease of the delocalization size due to polaron effects (Meier *et al.*, 1997; Freiberg *et al.*, 2003).

Because of the collective character of the excitation in photosynthetic complexes, the Förster theory cannot give an adequate picture of energy transfer. The generalized Förster theory (Novoderezhkin and Razjivin, 1994, 1996; Sumi, 1999; Scholes and Fleming, 2000; Jang *et al.*, 2004) considers energy transfer between clusters with an arbitrary degree of delocalization, but is restricted to weak inter-cluster interactions. In the standard Redfield theory (Redfield, 1957), all exciton couplings are taken into account explicitly, thus allowing a description of all types of exciton relaxation/migration within strongly coupled antenna complexes, including coupled dynamics of the populations and coherences between the exciton states. In this theory, the dynamics is described in the pure exciton basis, where the relaxation between exciton states is accounted for by including exciton–phonon coupling as an off-diagonal perturbation. The standard Redfield approach can be generalized by including strong coupling of excitations to a few vibrational modes. Relaxation in such a system can be described in the basis of electron-vibrational eigenstates. This approach has allowed us to describe the electron transfer coupled to coherent nuclear motion in the bacterial RC, discussed above (Novoderezhkin *et al.*, 2004), long-lived vibrational coherences in LH1 (Chachisvilis *et al.*, 1994; Novoderezhkin *et al.*, 2000) and coupled exciton-vibrational relaxation in LH1 (Novoderezhkin and van Grondelle, 2002).

In the modified version of the Redfield theory (Zhang *et al.*, 1998), the diagonal (in the exciton basis) part of the electron–phonon coupling is taken into account non-perturbatively, thus giving more realistic lineshapes and relaxation rates due to the inclusion of multiphonon processes (Yang and Fleming, 2002). The modified Redfield theory allows a quantitative treatment of spectra and dynamics in many systems: LH2/LH1 (Novoderezhkin *et al.*, 2006), FMO (Adolphs and Renger, 2006), LHCII (Novoderezhkin *et al.*, 2005c; Novoderezhkin and van Grondelle, 2010; Novoderezhkin *et al.*, 2011b; Renger *et al.*, 2011), PSII-RC (Novoderezhkin *et al.*, 2005c; Raszewski *et al.*, 2005; Novoderezhkin *et al.*, 2007a; Raszewski *et al.*, 2008; Novoderezhkin *et al.*, 2011b), the cryptophyte LH-complex PE545 (Novoderezhkin *et al.*, 2010) and PSII-core (Raszewski and Renger, 2008), including the conformational fluctuations of the single-molecule spectra observed for LH1/LH2 (Rutkauskas *et al.*, 2004; Novoderezhkin *et al.*, 2006; Rutkauskas *et al.*, 2006; Novoderezhkin *et al.*, 2007b), and later in LHCII (Kruger *et al.*, 2010).

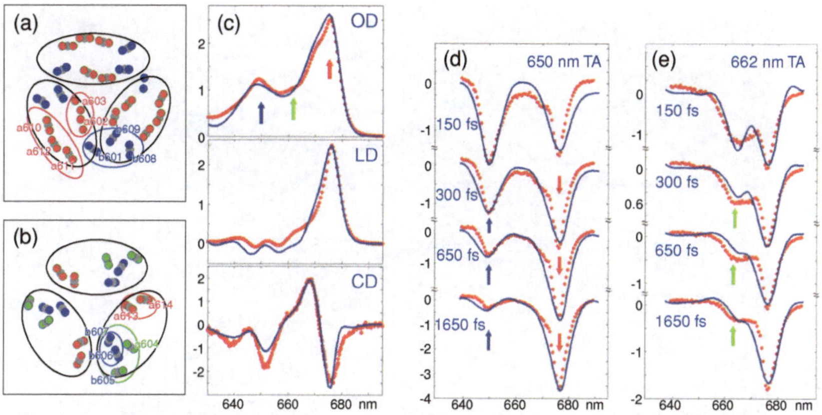

Figure 8.7 Arrangement of Chls within the LHCII trimer at the stromal (a) and lumenal (b) sides. Chls are represented by three atoms: the central magnesium atom and two nitrogen atoms. The connecting line between the two nitrogens defines the direction of the Q_y transition dipole. Red, Chl *a* nitrogen; blue, Chl *b* nitrogen; grey, magnesium; green, nitrogen of Chl *a*604 and Chl *b*605 (according to the structure reported by (Liu *et al.*, 2004). Clusters of Chls *a*, Chls *b*, and a mixed group containing long-lived intermediate sites (Chl *a*604 and Chl *b*605) are encircled by red, blue and green, respectively. c, d, e: Simultaneous fit of optical density or absorption (OD), linear dichroism (LD), circular dichroism (CD) and time-dependent transient absorption (TA) spectra using the modified Redfield approach. Experimental OD/LD spectra (van Amerongen *et al.*, 1994) and CD spectrum (Hemelrijk *et al.*, 1992) have been measured for the LHCII trimer at 77 K (c, red points). Experimental TA spectra are obtained upon 650 nm (d, red points), 662 nm (e, red points) excitation with 120 fs pulses and pump–probe delays of 150, 300, 650 and 1650 fs. Calculated spectra (c), (d), (e), blue lines) are obtained with the disordered exciton model for the whole trimer, where the unperturbed site energies within a monomeric unit have been adjusted in order to obtain the best simultaneous fit of all the data. Reproduced from (Novoderezhkin and van Grondelle, 2010) with permission from the PCCP Owner Societies.

Recently the theory was used to model the 2D-photon echo spectra in FMO (Cho *et al.*, 2005), B800–820 complex (Zigmantas *et al.*, 2006) and LHCII (Schlau-Cohen *et al.*, 2009). Notice that the present version of the modified Redfield theory is restricted to relaxation dynamics of populations and does not include one-exciton coherences. For a further discussion on the role of quantum coherence in photosynthetic light-harvesting and charge separation, we refer the reader to Chapters 6 and 7 by Engel *et al.* and Robentrost *et al.*, respectively, in this book.

An example of quantitative fit of the linear and non-linear spectral responses for LHCII using modified Redfield theory (Novoderezhkin and van Grondelle, 2010) is shown in Figure 8.7. The Chls *a* (red), Chls *b* (blue), and long-lived intermediate sites Chls *a*604, *b*605 (green) contribute to the 675 nm, 650 nm and 662 nm bands (shown by arrows in Figure 8.7c). Selective excitation of the 650 or 662 nm bands allows a visualization of the energy transfer dynamics from Chls *b* or intermediate

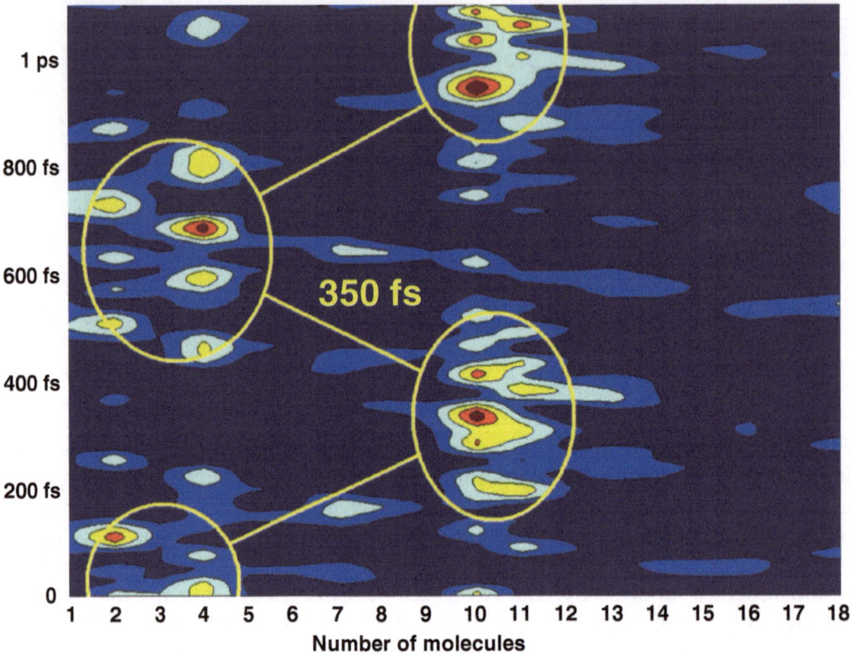

Figure 8.8 Oscillatory dynamics in the bacterial light-harvesting complex LH2, as revealed by analysis of single-molecule exciton spectra and femtosecond spectroscopy (for details see (Novoderezhkin *et al.*, 2006)). Dynamics of populations after impulsive excitation is shown. A colour scale is used to indicate the absolute values of the site population from zero (blue) to the maximal value (red). Coherence between the collective exciton states created by the impulsive excitation manifests itself as reversible oscillatory jumps (half-period of 350 fs) between small groups of molecules (encircled). Within each group, faster oscillations between individual sites (with a half-period of 60 fs) are discernible. Reprinted from (Novoderezhkin *et al.*, 2006), Copyright 2006, with permission from Elsevier.

sites to the Chls *a* with final localization of excitation at the peripheral Chl *a* cluster *a*610–612. The spectral signatures of this dynamics can be quantitatively reproduced by our model, as shown in Figure 8.7d, e. Details of the current model can be found in Novoderezhkin *et al.* (2005c), Novoderezhkin and van Grondelle (2010) and Novoderezhkin *et al.* (2011b).

8.3.2 Quantum coherence and photosynthetic light-harvesting

Dipolar couplings between localized excited states not only delocalize the excited state, also 'coherences' are generated, meaning that products such as $c_i c_j^*$ have a value that may give rise to oscillatory dynamics, depending on the excitation conditions. The existence of such coherences implies that the excitation 'remembers' where it has been. An example of such predicted coherent, oscillatory dynamics in LH2 (for one arbitrarily taken realization of the disorder) is shown in Figure 8.8

(van Grondelle and Novoderezhkin, 2010). It is clear that in this realization the excitation prefers to be localized either on sites 2–4 or on sites 10,11 of the LH2 ring and the excitation 'hops' between these two sites in about 350 fs. Once in position 2–4, the excitation oscillates between pigments 2 and 4 in about 60 fs; a similar dynamics is observed in position 10, 11.

Recently, coherent oscillatory behaviour could be visualized using 2DES. In the 2DES spectra cross-peaks are observed, that directly reflect the coherences between exciton states, while their oscillatory time dependence reveals the coherent dynamics (Engel *et al.*, 2007; Ginsberg *et al.*, 2009; Collini *et al.*, 2010). In an experiment at cryogenic temperatures (Engel *et al.*, 2007) long-lived coherences (660 fs) were observed to occur among the exciton states of the FMO complex. Recently (Collini *et al.*, 2010), in light-harvesting complexes of cryptophytes, long-lived coherences could be observed at room temperature, even between relatively weakly coupled states, suggesting an explicit role of the protein in maintaining these coherences (Chin *et al.*, 2013). Much effort is put into understanding the persistence of these quantum phenomena in such a noisy environment and to relate these 'non-trivial' quantum properties of photosynthetic pigment–proteins to the efficiency of the excitation energy transport process and energy transfer to the RC (Mohseni *et al.*, 2008; Olaya-Castro *et al.*, 2008; Plenio and Huelga, 2008; Caruso *et al.*, 2009; Ishizaki and Fleming, 2009a; Rebentrost *et al.*, 2009a, c; Fassioli and Olaya-Castro, 2010; Sarovar *et al.*, 2010). It is of interest to discover whether these properties of the natural system were designed, meaning optimized by evolution to enhance the efficiency of energy and/or electron transfer, or if they are simply a consequence of the dense packing of pigments in the pigment–protein complexes.

9

Electron transfer in proteins

SPIROS S. SKOURTIS

9.1 Introduction

Protein electron transfer (ET) reactions are central to biological function. They are important components of bioenergetic pathways (photosynthesis and respiration) and they are involved in biological signalling and in the generation and the control of disease (Marcus and Sutin, 1985; Bendall, 1996; Canters and Vijgenboom, 1997; Page *et al.*, 1999; Blankenship, 2002; Gray and Winkler, 2003, 2005). For a fundamental understanding of these biological processes it is necessary to study protein ET mechanisms at the molecular level. Protein ET physics is very rich because it involves charge transport through dynamic and responsive (to the transferring charge) molecular media organized in cellular molecular assemblies. A common feature among protein ET assemblies is that they are designed to move electrons to specific locations along transport pathways that partially suppress backward ET (Figure 9.1). In many cases the structures and dynamics of the protein ET complexes are such that ET takes place with high efficiency (Blankenship, 2002). Needless to say, an understanding of structural and dynamical effects on protein ET processes is very important for the development of new biomimetic electronic and energy-conversion materials with controlled functionalities (Jortner and Ratner, 1997; Balzani *et al.*, 2001; Adams *et al.*, 2003; Blankenship *et al.*, 2011). The field of biological ET (and in particular protein ET) is one of the oldest fields in molecular biophysics (Marcus and Sutin, 1985; Bendall, 1996; Page *et al.*, 1999; Jortner and Bixon, 1999; Kuznetsov and Ulstrup, 1999; May and Kühn, 2011; Balzani *et al.*, 2001; Blankenship, 2002; Gray and Winkler, 2003, 2005; Nitzan, 2006). This chapter covers only some of the basic concepts of biomolecular ET theory and a few examples of ET systems. The quantum nature of biological ET processes has long been recognized, but the field is continuously enriched by experimental

Quantum Effects in Biology, ed. Masoud Mohseni, Yasser Omar, Gregory S. Engel and Martin B. Plenio. Published by Cambridge University Press.

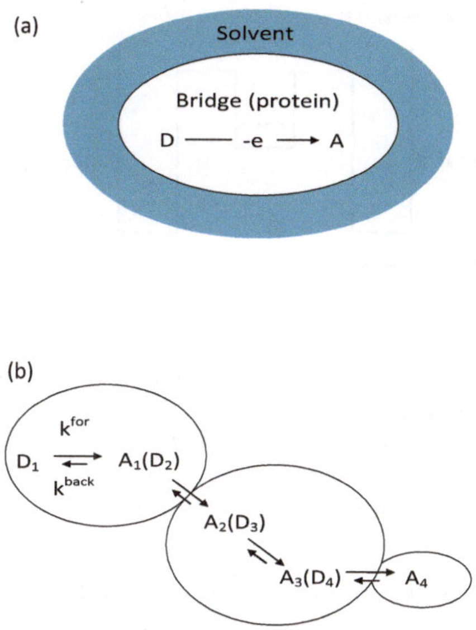

Figure 9.1 (a) A biomolecular ET reaction between an electron-donating moiety (electronic state D) and an electron-accepting moiety (electronic state A) that are connected by a protein bridge. In single-step tunnelling ET reactions, the bridge acts as a tunnelling barrier for the electron (see Figure 9.2). (b) Biological ET chains are often comprised of single-step tunnelling ET reactions between successive D–A pairs. The forward D-to-A ET reaction rate for each step is usually greater than the backward A-to-D ET rate.

discoveries that present new challenges for experiment, theory and computation (see Chapters 5, 10 and 12).

To start with I will briefly describe some general features of biological ET reactions. They often involve chains of electron (hole)-transfer steps in specialized molecular assemblies (Figure 9.1). Each step is the transfer of an electron (hole) from a localized donor (D) electronic state to a localized acceptor (A) electronic state, through an intervening molecular matrix (the bridge). The bridge between D and A is often protein (there are also protein/DNA and pure DNA ET systems) and it may include other cofactor molecules. From a practical point of view the D and A states of a charge-transfer step are initial and final electronic states that have long lifetimes and high populations, such that they can not be observed experimentally. The donor (acceptor) moieties of an ET biomolecular assembly can be metals atoms, aminoacids, other small organic molecules, or DNA bases. This variety of ET molecular assembly structures leads to a richness of transport mechanisms that includes coherent deep tunnelling, coherent resonant tunnelling and thermally activated hopping (Jortner and Bixon, 1999; Kuznetsov and Ulstrup, 1999; May and Kühn, 2011; Balzani *et al.*, 2001; Nitzan, 2006). Coherent deep tunnelling

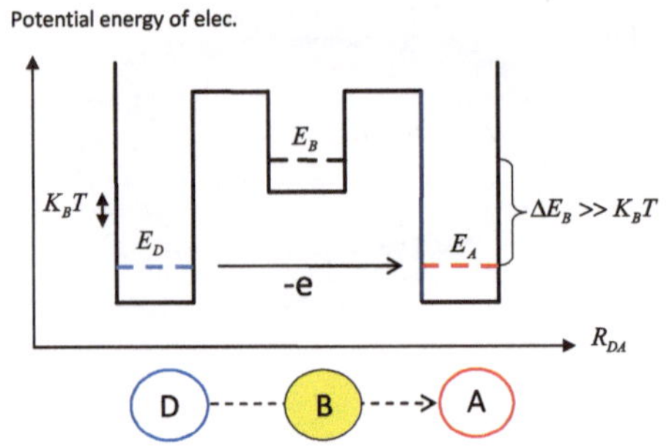

Figure 9.2 Relationship between D-to-bridge (A-to-bridge) energy gap and temperature for the deep tunnelling mechanism of ET. The figure refers to the D-to-A resonance conformation of the ET system (Q_{res} of Figure 9.3). The bridge is a tunnelling barrier for the transferring electron, i.e. $\Delta E_B \gg K_B T$. Through-bridge resonant tunnelling or hopping can be observed if $\Delta E_B \leq K_B T$.

is the transfer mechanism when the bridge medium between D and A presents a tunnelling barrier for the transferring electron, with a height $\Delta E_B \gg K_B T$ (K_B is Boltzmann's constant and T is the temperature, Figure 9.2). Since the electron tunnels through the bridge, it occupies the intermediate bridge electronic states with very low probabilities and for very short times. In this regime, electronic coherence in the bridge is maintained because the bridge medium is not perturbed by the transferring electronic charge and there is no electron–phonon energy exchange in response to the charge. The electronic coupling that induces electron transfer from D to A is described by a D-to-A tunnelling matrix element. According to the simplest (average-tunnelling-barrier) model, the tunnelling matrix element is

$$T_{DA} \propto \exp[(\sqrt{2m_e \Delta E_B}/\hbar) R_{DA}], \tag{9.1}$$

where ΔE_B is the average D/A-to-bridge energy gap and R_{DA} the D-to-A distance (Figure 9.2) (Jortner and Bixon, 1999; Kuznetsov and Ulstrup, 1999; May and Kühn, 2011; Balzani *et al.*, 2001; Nitzan, 2006). In coherent resonant tunnelling mechanisms, ΔE_B is small enough such that thermally induced structural fluctuations can bring one or more bridge electronic states into resonance with the D and/or A states ($\Delta E_B \leq K_B T$). In these mechanisms, the transferring electron occupies the resonant bridge states with high probability, but electronic coherence in the bridge is maintained because the occupation time of bridge states is too short for the bridge to respond. The thermally activated hopping regime also refers to cases where $\Delta E_B \leq K_B T$, but the bridge occupation time of the electron transferring from D is long enough such that the bridge relaxes to the electronic charge, trapping the electron in a bridge electronic state (B). For ET to take place from the B to the A state, the system must be thermally activated to a conformation that brings B into resonance with A. For biological ET the D–A distance for through-protein single-step tunnelling can be up to ∼10 Å. Hopping mechanisms comprised of many tunnelling steps support much greater distances (Gray and Winkler, 2009; Winkler and Gray, 2010) that in some cases reach the micron length scale (El-Naggar *et al.*, 2010).

What is it in the molecular architecture that determines transport mechanism? Usually deep tunnelling steps involve metal D and A moieties whose electronic energies are separated from the bridge electronic state energies (Page *et al.*, 1999; Gray and Winkler, 2003, 2005). In the context of a multi-step ET chain comprised of through-protein tunnelling steps (Figure 9.1b), the donor and acceptor moieties define the topology of the ET chain because they act as relay stations for the transferring electron (the electron cannot be localized in the bridge which is a tunnelling barrier). Through-protein resonant tunnelling and hopping mechanisms are encountered when the D and A states are aminoacids or organic molecules whose electronic states are similar to the protein electronic states (Jortner and Bixon, 1999; May and Kühn, 2011; Nitzan, 2006; Gray and Winkler, 2009; Winkler and Gray, 2010). This chapter focuses on ET reactions mediated by through-protein deep tunnelling, as these reactions are the elementary steps of most ET chains and they are very well understood experimentally and theoretically.

9.2 The rate for a single-step electron transfer reaction mediated by elastic through-bridge tunnelling

The measurable quantity in a biological ET reaction is usually the ET rate from D to A which is often probed by time-resolved optical spectroscopy (Balzani *et al.*, 2001). Electron transfer reactions in the deep tunnelling regime involve weak D–A electronic couplings (tunnelling matrix elements T_{DA}) and are therefore rate-limited by these electronic couplings. The simplest description of the D-to-A ET rate for such reactions is the 'non-adiabatic' Marcus-rate expression (Marcus and Sutin, 1985; Jortner and Bixon, 1999; Kuznetsov and Ulstrup, 1999; May and Kühn, 2011; Nitzan, 2006) which captures the essential physics. The ET rate is a product of a classical Boltzmann activation factor ρ_{FC}^{clas} for the conformational degrees of freedom of the ET system, and a tunnelling probability $(2\pi/\hbar)T_{DA}^2$ for the transferring electron,

$$k_{DA} = \frac{2\pi}{\hbar} T_{DA}^2 \, \rho_{FC}^{clas}. \tag{9.2}$$

What is the physical picture behind Equation (9.2)? Electron tranfser from D to A can take place only if the total energy of the system (ET molecule and

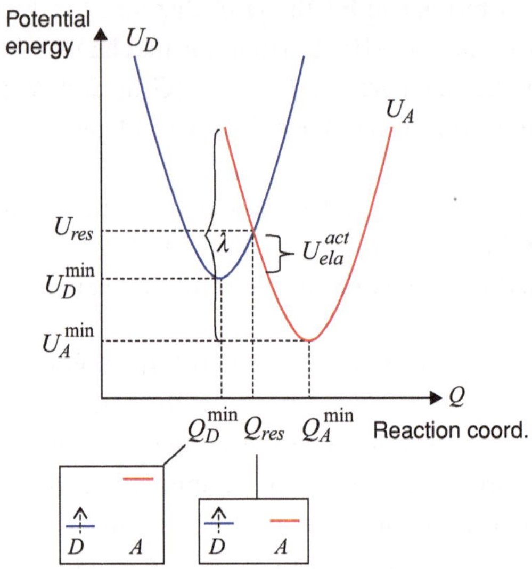

Figure 9.3 Potential energy (Born–Oppenheimer) surfaces of an ET system when the electron is at the initial (D) and the final (A) electronic states (U_D and U_A respectively). The surfaces are plotted as a function of a collective system coordinate Q that modulates the relative electronic energies of D and A (see boxes below plot). It is assumed that the collective coordinate is classical. ET can take place only when the potential energy surfaces cross at Q_{res}, i.e. when D and A electronic energies are resonant at $Q = Q_{res}$. Therefore, the D-to-A ET rate contains a Boltzmann activation factor to reach the resonance conformation (Equations (9.2), (9.3)). This figure does not show bridge-electonic-state surfaces. At the resonance conformation the transferring electron feels the bridge as a tunnelling barrier (see Figure 9.2) and the bridge Born–Oppenheimer surfaces have higher energies compared with U_D and U_A.

solvent) is conserved. If we assume that the conformational degrees of freedom of the ET system are classical, then energy-conserving ET can happen when the D and A electronic state energies E_D and E_A are resonant ($E_D = E_A$). This is because the timescale of the D-to-A electronic transition is much faster than the classical motions of the molecular system, such that during the D-to-A transition these motions are frozen and cannot exchange energy with the transferring electron. In general, there will be a set of system motions (denoted the reaction coordinate Q) that modulate the relative energies of the D and A electronic states, i.e. $E_D(t) = E_D(Q(t))$ and $E_A = E_A(Q(t))$. Thermal fluctuations of the reaction coordinate enable ET because they bring the system to a D-to-A resonance conformation Q_{res} for which $E_D(Q_{res}) = E_A(Q_{res})$. Figure 9.3 shows the potential energy (diabatic Born–Oppenheimer) surfaces of the ET system as a function of Q, when the electron is either in the D electronic state (U_D) or in the A electronic state (U_A). These surfaces will generally have different minimum-energy values and minimum-energy reaction-coordinate values, i.e. $U_D^{min} \neq U_A^{min}$, where $U_D^{min} = U_D(Q_D^{min})$, $U_A^{min} = U_A(Q_A^{min})$ and $Q_D^{min} \neq Q_A^{min}$. When the electron is in the D state, the reaction coordinate's dynamics is determined by the U_D energy surface and the system fluctuates around the reaction-coordinate conformation that minimizes U_D (i.e. around Q_D^{min}). Since $Q_D^{min} \neq Q_{res}$, the system has to wait for a thermal fluctuation $Q_D^{min} \rightarrow Q_{res}$ to bring the D and A energies to resonance. The classical (high-temperature) Boltzmann probability for this activation step is

$$\rho_{FC}^{clas} = \frac{1}{\sqrt{2\pi\sigma_{\Delta U}^2}} exp\left[-U_{ela}^{act}/K_B T\right] \tag{9.3}$$

(often denoted the classical Franck–Condon factor). In the equation above,

$$U_{ela}^{act} = U_{res} - U_D^{min} = \frac{\left(U_A^{min} - U_D^{min} + \lambda\right)^2}{4\lambda} \tag{9.4}$$

is the activation energy to reach the resonance conformation Q_{res} ($U_{res} = U_D(Q_{res}) = U_A(Q_{res})$). The subscript 'ela' stands for elastic tunnelling, which means that the tunnelling electron does not lose energy during tunnelling. The factor λ denotes the reorganization energy, given by $\lambda = U_D(Q_A^{min}) - U_D(Q_D^{min})$ or, equivalently, by $\lambda = U_A(Q_D^{min}) - U_A(Q_A^{min})$. This reorganization energy is often called the classical reorganization energy because it involves reorganization of low-frequency (classical) motions that are perturbed by ET and that contain delocalized (protein) and environmental (solvent) degrees of freedom. The name is used to distinguish it from inner-sphere reorganization energies which involve the reorganization of high frequency (quantum) vibrational modes perturbed by ET that often contain local (D/A and protein) degrees of freedom. $\sigma_{\Delta U}^2 = 2\lambda K_B T$ is the variance in the energy gap between U_D and U_A potential energy surfaces, given by $\sigma_{\Delta U}^2 =< \Delta U_{AD}^2 > - < \Delta U_{AD} >^2$, where $\Delta U_{AD} = U_A(Q) - U_D(Q)$ ($< .. >$ denotes the thermal average with respect to motion in the U_D surface).

Once the resonance conformation is reached, ET takes place by tunnelling with a probability $(2\pi/\hbar)T_{DA}^2$. The dependence of this probability on the square of the electronic coupling is valid for weak electronic coupling in the non-adiabatic limit. This dependence can be understood in terms of lowest-order perturbation theory with respect to the coupling, since the golden-rule expression for a transition rate is proportional to the square of the matrix element that induces the transition. It can also be derived from the Landau–Zener expression for a transition between two states with time-dependent energies that cross. This derivation is useful for understanding the terminology 'non-adiabatic' (Marcus and Sutin, 1985; Jortner and Bixon, 1999; Kuznetsov and Ulstrup, 1999; May and Kühn, 2011; Balzani *et al.*, 2001; Nitzan, 2006).

Consider the D–A electronic Hamiltonian,

$$\hat{H}(Q) = |D\rangle\langle D|E_D(Q) + |A\rangle\langle A|E_A(Q) + \{|D\rangle\langle A|T_{DA} + h.c.\}, \tag{9.5}$$

whose elements depend on the reaction coordinate Q. We assume that initially the electronic state is D and that the reaction coordinate is time-dependent, $Q = Q(t)$ (its dynamics is determined by U_D). At some point in time t_{res}, the D and A state energies cross, $Q(t_{res}) = Q_{res}$ such that $E_D(Q_{res}) = E_A(Q_{res}) = E_{res}$. The probability for a transition from D to A during this crossing event can be computed using the Landau–Zener approach, which assumes a linear dependence of the D–A energy gap on time around t_{res}, i.e. $E_{D(A)}(Q(t)) = E_{res} + \{dQ/dt\}_{t_{res}} \times \{dE_{D(A)}/dQ\}_{Q_{res}} \times (t - t_{res})$. The Landau–Zener probability for a transition from D to A is given by

$$P_{DA} = 1 - exp[(2\pi)^2\gamma], \tag{9.6}$$

where $\gamma = \tau_{LZ}/\tau_{Rabi}$ is the Landau–Zener parameter, written as a ratio of two times, the Rabi time and the Landau–Zener time. The Rabi time $\tau_{Rabi} = h/|T_{DA}|$ is taken from time-independent quantum mechanics. It is a measure of the time it takes in a static D–A system at resonance, i.e. when $|E_D - E_A| \leq |T_{DA}|$, for $P_{DA}(t)$ to become unity given that $P_{DA}(0) = 0$ ($P_D(0) = 1$). The Landau–Zener time,

$$\tau_{LZ} = \frac{|T_{DA}|}{|\{dQ/dt\}_{t_{res}}\{dE_D/dQ - dE_A/dQ\}_{Q_{res}}|} \tag{9.7}$$

is a measure of the time the D and A energies spend in the resonance region ($|E_D - E_A| \leq |T_{DA}|$). The non-adiabatic limit means that the D and A energies cross fast, such that they do not remain in the resonance region long enough for a complete D to A transition, i.e. $\gamma = \tau_{LZ}/\tau_{Rabi} \ll 1$. In this case, $P_{DA} \simeq (2\pi)^2\gamma \propto T_{DA}^2$. It is important to note that Equations (9.2) and (9.3) assume that the reaction coordinate is thermalized on the U_D surface (hence the classical Boltzmann distribution in Equation (9.3)). This is a good approximation for weak-coupling non-adiabatic ET, because the ET rate is slow with respect to thermalization times of molecular motions. Adiabatic ET reactions ($\gamma \geq 1$) are also encountered in biology for strong-coupling short-distance ET and they often involve fast ET rates (see Conclusions, Section 9.7).

A central prediction of Equation (9.2) is that the rate is maximum for $U_D^{min} - U_A^{min} = \lambda$, because it becomes activationless ($U_{ela}^{act} = 0$), and it decreases otherwise since it becomes activated. In particular, if $U_D^{min} > U_A^{min}$, the rate is activated for $U_D^{min} - U_A^{min} < \lambda$ (normal region) and for $U_D^{min} - U_A^{min} > \lambda$ (inverted region). The inverted region has been observed for small-molecule ET, where it is possible to modify $U_D^{min} - U_A^{min}$ by changing artificially the donor (acceptor) chemical groups. Activationless ET reactions are fast and are often used by biological systems, whenever it is necessary to rapidly transfer electrons (e.g. in the primary charge separation of photoysnthesis (Page *et al.*, 1999; Blankenship, 2002)).

9.3 Dependence of tunnelling on protein structure: tunnelling pathways and their interferences

An important effort in the field of bridge-mediated ET, especially in the coherent deep-tunnelling regime, is to understand how molecular conformation and molecular electronic structure influence the D–A electronic coupling. The dependence of the D–A tunnelling matrix element (T_{DA}) on molecular structure has long been established, both theoretically and experimentally for a large number of protein and small-molecule systems ((Winkler *et al.*, 1982; Beratan *et al.*, 1991) and (Skourtis and Beratan, 1999; Balzani *et al.*, 2001; Gray and Winkler, 2003, 2005) for reviews). Sophisticated techniques exist for the computation of T_{DA} and for its analysis in terms of bridge structure and of D-bridge and A-bridge connectivity ((Skourtis and Beratan, 1999; Newton, 2001; Stuchebrukhov, 2001; Rösch and Voityuk, 2004; Hsu, 2009) for reviews). All such methods start from the D-bridge-A molecular structure which is experimentally derived (from X-ray crystallography or NMR techniques). The electronic Hamiltonian of the system is built at the semi-empirical (e.g. Huckel, INDO/S) or ab initio level (HF, DFT), depending on the size of the system. T_{DA} is computed from the electronic Hamiltonian by partitioning the system into D, A and bridge subsystems. This partitioning is usually based on chemical and physical arguments. Once such subsystems are defined, one can use either effective Hamiltonian and Green's function techniques to compute bridge-mediated couplings between specific D and A electronic states. Alternatively one can start from the entire D–B–A system eigenstates and employ diabatization methods that derive D and A electronic states, localized on the D and A subsystems using some localization criterion (e.g. maximal charge localizations on the D and A subsystems). The Hamiltonian matrix element between such diabatic states is taken to be equal to T_{DA}. There also exist different methods for the structural analysis of T_{DA}, such as tunnelling pathways, tunnelling-current and pruning analysis ((Skourtis and Beratan, 1999; Newton, 2001; Stuchebrukhov, 2001; Rösch and Voityuk, 2004) for reviews). The structural analysis of T_{DA} ultimately leads to a description of how the bridge bonding topology influences the tunnelling paths of the electron from D to A. In general, increasing the number of bridge through-space jumps compared to through-bond jumps that the electron has to tunnel through reduces T_{DA}, because it introduces higher (through-space) tunnelling barriers in the electron's path (Beratan *et al.*, 1991).

To give an idea of how tunnelling pathways can be extracted from T_{DA}, I will describe the basics of computation and structural analysis methods that use

Green's function theory (Skourtis and Beratan, 1999). Consider the full electronic Hamiltonian of the system at $Q = Q_{res}$, which is usually expressed as a tight-binding effective one-electron Hamiltonian:

$$\hat{H}^{el} = \hat{H}^{DA} + \hat{H}^{br} + \hat{V}, \tag{9.8}$$

comprised of D–A, bridge and interaction parts. The D–A Hamiltonian at the D–A resonance conformation (Q_{res} in Figure 9.3) is

$$\hat{H}^{DA} = |D\rangle\langle D|E_{res} + |A\rangle\langle A|E_{res}, \tag{9.9}$$

where the direct coupling between $|D\rangle$ and $|A\rangle$ electronic states is neglected due to the large D–A distance. The bridge Hamiltonian is

$$\hat{H}^{br} = \hat{H}^{br}_{dia} + \hat{V}^{br}, \tag{9.10}$$

where

$$\hat{H}^{br}_{dia} = \sum_i |B_i\rangle\langle B_i|E_{Bi} \tag{9.11}$$

is the diagonal part and

$$\hat{V}^{br} = \sum_i \sum_{j>i} \{|B_i\rangle\langle B_j|V_{Bi,Bj} + h.c.\} \tag{9.12}$$

is the off-diagonal part. The interaction Hamiltonian between D(A) and bridge electronic states is

$$\hat{V} = \sum_i \{|D\rangle\langle B_i|V_{D,Bi} + |A\rangle\langle B_i|V_{A,Bi} + h.c.\}. \tag{9.13}$$

The $|B_i\rangle$ are often chosen to be localized bridge electronic states (e.g. two-centre bonding and antibonding orbitals), and the V_{ij} are electronic couplings between states (either intra-bridge or D-bridge and A-bridge).

The bridge-mediated D-to-A tunnelling matrix element (i.e. T_{DA} in Equation (9.5)) is given by

$$T_{DA} = \langle D|\hat{V}\hat{G}^{br}(E_{tun})\hat{V}|A\rangle, \tag{9.14}$$

where $E_{tun} = E_{res}$ and

$$\hat{G}^{br}(E_{tun}) = \left(E_{tun}\hat{I} - \hat{H}^{br}\right)^{-1} = \sum_k \frac{|\Psi^{br}_k\rangle\langle\Psi^{br}_k|}{E_{tun} - E^{br}_k}. \tag{9.15}$$

The $|\Psi^{br}_k\rangle$ and E^{br}_k denote the eigenstates and eigenenergies of $\hat{H}^{br}$. The above expression for the coupling (Equation (9.14)) is valid in the tunnelling limit, i.e. when $|\langle D|\hat{V}|\Psi^{br}_k\rangle|/|E_{tun} - E^{br}_k| < 1$ and $|\langle A|\hat{V}|\Psi^{br}_k\rangle|/|E_{tun} - E^{br}_k| < 1$ for

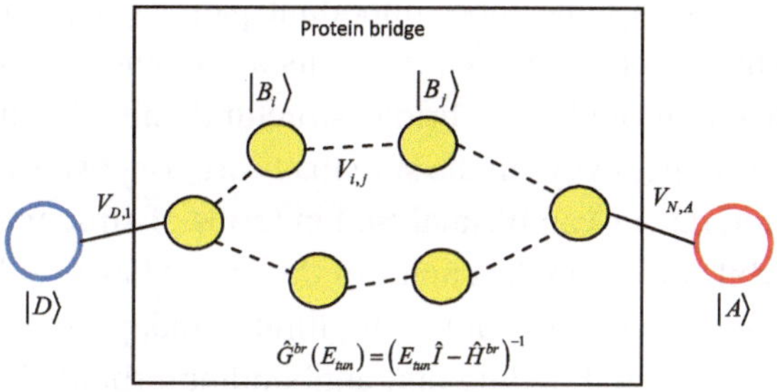

$$T_{DA} = \langle D|\hat{V}\hat{G}^{br}(E_{tun})\hat{V}|A\rangle$$

Figure 9.4 The bridge (protein)-mediated tunnelling matrix element between D and A electronic states can be expressed as a sum of interfering tunnelling pathways through localized electronic states of the bridge. This can be done by use of Green's function theory (Equations (9.14), (9.17), (9.18) and (9.19)). The method relates the tunnelling matrix element to protein-bridge secondary and tertiary structure.

all $|\Psi_k^{br}\rangle$ and E_k^{br} (Skourtis and Beratan, 1999). Having computed T_{DA} via Equations (9.14) and (9.15), analysis of the coupling value in terms of bridge structure can be done by expanding $\hat{G}^{br}$ in Equation (9.15) to different orders of the $V_{Bi,Bj}$ intra-bridge matrix elements. We define the diagonal bridge Green's function,

$$\hat{g}^{br}(E_{tun}) = \left(E_{tun}\hat{I} - \hat{H}_{dia}^{br}\right)^{-1} \tag{9.16}$$

and perform a Dyson expansion in $\hat{V}^{br}$, i.e. $\hat{G}^{br} = \sum_{N=1}^{\infty} \hat{g}^{br}(\hat{V}^{br}\hat{g}^{br})^{N-1}$. Substituting the Dyson expansion into Equation (9.14) gives

$$T_{DA} = \sum_{N=N_{min}}^{\infty} T_{DA}(N), \tag{9.17}$$

where

$$T_{DA}(N) = \sum_{Path_N} Path_N\left(D, B_{i_1}, B_{i_2}, \ldots, B_{i_N}, A\right) \tag{9.18}$$

and

$$Path_N\left(D, B_{i_1}, B_{i_2}, \ldots, B_{i_N}, A\right) = V_{D,B_{i_1}} \frac{1}{\Delta E_{i_1}} V_{B_{i_1},B_{i_2}} \cdots \frac{1}{\Delta E_{i_N}} V_{B_{i_N},A} \tag{9.19}$$

where $(\Delta E_{i_n} = E_{tun} - E_{B_{i_n}})$. Equation (9.19) defines a through-bridge tunnelling pathway from to D to A that visits N-bridge states (Figure 9.4). The tunnelling matrix element is formally a sum of an infinite number of such tunnelling pathways, (Equation (9.17)), but often a relatively small subset of pathways give most of the T_{DA} value. Each pathway in this subset is not necessarily of the same order in $\hat{V}^{br}$ as the other members of the subset, because the V_{B_i,B_j} have variable values.

Identifying the subset of the most important pathways enables the structural analysis of T_{DA}, since each pathway represents a sequence of tunnelling steps of the transferring charge through local bridge structural units (bonds and antibonds). Further, since different pathways in the sum may carry opposite signs, constructive and destructive inteferences can be analysed in terms of structure. In this methodology there is an arbitrariness in the choice of the bridge basis set $|B_i\rangle$, although the validity of Equation (9.17) in the tunnelling limit is independent of basis set. The choice of two-centre bridge bonding and antibonding orbitals for the $|B_i\rangle$ is useful, because it allows one to identify critical through-space jumps (low-magnitude V_{B_i,B_j}) that weaken the overall coupling strength and, thereby, to relate T_{DA} to protein secondary and tertiary structure. Another choice of basis set may be the occupied and unoccupied eigenstates of individual amino acids or the eigenstates of the entire bridge. The important point is that different parts of the bridge structure provide alternative tunnelling paths for the transferring electron, which interfere constructively and/or destructively. Because of the variability of protein structures, there exist few pathway and multiple pathway protein ET systems (Skourtis *et al.*, 2006; Beratan *et al.*, 2009; Skourtis *et al.*, 2010). The former are more sensitive to bridge structural changes (e.g. local mutations that introduce new through-space jumps into the few critical pathways). The success of structural analysis methods for T_{DA} is reflected by their ability to predict protein-bridge mutations that will reduce or enhance T_{DA} and thus the ET rate.

9.4 Tunnelling matrix element fluctuations in deep-tunnelling ET reactions

The Marcus rate expression (Equation (9.2)) takes into account the energy gap fluctuations that are caused by thermal molecular motions (and which provide activation events to D–A resonance conformations). For an ensemble of ET systems in the deep-tunnelling ET regime, Equation (9.3) gives the percentage of ensemble members for which D and A electronic-state energies cross (the D–A resonance sub-ensemble). An assumption behind Equation (9.2) is that the D–A tunnelling matrix element T_{DA} is the same for all members of the resonance sub-ensemble. However, biomolecules are floppy, with a large range of molecular-motion timescales and magnitudes. Therefore, each member of the resonance sub-ensemble generally sees a different bridge structure compared with other members and has a different T_{DA} value. In this situation the ET rate expression should be modified to account for T_{DA} fluctuation effects. In the simplest theory of coupling-modulated rates,

$$k_{DA} = \frac{2\pi}{\hbar}\langle T_{DA}^2\rangle \rho_{FC}^{clas}, \tag{9.20}$$

where $\langle \ldots \rangle$ denotes ensemble average ((Daizadeh *et al.*, 1997; Troisi *et al.*, 2003, 2004; Nishioka *et al.*, 2005; Skourtis *et al.*, 2005) and (Skourtis *et al.*, 2006; Beratan *et al.*, 2009; Skourtis *et al.*, 2010) for reviews). The above expression for the rate is valid in the slow coupling-fluctuation regime. We define the average timescale of significant coupling fluctuations (τ_{coupl}) as the shortest decay time of the T_{DA} correlation function, $< T_{DA}(t)T_{DA}(0) >$. We define the average amount of time that D and A remain in resonance as $\tau_{FC} = \hbar/\sigma_{\Delta U} \simeq \hbar/\sqrt{2\lambda K_B T}$ (the Franck–Condon time) (Troisi *et al.*, 2003; Skourtis *et al.*, 2005) ((Skourtis *et al.*, 2006, 2010) for reviews). In the slow-coupling fluctuation regime these timescales satisfy the relation $\tau_{coupl} \gg \tau_{FC}$, which means that for each D–A energy crossing event, the D and A states see a static T_{DA} value (not necessarily the same for all sub-ensemble members). Usually τ_{coupl} is determined by valence angle motions, because these motions destroy atomic orbital overlaps and modulate atomic orbital-to-orbital couplings (Skourtis *et al.*, 2005). The periods of valence angle motions are of the order of tens of fs, so that $\tau_{coupl} \geq 10$ fs. For solution-phase ET, $\tau_{FC} = \hbar/\sqrt{2\lambda K_B T} \simeq$ fs, because $\lambda \simeq 1$ eV (Skourtis *et al.*, 2005). Therefore, the slow-coupling fluctuation regime is quite common in molecular ET (Skourtis *et al.*, 2006, 2010).

In Equation (9.20), $\langle T_{DA}^2 \rangle = \langle T_{DA} \rangle^2 + \sigma_{DA}^2$. The small and large coupling-fluctuation regimes correspond to $\sigma_{DA} < \langle T_{DA} \rangle$ and $\sigma_{DA} > \langle T_{DA} \rangle$, respectively (Balabin and Onuchic, 2000; Troisi *et al.*, 2004; Skourtis *et al.*, 2005, 2006; Balabin *et al.*, 2008; Skourtis *et al.*, 2010). When $\sigma_{DA} > \langle T_{DA} \rangle$, the most populated low-energy structures of the ET-system ensemble do not provide the largest T_{DA} values. For significant ET to take place, the system has to access higher energy molecular conformations that enhance T_{DA} (as well as conformations that bring D–A resonance). Therefore, the coupling fluctuations σ_{DA} introduce an additional (to ρ_{FC}^{clas}) temperature dependence to the rate that modifies the rate's activation energy U_{ela}^{act} ((Skourtis *et al.*, 2006, 2010) for reviews). The occurrence of large coupling-fluctuations depends on the D–A distance R_{DA} and on the bridge medium (Balabin *et al.*, 2008). The large fluctuation regime $\sigma_{DA} > \langle T_{DA} \rangle$ sets in, on average, for $R_{DA} > R_c$, where $R_c \simeq$ 3–4 Å for water-mediated tunnelling and $R_c \simeq$ 6–7 Å for protein-mediated tunnelling with metal donors and acceptors (Balabin *et al.*, 2008). R_c is a statistically derived medium-dependent quantity which measures the bridge length beyond which T_{DA} fluctuations are likely to become important due to bridge motion. For example, for water-mediated tunnelling, $R_c \simeq$ 3–4 Å reflects the size of the non-covalently bound bridge units (i.e. the size of the water molecule). As discussed in the previous section, the structural analysis of T_{DA} is important for relating biomolecular structure to the ET rate. However, in the large coupling-fluctuation regime, applying structural analysis methods to the minimum energy (or crystallographic) ET-molecule structures cannot reveal the important tunnelling pathways for ET, because the important pathways are provided by higher-energy

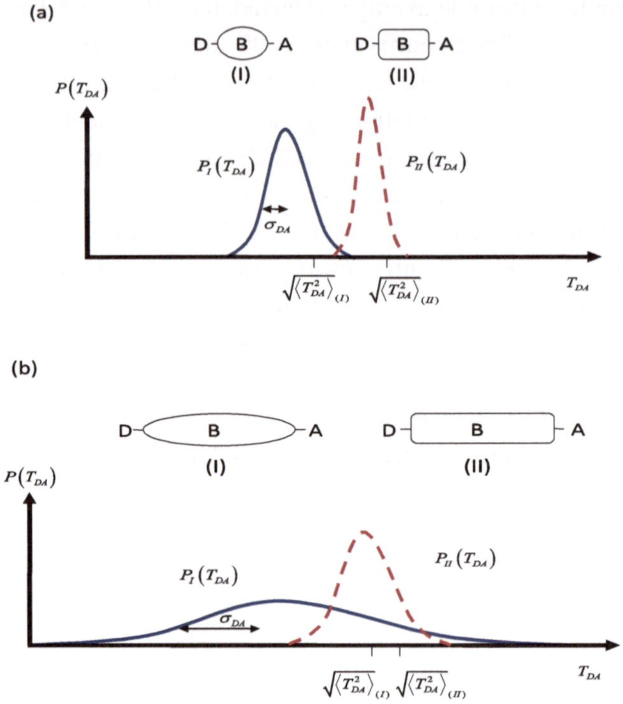

Figure 9.5 (a) Probability densities of T_{DA} for two distinct D-protein-A systems (I and II) with the same average D-to-A distances (and D/A moieties), the same activation energies, but different protein-bridge structures. The densities are centred at distinct T_{DA} values due to bridge structural differences between the two systems. The widths σ_{DA} of the probability densities are measures of the strengths of T_{DA} fluctuations for each of the systems. The small T_{DA} fluctuation regime shown in (a) means that the widths of $P_I(T_{DA})$ and $P_{II}(T_{DA})$ are not large enough to make $T_{DA}^{rms}(I) \simeq T_{DA}^{rms}(II)$. Therefore, the differences in the T_{DA} values of the two systems cause differences in the D-to-A rates through T_{DA}^{rms} (Equation (9.20)). (b) The case for which the σ_{DA}s become large (due to an increase in D–A distance, the floppiness of the bridge, or an increase in temperature). In this case, $T_{DA}^{rms}(I) \simeq T_{DA}^{rms}(II)$ and bridge structural fluctuations that cause fluctuations in T_{DA} wash out the differences of T_{DA}^{rms} and thus in the ET rates.

molecular conformations. Therefore, it is necessary to find these conformations (via molecular dynamics simulations, if possible), and then to do T_{DA} structural analysis on them, or to do a structural analysis of the statistical quantity $\langle T_{DA}^2\rangle$ that enters the rate. Methods that do a structural analysis of $\langle T_{DA}^2\rangle$ have appeared in the literature (Prytkova *et al.*, 2007; Nishioka and Kakitani, 2008).

In the large coupling-fluctuation regime ($\sigma_{DA} > \langle T_{DA}\rangle$), an important question is whether the coupling fluctuations wash out structural differences in T_{DA} among ET species (Balabin *et al.*, 2008; Skourtis *et al.*, 2010). A way to visualize this effect is to consider ensemble probability densities of T_{DA} (Figure 9.5)

((Balabin *et al.*, 2008) supporting information). Suppose two different D-bridge-A systems (ET species (I) and (II)) have identical average R_{DA} and U_{ela}^{act} values, such that differences in their ET rates may only be due to their bridge-mediated electronic couplings T_{DA}. If we plot the ensemble probability densities of T_{DA} for the two species, denoted $P_I(T_{DA})$ and $P_{II}(T_{DA})$, we will find that the densities have finite widths σ_{DA} (due to coupling fluctuations) and different averages $\langle T_{DA}\rangle$ (due to structural differences). Figure 9.5a shows the small coupling-fluctuation regime ($\sigma_{DA} < \langle T_{DA}\rangle$) where the probability-density widths σ_{DA} are small enough such that the rms coupling values $T_{DA}^{rms} = \sqrt{\langle T_{DA}^2\rangle}$ of the two species are distinguishable. In this case, differences in the electronic couplings will lead to distinct ET rates (Equation (9.20)). Figure 9.5b shows a case with $\sigma_{DA} > \langle T_{DA}\rangle$ such that the probability densities overlap substantially and $T_{DA}^{rms}(I) \simeq \langle T_{DA}^{rms}(II)$. In this case coupling fluctuations wash out the structural differences in the couplings and in the corresponding rates (an average-tunnelling-medium picture would then describe the couplings for both I and II, Equation (9.1)). This limit has not yet been encountered for protein ET in the deep-tunnelling regime (i.e. metal D and A (Balabin *et al.*, 2008)). To explore coupling fluctuation effects on rates one can try to enhance bridge motions (σ_{DA}) by increasing the temperature of the system. Another, more direct method is the IR-induced modulation of σ_{DA}, where specific bridge vibrations in regions critical to the coupling pathways are vibrationally excited by an IR pulse (Skourtis *et al.*, 2004; Skourtis and Beratan, 2007; Xiao *et al.*, 2009). This method has been realized experimentally for small-molecule ET systems, where it was shown that IR excitation of specific bridge vibrations changes the ET rate (Lin *et al.*, 2009).

9.5 Vibrational quantum effects and inelastic tunnelling

In deriving Equation (9.2) it was assumed that the reaction coordinate Q is classical. However, there may exist quantum degrees of freedom which are perturbed by ET, such as a high frequency vibrational mode with $\hbar\omega > K_B T$, whose equilibrium position changes upon ET from D to A. In this situation the reaction coordinate has to be extended to include the quantum degree of freedom (Jortner and Bixon, 1999; Kuznetsov and Ulstrup, 1999; May and Kühn, 2011; Nitzan, 2006). For simplicity, let us assume that there is one high frequency vibrational mode and denote by y the displacement of the mode from its equilibrium position when the electron is in $|D\rangle$. The mode Hamiltonian is

$$\hat{H}^{vi} = \frac{\hat{p}_y^2}{2m} + \frac{m\omega^2\hat{y}^2}{2} = \sum_{n_D=0}^{\infty} \epsilon_{n_D} |n_D(y)\rangle\langle n_D(y)|, \tag{9.21}$$

where $|n_D(y)\rangle$ denotes the vibrational (mode) eigenstates and $\epsilon_{n_D} = \hbar\omega(n_D + 1/2)$ the vibrational eigenenergies. When the electron transfers from D to A, the mode feels a force F_A. The equilibrium position of the mode thus shifts from $y_D^{eq} = 0$ to $y_A^{eq} = F_A/m\omega^2$. The Hamiltonian for the mode–electron interaction is given by

$$\hat{H}^{el-vi} = -F_A\hat{y}|A\rangle\langle A|. \quad (9.22)$$

With the electron in $|A\rangle$, the vibrational Hamiltonian is $\hat{H}^{vi} + \langle A|\hat{H}^{el-vi}|A\rangle$ and its eigenstates are $|n_A(y)\rangle$. They are related to $|n_D(y)\rangle$ by $|n_A(y)\rangle = |n_D(y - y_A^{eq})\rangle$ and their eigenenergies are $\epsilon_{n_A} = \epsilon_{n_D} - F_A^2/(2m\omega^2)$. If we assume that the mode is thermally equilibrated prior to ET, then the ET rate from D to A can be written as a thermally weighted sum of independent rates for vibronic (electronic-vibrational) transitions from initial vibronic states $|D; n_D\rangle = |D\rangle|n_D\rangle$ to final vibronic states $|A; n_A\rangle = |A\rangle|n_A\rangle$,

$$k_{DA} = \sum_{n_D} P_{n_D} \sum_{n_A} k(D, n_D \to A, n_A). \quad (9.23)$$

P_{n_D} is the thermal equilibrium probability of state $|n_D\rangle$ and $k(D, n_D \to A, n_A)$ is the rate for the vibronic transition $|D; n_D\rangle \to |A; n_A\rangle$,

$$k(D, n_D \to A, n_A) = \frac{2\pi}{\hbar} T_{DA}^2 \, (FC)_{n_D,n_A}^{qua} \rho_{FC}^{clas}(D, n_D \to A, n_A), \quad (9.24)$$

where $(FC)_{n_D,n_A}^{qua} = |\langle n_D|n_A\rangle|^2$ is a quantum Franck–Condon factor. It is the square of the overlap between the undisplaced and displaced mode wavefunctions, i.e. $|\langle n_D(y)|n_D(y - y_A^{eq})\rangle|^2$. Franck–Condon factors depend on the mode frequency and the (inner-sphere) reorganization energy, $\lambda^{in} = m\omega^2(y_A^{eq} - y_D^{eq})^2/2 = F_A^2/(2m\omega^2)$. For example, $(FC)_{n_D=0,n_A}^{qua} = (exp(-\lambda^{in}/\hbar\omega))(\lambda^{in}/\hbar\omega)^{n_A}/n_A!$. The term $\rho_{FC}^{clas}(D, n_D \to A, n_A)$ in Equation (9.24) is the Boltzmann activation factor for the classical reaction coordinate, given by Equation (9.3), with a modified activation energy,

$$U^{act}(D, n_D \to A, n_A) = \left((U_A^{min} + \epsilon_{n_A}) - (U_D^{min} + \epsilon_{n_D}) + \lambda\right)^2 / 4\lambda K_B T, \quad (9.25)$$

instead of U_{ela}^{act} (Equation (9.4)). For all channels with $\epsilon_{n_D} = \epsilon_{n_A}$ the activation energy is the same as for the classical reaction coordinate, i.e. $U^{act}(D, n_D \to A, n_A) = U_{ela}^{act}$. These channels describe elastic tunnelling with respect to this mode because the mode does not lose or gain energy due to ET. All channels for which $\epsilon_{n_D} \neq \epsilon_{n_A}$ describe inelastic tunnelling, because the perturbation of the mode upon arrival of the electron at A causes the exchange of energy (in multiples of $\hbar\omega$) between the mode and the electron. Inelastic channels have modified activation energies and can be activationless even if the elastic channels are activated. For example, suppose that $U_D^{min} - U_A^{min} = 3\lambda$, such that in Equation (9.4),

$U_{ela}^{act} = \lambda$. In this case the elastic rate is suppressed because $\rho_{FC}^{cla} \propto \exp(-\lambda/K_BT)$ and $\lambda \gg K_BT$. Consider an inelastic channel for which $\epsilon_{n_A} \simeq \epsilon_{n_D} + 2\lambda$. Then $U^{act}(D, n_D \to A, n_A) \simeq 0$ and this inelastic channel becomes activationless because the electron has given energy to the mode. If for this channel $(FC)_{n_D,n_A}^{qua}$ is not small then the corresponding inelastic rate will be significant. This type of inelastic ET enhances the ET rate in the inverted region where the elastic channels are suppressed because $U_D^{min} - U_A^{min} > \lambda$.

It is also possible that the quantum vibrational mode coupled to ET is a bridge-localized mode that is perturbed by the electronic occupation of a bridge electronic state $|B\rangle$ ((Onuchic and Da Gama, 1986; Medvedev and Stuchebrukhov, 1997; Troisi *et al.*, 2003; Skourtis *et al.*, 2004; Teklos and Skourtis, 2005; Nishioka *et al.*, 2005; Skourtis *et al.*, 2005; Skourtis and Beratan, 2007; Xiao *et al.*, 2009) and (Skourtis *et al.*, 2006, 2010) for reviews). Consider the simplest case of a single bridge electronic state $|B\rangle$ where $\hat{H}^{br} = E_B|B\rangle\langle B|$ and $\hat{V} = V_{D,B}|D\rangle\langle B| + V_{A,B}|A\rangle\langle B| + h.c.$ in Equations (9.10) and (9.13). When $|B\rangle$ is not occupied (i.e. the electronic state is either $|D\rangle$ or $|A\rangle$), the vibrational Hamiltonian is $\hat{H}^{vi} = \sum_n^{\infty} \epsilon_n |n(y)\rangle\langle n(y)|$, where $|n(y)\rangle$ denote the vibrational (mode) eigenstates and $\epsilon_n = \hbar\omega(n + 1/2)$ the vibrational eigenenergies. When the electron is at $|B\rangle$ it exerts a force F_B on the mode, i.e.

$$\hat{H}^{el-vi} = -F_B\hat{y}|B\rangle\langle B|. \tag{9.26}$$

The mode Hamiltonian then becomes $\hat{H}^{vi} + \langle B|\hat{H}^{el-vi}|B\rangle$, with eigenstates $|n(y - y_B^{eq})\rangle$ where $y_B^{eq} = F_B/m\omega^2$ and eigenenergies $\epsilon_n - F_B^2/(2m\omega^2)$. The total electron-vibrational Hamiltonian of the bridge is $\hat{H}_{el-vi}^{br} = U_B^{min}|B\rangle\langle B| + \hat{H}^{vi} + \hat{H}^{el-vi}$, with eigenstates $|B; n(y - y_B^{eq})\rangle = |B\rangle|n(y - y_B^{eq})\rangle$ (U_B^{min} is the minimum of the bridge-electronic-state Born–Oppenheimer surface).

As the electron tunnels through the bridge it may exchange energy with this bridge mode and cause a transition from an initial state $|i_B(y)\rangle$ to a final state $|f_B(y)\rangle$ of the undisplaced mode. Elastic tunnelling corresponds to $|i_B(y)\rangle = |f_B(y)\rangle$ and inelastic to $|i_B(y)\rangle \neq |f_B(y)\rangle$. If we assume that the mode is thermally equilibrated prior to ET, then the ET rate from D to A can be written as a thermally weighted sum of independent rates for vibronic transitions from initial vibronic states $|D; i_B\rangle = |D\rangle|i_B\rangle$ to final vibronic states $|A; f_B\rangle = |A\rangle|f_B\rangle$,

$$k_{DA} = \sum_{i_B} P_{i_B} \sum_{i_B} k(D, i_B \to A, f_B). \tag{9.27}$$

P_{i_B} is the thermal equilibrium probability of state $|i_B\rangle$ and $k(D, i_B \to A, f_B)$ is the rate for the vibronic transition $|D; i_B\rangle \to |A; f_B\rangle$, given by

$$k(D, i_B \to A, f_B) = \frac{2\pi}{\hbar}|\langle D; i_B|\hat{T}|A; f_B\rangle|^2 \rho_{FC}^{clas}(D, i_B \to A, f_B). \tag{9.28}$$

$\rho_{FC}^{clas}(D, i_B \rightarrow A, f_B)$ is the Boltzmann activation factor (Equation (9.3)) with an activation energy similar to (9.25) where ϵ_{n_D} is replaced by ϵ_{i_B} and ϵ_{n_A} by ϵ_{f_B}. $\langle D; i_B|\hat{T}|A; f_B\rangle$ is a vibronic bridge-mediated tunnelling matrix element that contains the vibronic bridge Green's function, i.e.

$$\langle D; i_B|\hat{T}|A; f_B\rangle = \langle D; i_B|\hat{V}\hat{G}^{br}_{el-vi}(E_{tun})\hat{V}|A; f_B\rangle, \tag{9.29}$$

where

$$\hat{G}^{br}_{el-vi}(E_{tun}) = \left(E_{tun}\hat{I} - \hat{H}^{br}_{el-vi}\right)^{-1} = \sum_n \frac{|B; n(y - y_B^{eq})\rangle\langle B; n(y - y_B^{eq})|}{E_{tun} - \left(U_B^{min} + \epsilon_n - F_B/(2m\omega^2)\right)}, \tag{9.30}$$

(here E_{tun} is the energy of the initial (D) and final (A) vibronic states at D-to-A resonance). Substituting Equation (9.30) into (9.29) gives

$$\langle D; i_B|\hat{T}|A; f_B\rangle = V_{D,B}V_{B,A}\sum_n \frac{\langle i_B(y)|n(y - y_B^{eq})\rangle\langle n(y - y_B^{eq})|f_B(y)\rangle}{E_{tun} - \left(U_B^{min} + \epsilon_n - F_B/(2m\omega^2)\right)}, \tag{9.31}$$

(Skourtis *et al.*, 2004, 2006; Skourtis and Beratan, 2007; Xiao *et al.*, 2009; Skourtis *et al.*, 2010). The above expression is a generalization of the elastic T_{DA} formulae (Equations (9.14) and (9.15)). The relevance of inelastic-tunnelling ET in biological processes is largely unexplored (for recent developments in olfaction see Chapter 12).

9.6 Biological ET chains with tunnelling and hopping steps through the protein medium

ET reactions mediated by electron tunnelling where the D and A moieties are metal atoms and the tunnelling barrier is protein are often the components of biological ET chains (Bendall, 1996; Page *et al.*, 1999; Blankenship, 2002; Gray and Winkler, 2003, 2005). The latter can be thought of as networks of D–A pairs connected by forward ($k_{D\rightarrow A}$) and backward ($k_{A\rightarrow D}$) tunnelling-mediated ET rates where, usually, $k_{D\rightarrow A} \gg k_{A\rightarrow D}$ (Figure 9.1b). In these rate networks the role of the protein medium is to provide a scaffold for the stabilization of the D and A locations in space, to influence the energy gap and reorganization energy of each D–A tunnelling step and to provide the tunnelling barrier (tunnelling matrix element value) for each tunnelling step. The advantage of having electron tunnelling as the basic ET transfer mechanism of the elementary transfer steps in biological ET chains is that tunnelling forces the electron to occupy only the D and A moieties of the rate network and to avoid the protein bridge. Therefore, biological ET chains are designed to transfer electrons to specific locations (the D/A moieties) and along specific directions (defined by the sequences D and A pairs).

There is compelling evidence that metal–protein–metal ET chains can support very long distance (micron scale) ET. This situation can arise in extra-cellular ET (EET) processes such as the respiratory chains of dissimilatory metal-reducing bacteria (DMRB) (Lovley and Phillips, 1988; Myers and Nealson, 1988; Nealson *et al.*, 2002; Gralnick and Newman, 2007). In DMRB, the final respiratory-electron acceptors of the respiratory chains lie outside the cell and are often environmental metal oxides. When there are no soluble electron acceptors in the immediate cell environment of a DMRB, the bacterium can grow very long pilus-like appendages, (bacterial nanowires), in order to find environmental metal oxides and to transfer its electrons to them (Reguera *et al.*, 2005; Gorby *et al.*, 2006; El-Naggar *et al.*, 2008, 2010; Malvankar *et al.*, 2011). Conductive probe atomic force microscopy has been used to measure the electron current along bacterial nanowires of the DMRB S. Oneidensis MR-1 (El-Naggar *et al.*, 2010). In El-Naggar *et al.* (2010), bacterial nanowires were connected to Au electrodes and it was shown that nA currents could be sustained over bacterial wire lengths of 0.6 microns under a bias voltage of 100 mV (resistivity 1 Ω cm). There is evidence that such transport is mediated by multi-heme cytochromes (Fe-containing proteins) because mutants of the bacterium which did not express two types of decaheme cytochromes were not conductive (El-Naggar *et al.*, 2010). Although the structures and assembly architectures of these proteins within the nanowires are still unknown (Fredrickson *et al.*, 2008), theoretical models of bacterial nanowire transport (Strycharz-Glaven *et al.*, 2011; Polizzi *et al.*, 2012; Pirbadian and El-Naggar, 2012) suggest that a multi-site hopping mechanism can explain the observed current magnitudes (El-Naggar *et al.*, 2010). If each each site-to-site hopping step is D–A tunnelling (e.g. Fe to Fe), then the site-to-site rates must lie in the fast limit of biological ET (rate^{-1} $\sim$ 1–10 ps). Such rates require short ($\sim$nm) D–A distances and relatively small reorganization energies (less than 1 eV). Therefore, the ET proteins responsible for EET must be closely packed within the bacterial nanowire. Another mechanism of transport that has been proposed for EET is band-like pi-stacking conductivity (Malvankar *et al.*, 2011). However, there is no theoretical analysis yet supporting this mechanism (Malvankar *et al.*, 2012; Strycharz-Glaven and Tender, 2012). The observed micron-sized transport distances of EET in bacterial nanowires raise new and exciting questions in the field of ET, as they force us to re-think charge transport optimization on a cellular length scale. Understanding EET is important for biogeochemistry and for energy-harvesting and bioelectrochemical devices (Hau and Gralnick, 2007; Logan, 2009).

Although metal–protein–metal ET chains are very common in biology, there are biological-ET-chain examples where the D and A are aminoacid moieties or other organic molecules rather than metals. In these cases, for many ensemble members, the D and A electronic states will have similar energies to the bridge electronic states (i.e. $\Delta E_B \leq K_B T$ or, more precisely, $\langle \Delta U \rangle \leq \sigma_{\Delta U}$, where ΔU refers to D–B and A–B energy gaps). Then, ET from D to A cannot be described as a single tunnelling step, since there will be a high number of ensemble members for which D (or A) become resonant to a protein bridge state (B). For these ensemble members the electron (hole) will hop from D to B and then from B to A. Therefore the T_{DB} and T_{AB} electronic couplings are important to the overall transfer, rather than T_{DA} (Skourtis and Onuchic, 1993; Skourtis *et al.*, 1993; Skourtis and Mukamel, 1995; Xie *et al.*, 1999; Skourtis *et al.*, 2001). In general, when $\Delta E_B \leq K_B T$, through-bridge hopping takes over and the fluctuations that minimize ΔE_B are more important to the overall rate than the fluctuations in couplings. It is well established that base-mediated hole hopping is important for long-distance hole transfer in DNA (Schuster, 2004). Hopping within protein bridges, with aminoacids acting as D and A, is encountered in biological systems such as DNA photolyases and cryptochromes (Sancar, 2003; Kao *et al.*, 2008; Brettel and Byrdin, 2010), in some ribonuclease reductases (Stubbe *et al.*, 2002), and in synthetically-modified azurin (Shih *et al.*, 2008; Gray and Winkler, 2009; Winkler and Gray, 2010). These systems support charge transport reactions involving hole hopping along tyrosyl or tryptophan aminoacids, which act as hole donors and acceptors (see Chapter 10 for relevanse to magnetoreception).

9.7 Conclusions

Because of the large variety of biomolecular ET architectures, the mechanisms of biological ET are variable, ranging from coherent deep tunnelling to thermally activated hopping. This chapter has focused on protein-mediated ET and describes ET reactions for which the timescale of ET is long compared to that of thermal equilibration of the ET reaction coordinate on the U_D energy surface. This assumption leads to the Boltzmann activation factors in the rate expressions (Equations (9.2), (9.24)). It has long been recognized that thermal equilibration of the reaction coordinate is not a valid assumption for fast ET reactions, for which the timescale of ET is similar to a timescale of relaxation of the reaction coordinate motions on the U_D surface. This is often the case with fast (ps) photo-excited ET, where ET competes with vibrational relaxation. Examples of such reactions are the primary charge separation in photosynthesis (Blankenship, 2002) and the flavin-to-Trp hole transfer step in cryptochromes and DNA photoylases (Sancar, 2003; Kao *et al.*, 2008; Brettel and Byrdin, 2010). These cases need theoretical and computational methods that go beyond the Marcus-type non-adiabatic rate expressions and that involve density matrix approaches with non-equilibrium vibrational dynamics (Warshel *et al.*, 1989; Schulten and Tesch, 1991; Xu and Schulten, 1994; Jortner and Bixon, 1999; Sumi and Kakitani, 2001; Warshel *et al.*, 2001), or direct propagation of the electronic and nuclear degrees of freedom (e.g. (Woiczikowski *et al.*, 2011)). However, in all types of ET reactions, the quantum-mechanical nature of the electron (and in some cases of the high-frequency molecular vibrations) play a central role in determining the ET rate. An interesting and largely unanswered question is how evolution has optimized biomolecular ET assemblies and whether this optimization involves the quantum or the classical components of the ET rate.

9.8 Acknowledgements

S. S. Skourtis acknowledges the hospitality of the Institute of Advanced Study, Hebrew University, during the writing of this chapter.

10

A chemical compass for bird navigation

ILIA A. SOLOV'YOV, THORSTEN RITZ, KLAUS SCHULTEN AND PETER J. HORE

10.1 Introduction

Migratory birds travel spectacular distances each year, navigating and orienting by a variety of means, most of which are poorly understood. Among these is a remarkable ability to perceive the intensity and direction of the Earth's magnetic field (Mouritsen and Ritz, 2005; Wiltschko and Wiltschko, 2006; Johnsen and Lohmann, 2008). Biologically credible mechanisms for the detection of such a weak field (25–65 μT) are scarce, and in recent years just two proposals have emerged as front-runners. One, essentially classical, centers on clusters of magnetic iron-containing particles in the upper beak, which appear to act as a magnetometer for determining geographical position (Kirschvink and Gould, 1981; Kirschvink *et al.*, 2001; Fleissner *et al.*, 2007; Solov'yov and Greiner, 2007; Walker, 2008; Solov'yov and Greiner, 2009a, b; Falkenberg *et al.*, 2010). The other relies on the quantum spin dynamics of transient photoinduced radical pairs (Schulten *et al.*, 1978; Schulten, 1982; Schulten and Windemuth, 1986; Ritz *et al.*, 2000b; Cintolesi *et al.*, 2003; Möller *et al.*, 2004; Mouritsen *et al.*, 2004; Heyers *et al.*, 2007; Liedvogel *et al.*, 2007b, a; Solov'yov *et al.*, 2007; Feenders *et al.*, 2008; Maeda *et al.*, 2008; Solov'yov and Schulten, 2009; Ritz *et al.*, 2009; Rodgers and Hore, 2009; Zapka *et al.*, 2009). Originally suggested by Schulten in 1978 (Schulten *et al.*, 1978) as the basis of the avian magnetic compass sensor, this mechanism gained support from the subsequent observation that the compass is light dependent (Wiltschko *et al.*, 1993) (for a review see e.g. (Wiltschko *et al.*, 2010)). The radical pair hypothesis began to attract increased interest following the proposal in 2000 that free radical chemistry could occur in the bird's retina, initiated by photoexcitation of cryptochrome, a specialized photoreceptor protein (Ritz *et al.*, 2000b; Solov'yov and Schulten, 2012; Solov'yov *et al.*, 2012, 2014).

Quantum Effects in Biology, ed. Masoud Mohseni, Yasser Omar, Gregory S. Engel and Martin B. Plenio. Published by Cambridge University Press.

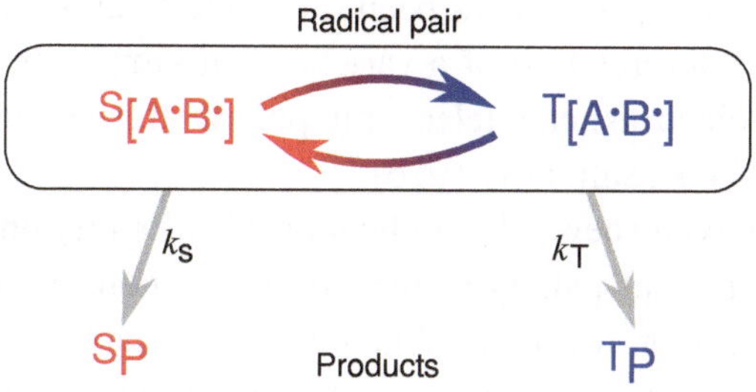

Figure 10.1 A simple radical pair reaction scheme.

The quantum evolution of highly non-equilibrium electron spin states of pairs of transient spin-correlated radicals is conjectured to change the yields of their reaction products in 'wet, warm and noisy' biological surroundings, even though the Zeeman interaction with the geomagnetic field is more than six orders of magnitude smaller than the thermal energy per molecule ($k_B T$). The classical thermodynamic effect of such minuscule interactions on the positions of chemical equilibria and the rates of activated reactions would be entirely negligible. The radical pair mechanism is the only well-established way in which an external magnetic field can influence a chemical reaction (Schulten *et al.*, 1976; Steiner and Ulrich, 1989; Brocklehurst, 2002; Timmel and Henbest, 2004; Rodgers, 2009).

The origin of the magnetic field effect (MFE) can be understood by reference to the simple reaction scheme shown in Figure 10.1: (1) A pair of radicals $A^{\bullet}B^{\bullet}$ is formed (e.g. by an electron transfer reaction) in an entangled state which may be either singlet (spin quantum number, $S = 0$) or triplet ($S = 1$), depending on the spin of the precursor molecule(s), which is conserved in the reaction. (2) The radical pair is able to recombine from both the S and T states to form chemically distinct products (^{S}P and ^{T}P in Figure 10.1) with rate constants k_S and k_T respectively. (3) S and T radical pairs coherently interconvert under the influence of local magnetic fields arising from hyperfine interactions of the electron spins with magnetic nuclei in the two radicals. As a consequence, the fractional yields of the two products are determined not only by k_S and k_T, but also by the extent and timing of the magnetically controlled S$\leftrightarrow$T interconversion step. (4) This step is also, crucially, enhanced or hindered by electron Zeeman interactions with an external magnetic field. Thus, the fractional yields of the two products and the lifetime of the radical pair become magnetic field dependent. If the radical pair is immobilized, the tensorial nature of the hyperfine interactions implies a directionality in the response to an external magnetic field, which could form the basis of a compass sensor (Schulten and Tavan, 1978; Timmel *et al.*, 2001; Cintolesi *et al.*, 2003; Lau *et al.*, 2010; Hill and Ritz, 2010; Solov'yov *et al.*, 2010). The theory of the radical pair mechanism is well developed and has been successfully used over the last 40 years for the quantitative interpretation of a variety of in vitro experimental data – not just MFEs, but also electron and nuclear spin polarizations (Muus *et al.*, 1977) and magnetic isotope effects (Salikhov, 1996).

There is no doubt whatsoever that radical pair MFEs rely on coherent quantum dynamics. When a radical pair is formed in a spin-conserving reaction from a singlet or triplet precursor, it is created in a non-stationary coherent superposition of the eigenstates of its spin Hamiltonian. As a consequence, the spin state of the radical pair oscillates coherently at frequencies and with amplitudes determined by the internal and external magnetic interactions. The frequencies typically fall in the 10^7–10^9 Hz range and can be significantly faster than the spin relaxation processes (often $<10^7$ s^{-1}) that cause decoherence and loss of spin-correlation.[1] In many cases, there is ample time for weak magnetic interactions to influence the spin dynamics before the radicals react, and therefore to affect the product yields. The clearest experimental demonstrations of this fundamentally quantum-mechanical behavior, without which there would be no significant response to an external magnetic field, are the observations of quantum beats in the recombination luminescence of radical ion pairs in non-polar solvents (Grigoryants *et al.*, 1995; Bagryansky *et al.*, 2000, 2007) and the detection by EPR (electron paramagnetic resonance) of zero-quantum coherences in radical pairs in photosynthetic reaction centers (Bittl and Kothe, 1991; Kothe *et al.*, 1991; Wang *et al.*, 1992; Dzuba *et al.*, 1996).

In most demonstrations of MFEs on radical pair reactions, the applied magnetic field is comparable to or stronger than the internal hyperfine interactions, producing significant changes in the spin energy levels of the radical pair and so altering the yields of reaction products. One might therefore expect MFEs to be rather slight when the electron Zeeman interaction is smaller than most of the hyperfine interactions, as is probably the case for an in vivo radical pair magnetoreceptor. In fact, a somewhat different mechanism, known as the 'low field effect' (LFE) (Brocklehurst, 1976; Werner *et al.*, 1977; Schulten, 1984; Timmel *et al.*, 1998; Eveson *et al.*, 2000), comes into play in this regime.

In the following, we first summarize the origin and properties of the LFE, then review the important physical and chemical constraints on a possible radical-pair-based compass sensor and discuss the suggestion that radical pairs in cryptochromes might provide a biological realization for a magnetic compass. We then summarize pertinent in vitro experimental data, and discuss their relevance to detecting the direction of the Earth's magnetic field. Finally, we review the current evidence supporting a role for radical pair reactions in the magnetic compass of birds.

[1] The term 'spin relaxation' is used here in the sense normally employed in magnetic resonance spectroscopy, to denote the return of the radical pair electron–nuclear spin system to thermal equilibrium as a result of the modulation of spin interactions by stochastic molecular motions.

10.2 Theoretical basis for a chemical compass

10.2.1 Theoretical model and origin of the low field effect

Following Timmel *et al.* (Timmel *et al.*, 1998), we first outline a simple theoretical model of radical pair spin dynamics focusing on the effects of weak applied magnetic fields. Imagine a pair of radicals produced at time $t = 0$ in a pure S state, e.g. by photoinduced inter- or intramolecular electron transfer. The density operator of the spin system is then simply proportional to the singlet projection operator, $\hat{P}^{\mathrm{S}}$,

$$\hat{\rho}(0) = \frac{\hat{P}^{\mathrm{S}}}{M} = \frac{1}{M}\sum_{\nu=1}^{M} |\mathrm{S}\nu\rangle\langle\mathrm{S}\nu| \,, \qquad (10.1)$$

where the sum runs over the M nuclear spin configurations, ν. The probability that the radical pair is found to be singlet at some later time is determined by the spin Hamiltonian $\hat{H}$, which we assume to be a time-independent sum of hyperfine ($\hat{H}_{\mathrm{HF}}$) and electron Zeeman ($\hat{H}_{\mathrm{Z}}$) contributions:

$$\begin{aligned}\langle\hat{P}^{\mathrm{S}}\rangle(t) &= \mathrm{Tr}\big[\hat{P}^{\mathrm{S}}\hat{\rho}(t)\big] = \mathrm{Tr}\big[\hat{P}^{\mathrm{S}}\mathrm{e}^{-\mathrm{i}\hat{H}t}\hat{\rho}(0)\mathrm{e}^{\mathrm{i}\hat{H}t}\big] \\ &= \frac{1}{M}\mathrm{Tr}\big[\hat{P}^{\mathrm{S}}\mathrm{e}^{-\mathrm{i}\hat{H}t}\hat{P}^{\mathrm{S}}\mathrm{e}^{\mathrm{i}\hat{H}t}\big]. \qquad (10.2)\end{aligned}$$

Defining $P^{\mathrm{S}}_{mn} = \langle m|\hat{P}^{\mathrm{S}}|n\rangle$ and $\omega_{mn} = \langle m|\hat{H}|m\rangle - \langle n|\hat{H}|n\rangle$, where $|m\rangle$ and $|n\rangle$ are eigenstates of $\hat{H}$, we have (Timmel *et al.*, 1998),

$$\langle\hat{P}^{\mathrm{S}}\rangle(t) = \frac{1}{M}\sum_{m=1}^{4M}\sum_{n=1}^{4M}\left|P^{\mathrm{S}}_{mn}\right|^2\cos(\omega_{mn}t)\,. \qquad (10.3)$$

In general, neither $\hat{H}_{\mathrm{HF}}$ nor $\hat{H}_{\mathrm{Z}}$ commutes with $\hat{P}^{\mathrm{S}}$, meaning that the spin system is formed in a non-stationary state. The coherent superpositions P^{S}_{mn} oscillate at frequencies ω_{mn}, causing S↔T interconversion (Timmel *et al.*, 1998). In the absence of an applied magnetic field, this coherent time dependence, driven by the hyperfine interactions, is typically in the MHz range. Assuming, for simplicity, that the S and T states recombine spin-selectively with the same first-order rate constant k ($k = k_{\mathrm{S}} = k_{\mathrm{T}}$ in Figure 10.1), the ultimate yield of the reaction product formed from the S state of the radical pair (the 'singlet yield') is given by

$$\begin{aligned}\Phi_{\mathrm{S}} = k\int_0^{\infty}\langle\hat{P}^{\mathrm{S}}\rangle(t)\,\mathrm{e}^{-kt}\,\mathrm{d}t &= \frac{1}{M}\sum_{m=1}^{4M}\sum_{n=1}^{4M}\left|P^{\mathrm{S}}_{mn}\right|^2 f(\omega_{mn}) \\ &= \frac{1}{M}\sum_{m=1}^{4M}\sum_{n=1}^{4M}\left|P^{\mathrm{S}}_{mn}\right|^2\frac{k^2}{k^2+\omega^2_{mn}}\,. \qquad (10.4)\end{aligned}$$

Thus, the contribution of a coherence P^{S}_{mn} to Φ_{S} is determined by its frequency relative to the recombination rate. If k is much larger than all $|\omega_{mn}|$, then all $f(\omega_{mn}) = 1$, and $\Phi_{\mathrm{S}} = 1$. For the hyperfine and Zeeman interactions to have a significant effect on the product yields (i.e. $\Phi_{\mathrm{S}} < 1$), the recombination must be slow enough that $k \leq |\omega_{mn}|$ for at least some (m, n), to allow time for the coherent spin dynamics to drive the spin system out of the S state.

The origin of the effect of a very weak applied magnetic field may now be seen (Timmel *et al.*, 1998). In zero-field, there will, in general, be some coherent superpositions that do not oscillate, because $\omega_{mn} = 0$. If an applied field lifts such a degeneracy, the coherence can then evolve and contribute to S↔T interconversion. Provided the splitting so produced (ω_{mn}) is not much smaller than *k*, this will result in a reduction in Φ_{S} (if the initial state of the pair is S). For a sufficiently long-lived radical pair, even a field much weaker than the hyperfine interactions can "unlock" zero-frequency coherences in this way. The energy-level shifts produced by the field are of the order of the electron Zeeman frequency, $\gamma_{\mathrm{e}} B_0/2\pi$, i.e. ~1.4 MHz for a 50 μT field. Such a weak field can therefore be expected to have a significant effect on Φ_{S} if the radical lifetime k^{-1} exceeds about 100 ns. An additional condition for a significant LFE is that the rate of electron spin relaxation R must also be slow enough ($R^{-1} \geq 100$ ns) to allow the electron–nuclear spin coherence to persist throughout the lifetime of the radical pair. Depending on the number and distribution of hyperfine couplings, changes in Φ_{S} of up to ~50% are predicted (Schulten *et al.*, 1978; Timmel *et al.*, 1998). For examples of observations of LFEs on chemical reactions in solution, see (Batchelor *et al.*, 1993; Saik *et al.*, 1995; Stass *et al.*, 1995a, b; Sacher and Grampp, 1997; Eveson *et al.*, 2000; Henbest *et al.*, 2006; Rodgers *et al.*, 2007).

10.2.2 Requirements for a magnetic compass

To form the basis of an effective compass magnetoreceptor, a radical pair reaction must satisfy a number of conditions (Rodgers and Hore, 2009), which fall into five broad overlapping areas: chemical, magnetic, kinetic, structural, and dynamic. (1) The radical pair must be formed in a coherent superposition of its electron–nuclear spin states and at least one of the S and T states should undergo a spin-selective reaction that the other cannot. (2) There should be suitable anisotropic hyperfine interactions. (3) The lifetime of the radical pair must be long enough to allow the weak magnetic field to affect the spin dynamics, and the rate constants k_{S} and k_{T} should not be too dissimilar. (4) The Zeeman interaction can only modulate the S↔T interconversion if inter-radical spin–spin (exchange and dipolar) interactions are sufficiently weak. (5) To deliver directional information, the radical pairs must be aligned and immobilized and the spin system should relax sufficiently slowly. These criteria are interlinked and not automatically mutually compatible. For example, (3) places an upper limit on the separation of the radicals, while (4) may require larger inter-radical distances. Both (3) and (4) may constrain the chemistry, e.g. by requiring the magnetically sensitive radical pair to be formed by sequential electron transfers rather than in a single step (Solov'yov *et al.*, 2007; Rodgers and Hore, 2009). Motional modulation of anisotropic hyperfine interactions is a major source of spin relaxation in radicals so that (2) may be incompatible with (5) unless the radicals are strongly immobilized.

Most of these factors and others have been investigated experimentally or theoretically (Timmel *et al.*, 1998; Solov'yov *et al.*, 2007; Efimova and Hore, 2008; Lau *et al.*, 2010; Hill and Ritz, 2010; Solov'yov *et al.*, 2010). Our impression is that there is no obvious "show-stopper" as far as a biological magnetic compass is concerned and that cryptochromes (see below) appear to be fit for this purpose (Solov'yov *et al.*, 2014). One great area of uncertainty, though, is how large the primary effect needs to be. In practice, isotropic LFEs in vitro rarely exceed 10% and are smaller still in magnetic fields approaching 50 μT. Anisotropic effects are likely to be no bigger. As so little is known about cryptochrome signalling, it is unclear whether magnetic responses at the 1–10% level would be sufficient. However, a few semi-quantitative conclusions can be drawn from the above list (Rodgers and Hore, 2009). For example, at 50 μT, the MFE, is expected to saturate when the spin coherence persists for roughly 1 μs, so that there would be little evolutionary pressure for the lifetime or the relaxation time to be much longer than this, and a sensitivity penalty if they were much shorter. Perhaps coincidentally, the lifetime of the magnetically sensitive radical pair in DNA photolyase (a close relation of cryptochrome, see below) is about 1 μs in vitro (Henbest *et al.*, 2008).

10.2.3 Cryptochrome magnetoreception

Ritz *et al.* (Ritz *et al.*, 2000b) proposed in 2000 that radical pairs formed photochemically in the protein cryptochrome could form the basis of the compass magnetoreceptor. No other candidate molecule has been put forward in the intervening years. Cryptochromes occur in several of the organisms for which magnetic field effects have been reported, including fruit flies, plants and migratory birds and have been shown to act as photoreceptors in a variety of species (Lin and Todo, 2005). In plants, they serve as photosensors for a number of developmental responses such as hypocotyl growth, leaf expansion, induction of flowering time and entrainment of the circadian clock. In insects, cryptochromes act as circadian photoreceptors.

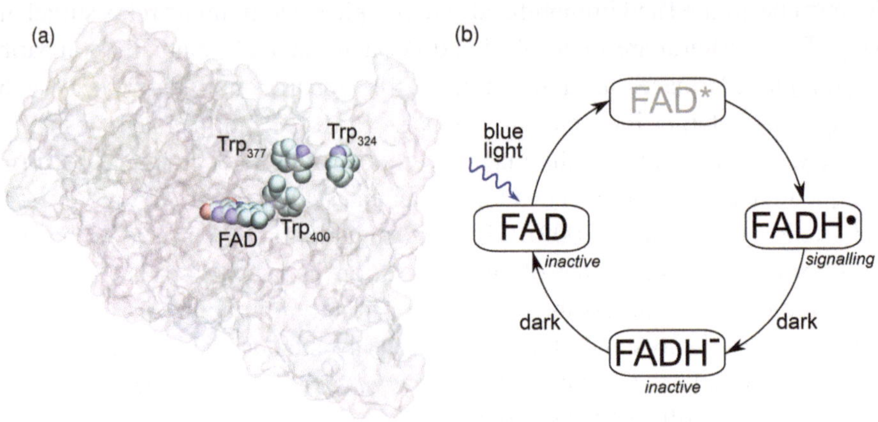

Figure 10.2 (a) Structure of cryptochrome, the protein implicated in avian magnetoreception. Cryptochrome internally binds the FAD (flavin adenine dinucleotide) cofactor which governs the functioning of the protein. The signalling state is achieved via a light-induced photoreduction pathway involving a chain of three tryptophan amino acids, indicated as Trp_{400}, Trp_{377} and Trp_{324} using the amino acid sequence numbers for *Arabidopsis thaliana* cryptochrome-1. (b) The cryptochrome photocycle. The signalling function of cryptochrome is controlled by the oxidation state of its flavin cofactor, which can exist in three interconvertible redox forms, FAD, $FADH^{\bullet}$ (or $FAD^{\bullet -}$), and $FADH^{-}$ (or $FADH_2$) (Bouly *et al.*, 2007; Kao *et al.*, 2008; O'Day, 2008). The FAD form is inactive and is thought to be the resting state of the protein in the dark. Blue light triggers photoreduction of FAD to establish a photo-equilibrium that favors $FADH^{\bullet}$ over FAD and $FADH^{-}$. The semiquinone radical $FADH^{\bullet}$ state is the signalling state of the protein. $FADH^{\bullet}$ can be further reduced to the inactive $FADH^{-}$ form. The $FAD \rightarrow FADH^{\bullet}$ and $FADH^{\bullet} \rightarrow FADH^{-}$ reactions may be affected by an external magnetic field. The excited state of the flavin cofactor, FAD*, is a short-lived intermediate in the photocycle.

Light-induced cryptochrome signalling appears to proceed via electron transfer involving a chain of three tryptophan amino acids (the Trp-triad) and the cofactor, flavin adenine dinucleotide (FAD) (Giovani *et al.*, 2003; Bouly *et al.*, 2007; Hoang *et al.*, 2008; Biskup *et al.*, 2009; Solov'yov *et al.*, 2014), shown in Figure 10.2a. Photo-excitation of the FAD in its fully oxidized state leads to the formation of three consecutive radical pairs by donation of an electron along the Trp-triad to the FAD to form the $FADH^{\bullet}$ radical, as illustrated in Figure 10.2b. It is this state that is thought to be responsible for biological signalling. Any factor that increases (decreases) the yield of this state of the protein should result in an increased (decreased) cryptochrome signal for a given light intensity. In principle, an external magnetic field could alter the yield of the signalling state via its effect on the flavin–tryptophan radical pair (Cintolesi *et al.*, 2003; Solov'yov *et al.*, 2007). In vitro, the $FADH^{\bullet}$ state of cryptochrome has a lifetime of about 1–10 ms with respect to reversion to the FAD state (Giovani *et al.*, 2003; Liedvogel *et al.*, 2007a; Biskup *et al.*, 2009). There has been speculation that this back-reaction might involve the superoxide radical $O_2^{\bullet-}$ (Ritz *et al.*, 2009), and could also therefore be modulated by the Earth's magnetic field (Hogben *et al.*, 2009; Solov'yov and Schulten, 2009). If external fields influence the rate of flavin reoxidation, they could alter the yield of the signalling state and so affect biological activity.

Experimental evidence provides some support for a magnetosensing role for cryptochrome. Growth of *Arabidopsis thaliana* seedlings in a 500 μT magnetic field has been reported to enhance cryptochrome activity, such that the plants responded as though they had been exposed to higher intensities of blue light than was in fact the case (Ahmad *et al.*, 2007). Magnetically enhanced cryptochrome activity was manifest in shorter hypocotyls and higher anthocyanin levels compared with control plants grown under identical blue-light intensities in weaker magnetic fields. However, none of these effects could be replicated in a subsequent study, which also failed to detect responses using substantially stronger magnetic fields where radical pair effects might be expected to be more pronounced (Harris *et al.*, 2009). Related effects have been found for the circadian clocks of fruit flies in which cryptochrome acts as a photoreceptor (Yoshii *et al.*, 2009). In response to blue light, cryptochrome activity increases the circadian period in *Drosophila melanogaster*, an effect that was found to be more pronounced in the presence of a weak magnetic field, indicating enhanced cryptochrome signalling. Cryptochrome knock-out mutants showed no magnetic field sensitivity, while flies overexpressing cryptochrome in the clock neurons showed enhanced magnetic responses compared to wild type. A recent investigation of behavioral responses of *Drosophila melanogaster* in applied magnetic fields has also implicated cryptochrome (Gegear *et al.*, 2008, 2010). In these experiments, flies were trained to associate the magnetic field with a food source, and learned to use it as an orientational cue; these responses were absent in cryptochrome-deficient flies.

In the context of avian magnetoreception, it is noteworthy that cryptochromes have been found in birds' retinas (Möller *et al.*, 2004; Mouritsen *et al.*, 2004). There are some genetic indications of an involvement of cryptochromes in magnetoreception in birds (Freire *et al.*, 2008), but the lack of transgenic birds has hitherto precluded more clear-cut evidence. Theoretical considerations also provide support for the cryptochrome hypothesis. For example, a lifetime as long as 1 μs can actually be achieved according to the theory of electron transfer reactions (Moser and Dutton, 1992) for an edge-to-edge inter-radical separation, r_e of 1.5 nm (Rodgers and Hore, 2009). This appears to be consistent with the crystal structure of *Arabidopsis thaliana* cryptochrome in which one finds $r_e = 1.47$ nm for the FAD cofactor and the terminal residue of the tryptophan triad (Brautigam *et al.*, 2004). A further, related kinetic constraint can be derived from the

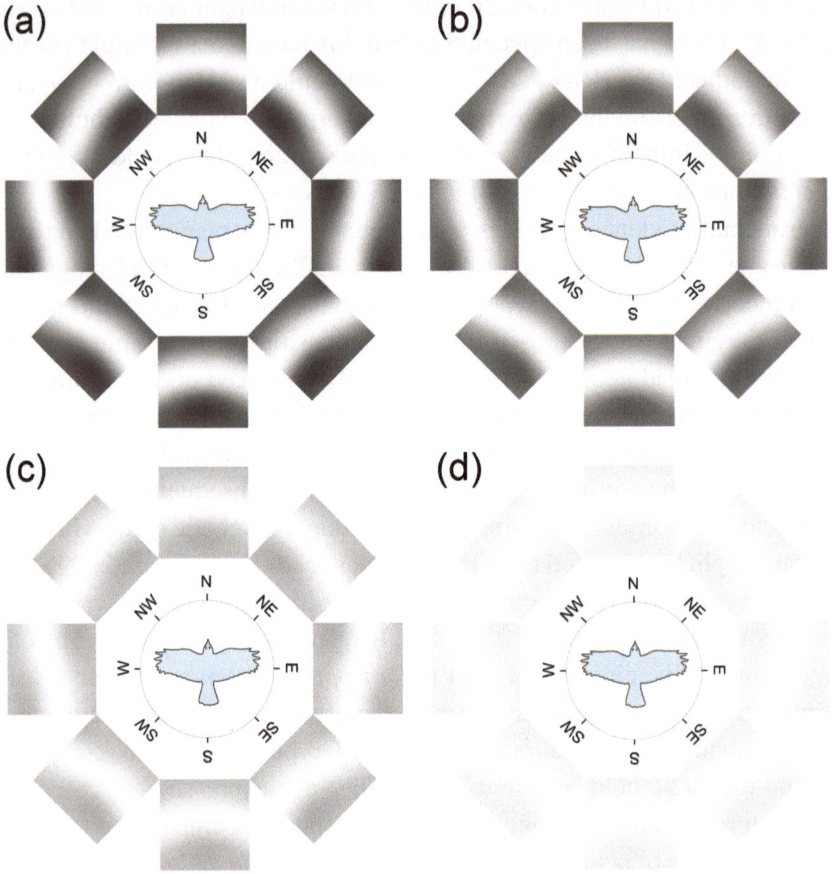

Figure 10.3 Schematic illustration of the visual modulation patterns that might be induced by the geomagnetic field for a bird flying in the eight cardinal directions (N, NE, E, SE, S, SW, W and NW). The geomagnetic field inclination angle is 66° (appropriate for Frankfurt am Main, Germany) (Wiltschko and Wiltschko, 1995b; Wiltschko *et al.*, 2008). The modulation patterns are calculated for different degrees of orientational disorder of the cryptochrome molecules within the receptor cells, characterized by the parameter ε_0: (a) $\varepsilon_0 = 100$ (highly ordered); (b) $\varepsilon_0 = 10$; (c) $\varepsilon_0 = 3$; (d) $\varepsilon_0 = 1$ (moderately disordered). For details, see (Solov'yov *et al.*, 2010).

reasonable assumption that the magnetically responsive radical pair should be formed in less than 1 ns (so as to have a high quantum yield and a pure initial spin state). Estimates based on Marcus theory (Marcus and Sutin, 1985) suggest that such a radical pair could be achieved if every electron transfer step involved in the formation of the pair had a donor–acceptor separation $r_e < 1.0$ nm, a condition which again is consistent with the FAD/Trp triad structure in *Arabidopsis thaliana* cryptochrome (Rodgers and Hore, 2009).

Finally, we turn to the degree of molecular ordering that would allow an array of cryptochromes to show a significant directional response to a 50 μT magnetic field and therefore to act as a compass sensor. Assuming that the magnetic signal-transduction mechanism is linked to the rhodopsin-mediated visual detection system, so that the bird literally sees a representation of the Earth's magnetic field, one can derive a filter function to model the transformation of the visual field produced by a cryptochrome-based magnetoreceptor (Solov'yov *et al.*, 2010). Figure 10.3 shows examples of visual modulation patterns simulated in this way for a bird flying horizontally in eight cardinal directions. The calculations underlying the modulation patterns in Figure 10.3 indicate that even modest uniaxial molecular alignment could be sufficient to yield a directional response suitable for compass detection (Lau *et al.*, 2010; Hill and Ritz, 2010; Solov'yov *et al.*, 2010).

10.3 In vitro magnetic field effects on radical pair reactions

10.3.1 Anisotropic magnetic field effects

Radical pairs with lifetimes in excess of 100 ns can in principle form the basis of a field-intensity sensor in the ~50 μT range. However, for a compass, the reaction must also respond to the direction of the field, implying that the radicals cannot be free to rotate rapidly. As Schulten originally argued, (Schulten *et al.*, 1978; Schulten, 1982; Schulten and Windemuth, 1986), radical pairs are expected to show anisotropic MFEs by virtue of the anisotropy of their hyperfine interactions, provided the radicals are both immobilized and aligned. The theory of anisotropic LFEs is essentially as presented above, with the exception that the spin Hamiltonian now depends on the direction of the applied magnetic field with respect to the molecular frame of the radical pair (Timmel *et al.*, 2001; Cintolesi *et al.*, 2003). As before, the electron Zeeman interaction can unlock zero-frequency coherences and so affect S↔T interconversion, but now the amplitudes of the relevant coherences (P^{S}_{mn}) will depend on the direction of the field. To make this more concrete, we consider briefly the simple case in which one radical has no magnetic nuclei and the other has an axial anisotropic hyperfine interaction with a single spin-1/2 nucleus. In the limit $k \ll \|\hat{H}_{\mathrm{Z}}\| \ll \|\hat{H}_{\mathrm{HF}}\|$, the singlet yield, calculated as above, is (Timmel *et al.*, 2001; Cintolesi *et al.*, 2003)

$$\Phi_{\mathrm{S}} = \frac{1}{4} + \frac{1}{8}\cos^2\psi = \frac{7}{24} + \frac{1}{24}(3\cos^2\psi - 1), \qquad (10.5)$$

where ψ is the angle between the magnetic field and the symmetry axis of the hyperfine tensor. In this case, Φ_{S} has a rather simple form that varies between 1/4 and 3/8. In zero-field, Φ_{S} is isotropic and equal to 3/8.

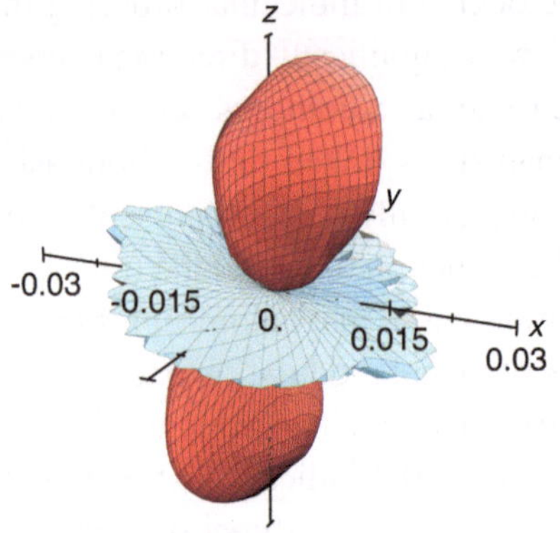

Figure 10.4 Anisotropy of the singlet recombination yield, Φ_S, calculated for the 8-nucleus flavin–tryptophan radical pair described in the text. The magnetic field strength was 50 μT and the recombination rate constant, $k = 2 \times 10^5$ s^{-1}. [Taken from (Cintolesi *et al.*, 2003).]

Any in vivo radical pair magnetoreceptor is unlikely to have magnetic responses as simple as the one-nucleus case just considered. Figure 10.4 shows the result of a simulation of the anisotropic part of Φ_S for a more realistic, multinuclear, flavin–tryptophan radical pair, [FH$^\bullet$ Trp$^\bullet$] (Cintolesi *et al.*, 2003). This calculation was performed for a spin system comprising two electrons and a total of eight nuclear spins (two ^{14}N and three ^{1}H in FH$^\bullet$, one ^{14}N and two ^{1}H in Trp$^\bullet$). The hyperfine tensors were taken from EPR and ENDOR data and DFT calculations, and the relative orientation of the two radicals is that of the FAD cofactor and Trp$_{306}$ in the crystal structure of *E. coli* DNA photolyase. The polar plot in Figure 10.4 shows the anisotropy of Φ_S for different directions of the static field, with red and blue representing values of Φ_S that are respectively larger and smaller than the spherical average. It is clear, at least for this particular radical pair, that the orientation dependence of the reaction yield is not very different from that in Equation (10.5). The approximate symmetry axis of Φ_S appears to be determined by the two nitrogens in the flavin radical which have strong, near-axial hyperfine tensors with almost collinear principal axes (Cintolesi *et al.*, 2003).

Recent proof-of-principle experiments have shown for the first time that a radical pair reaction can respond both to the presence of a magnetic field weaker than 50 μT and to the direction of a (slightly stronger) field. Maeda *et al.* (Maeda *et al.*, 2008) studied a carotenoid–porphyrin–fullerene triad molecule (CPF) in which two sequential photo-induced intramolecular electron transfer steps produce the magnetically sensitive radical pair (or more strictly, biradical) [C$^{\bullet+}$ P F$^{\bullet-}$] (Figure 10.5a, b). Both S and T states are able to recombine (at different rates, k_S and k_T). Transient absorption experiments show that in frozen isotropic solution

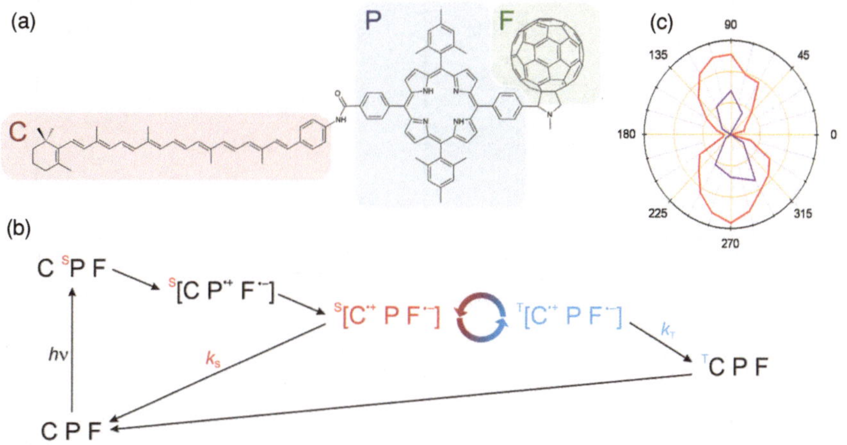

Figure 10.5 (a) Structure and (b) reaction scheme of the CPF (carotenoid–porphyrin–fullerene) triad used to demonstrate the principle of a chemical compass. ^{S}P and ^{T}C are excited singlet and triplet states of the porphyrin and the carotenoid, respectively. (c) Polar plot of the anisotropy of the magnetic field effect on the transient absorption of [C$^{•+}$ P F$^{•-}$] detected using an aligned sample (purple, 3.1 mT, 193 K) and by photoselection (red, 3.4 mT, 88 K). The maximum magnetic field effects in the two cases were ~1.5% and ~5%, respectively. The data for the aligned sample have been doubled for clarity. [Adapted from (Maeda *et al.*, 2008).]

the yield of [C$^{•+}$ P F$^{•-}$] on a sub-microsecond timescale changes by ~1.5% in a magnetic field of 40–50 μT (Maeda *et al.*, 2008). Anisotropic magnetic responses of similar magnitude were observed in ~3 mT magnetic fields using either an aligned sample in a liquid crystalline solvent or photoselection with polarized light (Maeda *et al.*, 2008). The orientation dependence of the radical pair signal (Figure 10.5c) is qualitatively consistent with Equation (10.5) and, like the avian magnetic compass, depends on the inclination rather than the polarity of the field. These measurements were made at low temperature to attenuate the effects of spin relaxation, which would otherwise destroy all coherences before the radical pair could recombine. The radical dynamics in a magnetoreceptor protein could well be more favourable in this respect. For example, flavin–tryptophan radical pairs in a cryptochrome retain their spin correlation for up to 10 μs at 1 °C (Biskup *et al.*, 2009).

10.3.2 Radiofrequency magnetic field effects

Radical pair reactions can also be affected by weak time-dependent magnetic fields. If close to resonance with an energy-level spacing produced by Zeeman and/or

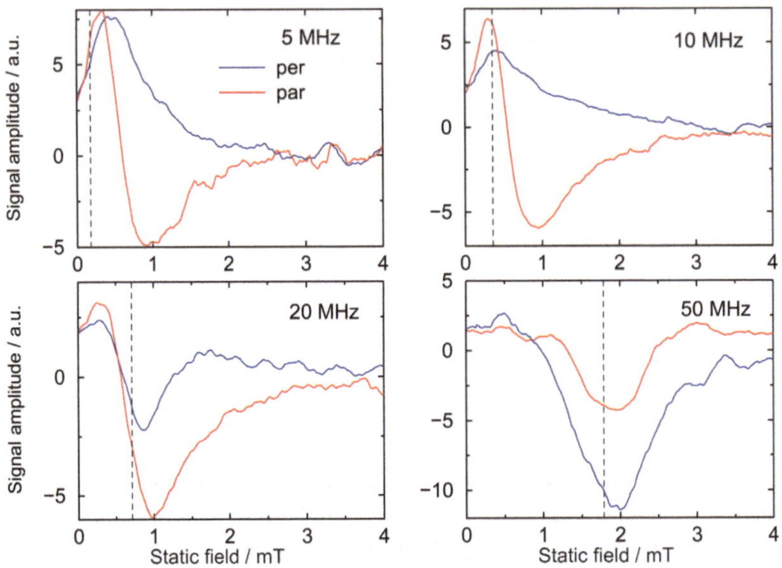

Figure 10.6 Effect of a ~300 μT radiofrequency field on the photochemical reaction of chrysene-d_{12} with 1,4-dicyanobenzene as a function of the strength of the applied static magnetic field. Red and blue lines are, respectively, for parallel and perpendicular static and RF fields. The radiofrequencies are as indicated. The dashed vertical lines indicate the static fields at which the RF radiation is in resonance with the electron Zeeman interaction. [Taken from (Henbest *et al.*, 2004).]

hyperfine interactions, an oscillating field $B_1(t)$ can modify the spin dynamics (Canfield *et al.*, 1994, 1995, 1995a; Timmel and Hore, 1996; Woodward *et al.*, 2001). As before, one of the conditions for a significant effect is that the Zeeman interaction, $\gamma_e B_1$, should at least be comparable to the recombination rate k. If a rotating frame transformation is appropriate, so that B_1 can be treated as independent of time, the theory is essentially as given in Equation (10.4) (Timmel and Hore, 1996). In general, though, a more sophisticated approach (e.g. perturbation theory (Canfield *et al.*, 1994, 1995a) or γ-COMPUTE (Henbest *et al.*, 2004; Rodgers *et al.*, 2005)) is required to account properly for the periodic time dependence of $H_Z(t)$.

Figure 10.6 illustrates some aspects of the effects of combined static and radiofrequency (RF) fields on the reactions of small organic radicals in solution. The reaction here is a photo-induced intermolecular electron transfer between the perdeuterated polycyclic aromatic hydrocarbon, chrysene (Chr-d_{12}), and the electron acceptor 1,4-dicyanobenzene (DCB). The singlet yield is monitored via exciplex fluorescence. The figure shows the influence of a ~300 μT RF field on the recombination of [Chr-$d_{12}^{\bullet+}$ DCB$^{\bullet-}$] as a function of the applied static field strength (B_0) for four radiofrequencies when the two fields are either parallel or perpendicular. Several features are apparent. The effect of the RF field depends strongly on its frequency; the parallel and perpendicular responses differ; the signals are weak when the RF is off-resonance with respect to both hyperfine and Zeeman interactions (static field $B_0 > 3$ mT); and at the higher frequencies, a dominant peak emerges close to the high-field EPR resonance condition, $\nu = \gamma_e B_0/2\pi$ or $\sim$0.036 mT MHz^{-1} (the 'Zeeman resonance'). Although not shown here, there are also differences between the responses to linearly and circularly polarized RF fields (Rodgers *et al.*, 2009; Wedge *et al.*, 2009). All of these effects, in a variety of radical pairs, with and without isotopic substitution, can satisfactorily be simulated, using independently determined hyperfine parameters and physically realistic estimates of the recombination rate constant k (Rodgers *et al.*, 2005, 2007, 2009; Wedge *et al.*, 2009). The observation of a pronounced Zeeman resonance provides scope for a general diagnostic test for the operation of the radical pair mechanism that does not require prior knowledge of the identity of the radicals or details of their chemistry. As illustrated in Figure 10.6 (and described in more detail elsewhere (Henbest *et al.*, 2004)), radical pair reactions that exhibit responses to a static magnetic field should also be sensitive to the frequency and direction of an additional RF field, and should exhibit a Zeeman resonance at a frequency that is not strongly dependent on the hyperfine interactions, provided that they are dominated by the static field. Such ideas lie behind the use of RF fields in behavioral experiments on migratory birds (see below) (Ritz *et al.*, 2004, 2009).

10.3.3 Magnetic field effects on a photoactive protein

Hitherto in this section, we have discussed exclusively MFEs on radical pairs that bear little structural similarity to cryptochrome. A demonstration that cryptochrome is sensitive, in vitro, to a 50 μT magnetic field at physiological temperatures would provide powerful support for the cryptochrome hypothesis. So far, this has not been achieved. However, there has been a study of *E. coli* DNA photolyase, a protein with high sequence homology to cryptochrome and similar photochemistry, in applied magnetic fields somewhat stronger than the Earth's (Henbest *et al.*, 2008). Laser flash photolysis of photolyase at 5 °C produced transient flavin and tryptophan radicals in yields that showed 3–4% responses to a 39 mT static magnetic field. These observations are consistent with a photochemical reaction scheme in which a primary radical pair comprising the deprotonated FADH$^\bullet$ radical and a tryptophan cation radical, [FAD$^{\bullet-}$ Trp(H)$^{\bullet+}$], is the magnetically sensitive species. The MFE arises from competition on a $\sim$1 μs timescale between spin-selective back electron transfer from the initially formed singlet state and deprotonation of the Trp(H)$^{\bullet+}$ radical, to form a non-magnetically sensitive secondary radical pair, [FAD$^{\bullet-}$ Trp$^\bullet$] with a lifetime of $\sim$1 ms. The MFE is thus manifest in the yield of a relatively long-lived FAD-radical state of the protein corresponding to the signalling state in cryptochrome (Henbest *et al.*, 2008). Although it has not yet been possible to extend these measurements to higher temperatures or lower fields, they do indicate that members of the cryptochrome/photolyase family satisfy many of the conditions necessary for a viable magnetoreceptor. As discussed in (Henbest *et al.*, 2008), there are several possible reasons why these responses are too small to be detected in an Earth-strength magnetic field and why they might be larger in specialized magnetoreceptor cells.

10.4 Evidence for a radical pair mechanism in birds

10.4.1 Lack of evidence for alternative mechanisms

Magnetoreception has long been postulated to be based on magnetite or other biogenic magnetic iron oxide particles. Simple detection of iron oxide contents in an animal is by itself not sufficient to indicate a role for iron oxides in magnetoreception, unless this is supported by corroborating behavioral observations. In birds, an iron oxide system has been found in the beaks in the vicinity of the ophthalmic nerve (Kirschvink *et al.*, 2001; Fleissner *et al.*, 2003, 2007; Solov'yov and Greiner, 2007, 2009b; Falkenberg *et al.*, 2010). However, under conditions in which birds show normal magnetic compass orientation in the seasonally appropriate migratory direction, their magnetic orientation responses are unaffected by anesthetization (Beason and Semm, 1996) of the beak or lesioning of the trigeminal branch of the ophthalmic nerve (Zapka *et al.*, 2009). These results show clearly that birds can detect the direction of a magnetic field without using the iron oxide system in the beak, thus indicating the existence of another, as yet undiscovered, magnetoreception system.

If this undiscovered system were based on iron oxide particles, one would expect that a strong magnetic pulse would re-magnetize or re-organize the magnetic material and therefore affect magnetoreception of an iron oxide based system. In a behavioral test, the bird beak system was anesthetized and a strong magnetic pulse applied, prior to testing magnetic compass responses. In these experiments, the birds showed unimpaired magnetic compass orientation, strongly suggesting that the undiscovered magnetoreception system is not based on a mechanism involving iron oxides (Wiltschko *et al.*, 2006). It is very likely that the beak iron oxide system plays some role in magnetoreception, but there is scant evidence suggesting that magnetoreception in birds can occur *only* with the help of iron oxide based mechanisms: birds can orient magnetically without using the only known iron oxide system in their beaks and application of a strong magnetic pulse, the standard indirect behavioral test for identifying an iron oxide based system, fails to indicate anyother iron oxide based system.

10.4.2 Neurobiology

The radical pair mechanism postulates that magnetic field effects are perceived as an indirect effect on light sensing. The most likely place for the receptors to be located would be in the eye(s), so as to harness the power and speed of the visual processing system. As mentioned above, the candidate photo-magnetoreceptor molecule cryptochrome has indeed been found in avian retinas. Thus, the question arises whether brain centers have been identified that receive visual inputs and are involved in processing magnetic information. Using genetic markers, a brain area termed Cluster N has been identified in European robins that is most active during magnetic compass orientation experiments at night (Mouritsen *et al.*, 2005), when European robins migrate, and much less active when the eyes are closed (Mouritsen *et al.*, 2005; Feenders *et al.*, 2008). Cluster N is part of the tecto-fugal visual processing pathway, and neuronal tracing has shown that it receives input from the eyes through only one synaptic transition (Heyers *et al.*, 2007; Liedvogel *et al.*, 2007b). European robins with bilateral Cluster N lesions cannot perform magnetic compass orientation (Zapka *et al.*, 2009), but are capable of sun and star compass orientation, demonstrating that Cluster N is involved in processing magnetic information. It is unclear whether this area is involved in processing compass information in birds whose compass operates during daytime and seems to show the same functional properties as the compass of night migrants (Wiltschko *et al.*, 2007).

10.4.3 Radiofrequency effects on magnetic orientation

An oscillating magnetic field with a frequency that matches an energy-level splitting between radical pair spin states is expected to affect S↔T interconversion, as in the in vitro experiments discussed above. Such fields could therefore change the sensitivity of a radical pair to the geomagnetic field. Analogous to the application of a strong magnetic pulse to modify the response of an iron oxide based compass system, one thus expects that the presence of a resonant oscillating field will modify the response of a radical pair based compass system, leading to re-orientation or disorientation in behavioral experiments when such a field is applied. Frequencies of resonances with typical hyperfine couplings and the free electron Larmor frequency fall into the range 1–100 MHz and one expects such fields to affect magnetic compass orientation. The lack of knowledge of the chemical nature of the hypothetical radical pairs in animal compass systems precludes more accurate predictions.

Figure 10.7 shows (a) the experimental arrangement used to investigate the effects of oscillating magnetic fields on the orientation of European robins and (b)

(a)

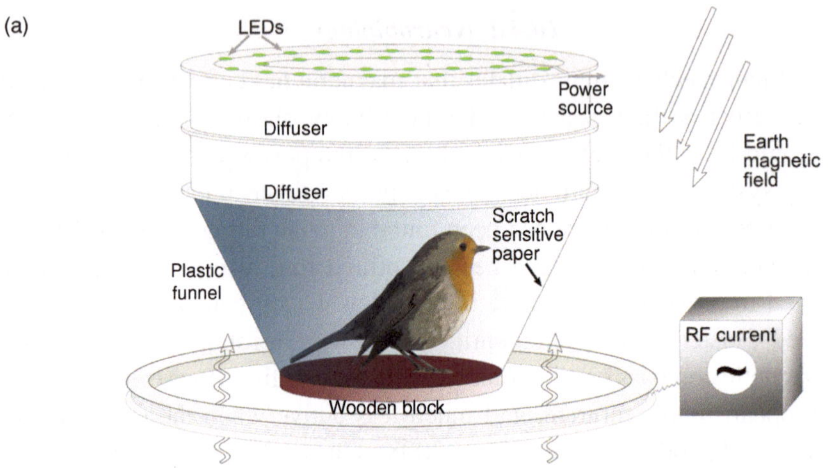

(b)

Intensity (nT)	46 μT static field						92 μT static field	
480	+	+	−	−	−	−		
150			+		+		+	
48				−	+		+	−
15				−	+			−
5				+				+
	0.01	0.03	0.65	1.315	2.63	7.0	1.315	2.63

Frequency (MHz)

Figure 10.7 (a) Schematic representation of the experimental arrangement used in Frankfurt to investigate the effects of radiofrequency magnetic fields on the orientation of European robins in the Earth's magnetic field. The birds' responses were recorded in funnel-shaped cages illuminated by diffuse light from above. In addition to the local geomagnetic field, an oscillating magnetic field was applied in each experimental condition. The funnels were lined with coated paper on which the birds left scratches as they moved. Analysis of the distribution of scratch marks allowed the birds' degree of orientation to be determined. For details see (Ritz *et al.*, 2004, 2009). (b) The outcome of the experiments using radiofrequency fields of the indicated intensity and frequency is presented as follows: circles with a minus denote disorientation, and circles with a plus denote statistically significant orientation in the appropriate migratory direction.

the results of experiments performed between 2004 and 2007 (Ritz *et al.*, 2004; Thalau *et al.*, 2005; Ritz *et al.*, 2009). In all conditions, the oscillating magnetic field was superimposed on a static magnetic field of either 46 μT (geomagnetic field) or an amplified static field of doubled intensity. The linearly polarized oscillating field was vertical, thus forming at the Frankfurt location a 24° angle with the static magnetic field. At an intensity of about 1% of the geomagnetic field, oscillating fields disrupt orientation of European robins at frequencies between from 0.65 up to 7 MHz, the highest frequency realized in the experimental setup. At frequencies below 30 kHz, the oscillating fields did not affect the robins' orientation. Bimodal orientation results at 0.1 and 0.5 MHz (not shown in the figure), suggested a transition region between oriented and disoriented behavior. These results suggest that the radical pair lifetime or the spin relaxation time, whichever is shorter, is in the range 2–10 μs (Ritz *et al.*, 2009). Clearly, an oscillating field with a period longer than the spin-relaxation time would be effectively static, and addition of a weak static magnetic field at 1% of the geomagnetic intensity is not expected to have a significant effect. Perhaps the most noteworthy feature of the oscillating field effects is that there is a dramatically stronger disruptive effect at 1.315 MHz, corresponding to the spin-only (i.e. $g = 2$) electron Larmor frequency in the geomagnetic field of 46 μT. At this frequency, a 15 nT RF field led to disoriented behavior, whereas about 30 times stronger fields were necessary to disorient birds at other frequencies. These observations suggest that one of the electron spins is magnetically isolated, i.e. that it is located on a radical with no hyperfine interactions (Ritz *et al.*, 2009). This suggestion is bolstered by the observation that doubling the static field intensity also doubles the frequency at which a 15 nT field leads to disorientation, as expected for the Zeeman resonance of a $g = 2$ radical.

A particularly strong disruptive effect of oscillating magnetic fields at the spin-only Larmor frequency has been observed in all species for which effects of oscillating fields on magnetic compass orientation have been found, namely in migratory European robins, non-migratory chickens (Wiltschko *et al.*, 2007) and Zebra finches (Keary *et al.*, 2009), as well as in cockroaches (Vacha *et al.*, 2009). This suggests that the magnetically sensitive radical-pair reaction has a similar chemical nature in different species. Radicals with an isolated electron spin are unusual in organic environments, as they need to be devoid of hydrogen or nitrogen atoms. The chemical nature of this postulated radical remains unknown. Superoxide and dioxygen have been suggested as possible candidates (Solov'yov and Schulten, 2009; Ritz *et al.*, 2009), but cannot be reconciled with known physical properties (Hogben *et al.*, 2009).

The existence of disruptive effects is a first indication supporting the radical-pair mechanism, but it is crucial that additional control conditions be tested to rule out that the change in orientation is due to an unrelated non-specific cause, e.g. a change in motivation due to the presence of the oscillating fields. Oscillating fields had no effect on the magnetic compass of mole rats, a blind, subterranean animal whose compass is probably based on iron oxide materials (Thalau *et al.*, 2006), indicating that effects of oscillating fields appear not to affect iron-based systems. A key control observation is that the angle of the oscillating fields with respect to the geomagnetic field determines whether birds are oriented or disoriented (Ritz *et al.*, 2004; Thalau *et al.*, 2005). Birds were disoriented when the oscillating fields formed a 48° (or 24°) angle with the geomagnetic field, but not when they were collinear with the geomagnetic field. The choice of 48° is particularly meaningful as a control condition, because at this angle, the oscillating field is applied at the same angle relative to the horizontal plane (in which the birds move during the experiments) as in the 0° condition. There is no reason why the birds' motivation should be affected differently by non-specific effects of oscillating fields of equal intensity, frequency and direction with respect to the horizontal. It appears much more likely that oscillating fields produce a resonance effect, in which case it is indeed expected that a collinear oscillating field will leave radical-pair reactions unaffected (Ritz *et al.*, 2009). The observation of oscillating magnetic field effects strongly supports the existence of a radical-pair mechanism, but the quantitative interpretation of the observed effects still raises a number of questions, discussed in more detail in (Ritz, 2011).

10.5 Conclusion

The last decade has seen a number of studies from different fields that support the photo-magnetoreceptor and cryptochrome hypotheses. Man-made radical-pair reactions have been designed that have proved to be sensitive to Earth-strength magnetic fields. Behavioral experiments using radiofrequency fields support the existence of a radical-pair mechanism in birds. Studies at the protein level suggest that cryptochromes and the closely related photolyases have properties conducive to magnetic sensing, such as formation of long-lived radical pairs. Magnetic field effects have been observed in several genetic organisms and were absent when cryptochromes were deleted. A visual brain area has been identified that is active during magnetic orientation behavior, and without which birds become disoriented in magnetic orientation experiments. At this point, the radical-pair hypothesis is not proven: fundamental questions remain in many relevant areas, as discussed in (Ritz, 2011; Solov'yov *et al.*, 2012). However, support for this hypothesis has strengthened significantly, in particular for migratory birds. If it can be shown conclusively that birds use a radical-pair-based compass, this would be a dramatic example of the use of a coherent quantum-mechanical process in a biological system and with clear biological relevance.

11

Quantum biology of retinal

KLAUS SCHULTEN AND SHIGEHIKO HAYASHI

11.1 Introduction

Retinal is a biological chromophore ubiquitous in visual receptors of higher life forms, but serving also as an antenna in light energy transformation and phototaxis of bacteria. The chromophore arises in various retinal proteins, the best known two being the visual receptor rhodopsin and the light-driven proton pump bacteriorhodopsin. The ubiquitous nature of retinal in photobiology is most remarkable as the molecule shows an extremely wide adaptability of its spectral absorption characteristics and a precise selection of its photoproducts, both properties steered by retinal proteins.

Rhodopsin (Rh) is a membrane protein of the rod cells in the retina of animal eyes and contains a retinal molecule as a chromophore surrounded by the protein's seven transmembrane helices (Khorana, 1992), as shown in Figure 11.1. Rhodopsin serves as the receptor protein for monochromic vision, in particular, for vision in the dark. Analogous retinal proteins, called iodopsins, exist in the cone cells of the retina and serve as receptor proteins for colour vision in daylight (Nathans *et al.*, 1986).

Retinal proteins also serve in certain bacteria as light-driven proton pumps that maintain the cell potential, as in case of bacteriorhodopsin (bR) (Schulten and Tavan, 1978), or as light sensors in bacterial phototaxis (Spudich and Jung, 2005).

All retinal proteins are structurally homologous, being composed of seven transmembrane helices with a retinal chromophore bound to a lysine side group. The photoactivation mechanisms of the proteins' retinal moieties are similar, but distinct from each other. The primary event of retinal photoactivation is a photoisomerization reaction (Birge, 1990). The photobiological mechanism of retinal has been fascinating experimental and theoretical researchers over many decades until

Quantum Effects in Biology, ed. Masoud Mohseni, Yasser Omar, Gregory S. Engel and Martin B. Plenio. Published by Cambridge University Press.

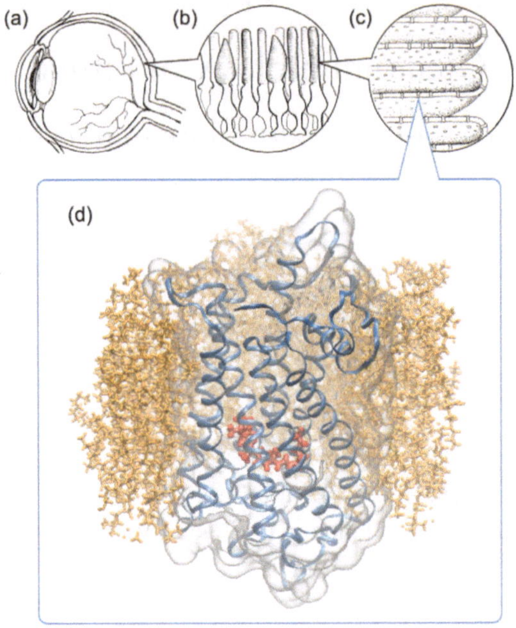

Figure 11.1 Visual receptors. (a) Cross-section of the eye with lens and iris on the left and the retina with nerve cells highlighted on the right. (b) The retinas of animals contain millions of rod and cone receptor cells. Shown are shapes and arrangement of rod and cone cells. (c) Discs in the cone and rod cells. The discs are hollow and made of membranes that are saturated with the light receptor protein, rhodopsin (Rh), in the case of rods, and several types of proteins, iodopsins, in the case of cones. In the figure Rhs are highlighted. (d) Light is absorbed by retinal (red), the chromatophore molecule situated inside Rh (blue tube and grey surface), which in turn is located inside a lipid (brown) membrane. Light-excited retinal undergoes an isomerization reaction triggering the visual signal; this chapter describes the quantum-mechanical nature of retinal's light absorption characteristics and photo reactions.

today. In this chapter, the quantum processes involved in the photoactivation of retinal in Rh and related proteins are presented.

11.2 Retinal in rhodopsin and bacteriorhodopsin

Figure 11.2 depicts the chemical structure of the retinal chromophore, mainly a polyene chain with six conjugated double bonds. One end of the polyene chain is covalently attached to a lysine residue through a protonated Schiff base (PSB); the other end forms a β-ionone ring. Upon photoabsorption, the polyene part of retinal undergoes isomerization towards an isomer different from that of the initial (reactant) state, i.e. one of the retinal double bonds (discussed further below) rotates

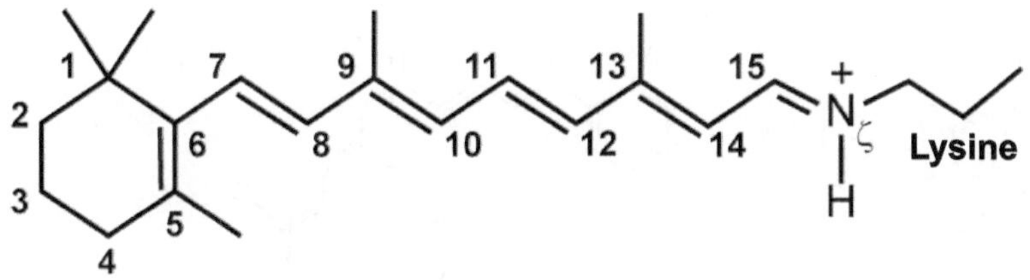

Figure 11.2 Chemical structure of the retinal chromophore linked through a protonated Schiff base to a lysine amino acid residue of the sensory protein rhodopsin; the nitrogen atom involved in the linkage is protonated. The polyene backbone of retinal ends on the right side in the protonated Schiff base and on the left side in a β-ionone ring. Shown is the conventional numbering of retinal carbon atoms. Bonds depicted by two parallel lines, e.g. the 11–12 bond, are the so-called double bonds that exhibit typically very high (compared with physiological thermal energy $k_B T$) torsional energy barriers preventing spontaneous $0° \rightarrow 180°$ rotation; bonds depicted by a single line, e.g. the 1–11 bond, are the so-called single bonds that exhibit typically lower torsional energy barriers.

by 180°. During the primary, extremely fast (50–500 fs) isomerization step, retinal, confined to its tight protein binding pocket, is forced to accommodate the 180° rotation around one bond with a minimal change of its overall shape, which it achieves through torsional counter-rotations involving its other bonds, though in doing so, ending up in a high energy conformation. Because of the high energy content, the torsional distortions of retinal seek to relax, which brings about an overall shape change of retinal that can be accommodated only if the surrounding protein alters its shape, too. This accommodation and subsequent protein processes coupled to it, for example motion of internal water and proton transfer, initiates a photocycle during which the retinal protein assumes a new functional state.

In the case of Rh, the protein enters an intermediate signalling state that activates a cascade of signalling reactions involving other proteins. Eventually, the cascade amplifies Rh signalling sufficiently for the interior electrical potential of the rod cell to change enough to induce firing of neurons, the so-called ganglion cells, linking the retina to a brain area relevant for vision. The description given here does not do justice to the complexity and impressive function of the retina.

In the case of bR, changes in the protein interior brought about by retinal photoisomerization induce the vectorial transfer of a proton from the interior to the exterior of the bacterial cell wall, in which bR resides, charging the bacterial cell energetically.

As shown in Figure 11.1, Rh is a visual pigment responsible in animal eyes for monochromic vision in the dark. In the human eye, three other visual pigments, iodopsins, also exist in the retina, namely in the so-called cone cells, and are responsible for colour vision. The visual pigments are members of the protein

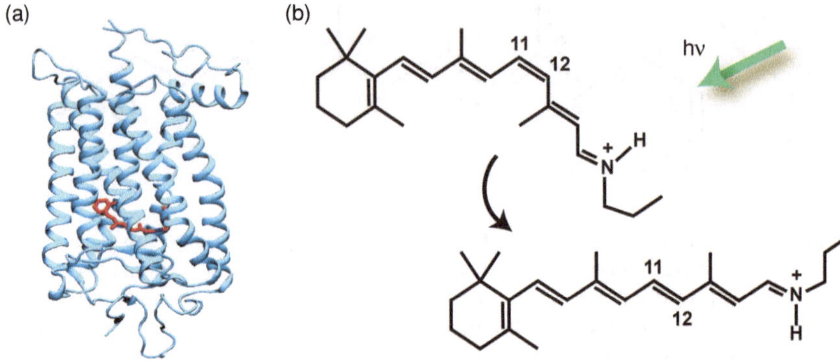

Figure 11.3 (a) Structure and photodynamics of rhodopsin. Seven transmembrane helices of the protein embed a retinal (red). (b) Photoreaction in Rh. Upon photoabsorption, retinal undergoes isomerization from 11-cis to all-trans. Retinal is shown bound as a protonated Schiff base to a lysine residue.

family of G-protein coupled receptors (GPCRs); photoactivation of the receptors leads to binding of G-protein to initiate a signalling transduction cascade (Shichida and Imai, 1998). Figures 11.1d and 11.3a depict the protein structure of Rh. In the resting dark state of Rh and the three human iodopsins, retinal is found in an 11-cis configuration. The primary photoreaction is a photoisomerization of retinal from its 11-cis state to its all-trans state, as shown in Figure 11.3b. The photoisomerization reaction in Rh is one of the fastest molecular reactions in nature; it completes within 200 fs (Schoenlein *et al.*, 1991; Kochendoerfer and Mathies, 1996; Polli *et al.*, 2010).

Photoabsorption spectra of Rh and of the three iodopsins reach over a wide range of the visible spectrum: the photoabsorption maximum of Rh is at 500 nm, i.e. at a wavelength where sunlight has maximum intensity on the surface of the Earth. Photoabsorption of the three human iodopsins covers 400–500 nm, 450–630 nm and 500–700 nm, with absorption maxima around 430, 550 and 570 nm. The absorption characteristics permit the iodopsins to discriminate colour hues, an impressive ability as most readers can judge. It is remarkable that all four human visual receptors employ 11-cis retinal, yet absorb 400–700 nm light; the quantum-mechanical mechanism underlying spectral tuning is explained further below.

Microbial rhodopsins are widely distributed in archaea, bacteria and unicellular eukaryotes, and function as photosynthetic light-driven ion pumps and photosensory receptors (Spudich and Jung, 2005). The proteins exhibit high sequence homology with one another, but no detectable homology with GPCR rhodopsins is seen. Recently, microbial rhodopsins have attracted much attention for their wide distribution in the huge euphotic zone of the deep sea (Béjà *et al.*, 2000; Kolber *et al.*, 2000). Microbial rhodopsins also caught the limelight as tools of a

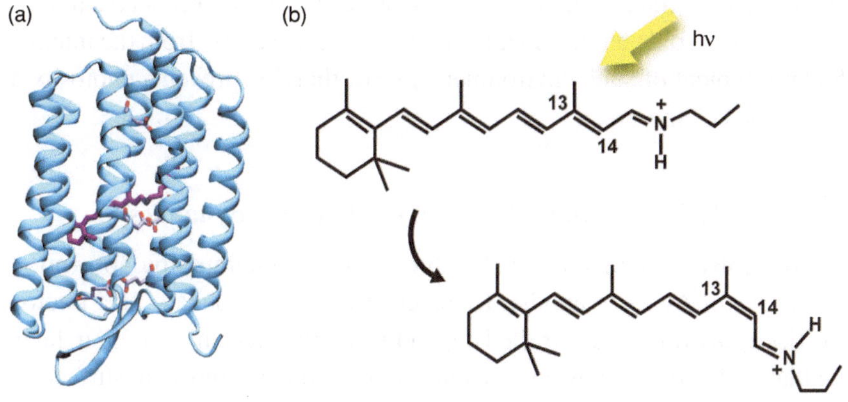

Figure 11.4 (a) Structure and photodynamics of bacteriorhodopsin. A PSB retinal chromophore (purple) is located in the protein. Carboxylate protein groups in the proton channel, involved in light-induced proton pumping, are also highlighted. (b) Photoreaction in bR. Upon photoabsorption, retinal undergoes an all-trans to 13-cis isomerization.

newly developed neuroscience technique called optogenetics; neural circuits are controlled in vivo by photoillumination through photosensitive ion channel and pump rhodopsins genetically expressed in neurons (Zhang *et al.*, 2007).

Bacteriorhodopsin (bR), an extensively studied archaeal rhodopsin, functions as a light-driven proton pump. Figure 11.4a depicts the structure of bR. A proton channel exists in the region between the transmembrane helices, involving carboxylate side-chain groups, the protonated Schiff base of retinal and water molecules. In the dark resting state of bR, retinal is found in its all-trans form, i.e. not in its 11-cis form as in the visual receptors. The primary photoreaction isomerizes the all-trans form to 13-cis, as shown in Figure 11.4b. The photoisomerization reaction in bR, as in Rh, is fast with a sub-picosecond time constant (Gai *et al.*, 1998). The photoisomerization triggers a photocycle that couples protein conformational changes to vectorial proton transport from the cytoplasmic side to the extracellular side of the bacterial cell membrane.

As in the case of visual receptors, absorption spectra of microbial rhodopsins are adapted to the bacterial habitat and to their particular function. For example, the spectra of proteorhodopsin of various aquatic species are adapted to the water depth in which a species typically lives. Another example is furnished by bacterial rhodopsins, sRI and sRII, which function as sensors for phototaxis in archaebacteria and possess different photoabsorption spectra due to their opposite functions. The absorption maximum of sRI, 587 nm, is close to the maxima of bacterial rhodopsins acting as ion pumps serving cell function and, thus, sRI features positive phototaxis guiding the bacteria to the suitable light environment. sRII, however, with an absorption maximum at 498 nm, features negative phototaxis and, mainly expressed in an aerobic environment situation, repels bacteria from the intense light at ~500 nm typical of such environments; sRII, thereby, prevents photo-oxidative damage to the cells.

11.3 Quantum physics of excited state dynamics

An electronically excited state of the chromophore generated by absorption of a photon decays one way or another to the electronic ground state. In a photoactive protein, the decay route is controlled depending on the protein's function, favouring a route that ends up in a new energy minimum on the ground state surface; in this minimum the protein is in its active state. Here we outline the quantum-mechanical description of the excited state dynamics and its control.

The excited state processes involve the nuclear motion of chromatophore and protein proceeding on the *Born–Oppenheimer (BO) or adiabatic* potential energy surfaces of the excited and ground electronic states, and also involve transition between the two surfaces. The description of the processes is based on the non-relativistic Hamiltonian of the system, $\hat{H}$, that is expressed as a differential operator with respect to coordinates of electrons, $\mathbf{r}$, and coordinates of nuclei, $\mathbf{R}$,

$$\hat{H}(\mathbf{r}, \mathbf{R}) = \hat{T}_N(\mathbf{R}) + \hat{H}_{BO}(\mathbf{r}, \mathbf{R}). \tag{11.1}$$

Here, $\hat{T}_N(\mathbf{R})$ is the nuclear kinetic energy operator and $\hat{H}_{BO}(\mathbf{r}, \mathbf{R})$ is the BO Hamiltonian, which consists of the electronic Hamiltonian and the nuclear repulsion energy,

$$\begin{aligned}\hat{H}_{BO}(\mathbf{r}, \mathbf{R}) = {} & \sum_{i}^{N_{elec}} \left(-\frac{\hbar^2}{2m_e} \nabla_i^2 \right) - \sum_{i}^{N_{elec}} \sum_{A}^{N_{nuc}} \frac{Z_A e^2}{4\pi \epsilon_0 r_{iA}} \\ & + \sum_{i}^{N_{elec}} \sum_{j>i}^{N_{elec}} \frac{e^2}{4\pi \epsilon_0 r_{ij}} + \sum_{A}^{N_{nuc}} \sum_{B>A}^{N_{nuc}} \frac{Z_A Z_B e^2}{4\pi \epsilon_0 R_{AB}},\end{aligned} \tag{11.2}$$

where N_{elec} and N_{nuc} are the numbers of electrons and nuclei, respectively, m_e is the mass of the electron, $\hbar = h/(2\pi)$ is the reduced Planck constant, Z_A is the atomic number of atom A, e is the elementary electric charge, and ϵ_0 is the vacuum permittivity. Electronic and nuclear motion are separated according to the *Born–Oppenheimer (BO) approximation*. In this approximation, electronic wavefunction and energy of the electronic states are determined by solving the Schrödinger equation for $\hat{H}_{BO}(\mathbf{r}, \mathbf{R})$:

$$\hat{H}_{BO}(\mathbf{r}, \mathbf{R})\Psi_I(\mathbf{r}, \mathbf{R}) = E_I^{BO}(\mathbf{R})\Psi_I(\mathbf{r}, \mathbf{R}), \tag{11.3}$$

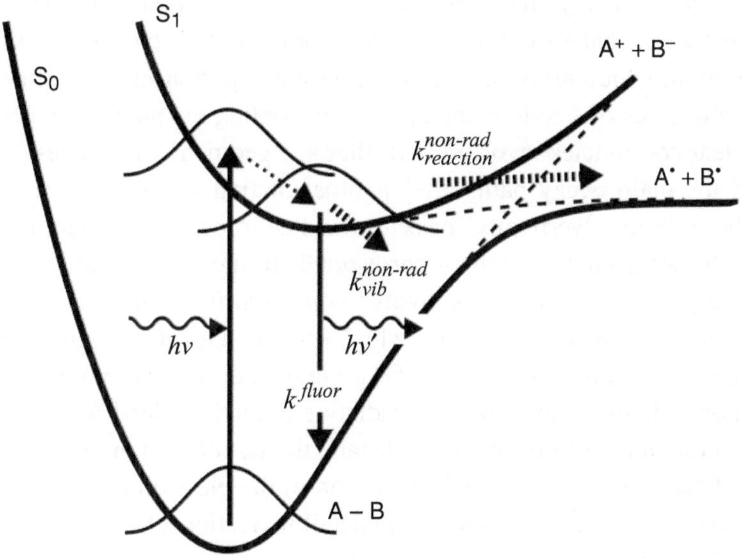

Figure 11.5 Competition of excited state decay paths in a photodissociation reaction of a diatomic molecule, e.g. NaI. The excited state decay paths seen in this simple photoreaction are representative of the more complex photoreaction of retinal in Rh. Thick solid and thin dashed lines represent adiabatic and diabatic potential energy curves, defined in Section 11.4. Upon photoabsorption to the FC region (solid upward arrow), a wavepacket describing the equilibrated nuclear motion on the ground state potential energy curve, labelled S_0, is lifted to the excited state potential energy curve, labelled S_1, and relaxes then to a nearby potential energy minimum (thin dotted arrow). Subsequently, the excited state population decays via three paths: fluorescence (solid downward arrow), non-radiative decay (thick dashed arrows) to vibrationally excited states of the ground state curve, and non-radiative decay to a photoproduct through an avoided energy curve crossing (discussed in Section 11.4).

where $\Psi_I(\mathbf{r}, \mathbf{R})$ and $E_I^{BO}(\mathbf{R})$ are the BO or adiabatic wavefunction and energy, respectively, of the Ith electronic state at a fixed nuclear geometry $\mathbf{R}$. The eigenvalue problem (11.3) is solved through methods of quantum chemistry.

In general, the excited state decay involves more than one decay path; the paths compete with each other kinetically (Atkins and de Paula, 2009). Figure 11.5 depicts, schematically, possible paths. Firstly, upon *photoabsorption,* the chromophore in the resting electronic ground state is promoted to the electronic excited state, according to the Franck–Condon (FC) principle, without changing the nuclear geometry, in what is referred to as a *vertical excitation.* The conformational space region in the excited state, reached after the vertical excitation from the ground state nuclear Boltzmann distribution, is called the *FC region.* The FC region is usually not at an energy minimum of the excited state potential energy surface and, hence, the vertical excitation generates vibrationally excited (i.e. hot) states of nuclear motion, which relax immediately, dissipating their energy into surrounding nuclear degrees of freedom and effectively cooling the nuclear motion viewed along nuclear coordinates involved in further key geometrical changes.

One of the main decay paths after photoexcitation is *fluorescence* where the chromophore decays 'vertically' back to the electronic ground state by emitting a photon, the wavelength of which corresponds to the energy gap bridged in the vertical decay. Other decay paths involve *non-radiative transitions* in which the system moves to a nuclear geometry, where excited state and ground state energy surfaces intersect, and transitions to the ground state without emitting a photon, i.e. in a non-radiative way. The non-radiative transitions lead to highly excited nuclear motion in the electronic ground state, the wavefunction of which matches best that of the pre-transition nuclear motion in the electronic excited state. The non-radiative transitions are described in detail in Section 11.5.

Non-radiative transitions that take place between electronic (excited and ground) states of the same spin symmetry are called *internal conversion*. In this case the transition is due mainly to so-called *non-adiabatic coupling* originating from the nuclear kinetic energy operator $\hat{T}_N(\mathbf{R})$. The system wavefunction in the BO-approximation is factored into an electronic wavefunction $\Psi(\mathbf{r}, \mathbf{R})$, the solution of Equation (11.3), and a nuclear wavefunction $\Phi(\mathbf{R})$. Considering at present just the electronic wavefunction, $\Psi_I(\mathbf{r}, \mathbf{R})$, which defines through Equation (11.3) the electronic state potential energy surfaces E_I^{BO}, the non-adiabatic coupling that is added to the wave equation for the nuclear motion $\Phi(\mathbf{R})$ is

$$\langle \Psi_J(\mathbf{r}, \mathbf{R}) | \hat{T}_N(\mathbf{R}) | \Psi_I(\mathbf{r}, \mathbf{R}) \rangle_r \simeq -i\hbar \sum_A^{N_{nuc}} d_{JI}^A(\mathbf{R}) \frac{\hbar}{iM_A} \nabla_A, \tag{11.4}$$

namely, a linear combination of operators acting on $\Phi(\mathbf{R})$. Here M_A is the mass of atom A. The $\mathbf{R}$-dependent coefficients of this operator,

$$d_{JI}^A(\mathbf{R}) = \langle \Psi_J(\mathbf{r}, \mathbf{R}) | \nabla_A | \Psi_I(\mathbf{r}, \mathbf{R}) \rangle_r, \tag{11.5}$$

are called the non-adiabatic coupling vectors (Groenhof *et al.*, 2009). In Equation (11.4), the second derivatives of the electronic wavefunction with respect to the nuclear coordinates, $\mathbf{R}$, are neglected. The role of the operator in Equation (11.4) is considered further in Section 11.5. In Equation (11.5), $\langle \cdots \rangle_r$ denotes integration over the electronic coordinates.

The non-radiative processes can be discerned indirectly by measuring the *quantum yield of fluorescence*, ϕ^{fluor}, which is defined as the ratio of the number of photons emitted to the number of photons absorbed. If non-radiative transitions do not take place, the number of photons emitted from the population of the excited state would be equal to the number of photons absorbed which produce the population; a measured decrease of the fluorescence quantum yield below unity is indicative of significant non-radiative decay. In this case, the actual value of the quantum yield permits one to estimate the speed of the non-radiative decay. For the sake of simplicity, let us assume that the decay process and fluorescence can be described by first-order kinetics with rate constants $k^{non-rad}$ and k^{fluor}, respectively. The quantum yield is then

$$\phi^{fluor} = \frac{k^{fluor}}{k^{fluor} + k^{non-rad}}. \tag{11.6}$$

One can conclude that an increased $k^{non-rad}$ leads to a decrease of the fluorescence quantum yield.

11.4 Regulation of photochemical processes for biological function

Photoinduced processes in proteins are regulated according to the protein's function. Functions are classified as photoactivation, as in rhodopsins, luminescence as in green fluorescent proteins, and light-harvesting as in some photosynthetic protein complexes.

For photoactive proteins, strong photoabsorption is a prerequisite to attain high sensitivity to incoming light. As photoabsorption cross-sections are proportional to the square of the transition dipole moment of the chromophore (Atkins and de Paula, 2009), an optically allowed electronically excited state that contributes a large transition dipole moment is needed. In this regard, the protonated Schiff base of retinal is ideally suited for visual photopigments. The respective characteristics of the optically allowed excited state of retinal is discussed in Section 11.6. In addition to the photoabsorption cross-section, absorption energy, i.e. the spectral maximum of absorption, is a key factor for visual photopigments. The mechanisms underlying tuning of absorption maxima are discussed in Section 11.7.

Control of excited state decay after photoabsorption is also crucial for photorelated protein function. For example, photoactivation requires very different decay channels from bio-luminescence. In the case of luminescence, the quantum yield of fluorescence, ϕ^{fluor}, is maximized; this also applies to chromophores involved in excitation energy transfer in light-harvesting systems. Both types of proteins minimize the role of non-radiative decay. In contrast, photoactivation requires the minimization of fluorescence and the channelling of non-radiative decay towards formation of ground state photoproducts distinct from the initial, i.e. dark, state.

In the case of photoactivative proteins, a large $k^{non-rad}$ is essential for keeping ϕ^{fluor} small, as can be concluded from Equation (11.6). As we noted above, the transition dipole moment of the chromophore in photoactive proteins must be large in order to assure strong absorption of incoming light. Hence, fluorescence from the optically allowed excited state is also fast, as the fluorescence rate is also proportional to the square of the transition dipole moment. In order to realize a large quantum yield for photoproduct formation, the respective non-radiative decay channel needs to be much faster in the protein than the fluorescence channel, the latter exhibiting a typical decay time, i.e. $1/k^{fluor}$, of 1 ns.

Disorganized motion on the excited state surface does not result in motion leading to a specific potential energy crossing point and subsequent product formation in the ground state. Accordingly, photoactive proteins need to trigger very fast specific motion on the excited state potential energy surface, leading to the suitable potential energy crossing point. Typically, the protein environment plays a role in funnelling the nuclear motion in the excited state along the desired photochemical reaction coordinate. As mentioned above and discussed in detail below, photochemical reactions of retinal in rhodopsins are extremely fast and well-steered, leading to very weak fluorescence and high quantum yields of photoproduct formation.

11.5 Potential energy crossing and conical intersection

Photochemical reactions constitute one of the non-radiative processes in chromophores. Figure 11.5 depicts a prototype photochemical reaction process, photodissociation of a diatomic molecule. This process is a one-dimensional analogue of the type of processes occurring in more complex molecules like retinal. In the case of a diatomic molecule, the ground state possesses a potential energy curve along the atom–atom distance, with a minimum that corresponds to a stable ionic bond; as the distance increases the potential energy increases steeply and undergoes a so-called *'avoided' energy crossing* (AEC) with a potential energy curve of the covalent excited state of the molecule. Separating the atoms beyond the avoided crossing in Figure 11.5 leads to dissociation into free neutral atoms.

In the one-dimensional reaction case depicted here, BO energy curves of two states with the same spatial and spin symmetry do not cross (von Neumann and Wigner, 1929). The non-adiabatic coupling term (11.5) contributes to the level repulsion, in particular in the AEC region, where the close energies of two BO states, I and J, give rise to strong non-adiabatic coupling, as one can see (note the energy difference in the denominator) from the readily derived expression for the non-adiabatic coupling vector,

$$\langle \Psi_J(\mathbf{r},\mathbf{R}) | \nabla_A | \Psi_I(\mathbf{r},\mathbf{R}) \rangle_r = \frac{\langle \Psi_J(\mathbf{r},\mathbf{R}) | [\nabla_A,\ \hat{H}_{BO}(\mathbf{r},\mathbf{R})] | \Psi_I(\mathbf{r},\mathbf{R}) \rangle_r}{E_I^{BO}(\mathbf{R}) - E_J^{BO}(\mathbf{R})}. \quad (11.7)$$

The commutator, $[\nabla_A, \hat{H}_{BO}(\mathbf{r}, \mathbf{R})]$, in this expression corresponds to multiplying the wavefunction $\Psi_I(\mathbf{r}, \mathbf{R})$ by function $[\nabla_A \hat{H}_{BO}(\mathbf{r}, \mathbf{R})]$, the square brackets $[\cdots]$ denoting that ∇_A only acts on $\hat{H}_{BO}(\mathbf{r}, \mathbf{R})$, but not beyond the brackets. $[\nabla_A \hat{H}_{BO}(\mathbf{r}, \mathbf{R})]$ represents the electrostatic force acting on atom A; since this force is a smooth, but non-constant, function of $\mathbf{R}$ in the AEC region, the coupling expressed through Equation (11.7) becomes very strong at the AEC due to the energy difference denominator in Equation (11.7) and noticing l'Hospital's rule.

The rate of non-radiative decay which competes against fluorescence, therefore, can proceed quickly along a non-adiabatic transition at an AEC. In general, the non-radiative decay rate, $k^{non-rad}$, involving two parallel channels described by rate constants $k_{vib}^{non-rad}$ (rate of transition from the vibrationally relaxed states from which fluorescence also occurs) and $k_{reaction}^{non-rad}$ (rate of transition through AEC for the photochemical reaction channel, see Figure 11.5) is approximately

$$k^{non-rad} = k_{vib}^{non-rad} + k_{reaction}^{non-rad}. \tag{11.8}$$

In polyatomic molecules, non-radiative decay through energy curve crossing requires consideration of more than one coordinate for the nuclear motion. Experimentally, many organic molecules have been found to exhibit remarkably small quantum yields of fluorescence, suggesting effective pathways for non-radiative decay through energy crossing. The observed non-radiative decays proceed on a timescale of picoseconds or shorter, which is comparable with the timescale of intramolecular vibrational relaxation after photoexcitation to the FC region. This fast non-radiative decay can be explained through non-adiabatic transition at so-called *conical intersections* (CIs) (Groenhof *et al.*, 2009). Figure 11.6a depicts, schematically, two BO potential energy surfaces around a CI. The two BO potential energy surfaces touch each other at the CI point, described here in a two-dimensional coordinate space, a linear subspace of the configuration space of $(3N_{nuc} - 6)$ degrees of freedom. The small energy gap of the BO energy surfaces near the CI point drastically accelerates the non-adiabatic transition, just as it does in the one-dimensional case shown in Figure 11.5.

The CI with the peculiar potential energy profile shown in Figure 11.6a is easily demonstrated as follows. For the sake of simplicity, let us consider a CI between the first excited state and the ground state. Firstly, the two adiabatic electronic wavefunctions, $\Psi_1(\mathbf{r}, \mathbf{R})$ and $\Psi_0(\mathbf{r}, \mathbf{R})$, defined by Equation (11.3), are transformed through a unitary transformation, $\hat{U}(\mathbf{R})$, to *diabatic* form where the non-adiabatic coupling, given by Equation (11.4), is minimized to a near zero value,

$$\tilde{\Psi}_{1,0}(\mathbf{r}, \mathbf{R}) = \hat{U}(\mathbf{R})\Psi_{1,0}(\mathbf{r}, \mathbf{R}), \tag{11.9}$$

$$\langle \tilde{\Psi}_1(\mathbf{r}, \mathbf{R}) | \hat{T}_N(\mathbf{R}) | \tilde{\Psi}_0(\mathbf{r}, \mathbf{R}) \rangle_r \simeq 0. \tag{11.10}$$

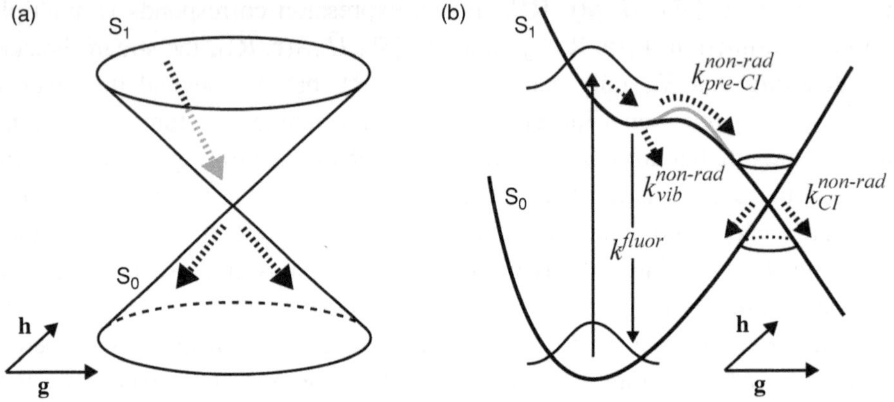

Figure 11.6 (a) Schematic representation of a conical intersection. A trajectory proceeding on the excited state BO potential energy surface (upper cone) undergoes an electronic transition near the conical intersection (CI) point at which apexes of two cones of the excited state and ground state touch in a coordinate space spanned by vectors **g** and **h** defined in Equations (11.12), (11.13). After the transition, the trajectory moves quickly on the ground state BO potential energy surface (lower cone) down the energy gradient. (b) Competition of excited state decay paths. Since the rate of non-radiative transition at the CI is fast, the rate of the overall non-radiative transition is determined mainly by the rate at which trajectories reach the CI region.

Here, $\tilde{\Psi}_1(\mathbf{r}, \mathbf{R})$ and $\tilde{\Psi}_0(\mathbf{r}, \mathbf{R})$ represent the diabatic electronic wavefunction of the excited state and the ground state, respectively. For a one-dimensional reaction of a diatomic molecule, diabatic wavefunctions that exactly eliminate the non-adiabatic coupling can be determined readily. However, it is not possible to find such a unitary transformation between two states eliminating the non-adiabatic coupling for polyatomic molecules, since all coordinate components of the non-adiabatic coupling vector, Equation (11.5), need to vanish in this case. However, one can choose a reaction coordinate along which the non-adiabatic coupling is minimized, determining in this way the unitary transformation.

The Hamiltonian matrix $\hat{H}_{BO}(\mathbf{r}, \mathbf{R})$ in the basis of the diabatic electronic wavefunctions is no longer diagonal. We express this matrix,

$$\hat{H}_{BO}(\mathbf{r}, \mathbf{R}) = \begin{pmatrix} \tilde{W}_1(\mathbf{R}) & \tilde{V}(\mathbf{R}) \\ \tilde{V}(\mathbf{R}) & \tilde{W}_0(\mathbf{R}) \end{pmatrix}, \tag{11.11}$$

where the diagonal elements, $\tilde{W}_{1,0}(\mathbf{R})$, can be regarded as potential energies of the so-called diabatic states. Unlike adiabatic potential energies, diabatic ones can cross. In a one-dimensional reaction of a diatomic molecule, the off-diagonal element $\tilde{V}(\mathbf{R_c})$ at the crossing point, $\mathbf{R_c}$, in general, is non-zero unless the spatial and spin symmetries of the two electronic states don't match. Thus the crossing of the energy curves of the adiabatic states with the same spatial and spin symmetries is 'avoided', with a minimal energy gap of $2\tilde{V}(\mathbf{R}_c)$.

In the case of diabatic potential energy surfaces existing in a configurational space of dimension higher than two, one can find a point, $\mathbf{R}_{\mathrm{CI}}$, where the potential energies cross, i.e. where $\tilde{W}_1(\mathbf{R}_{\mathrm{CI}}) - \tilde{W}_0(\mathbf{R}_{\mathrm{CI}}) = 0$ and $\tilde{V}(\mathbf{R}_{\mathrm{CI}}) = 0$ hold simultaneously. Since diagonal elements are degenerate and off-diagonal elements vanish at the crossing point, energies in the adiabatic representation obtained by the inverse unitary transformation are also degenerate, i.e. it holds that $E_1^{BO}(\mathbf{R}_{\mathrm{CI}}) - E_0^{BO}(\mathbf{R}_{\mathrm{CI}}) = 0$. Taylor expansion of the adiabatic potential energy surfaces around $\mathbf{R}_{\mathrm{CI}}$, the CI point, determines two directions $\mathbf{g}$ and $\mathbf{h}$ in the configuration space along which the adiabatic potential energies separate:

$$\mathbf{g} \equiv \nabla_{\mathbf{R}}[E_1^{BO}(\mathbf{R}) - E_0^{BO}(\mathbf{R})]|_{\mathbf{R}=\mathbf{R}_{\mathrm{CI}}}, \tag{11.12}$$

$$\mathbf{h} \equiv \nabla_{\mathbf{R}}\langle\Psi_1(\mathbf{r},\mathbf{R})|\hat{H}_{BO}(\mathbf{r},\mathbf{R})|\Psi_0(\mathbf{r},\mathbf{R})\rangle_r|_{\mathbf{R}=\mathbf{R}_{\mathrm{CI}}}. \tag{11.13}$$

Here, $\nabla_{\mathbf{R}}$ indicates the gradient operator with respect to all nuclear coordinates and $\mathbf{g}$ and $\mathbf{h}$ are called *gradient difference vector* and *derivative coupling vector*, respectively. The cone shape of the potential energy profile, therefore, is represented in a two-dimensional space spanned by $\mathbf{g}$ and $\mathbf{h}$. It is noteworthy that the zero-dimensional CI point in the two dimensional space of $\mathbf{g}$ and $\mathbf{h}$ constitutes a CI seam in the remainder of the $(3N_{nuc} - 8)$-dimensional configurational space; this is a ubiquitous feature of the conical intersection for polyatomic molecules.

The transition rate of non-radiative decay through the CI region is not properly accounted for by a perturbative approach, such as Fermi's golden rule, because of a typically large non-adiabatic coupling and the close proximity of two energy levels. Furthermore, in such a strong coupling case, the transition rate depends explicitly on the nuclear motion on the BO potential energy surfaces. In principle, nuclear wavepacket motion representing the molecular dynamics and the electronic transition need to be described by the time-dependent Schrödinger equation for nuclear motion with multi-electronic state coupling of the non-adiabatic interaction given by Equation (11.4). However, for polyatomic molecules the needed calculations quickly become too formidable a task, but a semi-classical treatment is feasible. In such a treatment, the nuclear motion on a single potential energy surface is described, classically. In the adabatic representation, a trajectory of the nuclear motion on the Ith BO potential energy surface is calculated by numerically solving Newton's equation,

$$M_A\ddot{\mathbf{R}}_A^I(t) = -\frac{\partial E_I^{BO}(\mathbf{R})}{\partial(\mathbf{R}_A)}, \tag{11.14}$$

where $\mathbf{R}_A^I(t)$ expresses a trajectory of the Ath atom on the Ith BO surface. Then, along the classical trajectory, the transition between two electronic states is represented by a transition between two BO surfaces induced by the non-adiabatic coupling. The relevant coupling element between two BO electronic states in a semi-classical representation is

$$\langle \Psi_J(\mathbf{r}, \mathbf{R}) | \hat{T}_N(\mathbf{R}) | \Psi_I(\mathbf{r}, \mathbf{R}) \rangle_r \simeq -i\hbar \mathbf{d}_{JI}(\mathbf{R}) \cdot \dot{\mathbf{R}}(t), \tag{11.15}$$

where the quantum-mechanical momentum operator in Equation (11.4) has been replaced by the corresponding classical velocity. The non-adiabatic coupling is then expressed through the scalar product of the non-adiabatic vector already introduced, $\mathbf{d}_{JI}(\mathbf{R})$, and the velocity of nuclear motion.

For a one-dimensional reaction, the transition probability of an AEC can be evaluated by the Landau–Zener (LZ) theory, which provides a simple formula for the transition probability (Zener, 1932). In this theory, three assumptions are introduced: (i) the velocity, $\dot{R}(t)$, is constant, (ii) the energy difference between the two diabatic states, $\Delta\tilde{W}(R(t)) = \tilde{W}_1(R(t)) - \tilde{W}_0(R(t))$, varies linearly with time along the trajectory, and (iii) the off-diagonal coupling element between the two diabatic states is constant, i.e. $\tilde{V}(R(t)) = \tilde{V}$. With these assumptions, the transition probability between two adiabatic states, P_{LZ}^{ad}, is (Desouter-Lecomte and Lorquet, 1979)

$$P_{LZ}^{ad} = \exp\left[-\frac{\pi}{4\hbar} \left| \frac{\Delta E^{BO}(R_c)}{\dot{R} d_{10}(R_c)} \right| \right], \tag{11.16}$$

where $\Delta E^{BO}(R_c) = E_1^{BO}(R_c) - E_0^{BO}(R_c) = 2\tilde{V}$, is the energy difference between the two adiabatic states at the crossing point, R_c. As seen in Equation (11.16), the probability of the transition between two adiabatic states becomes large when $\Delta E_{BO}(R_c)$ is small, the velocity is large and the non-adiabatic coupling element, $d_{10}(R_c)$, is large.

For the transition between multi-dimensional potential energy surfaces of a polyatomic molecule, the one-dimensional description by LZ theory may lead to difficulty in evaluating the transition probability due to the complex character of the potential energy surfaces and the non-adiabatic coupling. In particular, in the CI region, the off-diagonal coupling element between diabatic states, $\tilde{V}(\mathbf{R})$, can depend strongly on the nuclear coordinates, which violates the assumption underlying LZ theory. Nevertheless, the simple formulation of LZ theory provides valuable insight into the electronic transition of an AEC, even for complex systems.

The multi-electronic state dynamics of polyatomic molecules is often treated by Tully's fewest-switches surface hopping algorithm (Tully, 1990). According to this algorithm, the probability for hopping between surfaces is determined by numerically solving a time-dependent Schrödinger equation for the population of the adiabatic electronic states. In the case of a process including two electronic states, the respective Schrödinger equation is

$$i\hbar \begin{pmatrix} \dot{c}_1(t) \\ \dot{c}_0(t) \end{pmatrix} = \begin{pmatrix} E^1_{BO}(\mathbf{R}(t)) & -i\hbar \mathbf{d}_{10} \cdot \dot{\mathbf{R}}(t) \\ -i\hbar \mathbf{d}_{01} \cdot \dot{\mathbf{R}}(t) & E^0_{BO}(\mathbf{R}(t)) \end{pmatrix} \begin{pmatrix} c_1(t) \\ c_0(t) \end{pmatrix}, \tag{11.17}$$

where $c_{1,0}(t)$ represents expansion coefficients of the two adiabatic electronic states. The off-diagonal elements for the coupling between the adiabatic states is the semi-classical non-adiabatic coupling, given by Equation (11.15).

The Schrödinger equation is solved along a classical trajectory $\mathbf{R}(t)$ on the potential energy surface of the representative state, the population of which is supposed to be dominant. For example, initially the 1st adiabatic state may be populated. The probability of hopping to the 0th state, P^{ad}_{Tully}, in a time step of the integration of the Schrödinger equation is

$$P^{ad}_{Tully} = \max\left[0, \frac{-\dot{\rho}_{11}\Delta t}{\rho_{11}}\right], \tag{11.18}$$

where Δt is the time-step and $\rho_{11} = |c_1(t)|^2$ is the population of the 1st adiabatic state. The decision for a hop is made by comparing the transition probability (11.18) with a random number. After the transition occurs, the trajectory is calculated on the 0th state and the population change is also evaluated by the Schrödinger equation with the trajectory. Statistical sampling of the trajectory provides one with ensemble averages of the population decay. Extension to multi-electronic transitions of greater than two is straightforward.

A more general and detailed description is found in (Tully, 1990; Groenhof *et al.*, 2009). It is known that Tully's surface hopping approach has defects for re-transition from the lower to the upper state due to inconsistency of the mixed quantum-classical treatment. However, re-transition by AEC in the CI region for a polyatomic molecule is rare, since the conical shape of the potential energy surface of the lower state facilitates intramolecular vibrational relaxation which prevents the trajectory from re-entering the CI region.

The CI serves as the predominant channel for non-radiative decay. The excited state population that falls into the potential energy funnel of the CI region cannot escape from the region, like an ant entrapped in a doodlebug's pit; instead, it quickly undergoes non-adiabatic transition to the ground state. The remarkable efficiency for the non-adiabatic transition through a CI together with its ubiquitous feature has brought about a paradigm shift for the mechanisms underlying non-radiative decay in polyatomic molecules. It is now recognized that the rate of a very fast non-radiative decay is determined not by how fast the non-adiabatic transition proceeds at the energy crossing region, but by how fast the excited state population reaches the CI region, as depicted schematically in Figure 11.6b.

We like to demonstrate the stated behaviour through a simple kinetic description. Let us define by $k_{pre-CI}^{non-rad}$ and $k_{CI}^{non-rad}$ the rate of reaching the CI region in the first place and the rate of transition at the CI region, respectively. Since the two events take place sequentially, the overall rate of non-radiative decay for the photochemical reaction through CI, $k_{pre-CI}^{non-rad}$, is given by the approximate expression,

$$\frac{1}{k_{reaction}^{non-rad}} \approx \frac{1}{k_{pre-CI}^{non-rad}} + \frac{1}{k_{CI}^{non-rad}}. \tag{11.19}$$

Obviously, for $k_{CI}^{non-rad} \gg k_{pre-CI}^{non-rad}$, holds, $k_{reaction}^{non-rad} \simeq k_{pre-CI}^{non-rad}$. Thus, according to Equations (11.8) and (11.19), the process competing against fluorescence is the motion towards a CI region and not the transition at a CI.

The excited state process before reaching the CI region, therefore, is the primary target for control through the protein environment for a photoactive protein to realize its function effectively. In the case of rhodopsins, as seen below, the molecular dynamics before reaching the CI region is crucial for the sensory efficiency of the receptor. Crucial also is the branching ratio among possible chemically different photoproducts. Molecular dynamics simulations describing the approach to CI regions and branching ratios after crossing to the ground state are extremely helpful in revealing the role of the protein environment in steering the two processes (approach to CIs and branching) and, presently, the only means of obtaining a closer insight. Such simulations of excited state conformational dynamics are extremely challenging, but possible (Hayashi *et al.*, 2003; Frutos *et al.*, 2007; Hayashi *et al.*, 2009; Polli *et al.*, 2010).

11.6 Electronic structure of protonated Schiff base retinal

The common chromophore of retinal proteins is retinal bound as a PSB to lysine. Retinal consists of a linear polyene chain with six conjugated double bonds. The long π-conjugation of the polyene chain, together with a positive charge of the Schiff base group located at one end of the polyene, gives rise to strong optical absorption in the visible region of the spectrum. The absorption is due to a so-called π–π^* excitation.

The electronic structure of linear polyenes has been the subject of photophysical studies over many decades. The optically allowed π–π^* absorption band of a linear polyene with six conjugated double bonds arises in an ultra-violet region ($320 \sim 370$ nm). The corresponding excited state possesses ^1B_u symmetry and is mainly characterized as a single electron excitation from the highest occupied molecular orbital (HOMO) to the lowest unoccupied one (LUMO). In the vicinity of the ^1B_u state, an optically forbidden π–π^* state of ^1A_g symmetry exists, along

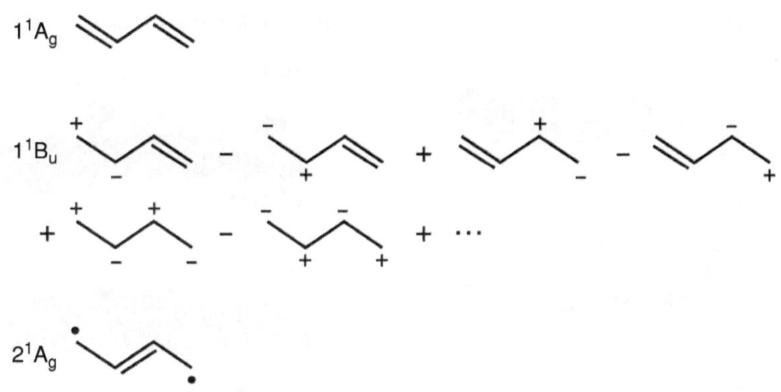

Figure 11.7 Valence bond structures of butadiene in the 1^1A_g ground state, in the 1^1B_u excited state and in the 2^1A_g excited state. Butadiene is the simplest polyene, involving only two double bonds, i.e. four π-electrons. The 1^1A_g ground state is represented by the familiar, two double bond, covalent valence bond (VB) structure. The optically allowed 1^1B_u state of butadiene is represented by a linear combination of ionic VB structures, the optically forbidden 2^1A_g state of this polyene is mainly represented by a VB structure, where two radicals at the two ends of the polyene exhibit a singlet coupling.

with several other forbidden states. The electronic structure of the ${}^1A_g$ excited state is described by multi-configurational excitations that involve two triplet excitations coupled to an overall singlet spin state (Schulten and Karplus, 1972). The ${}^1A_g$ state and its sister triplet–triplet states have been investigated computationally in Tavan and Schulten (1986, 1987) and Ritz *et al.* (2000a); an experimental-theoretical review can be found in Hudson *et al.* (1982).

The polyene excited states can also be interpreted through a valence bond (VB) representation (Schulten and Karplus, 1972). Figure 11.7 depicts, schematically, the main VB electronic cofigurations of the ${}^1B_u$ and ${}^1A_g$ electronic states. The ${}^1B_u$ state is expressed by a linear combination of polar structures. In each of the polar structures, π-electrons are shifted between the carbons constituting a double bond and, thereby, polarize the double bond. The ${}^1B_u$ state is often called *ionic* because of the polar nature of the VB configurations, although the polarization does not explicitly appear in the one electron density of the total wavefunction due to cancellation by the linear combination. The ionic character is responsible for the large transition dipole moment that endows the ${}^1B_u$ state with strong photo-absorbance. The polarization of π-electrons in each VB configuration induces counter-polarization of a σ-bond which compensates the π-polarization. Because of this σ, π-polarization, one needs to take into account explicitly dynamic electron correlation between the σ- and π-orbitals for an accurate computation of the electronic energy of the ${}^1B_u$ state. In the case of the ${}^1A_g$ excited state, as well

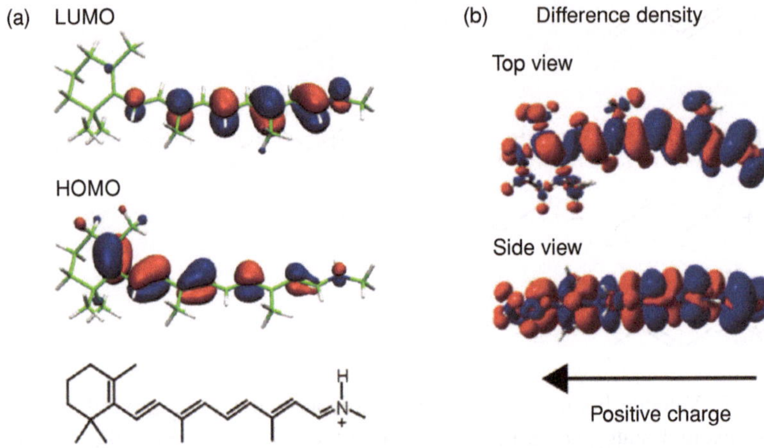

Figure 11.8 (a) Highest occupied molecular orbital (HOMO) and lowest unoccupied molecular orbital (LUMO) of PSB retinal. Depicted is the chemical structure of PSB retinal and superimposed on it, the molecular orbitals coloured blue and red according to the sign of the orbital wavefunctions; the characteristic blue-light lobes stem in (a) from the π-electron atomic orbitals. (b) Difference in electron density between the excited state and the ground state (Fujimoto *et al.*, 2010). Red and blue lobes represent in (b) a decrease and increase of the electron density, respectively. Blue and red lobes are bigger on the PSB side (right) and the β-ionone ring side (left), respectively, indicating an overall shift of positive charge along the polyene chain from the PSB side to the β-ionone ring side.

as the ground state, which also possesses ^1A_g symmetry, electronic structures are expressed as a linear combination of covalent Kekule-like and Dewar-like VB structures, as shown in Figure 11.7, indicating a covalent character for both states.

The electronic excitations of the PSB retinal chromophore in Rh can be delineated from polyene electronic excitations described above. However, the positive charge introduced in the conjugated π electron system through the protonation of the Schiff base linkage to lysine remarkably alters the electronic structure (Schulten *et al.*, 1980). In particular, the absorption spectrum of the optically allowed π–π^* excitation of PSB retinal is considerably red-shifted from that of a polyene with the same number of conjugated double bonds, coming to rest in the visible region, which is appropriate for the biological function. The positive charge on the PSB induces localization of π orbitals, as shown in Figure 11.8a. One can discern in the figure that retinal's HOMO and LUMO exhibit larger distributions in the β-ionone ring half and in the PSB half, respectively. A HOMO–LUMO single excitation which mainly characterizes the optically allowed ^1B_u-like state involves, therefore, a positive charge transfer from the PSB half to the β-ionone ring half along the polyene chain, as shown in Figure 11.8b. The localization of HOMO and LUMO also contributes to the red-shift of the optically allowed absorption band compared with that of the same length polyene because of reduction of electron repulsion. Finally, the positive charge breaks the polyene D_{2h} symmetry in PSB retinal; accordingly, the electronic characters of the ^1B_u-like and ^1A_g-like excited states become mixed, also endowing the ^1A_g-like excited state of retinal with non-zero optical absorbance, albeit only a weak one.

11.7 Mechanism of spectral tuning in rhodopsins

Control of photoabsorption, namely of the wavelength position of the absorption maximum, λ_{max}, is functionally important for retinal proteins, as discussed above. This control is referred to as spectral tuning. Since all retinal proteins possess a PSB retinal chromophore, i.e. their chromophores are identical (though in some animal species retinal is replaced, to reach longer wavelength absorption, by retinal$_2$ with π-conjugation extended in the β-ionone ring by one double bond, a case not considered here), interaction of the chromophore with the protein environment is responsible for the needed variation of absorption maxima. As the excitation energy is defined as the energy difference between the B_u-like excited state and the ground state, spectral tuning is based on the different effect of the protein environment on the energies of the two states. Indeed, the molecular mechanism of colour tuning in rhodopsins exploits differences in the electronic character between the states, which has been described in Section 11.6. The spectral shift originates mainly from conformational change of the chromophore and from electrostatic interaction between the chromophore and polar groups in the surrounding protein.

Conformational changes of the PSB retinal chromophore are induced by confinement due to steric interactions in the binding pocket, in which the chromophore is situated, and due to hydrogen bonding of retinal's PSB to polar amino acid side groups. The polyene chain of PSB retinal can be deformed torsionally in the ground state; however, π–π^* excitation significantly alters the torsional flexibility of retinal. One can recognize in Figure 11.8a that the HOMO exhibits bonding and anti-bonding phases at double and single bonds in the polyene chain, respectively, and that the phase behaviour of the LUMO is the opposite of that of the HOMO, bonding and anti-bonding phases for the LUMO appearing at single and double bonds, respectively. Hence, torsions around double and single bonds cause a decrease and an increase in the HOMO–LUMO energy gap, respectively, leading to respective red and blue shifts of the absorption maxima of the π–π^* excitations.

We illustrate the role of torsion on the HOMO–LUMO energy gap and, thereby, on the λ_{max} of retinal for the C_6–C_7 bond (for the numbering of carbon atoms see Figure 11.2). Torsion around this bond has little effect on the HOMO energy as the orbital exhibits an anti-bonding phase for this bond; however, the same torsion increases the LUMO energy as the latter orbital exhibits a bonding phase for the C_6–C_7 bond. In spite of the preference for planarity as part of retinal's π-conjugated bond system, the C_6–C_7 bond in retinal rotates rather readily due to steric interactions of the H_7 and H_8 atoms with methyl groups at positions C_1, C_5 and C_6. Quantum chemistry calculations demonstrate a steep increase in excitation energy with torsion around the C_6–C_7 bond (Wanko *et al.*, 2005). Thus, torsion around the C_6–C_7 bond can control retinal's absorption maximum in rhodopsins; in fact, the C_6–C_7 torsional angles of the chromophore in bR and Rh differ remarkably (deviation from planarity is $\sim 10°$ and $\sim 50°$ in bR and Rh, respectively) and explain the blue shift from the bR to the Rh spectral maximum (Fujimoto *et al.*, 2007).

Electrostatic interaction between retinal chromophore and protein polar and charged amino acids also plays a central role in spectral tuning. As described above (see also Figure 11.8b), PSB retinal excitation involves a charge shift along the polyene backbone. Hence, an electrostatic field along the polyene chain due to surrounding polar groups can shift the PSB retinal spectrum. Since upon photoexcitation the positive charge of the Schiff base moves towards the β-ionone ring, the excitation energy increases if in a retinal protein the electrostatic potential becomes more negative near the Schiff base and more positive near the β-ionone ring.

In retinal proteins, carboxylate groups such as Asp85 and Asp212 for bR and Glu113 for Rh serve as counter anions of the PSB and provide large contributions to the spectral shifts. In chloride pump proteins, i.e. halorhodopsin (Scharf and Engelhard, 1994) and a D85T mutant of bR (Sasaki *et al.*, 1995), a chloride anion bound in the vicinity of the PSB also affects the absorption maximum strongly; removal of the bound chloride anion results in a spectral red shift. In addition to the counter ion groups, polar groups surrounding the chromophore, such as hydroxyl groups of serine and threonine as well as water molecules, can contribute to spectral shifts. Among bacterial retinal proteins, bR and sRII exhibit a remarkable difference between their absorption spectra, i.e. $\lambda_{max} = \sim 570$ nm for bR and $\lambda_{max} = \sim 500$ nm for sRII, whereas X-ray crystallographic studies reveal only a slight difference in the chromophore structure. Comprehensive mutagenesis studies for sRII (Shimono *et al.*, 2000) and theoretical investigations (Hayashi *et al.*, 2001; Ren *et al.*, 2001; Hoffmann *et al.*, 2006) have shown that a major part of the spectral shift can be attributed, in the case of these pigments, to a difference in electrostatic interaction with the surrounding polar and charged amino acids.

As mentioned above, the retina of the human eye contains three colour receptors, namely, human red (HR), human green (HG) and human blue (HB) receptors. An explanation of the receptors' spectral tuning is of great interest. Unfortunately, biochemical and structural information on the colour receptors is sparse, in particular, no iodopsin structure has been determined yet. However, biochemical and mutagenesis studies have revealed that polar protein groups and a chloride ion at a binding site on the extracellular side of the protein give large contributions to the spectral shifts (Shichida and Imai, 1998). Computationally, an attempt by means of quantum chemistry calculations (Fujimoto *et al.*, 2009) has been made with model protein structures for the colour receptors constructed based on the available Rh structure. The latter study suggests that electrostatic interaction contributes to spectral shifts between the three receptors; a chromophore structural difference appearing in HB appears to produce the significant blue shift for this receptor.

11.8 Photoisomerization of retinal in rhodopsins

Photoabsorption of retinal initiates dynamics on the excited state surface that leads to strong torsional motion in the polyene chain moiety. As described above, the S_1 state responsible for photoabsorption and the photochemical reaction possesses ionic valence-bond configurations. Since these ionic configurations prefer reduction of overlap between 2p orbitals constituting the π molecular orbitals at a double bond of the polyene, one of the polyene double bonds undergoes torsion towards a 90° conformation. A remarkable charge separation, the so-called sudden polarization (Salem, 1979; Bonačić-Koutecky *et al.*, 1984) occurs in the twisted conformation, which can make energy levels of excited state and ground state approach each other to form a CI. In the case of a PSB polyene, the ground state positive charge is localized in the Schiff base half of retinal, and the excited state positive charge in the β-ionone ring half, the two halves being separated by the isomerizing bond (González-Luque *et al.*, 2000). Through the conical intersection arising for perpendicular torsion of the bond, the excited state population decays to the ground state where the perpendicular geometry relaxes to planar, the relaxation involving branching either back to the original isomeric state of the double bond or to the alternative isomeric state.

Photoisomerization reactions in rhodopsins are known to be the fastest molecular reactions in nature. The measured time constants for excited state decay in proteins lie in the range 50–500 fs. The fast decay diminishes ϕ^{fluor} (Equation (11.6)), almost completely avoiding decay through fluorescence, and provides a high photoactivation quantum yield of more than 0.6, this value being determined by the branching ratio of completion or lack of completion of the isomerization reaction after crossing through the CI. In fact, ϕ^{fluor} has been measured experimentally to be only $\sim 10^{-5}$ (Kochendoerfer and Mathies, 1996).

Actual photoisomerization in the various rhodopsins exhibits strict stereoselectivity. For example, in bR, photoisomerization of the chromophore, which is in the all-trans state in the resting ground state, produces only the 13-cis isomer or returns to the all-trans state, even though other double bonds in the polyene chain are, in principle, capable of undergoing isomerization. The extremely fast reaction rates and the strict stereoselectivity are due to steering by the protein environment. In fact, the photoisomerization reaction time constant of retinal in methanol solvent is 2.5–4 ps (Logunov *et al.*, 1996) and photoisomerization from the all-trans configuration in this solvent is not stereoselective: photoproducts include 9-cis, 11-cis and 13-cis isomers with quantum yields of 0.02, 0.14 and 0.01, respectively (Koyama *et al.*, 1991). Steering by the protein environment towards the proper photoproducts speeds up the reaction, by a factor of 10 to 100, and increases the quantum yield, thus, furnishing photoreceptors with high sensitivity to incoming light.

The stated behaviour can clearly be seen in the photoreactions of Rh and bR. The photoisomerization of Rh taking place from 11-cis retinal to all-trans retinal proceeds within an extremely short time of ~50 fs and with a high quantum yield, namely 0.68, without by-products other than the 11-cis reactant form (Kochendoerfer and Mathies, 1996; Polli *et al.*, 2010). In bR, the photoisomerization from all-trans to 13-cis retinal occurs witin 240–500 fs (Gai *et al.*, 1998), slightly slower than the reaction in Rh, and with a quantum yield of 0.6. Interestingly, kinetic behaviour differs remarkably between Rh and bR. Pump–probe experiments on Rh (Wang *et al.*, 1994; Polli *et al.*, 2010) revealed coherent (or synchronous in a classical mechanical sense) wavepacket dynamics from the FC region to the photoproduct, even in the presence of thermal noise in the surrounding protein environment. In contrast, pump–dump–probe experiments on bR (Ruhman *et al.*, 2002) saw photoproduct formation proceeding in an incoherent, stochastic manner (asynchronous in a classical mechanical sense), although synchronous vibrations were observed in the excited state (Kobayashi *et al.*, 2001). Apparently, in the case of Rh, excited state motion occurs synchronously along a single path, whereas in bR, excited state dynamics is funnelled briefly into a multi-dimensional basin of attraction, from where the system escapes asynchronously through a relatively narrow exit that leads to photoproduct formation. The difference in kinetic behaviour between Rh and bR gives a clue to understanding the physical mechanism underlying photoreaction steering in rhodopsins.

A key role of the protein environment is to achieve an optimal branching ratio by targeting, during excited state motion, the CI with 90° rotation of the right double bond and, after crossing through the CI, by completing bond rotation to reach the desired isomeric state. Figure 11.9 depicts, schematically, the potential energy profile of the isomerization reaction. Upon photoabsorption, the reactant isomer in the resting ground state is vertically excited to the FC region in the excited state where the isomer is located at the top of a potential hill with a single

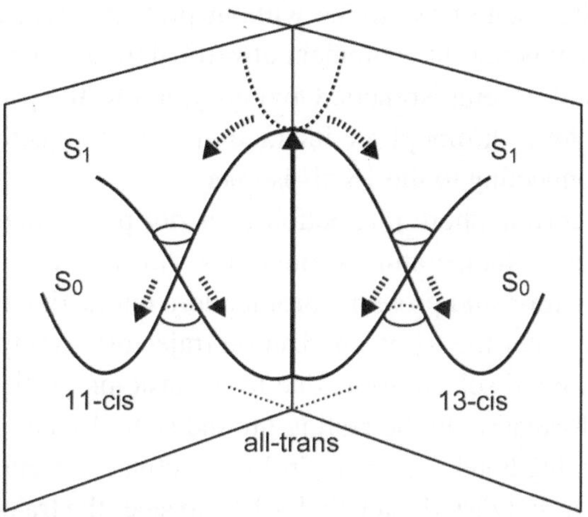

Figure 11.9 Potential energy profiles governing photoisomerization of PSB retinal from an all-trans configuration to two possible product isomeric states, 11-cis and 13-cis. The branching ratio of the product isomers is determined by dynamics in two regions, the FC and the CI region.

peak. From the FC region, downhill paths connecting to different isomers through alternative CIs are possible. One can discern from Figure 11.9 that the branching ratio is determined in two regions, namely, in the FC and the CI regions. Firstly, dynamics in the FC region determines stereoselectivity, i.e. the trajectory on the excited state energy surface must be steered by the protein environment towards the CI corresponding to 90° torsion around the selected bond. Once a trajectory falls into any one of the branching paths, it never comes back out of the CI to be able to reach the CI of an alternative bond, as the crossing at a CI is too fast. Secondly, the protein environment must assure that, after crossing at the right CI, when the torsion of the selected double bond is at a 90° angle, bond rotation has a good chance of completing to form the new isomeric state of retinal; the protein environment and chromophore properties must prevent too many CI crossings falling back to the reactant isomeric state.

As mentioned above, experimental evidence suggests that, in the protein environment of Rh and bR, there are no by-product isomers other than the product and reactant generated by the photoreaction. Hence, the dynamics at the FC region is strictly regulated for selectivity of the isomerizing bond. How is such strict selectivity achieved? Recent molecular simulation studies have suggested a striking contrast between the mechanisms in Rh and bR (Hayashi *et al.*, 2003, 2009).

In bR, interaction of the chromophore with the protein surroundings which determines bond selectivity is exerted on the dynamics in the excited state. A simulation of excited state trajectories without protein surrounding showed that thermal fluctuation existing at the moment of vertical transition by photoabsorption leads to formation of several isomers. Obviously, in bR, the paths to 'wrong' CIs are blocked by protein–chromophore interaction and all trajectories are funnelled to the proper CI connecting to the 13-cis isomer.

In Rh, protein–chromophore interaction does not play a major role in excited state dynamics. Room temperature simulations showed that thermal fluctuation before vertical transition and protein–chromophore interaction in the excited state do not significantly alter the Rh isomerization trajectories (Hayashi *et al.*, 2009). This indicates that bond selectivity is due to intrinsic properties of 11-cis retinal in its reactant conformation in the resting ground state. In fact, the dihedral angle around the isomerizing bond, $C_{11}{=}C_{12}$, in 11-cis retinal's reactant conformation is found to be sufficiently twisted, namely by 17°, to steer the trajectory upon excitation instantaneously towards the product isomer. In addition, concerted co-rotations of $C_9{=}C_{10}$ and $C_6{-}C_7$ also promote isomerization around $C_{11}{=}C_{12}$ (Frutos *et al.*, 2007; Hayashi *et al.*, 2009; Polli *et al.*, 2010).

The 'blocking-and-funnelling' and 'hands-off' roles of bR and Rh, respectively, during retinal excited state dynamics explain the observed photochemical kinetics very well. In bR, chromophore–protein interaction blocking wrong isomerization paths perturbs the reaction dynamics, leading to asynchronous motion after vertical excitation and decreasing the escape rate from the FC region. On the other hand, the hands-off mechanism of Rh avoids collision between isomerizing chromophore and protein environment, thus, maximizing the reaction rate.

Once a branching path from the FC region towards a product conformer is selected in Rh or bR, the trajectory along the branching path undergoes, in the CI region, an electronic transition to the ground state; subsequently, the trajectory undergoes another branching, namely, between reactant and product isomeric states. In both Rh and bR, the branching ratios of product and reactant isomers are larger than 0.6, indicating a preference for the product isomers. This preference can be explained by a simple semi-classical consideration for a one dimensional reaction case (Weiss and Warshel, 1979). Figure 11.10 depicts a schematic mechanism for the branching ratio in an AEC region. Firstly, a trajectory enters into the AEC region from the FC region with momentum towards the product isomer. Passing the AEC region, part of the population associated with the trajectory undergoes electronic transition to the ground state. Because of the momentum moving in the excited state towards the product isomer, the transition produces a trajectory that also has momentum towards the product isomer in the ground state. The momentum transfer to the ground state can be explained in the framework of LZ theory, Equation (11.16). The population not crossing to the ground state leaves the AEC

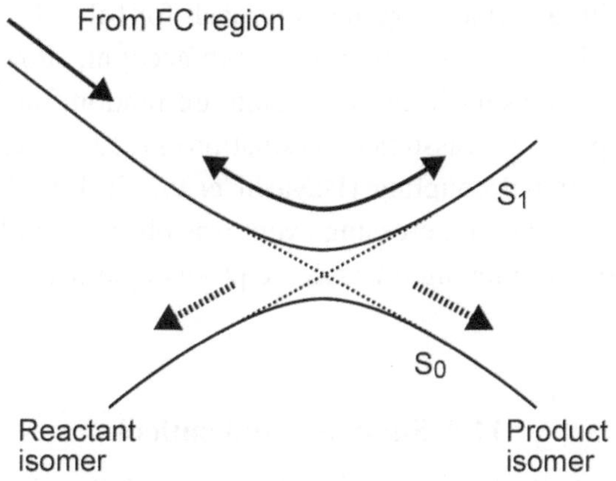

Figure 11.10 Mechanism of population branching at a one-dimensional avoided energy crossing. Solid and dotted lines represent adiabatic and diabatic potential energy curves. A trajectory coming from the FC region in the excited state exhibits oscillation on an adiabatic potential energy curve around the avoided energy crossing (AEC) point. Population associated with the trajectory undergoes transition to the electronic ground state during each passing of the AEC. The momentum at the AEC determines the isomeric state produced in the ground state.

region and then re-enters the AEC region with opposite momentum. The recrossing event gives rise to a ground state trajectory towards the reactant isomer. Further crossing events take place, with alternating directions of momentum until eventually all of the excited state population is divided into populations of reactant and product conformers in the ground state.

For the process outlined, the branching ratio, Φ, can be determined with the assumption of a constant transition probability, θ, at each crossing event. Summing up of all crossing events yields

$$\Phi = (1-f)[\theta + \theta(1-\theta)^2 + \theta(1-\theta)^4 + \cdots] = (1-f)/(2-\theta), \quad (11.20)$$

where f represents a population lost before the trajectory enters the AEC region due to fluorescence and non-adiabatic transitions, which is expected to be negligible in the case of Rh and bR because of their fast photoisomerization rates. The equation suggests that the branching ratio approaches one for $\theta \to 1$ and the lower bound of 0.5 for $\theta \to 0$. Branching ratios greater than 0.6, as observed for Rh and bR, are achieved for $\theta > 0.33$. One can discern from Equation (11.20) that the dominating contribution to the branching ratio stems from the term that corresponds to the first crossing event.

In the case of isomerization in the protein environment, the trajectory proceeds on a multi-dimensional potential energy surface. A quantum chemical molecular dynamics study with a surface hopping approach for bR, which took all degrees of freedom involved in the reaction process into account, showed that recrossing events after the first crossing in the CI region are random rather than exhibiting regular alternate repeats, as is assumed in Equation (11.20) due to multidimensional intramolecular vibrational coupling (Hayashi *et al.*, 2003). Nevertheless, in the multidimensional case the first crossing event was observed to lead exclusively to a ground state trajectory moving towards the product isomer, as seen in the model discussed above.

11.9 Summary and outlook

In this chapter, the mechanisms of spectral tuning and photoisomerization in retinal proteins were discussed in conceptual terms. Through many experimental and theoretical studies carried out over several decades, understanding of the quantum physics and quantum chemistry of the primary photoactivation events in rhodopsins has been significantly advanced. Because of the relative simplicity of the structures of rhodopsins and the rather straightforward reaction dynamics involved in photoactivation, rhodopsins have been one of the most extensively studied systems targeted for elucidation of quantum effects in biological function. There remain, however, important issues that are not yet understood and need to be addressed in future studies. For example, the photoisomerization reactions have been suggested, experimentally and theoretically, to involve not only the rotational motion around the isomerizing bond, but also other molecular motions, such as in-plane relaxation of the bond-alternation of the polyene, hydrogen-out-of-plane motion and C–H bond bending motion, as well as concerted co-rotations in the polyene chain (Kobayashi *et al.*, 2001; Hayashi *et al.*, 2003; Kukura *et al.*, 2005; Frutos *et al.*, 2007). The coupling of these various motions to photoisomerization still needs to be fully investigated. Molecular dynamics simulations at electronic and atomic resolution, becoming ever more feasible today, promise more insight into the quantum physics of retinal proteins, for example quantum coherent control of photoisomerization, as demonstrated recently, experimentally, in the case of bR (Prokhorenko *et al.*, 2006).

Eighty years ago, George Wald discovered the role of retinal in animal vision (Wald, 1933). Retinal has been extensively studied ever since, attracting, in particular, theoretical and computational scientists to biology due to the molecule's highly correlated σ, π-electron system. Retinal's remarkable role in vision, achieved through its strong light absorption, wide range of λ_{max} values tuned readily through changes of its protein environment, and capability for highly specific isomerization reactions, is still admired today and still the subject of intense investigations.

11.10 Acknowledgement

The authors are grateful to Kazuhiro Fujimoto, Ryoichi Kida, Olga Svinarski and Bo Liu for assistance with drawing of figures. This work was supported by National Science Foundation grants MCB0744057 and PHY0822613 as well as by National Institutes of Health grant P41-RR05969 (to K. S.).

12

Quantum vibrational effects on sense of smell

A.M. STONEHAM, L. TURIN, J.C. BROOKES AND A.P. HORSFIELD

12.1 Phonon assisted tunnelling in olfaction

Human vision is impressive, but the front-end mechanism – how the photons incident on the eye are detected – is quite well understood: they are absorbed by rhodopsin causing electronic transitions that lead, via a sequence of amplification steps, to a signal sent to the brain. Olfaction is somewhat more puzzling. Unlike photons, which differ only in wavelength, olfaction must detect and, harder still, discriminate between thousands of molecules (odourants) with their different physical and chemical properties. What is more, the repertoire of odourants is not fixed: newly synthesized odourants can be smelled immediately. It has been found empirically that to be detectable, the odourant molecules must of course be volatile, and typically contain fewer than 16 carbons: unless exotic heavy atoms are present, this means odourants have a maximum weight of 240 daltons (about 50 atoms). Empirically again, no two odourants (excluding enantiomer pairs) have ever been found to smell exactly identical.

How is odour character written into a molecule? A century of synthetic chemistry and hundreds of thousands of synthesized molecules have failed to provide an answer. There are some regularities in structure–odour relations, but none amount to a theory endowed with predictive power. Reviews of the field have for a long time consisted largely of lists of molecules grouped by odour character. More recent reviews suggest that a theory of structure–odour relations is not just around the corner and may, in fact, be impossible (Sell, 2006). While shape must help to determine the affinity of the odourant for the receptor, and so contribute to any molecular recognition event, rules based on 'lock and key' ideas are falling out of favour and do not obviously answer the question 'What turns the key?'.

Quantum Effects in Biology, ed. Masoud Mohseni, Yasser Omar, Gregory S. Engel and Martin B. Plenio. Published by Cambridge University Press.

Yet somehow the odourant must initiate a signal to the brain. However smart the brain is, it needs information to work on: it must get a different set of signals from different odourants. The task then is to understand the selective and sensitive activation step. The following observations, which are inconsistent with shape-only theories, provide clues:

1 Humans can easily detect the presence in an odourant of certain functional groups, whatever their molecular context. The best known example of this is -SH, which imparts to whatever molecule it is attached the eponymously named character known as sulfuraceous. This is true for all thiols, and is a very challenging molecular recognition problem. A shape-based system would not be expected to discriminate perfectly between thiols and, say, alcohols at all concentrations and with no error. In other words, alcohols should smell sulfuraceous at high concentrations, and thiols should sometimes be perceived as alcohols. Neither is true. Recognition of chemical groups (some expertise is needed) extends to amines, imines, nitriles, isonitriles, nitro, aldehydes and, with less reliability, possibly ethers (Klopping, 1971). Accordingly, many of these groups have given their name to odour characters: aldehydic, ethereal, etc.

2 The majority of enantiomer pairs have very similar, or in many cases identical, smells. Since receptors are chiral, this is a direct challenge to the notion that shape governs odour character. In general, molecular recognition is of course very sensitive to chirality. Receptors and enzymes generally have different affinities for enantiomers.

3 Replacement of carbon by silicon, germanium and tin – column 4A elements with tetrahedral bonding – in a variety of odourants almost invariably leads to large changes in odour character, despite modest changes in molecular shape (Wrobel *et al.*, 1982; Wannagat *et al.*, 1993, 1994).

4 Replacement of hydrogen by deuterium is perceived as a different odour character by fruit flies and humans (Franco *et al.*, 2011; Gane *et al.*, 2013). Establishing whether isotopes are perceived differently by animals and humans has turned out to be surprisingly difficult. The main obstacle to convincing results is the purity of reagents. Suppose one wishes to determine whether an animal responds differently to a molecule and, say, its deutereated isotopomer. It is not sufficient to merely test its ability to distinguish one from the other, because in most cases the synthetic pathways of the two isotopomers will have been different, and therefore the impurity profiles of the two samples will be different as well. Olfaction is highly non-linear, capable of simultaneously detecting majority and minority (less than 0.1 %) components. This makes the results essentially uninterpretable. There are two ways out of this difficulty. One is to strive for maximum purity, using odourants purified in a gas chromatograph (GC) or a high-pressure liquid

chromatograph (HPLC). The advantage of GC is that the pure odourant 'peak' exiting the purification column is carried by a stream of odourless gas, usually nitrogen. The disadvantage is the small quantity that can be loaded onto the column. HPLC allows purification of larger amounts, at the expense of obtaining them in a carrier solvent which, however, can be the same for both isotopomers and therefore need not introduce problems of interpretation. This approach has now been used successfully by Gane *et al.* (Gane *et al.*, 2013). The other approach to the problem, taken by Franco *et al.* (Franco *et al.*, 2011) is to ask whether the presence of deuterium confers on the molecules a specific odour character which, like other functional groups, can be recognised independent of context. Accordingly they trained *Drosophila* fruit flies to avoid either the hydrogen or the deuterium isotopomer of one odourant pair and found that the flies avoided the corresponding isotopomer in another pair of odourants. Further, once trained to seek out or avoid deuterium, the flies would seek out or avoid a chemically unrelated functional group, nitrile $-C \equiv N$, whose stretch frequency is very close to that of $-C-D$. These results have been successfully analysed using the vibrational theory by Bittner *et al.* (Bittner *et al.*, 2012).

Some of these observations are not new and, indeed, were often remarked upon by chemists interested in olfaction in the century or so since organic chemistry began. Indeed, physical (as opposed to chemical) theories of olfaction were popular, until Pauling suggested in 1946 that shape-based molecular recognition governed olfaction, as it does much of the rest of biology. One such physical theory, the vibrational theory (Dyson, 1938), posited that the nose was a vibrational spectroscope which detected and identified molecules by their vibrational spectra. Such a mechanism, were it possible, would have attractive features: it would be applicable to all molecules, even novel ones; it would provide direct information about the chemistry of the molecule, including its functional groups; if the 'fingerprint' region of the spectrum could be measured with sufficient accuracy, it would result in a unique odour character for each molecular structure. The attractiveness of this notion can be gauged by the fact that it survived for four decades, despite one major flaw: the lack of a plausible mechanism.

One such was found by Turin (Turin, 1996) when he realized that a non-optical nanoscale mechanism discovered in 1966 by Jaklevic and Lambe (Lambe and Jaklevic, 1968) was adaptable *mutatis mutandis* to a spectroscope made of proteins. This is where mechanics come in. The main ideas are that energies are quantized, and light particles (e.g. electrons) can tunnel through barriers. Whilst a classical harmonic oscillator can give up or take up energy in any amount, however small, a harmonic oscillator can only give or take energy in discrete quanta. The quanta have size hf, where h is Planck's constant and f is the frequency of oscillation of the oscillator. Thus it is possible to determine the frequency of vibration by looking for energy packets of size hf. Selection rules, coherence (important in photosynthesis) and entanglement (discussed by some authors in relation to consciousness) play very small parts.

Turin's proposal (Turin, 1996) put together the pieces needed for an olfactory receptor to identify a molecule by its vibrations, using tunnelling electrons to respond to specific frequencies. This theory was subsequently assessed critically and worked out in greater detail by Brookes *et al.* (Brookes *et al.*, 2007), and later confirmed by Solov'yov *et al.* (Solov'yov *et al.*, 2012). Electrons are required to tunnel between two sites that differ in energy by a given amount. This is only possible if the electron can either gain or lose this energy by giving it to something else, in this case the quantized vibrations of a molecule.

If molecular vibrational energies *alone* determine odour, then there are some clear predictions: there should be isotope dependence of scent; enantiomers (left- and right-handed chiral molecules) should smell the same;[1] there should be links between infrared spectra and odour (this has been verified for hydrogen sulphide and decaborane (Brookes *et al.*, 2007)). Further, for a model to be realistic, then tunnelling rates and other physical parameters must be consistent with standard observed biological structural information, and the processes must operate at ambient temperatures. Finally, if Nature has used this mechanism (or similar ones) more than once, then we might be led to an understanding of other selective responses of receptors to small molecules, like hormones, steroids or neurotransmitters.

12.2 Important processes and timescales

Here we take an overview of olfaction, to show where ideas fit in. There is a sequence of processes leading to recognition (i) odourant molecules begin in the atmosphere; (ii) they then enter the nasal cavity; (iii) they bind at some ensemble of receptors, leading to (iv) selective actuation of one or more receptors, generating signals that are amplified and transmitted to the brain, (v) where they are analysed. Possibly some analysis may occur before the brain is reached. The system as a whole imposes some constraints, but it is process (iv) that we assess thoroughly and propose must be quantal.

We have incomplete knowledge of the structure and operation of receptors. We know the sequence of amino acids that they are made from (primary structure), but not their correct 3D arrangement (secondary structure). However, olfactory receptors are probably like many other receptors, and so hydrophobic and chiral,

[1] The theory of Turin involves electron tunnelling and so is sensitive to geometry as well as vibrational frequency.

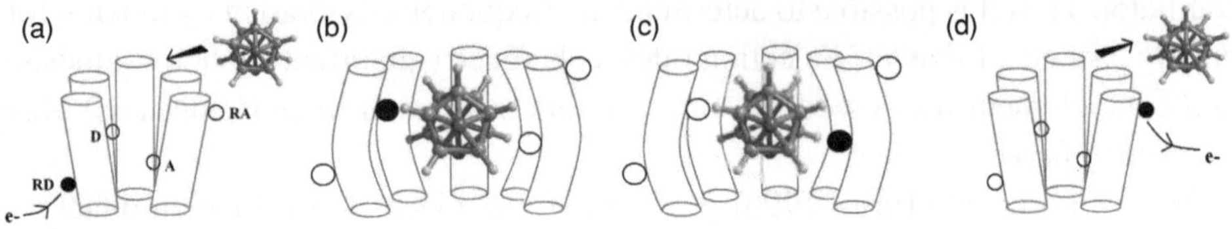

Figure 12.1 A cartoon for the proposal of electron transfer in the nose. Only five transmembrane helices (of the seven in total) for the olfactory receptor are shown (cylinders) here for clarity. (a) The odourant (in this case a carborane) approaches the receptor, meanwhile an electron diffuses to position RD on a helix. (b) The odourant docks at the ligand binding domain, the overall configuration of receptor and odourant changes, meanwhile the electron tunnels within the protein to D and it spends some time there. (c) The electron jumps from D to A causing the odourant to vibrate vigorously. (d) The odourant is expelled from the ligand binding domain and the electron tunnels within the protein to site RA. Signal transduction is initiated with the G-protein release. From the PhD Thesis of J. C. Brookes. Used with permission.

with significant thermal shape fluctuations allowing the receptor to change shape to some degree when binding a scent molecule.

We can still say something about the odourant recognition process, however. Experimentally, the overall process of olfaction occurs over milliseconds, decidedly slow compared with most processes at the molecular scale (see Table 12.1 in Section 12.3). This provides one benchmark against which to check the key physics of Turin's theory.

The key steps in the Turin theory for one receptor are illustrated in Figure 12.1. We now discuss them in detail. In the model of Figure 12.1 the likely rate-determining steps involve transport of an electron to the donor (D) or removal of an electron from the acceptor (A). The first requirement is a source of electrons or holes to allow this charge flow to take place. Producing this initial state requires some input of energy, though this is not expected to be problematic, as voltages of order 0.5 V are certainly available in cells. The precise biological origin is not known, but may well consist of reducing (oxidizing) species (X) in the cell fluid. These molecules diffuse through the aqueous medium and arrive with an average interval of τ_X. Using a standard approach for computing reactant collision rates in solution from the diffusion equation and the Stokes–Einstein relation for the diffusion coefficient (Atkins and de Paula, 2002), we get $\tau_X = 3\eta/2n_X k_B T$, where η is the viscosity of water, n_X is the concentration of X, k_B is Boltzmann's constant and T is the temperature. This result is independent of the nature of X or the receptor. Since n_X will probably lie in the range 1 μM to 100 μM, we get a range of values for τ_X of 10 μs to 1 ms.

The charge now has to cross from molecule X to the receptor molecule, a process that can be described by Marcus theory (Marcus, 1964; Ulstrup, 1979; Bendall,

1996) and characterized by a time τ_I. In proteins, times range from about 1 ms down to about 1 μs (Gray and Winkler, 2005). The injected charge has to propagate through to the donor (D). The route is not known, but probably involves hopping transport. Thus the journey time is likely be in the ms to μs range, as for charge injection. The next step is inelastic tunnelling from D to A (the acceptor), and it is this charge movement that actuates the receptor. From A the charge must now reach the mechanism that releases the G-protein, which in turn initiates the signal that is sent to the brain. Again, we do not know the route taken but it is likely to involve charge hopping. So the characteristic time τ_R will probably be in the range ms to μs. Thus, overall charge injection and extraction together are likely to occur on typical biological timescales of μs to ms.

Note that the electron path need not be through the odourant, as shown in Figure 12.1. All that is needed is rapid movement of a charge near the odourant; the electron might pass through one protein helix, for example. The physics involved is largely unchanged. Perhaps the main advantages of intra-chain charge transfer is that there is no need for any long-range motion of charge to reset the donor and acceptor to their original states, and it eliminates the electron tunnelling rate as a quantity contributing to selectivity, producing a signal that is easier to interpret. It is possible, for example, that the original intra-chain electronic states will be recovered simply by the odourant leaving the receptor, but that is pure speculation. However, decisions as to the precise charge transfer will not be possible until there is full structural information, and especially guidance about likely donor/acceptor units at the receptor binding site. At the time of writing, the olfactory receptors remain one of the elusive class of membrane protein receptors yet to be crystallized.

12.2.1 The electron-odourant force

The electron is charged as are atoms in odourants, thus there can be a force between them. When the electron jumps from D to A, the force it exerts on the odourant changes. This change in force is felt by the odourant as a kick, which then causes it to oscillate. This is the mechanism that couples the electron motion to the odourant vibration. For the Turin mechanism to work the kick has to be large enough to allow the electron to lose energy efficiently during the jump. Thus the size of this force is a key quantity in the theory.

The theory for the rates of processes in which one or more phonons are excited was developed by Huang and Rhys in 1950 (Huang and Rhys, 1950). Their work, originally for optical transitions, showed that the electron hopping rate depends on a dimensionless parameter, the Huang–Rhys factor (S), which is the ratio of the reorganization energy λ to the vibrational energy hf, where λ is proportional to the square of the change in force. Note that the relevant force here is that projected onto

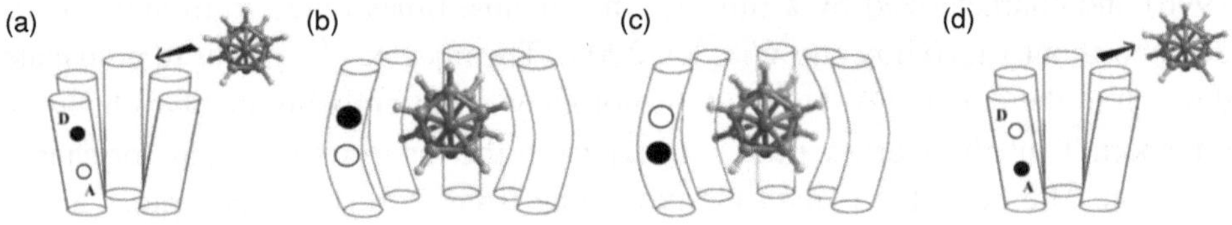

Figure 12.2 A cartoon illustrating intra-protein electron transfer. Only five transmembrane helices for the olfactory receptor are shown (cylinders) here for clarity. (a) The odourant approaches the receptor, meanwhile an electron is present at donor site D. (b) The odourant docks at the ligand binding domain, the overall configuration of receptor and odourant changes. (c) The electron jumps from D to A, causing the odourant to vibrate vigorously. (d) The odourant is expelled from the ligand binding domain. Adapted from the PhD Thesis of J. C. Brookes. Used with permission.

the vibrational mode we wish to excite. Marcus derived essentially the same theory in the context of redox reactions (Marcus, 1964). Other vibrations in the protein that are lower in frequency or too far away from the binding sites to contribute to the value of S of the odourant, add to the broadening of the response and so interfere with recognition (Brookes *et al.*, 2007).

We can use state-of-the-art electronic structure calculations to make rough estimates of the Huang–Rhys factor, and hence of the probabilities of vibrational excitation. We do this simply by exposing a model of the odourant to the field of a point charge placed at two sites, and directly compute the change in force along the relevant mode. Infrared active phonons, where there is a change in dipole moment, can be detected by this mechanism. We now describe the calculation for a simple model system. Let us represent the odourant by a charge q with mass M that is able to move along the x-axis and bound harmonically to the centre of a cavity. The harmonic vibration angular frequency is ω, and the force constant is $K = M\omega^2$. We further assume that the electronic charge moves from one point D on the wall of the cavity to another point A, again on the wall. We can choose these points to correspond to the inter-molecular (see Figure 12.1) or intra-molecular (see Figure 12.2) tunnelling cases. The sudden move of the electron from D to A changes the electric field at the odourant dipole, causing it to change vibrational state. We can now estimate the Huang–Rhys factor (S) as a function of the relative orientations of jump path and the oscillator axis, the magnitude of the dipole moment and the vibrational frequency.

We represent the applied force as an electric field having a projection E in the x direction, so it has magnitude $F = Eq$ on the oscillator charge. The relaxation energy that is relevant for the Huang–Rhys factor calculation is $\lambda = F^2/2K = E^2q^2/2K$. The Huang–Rhys factor itself is $S = \lambda/\hbar\omega$ or $(E^2q^2/2K)/\hbar\omega$. Since the force constant K is $M\omega^2$, then $S = E^2q^2/2M\hbar\omega^3$. This result is consistent with more formal theory of infrared absorption, compare Sect. 11.9.2 of Stoneham (1975).

It now remains to calculate E for given positions of D and A, and also to decide what effective mass M and charge q is appropriate. The effective charges need careful discussion in any realistic case (see, e.g. Catlow and Stoneham (1983)), but need not concern us so much in a model calculation. When the oscillator is at the origin pointing along the x-axis, we may take the donor D to be as at a site located at $\vec{R}_D$ and the acceptor A at $\vec{R}_A$. The change in force on transfer of an electron is

$$F = \frac{eq}{4\pi\epsilon_0\epsilon}\left(\frac{\vec{R}_D}{R_D^3} - \frac{\vec{R}_A}{R_A^3}\right)\cdot\hat{\imath}, \tag{12.1}$$

with $\hat{\imath}$ a unit vector along the x-axis and ϵ the relevant dielectric constant. This can be evaluated easily in the model case. It is easily seen without explicit calculation that the relevant Huang–Rhys factor falls off with increasing distance of D or A from the oscillator axis.

For the electron to move from D to A, energy must be conserved. Since D and A differ in energy, the electron must lose this amount to its environment. It can do this in one of two ways: either by exciting the odourant, or by exciting other modes in the neighbourhood (or a combination of the two). The first instance corresponds to receptor discrimination of an odourant and the second does not. For this mechanical mechanism to explain how humans smell, the discriminatory channel must make the larger contribution. Conveniently, the probability for both events, discriminatory and non-discriminatory, can be calculated with Fermi's golden rule. The Huang–Rhys/Marcus formula regards the single frequency case, and we consider in the olfaction model just one phonon of excitation. If there is a large change in the charge distribution, and thus a large Huang–Rhys factor, then we predict a strong odourant signal for that phonon. Thus odourant recognition may be quantified and predicted using Huang–Rhys factor spectra.

Huang–Rhys theory works classically as well as quantally, but the quantum version is essential here: (i) energy can only be given or received in fixed units; (ii) the ratios of the zero phonon, one phonon, two phonon, components are fixed quantally; and (iii) the significant oscillators are all in their ground vibrational states, i.e. at room temperature the critical mode in the odourant is not excited.

12.3 Quantum rate equations

Calculation of the elastic and inelastic tunnelling rates now follows straightforwardly from the considerations above (Brookes *et al.*, 2007). The times characterizing elastic (τ_{T0}) and inelastic (τ_{T1}) tunnelling from D to A are the key quantities.

Table 12.1 *Estimated values for the parameters needed to compute τ_{T0} and τ_{T1} (Brookes et al., 2007). Note here we use $S = 0.1$, which is more realistic than our previous value of $S = 0.01$. We discuss the likely sensitivities of the various parameter values in the text.*

Quantity	$\hbar\omega$	S	λ	$\|t\|$
Value	200 meV	0.1	30 meV	1 meV

We treat D and A as single molecular orbitals with energies ε_D and ε_A, coupled to each other by a weak hopping integral t, but not coupled to other electronic states. Since the hopping between D and A is slow on electronic timescales, the remaining electronic couplings must be very weak to prevent electron leakage. However, D and A will be coupled to oscillators in the odourant, receptor protein and the wider environment.

The transfer rate from D to A can be computed using the Marcus–Levitch–Jortner equation (Walker *et al.*, 1992). We consider one odourant oscillator with angular frequency ω which couples to the electron when on D and A, with the relevant measure being the Huang–Rhys factor S. The environment is treated as many oscillators which we characterize by a single number, the reorganization energy λ (Flynn and Stoneham, 1970; Song and Marcus, 1993). Taking the background fluctuations to be of low frequency, we get the following expression:

$$\frac{1}{\tau_{Tn}} = \frac{2\pi}{h} t^2 \frac{\sigma_n}{\sqrt{4\pi k_B T \lambda}} \exp\left(-\frac{(\varepsilon_n - \lambda)^2}{4k_B T \lambda} \right), \tag{12.2}$$

where $\sigma_n = \exp(-S) S^n / n!$ and $\varepsilon_n = \varepsilon_D - \varepsilon_A - n\hbar\omega$.

12.4 Putting in numbers

To test the theory we now compute the values of the rates it predicts, and ask the following questions: Do the numbers make sense when compared with experiment and with other biophysical values? Are very special values needed, or is the theory robust?

The non-radiative transition rates are given by Equation (12.2) for the two channels with $n = 0$ for the non-discriminating channel where the odourant is not excited and all energy is taken up by host vibrations, and $n = 1$ for the discriminating channel, where the odourant takes up this energy. Typical values of the important parameters, given in Table 12.1 and first shown in Brookes *et al.* (2007), give $\tau_{T0} = 87$ ns and $\tau_{T1} = 0.15$ ns. These easily satisfy the condition that $\tau_{T1} \ll \tau_{T0}$, and the discriminating inelastic channel dominates. This is, of course, the opposite of what is found for inelastic tunnelling involving metal electrodes with their continua of initial or final electronic states.

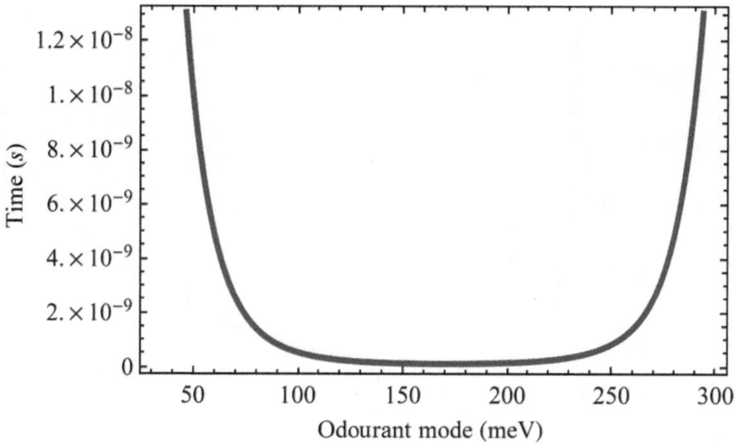

Figure 12.3 A plot to show the time (s) for an inelastic transmission (blue) for $\varepsilon_D - \varepsilon_A = 200$ meV, and varying the values of the odourant's mode of vibration $\hbar\omega$.

How reliable are the numbers in Table 12.1? For a particular odourant mode of vibration, $\hbar\omega$ can be easily and accurately calculated using first principles electronic structure methods. The Huang–Rhys factor (S) can be calculated as described above (see Section 12.2.1). The reorganization energy, however, is not so easily calculated, so we make an estimate based on what is known of the olfactory receptors. If the environment were strongly coupled to the mobile electron, the environment modes could take up most of the electronic energy and discrimination based on the odourant mode would be ineffective. Yet the hydrophobic nature of the odourant binding site ensures that the reorganization energy is low and there is weak coupling to the host modes.

The electron tunnelling matrix element (t) can be estimated as follows. We assume the odourant (M) contacts both D and A but interacts with them only weakly with hopping integral v. By considering the resulting admixtures of an M state with energy ε_M with those of D and A we obtain an effective hopping integral between D and A ($t = v^2/(\varepsilon_M - \varepsilon_A)$). If ε_M corresponds to a LUMO while ε_D and ε_A correspond to HOMOs, then the difference $\varepsilon_M - \varepsilon_A$ can be as large as 10 eV. The hopping integrals can be estimated for known molecular structures. Whilst the odourant structure is known, the donor and acceptor structures interacting with it are unknown, and we have to make an educated guess. If the bonds between M, and D and A are no stronger than hydrogen bonds, we can put a rough upper bound on the associated hopping integrals of order 0.1 eV, and hence obtain $t \sim 1$ meV. Our final conclusions are not sensitive to this value. However, the rate has some quite important dependencies on other parameters in Table 12.1, as represented graphically in Figures 12.3, 12.4 and 12.5. Figure 12.3 shows the

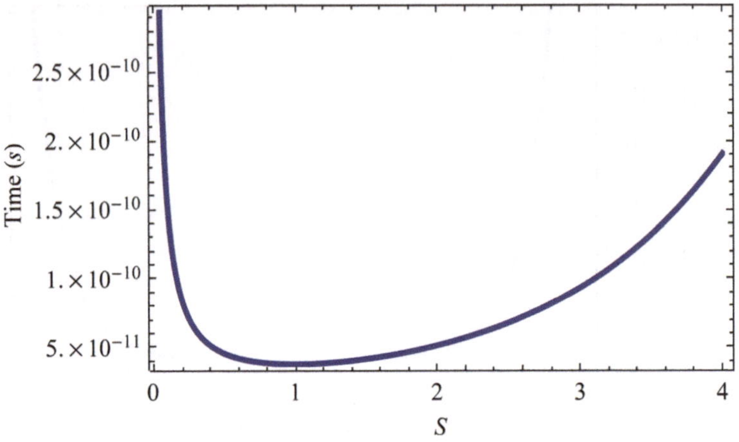

Figure 12.4 A plot to show the time (s) for an inelastic transmission (blue) for $\varepsilon_D - \varepsilon_A = 200$ meV $= \hbar\omega$ at resonance, and varying Huang–Rhys factor S.

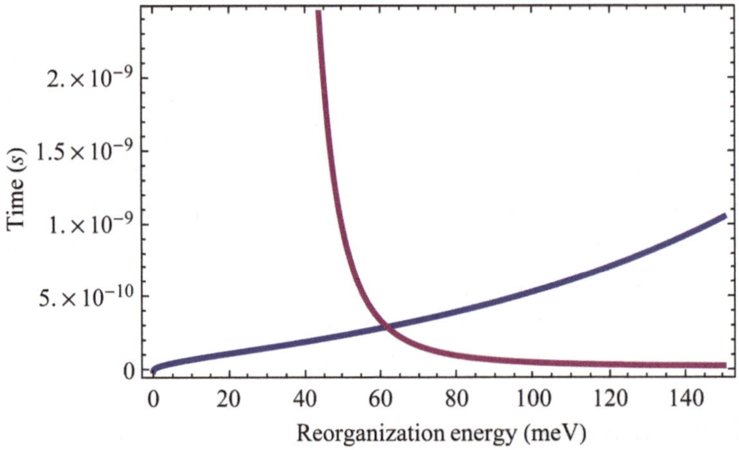

Figure 12.5 A plot to show the inelastic (blue) versus the elastic (red) transmission for the parameters in Table 12.1, for $\varepsilon_D - \varepsilon_A = 200$ meV $= \hbar\omega$ at resonance, but varying the reorganization energy (λ) of the environment. From the PhD Thesis of J. C. Brookes. Used with permission.

selectivity for a given odourant mode (200 meV) decreases drastically for $\hbar\omega <$ 100 meV and $\hbar\omega >$ 230 meV. Figure 12.4 indicates that for a strong odourant signal a Huang–Rhys factor (S) of 0.05–0.25 is desirable. Lastly, Figure 12.5 indicates the sensitivity the rate has on the reorganization energy; it shows a plot of the characteristic times for the channels with and without odourant vibrational excitation, as a function of λ. The figure shows that tunnelling dominated by environment modes would occur for $\lambda >$ 62 meV.

Our analysis gives the crucial result that the discriminating case (odourant with the right frequency) has a nearly 600 times higher rate of transmission than that of the non-discriminating case. This holds true in the harmonic approximations when background oscillations have low frequencies and couple weakly to the electron transfer, when there is a low reorganization energy, and when one phonon of odourant vibration is excited. Under these conditions the inelastic channel is preferred and the odourant with the 'right' frequency will be detected over the 'wrong' one.

12.5 Can we make predictions?

Olfaction is notorious for the problems associated with producing consistent and reliable experimental data. The problems include removing trace contamination, and overcoming subjective differences in the interpretation of odour. However, there are cases where there are systematic trends, and where we may test our ideas. We focus on the odourant and the question of whether odour can be correlated with vibrational frequency and coupling to odourant charges. The boranes provide an interesting example, since they can smell sulfuraceous despite containing no sulfur. We have computed the vibrational spectra of H_2S and four boranes (decaborane, m-, o- and p-carborane). The boranes are structurally similar, but all quite distinct from H_2S. However, H_2S and decaborane smell sulfuraceous, while the carboranes smell camphoraceous. Using *Gaussian03* we computed vibrational frequencies and infrared (IR) couplings, defined as $|dp/dQ_i|^2$ with p the dipole moment and Q_i a displacement along normal mode i. The sulfuraceous smell of H_2S is associated with vibrations in the region of 2600 cm^{-1}. In this region decaborane has IR couplings that are one to two orders of magnitude greater than the carboranes. Assuming that the IR couplings are a good estimate of the electron–oscillator coupling in an olfactory receptor, this could explain sulfuraceous and less sulfuraceous odours.

12.6 Extensions of the theory for enantiomers

A common reason to reject vibration based theories of olfaction is the existence of many enantiomer odourants (mirror image molecules, e.g. left- and right-handed chiral molecules) that have the same vibrational frequencies but smell different. Mirror image molecules exemplify the importance of shape in receptor detection, while still leaving the rules of shape selectivity obscure. They clearly show that the positions of atoms matter, though it is unkown why. Shape factors alone do not correctly predict the activity of odourants (Sell, 2006), but it is also true that vibrational (IR) features alone are also not always successful in predicting organoleptic properties (Raman and Gutierrez-Osuna, 2009), though in some cases they do very well. The theory of scent determination via vibrations cannot thus be completely divorced from shape requirements and this is interestingly exemplified by enantiomer pairs of odourant molecules. In some, if not all cases, actuation is likely determined by a combination of these factors. Note that the above rate equation determines discrimination based on an odourants critical mode of vibration, but this in turn is sensitive to the geometry of the molecule, and further, the positioning of D and A relative to the molecule. These geometrical contributions are accounted for in the calculation of S and t. We propose then that predictions based on the rate equation embrace the necessary contributions from geometry and energy.

Odourants are not the only small molecules that interact unpredictably with large proteins; steroid hormones, anaesthetics, neurotransmitters, to name a few, are examples of ligands that interact specifically with special receptors to produce important biological processes. Steroids, in particular, exhibit similar curiosities to odourants. The estrogen receptor for one is notably promiscuous and responds to its physiological 17β-estradiol, but also many non-steroidal stimulants such as 1,2-diarylethanes and ethylenes, flavones/isoflavones, macrolactones, aryl/alkyl phenols and halogenated carbocycles (Oettel and Schillinger, 1999). Such a group represents quite wide structural diversity. Attempts have been made to identify structure–activity relations, but it seems unlikely that all these different 'keys' are fitting the same 'lock'; even subtle changes to a ligand's stereochemistry can produce quite large bio-effects. This is not only exemplified by mirror image odourants, as discussed above, but by very small conformational changes, such as the α/β positioning of a C–H bond, for example.

Conventional inelastic electron tunnelling spectroscopy can determine the orientation of a molecule (Kirtley *et al.*, 1976). In similar ways, the tunnelling model is sensitive to positioning of the odourant in a binding site. This electron transfer model depends on a Huang–Rhys factor, which in turn depends on the orientation of the critical oscillating mode of vibration. Further, it is emerging in many electron transfer experiments that there is great sensitivity to small changes in geometry (of order picometres) and in conformational dynamics of the ligand (Sigala *et al.*, 2008). It may be that electron transfer in the nose is also sensitive to conformational mobility (Lu *et al.*, 2007; Brookes, 2010). It is interesting to hypothesize, that given that the positions of atoms can certainly be detected by the electron, perhaps any flexibility in the odourant may promote or demote the electron transfer in an actuating step. Recent advances in computational biology are contemplating a more dynamical world of the fluctuating protein; but it may be that the fluctuations of a 'key' ligand are the important ones. This seems unlikely in a classical model, but processes potentially reveal explanations for the seemingly impossible processes of receptor–small molecule recognition.

13

A perspective on possible manifestations of entanglement in biological systems

HANS J. BRIEGEL AND SANDU POPESCU

13.1 Introduction

In this chapter, we will focus on the phenomenon of quantum *entanglement*. Entanglement is a special property that some states of two or more quantum particles may possess. When particles are in an entangled state they are correlated. But entanglement is *far more than mere correlation*. Entanglement is a type of correlation that macroscopic objects cannot have, even in principle; it is what we call a 'non-local' correlation.

Entanglement, and the non-locality that derives from it, is by now considered to be, arguably, the most important aspect of quantum mechanics. Philosophically it leads to many puzzles and to a complete change in our view of nature. At the same time, the practical implications of this type of correlation are very significant. In particular, they can lead to the dramatically enhanced information processing capabilities that quantum systems possess. The whole field of quantum information processing (Steane, 1998; Bennett and DiVincenzo, 2000; Zoller *et al.*, 2005) has entanglement at its core.

Given the above, it is natural to ask whether biological systems make use and take advantage of the enhanced possibilities that entanglement offers. It seems natural to expect this, especially at biological molecular level, where the laws of quantum mechanics apply.

Entanglement between two isolated particles is easy to obtain: almost every time two particles interact they get entangled. But entanglement is very fragile. When two entangled quantum systems interact with a third, they easily become entangled in a tri-partite way; however, any entanglement between any two of them is reduced or disappears altogether. That is, one can see the interesting non-local correlations only if one now looks at correlations between all three parties; but not only between

Quantum Effects in Biology, ed. Masoud Mohseni, Yasser Omar, Gregory S. Engel and Martin B. Plenio. Published by Cambridge University Press.

two. In this sense, entanglement, like love, is monogamous. When two parties are fully entangled with one another, they cannot be entangled with a third; if they are entangled with the third, in a quantum love triangle, the entanglement of the original pair is lost. Even the cumbersome – but very interesting – entanglement of the triplet is lost when a fourth appears, and is replaced by an even more intricate four partite entanglement. And so on.

For the above reason, most experiments to date that exhibit entanglement try extremely hard to isolate the quantum systems under investigation. Otherwise entanglement spreads over more and more particles and nothing is left between the ones we are interested in. But while this can be done in physics laboratories, keeping particles in vacuum, cooling them to extreme temperatures and employing many other precautions to isolate them from the electric and magnetic fields of nearby objects, this is obviously impossible in biological systems. Biology is warm and wet. Most processes take place in a watery solution, and molecules continuously bounce into each other. So one would expect that biologically there can be no entanglement between any relevant degrees of freedom but get dissipated throughout the environment. It is for these reasons that until very recently the possibility of the existence of entanglement in biological systems was dismissed out of hand. However, it has recently been understood that, while such a *static* way of maintaining entanglement like simple isolation is impossible, *active* ways of maintaining it may be possible. The central idea is the fact that biological systems are *open, driven systems, far from thermal equilibrium* and that such systems are capable of *quantum error correction.* Loss of entanglement between two given systems is an 'error'; correcting it means maintaining entanglement. Discussing these recent ideas is the subject of this chapter.

13.2 Entanglement

In this section, we briefly introduce the essential concepts and definitions relating to entanglement. The related issue of quantum non-locality and the kind of correlations exhibited by entangled states will be discussed in the next section. For a more detailed and quantitative discussion we refer the reader to textbooks and reviews, see e.g. (Lo *et al.*, 1998; Bouwmeester *et al.*, 2000; Plenio and Virmani, 2007; Horodecki *et al.*, 2009).

13.2.1 Qualitative discussion

Consider two quantum systems (such as atoms, molecules, spins, etc.). For the purpose of the present discussion, their precise nature is irrelevant; we will call them, in an abstract way, systems A and B. Suppose system A is in state $|0\rangle_A$ and system B is in state $|0\rangle_B$. We can then write the common state of the two systems as

$$|\Phi\rangle_{AB} = |0\rangle_A|0\rangle_B. \tag{13.1}$$

Consider now another situation, in which system A is in state $|1\rangle_A$ and system B is in state $|1\rangle_B$, orthogonal on $|0\rangle_A$ and $|0\rangle_B$ respectively. In this case we can write the common state of the two systems as

$$|\Phi'\rangle_{AB} = |1\rangle_A|1\rangle_B. \tag{13.2}$$

But the rules of quantum mechanics say that whenever two states, such as $|\Phi\rangle_{AB}$ and $|\Phi'\rangle_{AB}$, are possible, so is any *superposition* of them, such as

$$|\Psi\rangle_{AB} = \frac{1}{\sqrt{2}}|\Phi\rangle_{AB} + \frac{1}{\sqrt{2}}|\Phi'\rangle_{AB}. \tag{13.3}$$

This is exactly the same principle as the one that applies to a single quantum system and says that, for example, since system A can be in states $|0\rangle_A$ and $|1\rangle_A$, then $\frac{1}{\sqrt{2}}|0\rangle_A + \frac{1}{\sqrt{2}}|1\rangle_A$ is a possible state.

Using the explicit forms of states $|\Phi\rangle_{AB}$ and $|\Phi'\rangle_{AB}$, the state (13.3) can be written as

$$|\Psi\rangle_{AB} = \frac{1}{\sqrt{2}}|0\rangle_A|0\rangle_B + \frac{1}{\sqrt{2}}|1\rangle_A|1\rangle_B. \tag{13.4}$$

This is an example of an *entangled state*. What we can immediately see is that in this joint AB state system A does not have a (pure) state of its own, neither has system B. That is, we cannot say that system A is in some state $|\psi\rangle_A$ and system B in some state $|\varphi\rangle_B$. Indeed, one can easily check that $|\Psi\rangle_{AB}$ cannot be factorized, that is,

$$|\Psi\rangle_{AB} \neq |\psi\rangle_A|\varphi\rangle_B. \tag{13.5}$$

Instead, the state of system A and that of system B are correlated, in a very peculiar way.

First of all, there is one immediate way in which they are correlated. Since $|0\rangle_A$ and $|1\rangle_A$ are orthogonal we can perform a measurement to distinguish them, i.e. we can make a measurement to check whether system A is in state $|0\rangle_A$ or $|1\rangle_A$. The probability of finding either result is 1/2. Now, if this measurement tells us that system A in state $|0\rangle_A$ then, according to quantum mechanics the original state (13.4) collapses onto state $|0\rangle_A|0\rangle_B$, so now system B is in state $|0\rangle_B$. More exactly, if we now measure system B to check whether it is in state $|0\rangle_B$ or $|1\rangle_B$, we find it with certainty in $|0\rangle_B$. On the other hand, if we find system A in $|1\rangle_A$ we know system B is in state $|1\rangle_B$. That is, if we now measure system B to check whether it is in state $|0\rangle_B$ or $|1\rangle_B$, we find it with certainty in $|1\rangle_B$.

But this is not all. Suppose that instead of performing the measurements described above, we had performed some other measurements on A and B. The results of these other measurements will also be correlated in some way. It is the pattern of these correlations between the results of all possible measurements that characterizes the entanglement.

13.2.2 Entanglement: general definitions

Before discussing the pattern of correlations that characterize entangled states, we shall give the general definitions of what entanglement is.

The case of pure states

Definition 1: Direct product Consider two quantum systems, A and B, in a pure state. A state that can be written in the form

$$|\Psi\rangle_{AB} = |\phi\rangle_A |\varphi\rangle_B, \tag{13.6}$$

where $|\phi\rangle_A$ is an arbitrary state of system A and $|\varphi\rangle_B$ is an arbitrary state of system B, is called a *direct product* and it is NOT entangled.

Definition 2: Entangled state Any state $|\Psi\rangle_{AB}$ that is not a direct product is called *entangled.*

Note that the definition of an entangled state is given indirectly, by reference to direct product states. Note also that in order to be entangled it is not sufficient that a state does not have the form (13.6), but that it cannot be written in that form by any appropriate choice of basis. Indeed, consider the state

$$\frac{1}{2}|0\rangle_A|0\rangle_B + \frac{1}{2}|1\rangle_A|1\rangle_B + \frac{1}{2}|0\rangle_A|1\rangle_B + \frac{1}{2}|1\rangle_A|0\rangle_B. \tag{13.7}$$

This state is actually a direct product since it can be written in the form (13.6) with $|\phi\rangle_A = \frac{1}{\sqrt{2}}|0\rangle_A + \frac{1}{\sqrt{2}}|1\rangle_A$ and $|\varphi\rangle_B = \frac{1}{\sqrt{2}}|0\rangle_B + \frac{1}{\sqrt{2}}|1\rangle_B$.

Verifying whether or not a given pure state is entangled or not is a trivial question. (This is in sharp contradiction with the case of mixed states discussed below.)

The definitions above generalize to the case of many quantum systems with

$$|\Psi\rangle_{AB...N} = |\phi\rangle_A|\varphi\rangle_B \ldots |\xi\rangle_N \tag{13.8}$$

being a direct product and all other states being entangled.

The case of mixed states

Definition 3: Separable states A mixed state ρ_{AB} that can be written as a mixture of direct products is a *separable mixed state*. That is, ρ_{AB} is separable if it can be written in the form,

$$\rho_{AB} = \sum_i p_i |\phi_i\rangle_A \langle\phi_i| \otimes |\varphi_i\rangle_B \langle\varphi_i|, \tag{13.9}$$

where p_i are probabilities ($p_i \geq 0$ and $\sum_i p_i = 1$), $|\phi_i\rangle_A$ are arbitrary states of system A, not necessarily orthogonal on one other and similar for $|\varphi_i\rangle_B$ and where the summation index i is $1 \leq i < \infty$.

Definition 4: Entangled mixed states Any mixed state that is not separable is entangled.

The definitions above extend to more than two systems in a similar way to the case of pure states.

13.2.3 Mixed-state entanglement

As opposed to the case of pure states, it is, in general, very difficult to establish whether or not a mixed state is entangled. In fact there is no known general algorithm for establishing this. The difficulty comes from the fact that the number of direct product states in the mixture (13.9) may be infinite and the coefficients p_i are restricted to being positive. In particular also note that, counterintuitively, mixtures of entangled states may sometime result in a separable state.

The above being said, there are some (quite important) cases in which we can easily decide whether a state is entangled or not:

- The case of two qubits is completely solved. ('Qubit' is the generic name for a two-state quantum mechanical system).
- There are instances in which we know explicitly the decomposition of a mixed state into direct products; hence the state is separable.
- There exist certain quantities that are 'entanglement witnesses'. If they have a certain value we conclude that the state is entangled; otherwise we cannot conclude anything. The list of these witnesses is ever increasing.

For more detailed information we refer the reader to recent reviews, e.g. (Plenio and Virmani, 2007; Horodecki *et al.*, 2009).

13.3 Non-local correlations

As we mentioned earlier, the most important aspect of entangled states is the type of correlations between the outcomes of different measurements performed on them. To start with, it is important to notice that the correlations characterize the state itself, not the interaction between the particles. That is, once the state is established – which is due to interactions between the particles – interaction is no longer needed to generate the correlations. So the best way to understand the nature of correlations is to imagine the two entangled particles to be completely separated in space and no longer interacting. (Of course, when particles continuously interact, as is probably the case in all relevant biological contexts, the state evolves in time and the correlations between the particles may change.)

Consider two experimentalists, Alice and Bob, situated far away from each other. They perform various measurements on particles that come from a common source and are prepared in an entangled state, one particle from each pair going to Alice and one to Bob. The experiments are 'space-like separated'. That is, the measurements take a short time compared with the time required for light (and hence for any signal) to propagate from Alice to Bob. Furthermore, by prearrangement, the measurements are timed in such a way that Alice's measurement finishes before she could receive any signal from Bob about the measurement he performed (i.e. what measurement he made and what the result was), and similarly all information from Alice about the measurement she performed can reach Bob only after he finishes his measurement. Nevertheless, their results turn out to be correlated (although this can only be found out later, when Alice and Bob are able to compare their results). The fact that the results are correlated is not a great surprise – after all, the particles came from a common source. What is astonishing however is that they are correlated in such a way that if we want to establish such correlations with any classical devices, they have to communicate with each other. Because of the timing of the measurements, this communication would have to be super-luminal. We refer to such correlations as non-local.

To describe the notion of non-local correlations, let us start with a game having nothing to do with quantum mechanics. It is a game played by one team of two players and two referees, Alice and Bob are situated far from each other. The game consists of many rounds. The two players meet before each round and agree on how to play in that round. After this meeting the players are separated – Player 1 goes to Alice while Player 2 goes to Bob. Alice asks Player 1 one question that she chooses at random between two possible questions. The questions are: 'Tell me a value for A_1' and 'Tell me a value for A_2'. There are two possible answers to each question, say 0 and 1. Similarly Bob asks Player 2 a value for B_1 or B_2. The players can give any answers they want.

The team wins in the following situations:

- If Alice asked A_2 and Bob asked B_2 the team wins if the players give different values: Player 1 says 0 and Player 2 says 1 or vice versa.
- In all other cases, i.e. A_1 and B_1 or A_1 and B_2 or A_2 and B_1 the team wins if both players give the same answer.

The players know the rules of the game; they only do not know which question will be asked in each round.

The most important aspect of the game is that after the players split and one goes to Alice and the other to Bob, they are *not allowed to communicate* to each other. That is, Player 1 cannot tell Player 2 something like, 'Alice asked me a value for A_1, I said 0, you take care what you do.' The players can decide on a common strategy of how to answer, but they can do this only before they go to the referees, i.e. before they find out what the questions are.

If the players just give random answers the probability of success is $1/2$. Pre-arrangement, that is, agreeing on a common strategy, allows the team to do better. The question is to find the strategy that maximizes the winning probability.

Let us start by asking if the team can win with certainty.

Since the players do not know what question they will be asked, they have to prepare in advance for all possibilities. Suppose, without any loss of generality, that in the first round the players decide that Player 1, if asked A_1 will answer '0'. For what follows it is important to keep in mind that, when in front of Alice, Player 1 has no information about what happens to Player 2, so he will answer the question A_1 with '0' regardless of what Bob asks. Similarly, Player 2 acts without knowing what Alice does.

Now, given the above decision for Player 1, in order to win when the questions are A_1 and B_1, they have no choice but to decide that Player 2 will give the answer '0' if asked B_1.

In order to win also in the case where the questions are A_1 and B_2, Player 2 must answer '0' to B_2 as well.

Next, in order to ensure a win in the case of A_2, B_1, since the answer of Player 2 to B_1 is '0', the answer of Player 1 to A_2 must also be '0'.

So far so good – the team wins if the questions are A_1 and B_1 or A_1 and B_2 or A_2 and B_1. But at this moment they have no more free parameters: they were forced to decide to answer by '0' all questions. Which means that if they are asked A_2 and B_2 they both say '0', which is a loss according to the rules of the game. It is important to realize that there is not much else what the players can do. The only other possibility would have been for Player 1 to answer A_1 by '1'. A similar analysis as above shows that again they cannot win in all four cases.

We thus conclude that the best that the players can do is to win the game in $3/4$ of the cases. Anything more can only be achieved if the two players communicate *after* finding out what the questions are and coordinate their answers.

Definition: Locality The fact that the players are not allowed to communicate with each other when in front of the referees is called 'locality', meaning that when in front of the referees the players have to rely only on the information that is locally available to them (such as their memory of the pre-arrangement).

Definition: Bell inequality The upper bound on the winning probability comes from the no-communication constraint and it is generally called a Bell inequality, after its discoverer John Bell. A 'violation of Bell's inequality' implies, in the context of our game, communication.

Coming back to quantum mechanics, exactly the same game can be played here; The 'players' are the quantum particles. The 'questions' that Alice and Bob ask are measurements performed on the particles. The 'answers' that the particles give are the outcomes of the measurements. It is important to note that the actual nature of the measurements and of their outcomes is irrelevant; we can always call them abstractly as measurements $A_1, \ldots, B_2$ and outcomes '0' and '1'. Finally, since the measurements are space-like separated, no signal should be able to reach from one particle to the other during the measurements, hence particles are not able to communicate with each other.

John Bell's great discovery is that situations can be arranged in which the particles win the game in *more* that 3/4 of the cases. This means that what the quantum particles do in this experiment is equivalent to classical particles that communicate with each other faster than light.

For the sake of argument, we will entertain such a "classical" picture in the remainder of this section. This will allow us to give concise definitions of various concepts related to non-locality.

Example A particular situation is the following. Two qubits are prepared in the entangled state:

$$|\Psi\rangle_{AB} = \frac{1}{\sqrt{2}}|0\rangle_A|1\rangle_B - \frac{1}{\sqrt{2}}|1\rangle_A|0\rangle_B, \tag{13.10}$$

and the measurements Alice and Bob make are of the projection operators,

$$A_1 = \frac{1}{2}\big(|0\rangle_A + |1\rangle_A\big)\big({}_A\langle 0| +_A \langle 1|\big) \tag{13.11}$$

$$A_2 = \frac{1}{2}\big(|0\rangle_A + i|1\rangle_A\big)\big({}_A\langle 0| - i_A\langle 1|\big) \tag{13.12}$$

$$B_1 = \frac{1}{2}\big(|0\rangle_B + e^{\frac{i\pi}{4}}|1\rangle_B\big)\big({}_B\langle 0| + e^{\frac{-i\pi}{4}}{}_B\langle 1|\big) \tag{13.13}$$

$$B_2 = \frac{1}{2}\big(|0\rangle_B + e^{\frac{-i\pi}{4}}|1\rangle_B\big)\big({}_B\langle 0| + e^{\frac{i\pi}{4}}{}_B\langle 1|\big). \tag{13.14}$$

Each projection operator can yield two possible outcomes, namely the eigenvalues 0 or 1.

Using the standard quantum-mechanical rules, we find that the probability $P(1, 1|A_1, B_1)$ for obtaining the pair of outcomes 1 and 1 when measuring A_1 and B_1 is simply the expectation value of the product of these two projectors, namely

$$P(1, 1|A_1, B_1) = \langle\Psi|A_1 B_1|\Psi\rangle = \frac{2+\sqrt{2}}{8}; \tag{13.15}$$

while the probability of obtaining the pair of outcomes 0 and 0 is the expectation of the product of the complementary projectors,

$$P(0, 0|A_1, B_1) = \langle\Psi|(1 - A_1)(1 - B_1)|\Psi\rangle = \frac{2+\sqrt{2}}{8}. \tag{13.16}$$

Hence the probability of winning when the measurements A_1 and B_1 are performed is equal to

$$P(\text{win}|A_1, B_1) = P(1, 1|A_1, B_1) + P(0, 0|A_1, B_1) = \frac{2+\sqrt{2}}{4}. \tag{13.17}$$

In a similar way, we find that $P(\text{win}|A_1, B_2) = P(\text{win}|A_2, B_1) = \frac{2+\sqrt{2}}{4}$. For the case when A_2 and B_2 are measured, winning happens when the outcomes of the measurements are different. We obtain

$$\begin{aligned} P(\text{win}|A_2, B_2) &= P(0, 1|A_2, B_2) + P(1, 0|A_2, B_2) \\ &= \langle\Psi|A_2(1 - B_2)|\Psi\rangle + \langle\Psi|(1 - A_2)B_2|\Psi\rangle = \frac{2+\sqrt{2}}{4}. \end{aligned} \tag{13.18}$$

Adding all together, and taking into account that the probability for Alice and Bob to perform a certain pair of measurements is 1/4 (since they chose their measurements at random) we obtain

$$\begin{aligned} P(\text{win}) &= \frac{1}{4}P(\text{win}|A_1, B_1) + \frac{1}{4}P(\text{win}|A_1, B_2) + \frac{1}{4}P(\text{win}|A_2, B_1) \\ &\quad + \frac{1}{4}P(\text{win}|A_2, B_2) \\ &= \frac{2+\sqrt{2}}{4} > \frac{3}{4}. \end{aligned} \tag{13.19}$$

One may immediately worry that this faster than light communication between the particles violates special relativity and hence there must be a mistake somewhere. There is not. The point is that quantum mechanics is a fundamentally indeterministic theory. That is, in quantum mechanics measurements performed on identically prepared systems may give different outcomes and only the probability of these outcomes is prescribed. Crucially, this indeterminism is fundamental. That is, it is *not* due to some random experimental errors (which affect every measurement in classical mechanics as well), but to the very nature of the theory. Random experimental errors can be reduced by improving the measurement; the quantum uncertainty cannot be reduced. Now, in all such fundamentally indeterministic theories the consistency with special relativity is subtle.

What happens in this game is that the *particles* may communicate with each other faster than light, but the *experimentalists* (Alice and Bob) cannot use these inter-particle communications to communicate with each other faster than light. If they could, we would immediately run into all the well-known paradoxes of special relativity – communication backwards in time, killing ones grandparents, etc. But they can not. The reason is that Alice and Bob have only limited control over their particles; they may choose what measurement to perform, but they cannot force the particles to give a particular outcome. Particles can 'hide' their faster than light communication under the umbrella of uncertainties. That is, the fact that the particles must have communicated with each other can only be observed when one looks at the *correlations* between the results of Alice and Bob's measurements. But for this Alice and Bob must transmit their results from one to the other, a process that occurs with, at most, the speed of light. On the other hand, immediately after she performed her measurements Alice only knows the probability of her results, not their correlations with Bob's, and similarly for Bob. Crucially, Bob cannot affect Alice's probabilities by choosing to perform different measurements on his particle, so he has no way to communicate anything to Alice right away. In the example above, when Alice measures A_1 the probability of obtaining '1' is $1/2$, regardless of whether Bob decides to measure B_1 or B_2. Indeed the probability $P_{\text{Alice}}(1|A_1, B_1)$ for Alice to obtain '1' when Bob measured B_1 is identical with her probability of obtaining '1' when Bob measured B_2,

$$\begin{aligned} P_{\text{Alice}}(1|A_1, B_1) &= P(1,0|A_1, B_1) + P(1,1|A_1, B_1) \\ &= P_{\text{Alice}}(1|A_1, B_2) = P(1,0|A_1, B_2) + P(1,1|A_1, B_2) \\ &= \frac{1}{2}, \end{aligned} \tag{13.20}$$

where we have used (13.15), and similar expressions for the other probabilities.

These types of correlations are said to be non-local. A very rough definition is the following.

Definition: Non-local correlations Correlations between space-like separated events (such as the outcomes of various measurements) which cannot be produced by pre-arrangement but, in a classical picture, require faster than light communication, at the same time do not allow the experimentalists to signal to one another, since the probabilities of the events occurring at one location are independent of the actions of the other experimentalist.

More generally, we can describe any experiment performed by Alice and Bob in the following abstract way. Alice and Bob may choose to perform in each round one of a number of possible measurements. The measurements are conventionally labelled A_x, with $x = 1, 2, \ldots$ and B_y, with $y = 1, 2, \ldots$ respectively. This notation by itself does not tell us what the measurements are, but simply which one out of the set of possible measurements Alice and Bob performed. The actual measurement that is labelled A_1 may be, for example, 'measure the electron spin along the vertical axis'. Similarly, let $a = 1, 2, \ldots$ and $b = 1, 2, \ldots$ be the labels we associate with the outcomes of Alice and Bob's measurements, respectively. Again, a and b need not represent the actual values produced by the measurements, which may be so-and-so many metres, joules, etc. but are abstract labelling of the outcomes. That is, we may decide to call the outcome '7 metres' as 'outcome number 1', the '19 metres' as 'outcome number 2' and so on. The whole experiment is then described by $P(a, b|x, y)$, the joint probabilities of Alice obtaining outcome a when she performed measurement A_x and for Bob obtaining answer b when he performed measurement B_y.

In full generality, a 'local' (i.e. no-communication) experiment can be described as follows. In each round the two players (or the particles that will be subjected to measurements) meet and agree on an answering strategy. This strategy could be one out of many; we denote it by the index λ. Once they separate and arrive at Alice and Bob who ask them questions/perform measurements, they behave according to the agreed strategy, but are not allowed to communicate with each other. Hence the probability $P(a, b|A_x, B_y; \lambda)$ of yielding answers a and b when the questions/measurements were A_x and B_y and when the strategy was λ, is given by

$$P(a, b|A_x, B_y; \lambda) = P(a|A_x; \lambda)P(b|B_y; \lambda). \tag{13.21}$$

The probability is the product of two local probabilities because, in a particular round what Player 1 does in front of Alice is independent from what Player 2 does in front of Bob. Their behaviour though, depends on the pre-agreed strategy λ. When the game is played many times the players choose different strategies λ according to some probability distribution $p(\lambda)$. Hence, the overall probability distribution for a 'local' game is

$$P_{\text{local}}(a, b|A_x, B_y) = \int p(\lambda)P(a|A_x; \lambda)P(b|B_y; \lambda)d\lambda, \tag{13.22}$$

where for full generality we allowed λ to be a continuous variable.

Definition. Local correlations Any correlations that can be described in the form (13.22) are called 'local correlations'.

Definition. Hidden variable The strategy λ is called a 'hidden variable' since we only have access to the overall distribution $P(a, b|A_x, B_y)$, but we have no knowledge of the specific strategy used in each round.

The other notion we discussed above is that of 'no-signalling'.

Definition. Non-signalling A probability distribution is called non-signalling if

$$\sum_b P(a, b|A_x, B_y) \text{ is independent of } B_y, \tag{13.23}$$

$$\sum_a P(a, b|A_x, B_y) \text{ is independent of } A_x. \tag{13.24}$$

We are now in the position of stating the main definition of this section.

Definition. Non-local correlations A non-signalling correlation that is not of the form (13.22) is called non-local.

13.3.1 Entanglement: conclusions

To summarize, entangled states have the property that, if appropriate measurements are performed on the particles, they generate non-local correlations. This happens even if there is no interaction between the particles at the time of measurement (interaction was only needed to established the entangled state in the first place). More precisely, *all* entangled pure states can generate non-local correlations if appropriate measurements are performed. Also, many mixed entangled states lead to non-local correlations; it is still an open question whether all entangled mixed states have this property. On the other hand, in most biological settings we do not expect the entangled particles to be separated in space and non-interacting with each other, as in the examples above. So the physical effects that they present will be far more complex, and not necessarily describable in the relatively simple language of non-local correlations discussed above. Non-local correlations as described above are only one of the non-local aspects of entanglement. Discussing other non-local aspects of entanglement goes beyond the scope of this book. Yet, it is essential to understand non-locality in its fundamental setting, as described above, to appreciate what it is all about, since it is at the core of all other non-local phenomena.

As far as we are concerned here, the main message is that whenever quantum states are entangled, they have non-local properties and these properties, make them qualitatively different from non-entangled states. In particular, non-locality is behind the significantly increased information processing power of the entangled states and is at the basis of the entire field of quantum information. But this is just the tip of the iceberg, and research in this area is still just beginning.

In conclusion, it is obvious that entanglement is an essential property of quantum mechanics. Arguably, on the scale of importance, quantum coherence is the first step; entanglement is the top. Hence studying entanglement in biological systems is of extreme importance.

13.4 Entanglement in biology

That entanglement and, more generally, coherent quantum effects exist – at some level – in all systems (including biological ones) is quite clear. After all the laws of quantum physics enter at the level of quantum chemistry, determining the structure and energy spectra of the molecules and their interactions. Coming to biology, most scientists would probably share the view that quantum-mechanical effects play a role, but only an indirect one. Quantum mechanics would thus be responsible for the molecular basis or substrate, while the biological functionality of the molecules can be explained by classical statistical physics, combined with the principles of molecular Darwinism (Hughes, 1999).

This makes it clear that, even before asking whether entanglement exists or not, we need to better define what we are actually talking about; that is, what kind of entanglement? In the following, we would like to distinguish three different kinds of entanglement that we expect to play a role in biology; the same classification also holds for any quantum coherent processes that may occur in biological systems.

One has to say from the beginning that the boundaries between theses different types of entanglement are fuzzy. Classification of some phenomena is clear-cut, while for others one may argue whether it is of type 1 or type 2, etc.; nevertheless, we believe this classification to be essential when proceeding to study the possibility of biological quantum effects.

The three types of entanglement are:

- entanglement of basic constituents;
- dead entanglement; and
- live entanglement.

Entanglement of basic constituents

Think of any atom or molecule; these elementary systems contain a lot of entanglement. Indeed, all their electrons are entangled, the protons and the neutrons in the nucleus are entangled and so on. The existence of this sort of entanglement is obvious and in many cases trivial.

As always, the boundary cases may be more interesting. For example, while the existence of a delocalized electron state in a benzene molecule is also quite trivial, finding delocalized states that extend over much larger molecules may have important consequences.

Dead entanglement

Dead entanglement occurs in molecules that have biological origin or occur in biological cells. However, the occurrence of this kind of entanglement does not require metabolic processes to function. As such, these molecules can be taken out of the cell, and they will continue to work.

Here we are talking about systems that are generally in thermal equilibrium. When some appropriate external perturbation comes, it takes them out of equilibrium, some coherent phenomenon takes place and it quickly dies out. A paradigmatic example of this, outside the biological context, would be a piece of an optical fibre. When a photon comes it propagates through the optical fibre and gets out at the other end. However, during the rest of the time, the piece of optical fibre just stays there, say on an optical table, and nothing happens to it. A similar example, taken from biology, occurs in photosynthesis. Here, in a complicated multi-molecular structure called a light-harvesting complex (LHC), chlorophyll molecules absorb energy (in the form of a photon) which then propagates (in the form of an exciton) from one part of the structure to another part, until it reaches the reaction centre. Again, in the absence of the photon, the LHC is at equilibrium and nothing interesting happens to it.

There are some key words and properties that we generally expect to be associated with this type of entanglement: *Incidental*, *Side effect*, *Short time*, *May have biological functionality*, and *May be evolutionary selected.*

Incidental The main characteristic of the process may not require entanglement, or coherence, but in a particular implementation of the process they may just occur. For example, when a photon propagates through an optical fibre, maintaining polarization coherence is not necessary. It so happens, however, that present day optical fibres are so good that polarization coherence is maintained. But as far as functionality is concerned, which is transmitting the light, this is not important.

Side effect Entanglement may be just some sort of side effect of the process. That is, it may always accompany a given process, but not play any role.

Short time Generally these phenomena are short time because the external perturbation produces some modification, but then the environment immediately brings it down to equilibrium. These are phenomena that may typically take pico- or femto-seconds. As a matter of fact, if you work on a very short timescale, there is always some quantum coherence, because it requires some time to die out, to de-cohere.

On the other hand, one must also be aware that although the absolute timescales involved in these phenomena are very short, this does not necessarily mean that entanglement/coherence does not play a significant role. Indeed, a relevant timescale is that of the duration of the process itself. If entanglement/coherence is present during the whole process, then it may play a significant role; otherwise its role is most probably irrelevant.

Biological functionality and evolution In some instances, the entanglement *may* have some biological functionality, and it *may* have been that this type of quantum coherence was evolutionary selected. It is also possible that even though the entanglement is just a side effect without biological functionality, the primary effect that leads to it was biologically selected. That is, other, important, things were selected and evolved, and with them the entanglement, but just as an accompanying effect. As an example, we can give here photosynthesis again.

The key word here is 'may'. That is, in the case of this type of entanglement/quantum coherence, while evolution may occur, it is not a *sine-qua-non* condition for its very existence.

By no means are 'dead' entanglement/quantum coherence non-interesting phenomena. Quite the opposite. They are extremely interesting and very complicated. Although conceptually they are the same, there is an enormous difference between, say, propagation of a photon through an optical fibre and propagation of an exciton through the light-harvesting complex. Showing that even such 'simple' quantum effects actually take place in biological systems is a great challenge. In fact, at present all the experimental work on biological quantum effects is focused exactly on this type of process.

Live entanglement

The defining property of this type of biological entanglement is that it exists only while metabolic processes take place. In other words, it exists only as long as the system is actively maintained far from thermal equilibrium, i.e. in **open, driven systems far from equilibrium**. When the metabolism stops and the system reaches equilibrium, this type of entanglement disappears.

The key properties that we expect this class of phenomenon to possess are: *It is persistent, It is dynamically controllable, It has biological functionality, It is evolutionary selected.*

Persistency By their very nature, these processes are such that as long as they are active, entanglement/coherence is maintained. This is in fact the purpose of the entire process. From this point of view they are fundamentally different from dead-entanglement phenomena, in which the entanglement appears only as a transitory phenomenon.

Dynamical control In the case of dead entanglement, such as, if confirmed, propagation of an exciton through a photosynthetic light-harvesting complex, LHC, the details of the process are governed by the structure of the molecular structure itself. This is a structural and therefore relatively static parameter. On the other hand, in processes that are driven far from equilibrium, changes in the way in which the process is driven can immediately alter the characteristics of the process, and hence of the entanglement.

Biological functionality and evolutionary selection These are, of course, much more complicated processes than the ones that give rise to dead entanglement. Unless the live entanglement has biological functionality, evolution most probably could not have arrived at it.

Again, all of the above are properties that we expect this class to possess. This is not to say, of course, that there could not be exceptions. For example, it is not impossible that some process leads to persistent entanglement as a side effect with no biological functionality. All we say is that this seems to us highly improbable.

To conclude this section, we want to emphasize that we do not have any experimental evidence, at present, that this type of entanglement (or similarly, quantum coherence) exists. The very scope of the present research in this area is to investigate the possibility that such entanglement actually exists.

We would also like to mention that we are by no means suggesting the possibility of entanglement on a very large scale – such as super-positions of brain states leading possibly to quantum computation in the brain, etc. This seems to us virtually impossible, and here we fully agree with the sceptical view expressed in (Wiseman and Eisert, 2007) (see also (Koch and Hepp, 2006)). What we are interested in is persistent and controllable entanglement with presumably biological function, at the level of bio-chemical processes.

13.5 Open driven systems and entanglement

As we discussed in the introduction, maintaining entanglement between two specific quantum systems is a very difficult task. Nevertheless, this task might be accomplished by *actively* stabilizing it. The possibility of such active stabilization derives from the fact that living systems are open, driven systems, that operate far from thermal equilibrium. Here we will comment in more detail about this idea.

Living organisms depend on permanent consumption of energy in the form of food (or photons, as in photosynthetic organisms). Different from, for example, some solid-state material, a living cell cannot be described as an isolated system. It continuously exchanges particles with its environment, and with them energy and entropy. In the language of thermodynamics, biological systems are open driven (quantum) systems, whose steady state is far away from the thermal equilibrium. As we mentioned in the introduction, this fact has major implications for the issue of the presence of entanglement.

To begin with, the fact that in open driven systems entanglement can exist at room temperature, is absolutely clear, once one realizes that every quantum physics laboratory is such a system. So, having established that, as a matter of principle, controllable entanglement can exist in room temperature systems, the question is only one of scale and complexity. Do we need sophisticated lasers and large fridges (which work at room temperature but cool a subpart of them) or can they be present at the small scale of a living cell? After all, there are numerous studies that show that analogues of large scale, man-made machines (ratchets, rotors, etc.) do actually exist on a bio-molecular level (for a review see for example, Browne and Feringa (2006)). In what follows, we will give a number of specific examples that seem suitable to be scaled down.

Before going to specific examples however, it is important to understand why open driven systems make a difference. The reason is that such systems can perform error correction. Decoherence introduces noise into systems and increases their entropy. On the other hand, open driven systems have, by definition, access to a source of free energy and can use it to get rid of the errors. In fact for the issue we consider, namely merely producing and stabilizing a particular entangled state, one does not even need full quantum error correction, i.e. an error correction protocol that can stabilize any arbitrary superposition of states in a given Hilbert subspace (Calderbank and Shor, 1996; Steane, 1996). The error correction we require here is, computationally speaking, trivial – the stabilization of a single state. Nevertheless, entropically the task is similar, i.e. the entropy continuously produced by noise has to be removed from the system.

Coming back to biology, we note that probably the most striking characteristic of biological systems is that they are error correcting systems – a dead animal starts decomposing in a matter of hours, so it must be that while living there are continuous error correction processes going on which maintain the body. So once we realize that error correction takes place in any living organism, whether or not it is enough to stabilize entangled states becomes a matter of scale, not one of principle.

In the following, we shall give three simple examples of driven systems that are capable of generating entanglement. The first example refers to a standard experimental set-up based on parametric down conversion, as used in many modern quantum optics laboratories. We use it to illustrate some of the issues addressed in the introduction, such as scalability to molecular dimensions and the distinction between live and dead entanglement. In the second example, we describe a molecular gas in which entanglement is generated due to molecular collisions. We use this example to illustrate the important issue of monogamy of entanglement, and what can be done to prevent entanglement from (dissipating or) spreading into an environment of many interacting particles. In the third example, we describe classical conformational motion as a mechanism that is capable, in principle, of generating entanglement inside a noisy and de-coherent environment. The example introduces a simple molecular machine that could be viewed as an 'entanglement generator' in the context of biological (living) systems. After these examples, we shall connect our discussion with concrete biological systems for which the occurrence of genuine quantum effects has been discussed in the literature. These are the problem of light-harvesting in photosynthesis, and the radical pair mechanism as a potential molecular process behind avian magneto-reception.

13.5.1 Simple models I: parametric down conversion

As a first example, consider the famous procedure used to produce entanglement in present day optical labs, namely parametric down conversion. A typical parametric down conversion experiment uses a laser to produce a stream of photons that are directed towards a non-linear crystal. Upon impinging onto the crystal, some of these photons generate pairs of polarization-entangled photons. The entanglement in that experiment is totally dependent on the fact that we are dealing with an open driven system. The laser is powered by a power source; when the source is turned off the entanglement disappears.

The parametric down conversion experiment which we have just described requires complex equipment including a laser and a power source. Is such complicated equipment necessary? The answer is no – there is no fundamental principle of nature that requires complex equipment. The following simple device (see Figure 13.1) could do the same thing.

Suppose we have a thermo-electric element – two simple wires, each made of a different metal – that is connected to a simple light emitting diode LED. The LED itself is a very simple device as well – two semiconducting crystals, each with a different impurity, joined together (a so-called 'n-p junction'). If one of the joints of the thermo-electrical element is heated relative to the other, e.g. by a simple flame, the LED will produce a light beam. When the output of the LED is directed to a non-linear crystal, entangled photons are produced by parametric down conversion.

Note that there is something quite remarkable about this device. (a) It is very simple. (b) The entire system is entirely driven by a temperature difference, thus,

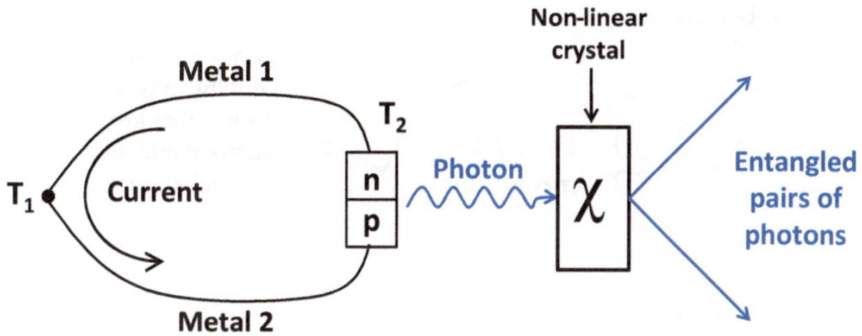

Figure 13.1 Simple entanglement-generating device (see text).

the source of free energy is not coherent, and neither is the transmission of current in the wires a coherent process (see also Plenio and Huelga (2002)). Furthermore, (c), the junction and the crystal are both at room temperature. This simple device is able to create something highly quantum mechanical, such as entanglement. How can that be? Of course, it has to do with the coherent interaction that goes on inside the crystal, when a photon is absorbed and converted into a pair of photons with lower energy. The main point to be observed is that this process is embedded into a hot and rather noisy environment, and it is only due to the continuous pumping of the system that the entanglement-generating interaction in the crystal can be exploited.

We may further imagine that the photons in the entangled pair are directed to two atoms that absorb them (such experiments have been carried out). Upon absorbing the photons, the atoms become entangled. If the atoms are subjected to noise, the entanglement will subsequently be destroyed; whether any entanglement can survive for a longer time is a question of how quick is the de-coherence versus the rate of pumping new entanglement into the atoms by subsequent entangled photon pairs.

A molecular analogue (see Figure 13.2) would be a chemical process where a large molecule (such as a protein) serves as a catalyst for an exothermic chemical reaction. We assume that the chemical reaction takes place at a certain functional centre ('docking site') of the catalyst molecule. The chemical process may lead to a transfer of free energy along the molecule; the free energy transfer may be conveyed by various channels, for example by excitons, phonons or electric current (electron displacement). This energy transport over the molecule need not be coherent and the molecule may be at room temperature. What matters is that, at other sites of the catalyst molecule, the energy may be converted by some non-linear process, similar to the one that occurs in the non-linear crystal of the parametric down conversion, into two entangled modes that pump entanglement into, say, some receiver atoms.

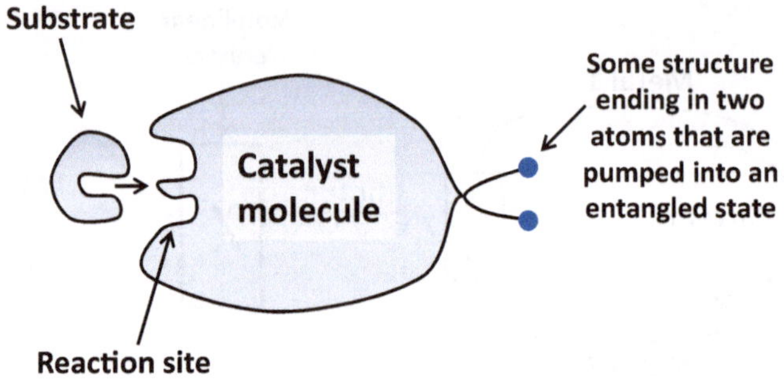

Figure 13.2 Hypothetical molecular device that is capable of creating entanglement by a molecular pumping process. The substrate undergoes an exothermic reaction and thereby delivers free energy into the catalyser molecule.

The important point is here that the two atoms, depicted in blue in Figure 13.2, will not be entangled if we stop driving the system, i.e. if we stop the supply of reactant chemical. This emphasizes the difference between elementary (static) entanglement, as between electrons in an atom, and the dynamic entanglement that is generated only when the system is driven away from thermal equilibrium.

13.5.2 Simple models II: a gas of molecular spins

Our second example is not derived from a macroscopic device, but from the study of entanglement in a gas-type scenario. It illustrates the property of monogamy of entanglement and the problem of its uncontrolled spreading in an environment of interacting particles.

Let us consider a gas of particles with the following properties. The particles move on classical trajectories but, at the same time, they have some internal structure which is described by quantum mechanics (see Figure 13.3). As a concrete realization, we may imagine molecules carrying a nuclear or an electron spin, as in the example of the radical pair mechanism discussed in Section 13.5.4, or neutral atoms with two internal hyperfine states $|0\rangle$ and $|1\rangle$, which are studied in the context of ultra cold collisions. When two particles come close to each other, an (dipole or collision-type) interaction is effectively switched on, which will entangle the particles with respect to their internal degrees of freedom.

Such models of spin gases have been studied in the field of quantum information, in particular in the context of multi-partite entanglement and non-Markovian decoherence (see for example, Calsamiglia *et al.* (2005); Hartmann *et al.* (2005)). The reason why they are interesting is that, for certain types of interactions, the generation and spreading of entanglement between the gas particles can be studied quite explicitly. If one views the entire gas as 'the system', it generates a rich

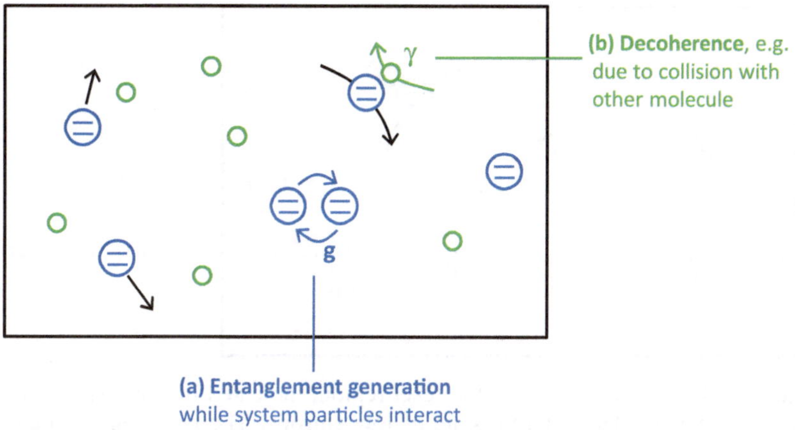

Figure 13.3 (a) Illustration of a spin gas (blue): particles move on classical trajectories, but each particle carries an internal quantum-mechanical degree of freedom, such as a spin. Upon collision, the spin degrees of freedom get entangled. (b) Collisions of the gas particles with other particles (buffer gas, green) lead to de-coherence.

class of entangled states, including states with known applications in quantum computation (Hein *et al.*, 2006). To exploit this entanglement, one would however need to keep track of all particles and their entire interaction history.[1] In most situations, the system under control will consist of only a small number of particles, while the larger number of particles is not controlled and represents degrees of freedom of the environment. In Figure 13.3 we have illustrated such a situation, with 'system particles' indicated in blue and 'environment particles' indicated in green. In such a situation, as is also typically the case in a biological environment, any entanglement that may initially exist between the system particles will quickly spread to the environment. On the reduced state space of the system particles, the entanglement will eventually disappear, a process which is also called de-coherence, leaving one with classical correlations between the system particles. It thus seems that, in such a gas-type scenario, no useful entanglement (i.e. between selected degrees of freedom) can survive.

However, as was shown in Hartmann *et al.* (2006), there is a simple mechanism that can sustain the entanglement between the system particles (blue) and balance the effect of de-coherence, without generating entanglement by itself! Imagine that, when particles enter a certain region in the cell, indicated by a red spot in Figure 13.4, they are reset in some pure state $|\chi\rangle = \alpha|0\rangle + \beta|1\rangle$. This could be

[1] This can indeed be achieved under very special conditions as they are realized in modern experiments using ultra cold collisions in atomic lattice gases.

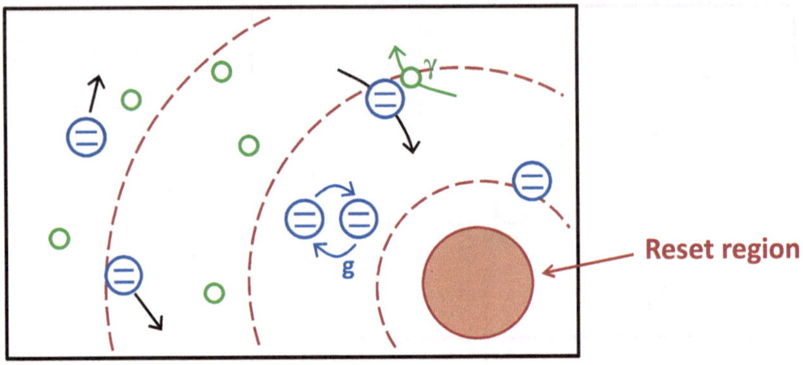

Figure 13.4 Particles that enter the reset region, indicated by a red spot, will be reset in some standard internal state of low entropy. The reset region could be some optically active region (light spot), or a structure inside the cell to which the particles dock. Particles that leave the spot will be reset in a pure state $|\chi\rangle = \alpha|0\rangle + \beta|1\rangle$. As a consequence, entanglement will build up around the spot and persist – in dynamic equilibrium – despite the presence of de-coherence.

realized by various mechanisms, for example by an interaction with some local structure in the cell, to which the particles 'dock' temporarily.

Such a reset mechanism will have two effects. First it will destroy any existing entanglement between this particle (i.e. its spin) and the other system particles (indicated blue in Figure 13.4) in the gas, but it will also destroy this particle's entanglement with the environment (indicated green in Figure 13.4). At first sight, this does not appear to be very constructive, but it has the effect that the particles that leave the spot are in a pure state, and, if two of such particles collide later on (e.g. in the vicinity of the spot), they will be capable of creating fresh entanglement! A detailed analysis shows that indeed, for a specific choice of parameters, the gas can have a steady state, where entanglement persists between the blue particles. Depending on the mean free path of the particles, the regions of persisting entanglement will be confined to the vicinity of the reset region, or they may extend over the entire gas cell.

While the details of this process are not of interest here, it should be mentioned that there are two regimes, one of vanishing and one of non-vanishing entanglement, and there is a sharp transition between them. Entanglement can be sustained if both the coherent interaction strength and the reset rate are sufficiently large. Furthermore, the particles need not be reset to a pure state; it suffices if they leave the reset region in a mixed state of sufficiently low entropy (high purity) (Hartmann *et al.*, 2006, 2007).

This illustrates, in the context of an explicit quantum many-body model, that the possibility of sustaining entanglement is not confined to systems that are well isolated from the environment and at low temperatures. The notion of a quantum reset mechanism (Hartmann *et al.*, 2006) allows entanglement to be maintained in a dynamic equilibrium state, even in an open system with a noisy environment. The property of such a mechanism is that it removes the entropy of part of the system, resetting particles in a sufficiently pure state ready to become entangled again.

While the spin gas represents an abstract model, it is clear that the set-up is reminiscent of biological systems where various analogues of reset processes exist (e.g. in the context of bio-molecular synthesis). It is thus, again, not a question of principle, but of scale, whether there are processes on a molecular scale capable of generating and sustaining entanglement. We shall have a closer look at this possibility in the next section.

13.5.3 Simple models III: conformational changes and time-dependent Hamiltonians

In the introductory part we pointed out that, when discussing entanglement in biology, one should describe more carefully what one is actually looking for. In a certain way, quantum mechanics plays a fundamental role in biology, insofar as the structure and shape of bio-molecules is determined by the laws of quantum physics (and chemistry). In other words, the entire molecular substrate on which biological processes operate is described by quantum mechanics. If we are specifically looking for entanglement, we will find plenty of it on this level and this is what we called 'entanglement of basic constituents.' But this situation is similar to atomic or condensed matter physics and there is nothing specifically relevant to biological function. This kind of entanglement will exist both in dead and in living bodies.

A more interesting question is whether we can identify situations, most likely at the level of molecular biology, where entanglement exists only because and while some metabolic processes take place. What distinguishes biology from condensed matter physics is the fact that biological systems operate, per definition, away from thermal equilibrium. On the level of molecular biology, molecular motion plays an important role and controlled conformational changes are indeed an omnipresent feature of bio-molecular processes. Protein function, for example, requires conformational motion (Frauenfelder *et al.*, 1999; Alberts *et al.*, 2008) and so do all kinds of molecular machines. Many of these conformational changes are reversible (under a supply of free energy) and lead to quasi-periodic processes.

The question is how to describe these processes. In molecular dynamics simulations, for example, one treats the problem essentially classically and computes an (approximate) electrostatic potential and force field associated with a given molecular conformation. While this method is sufficient for many applications, it cannot resolve finer structural or dynamical details. In some situations, however, a more

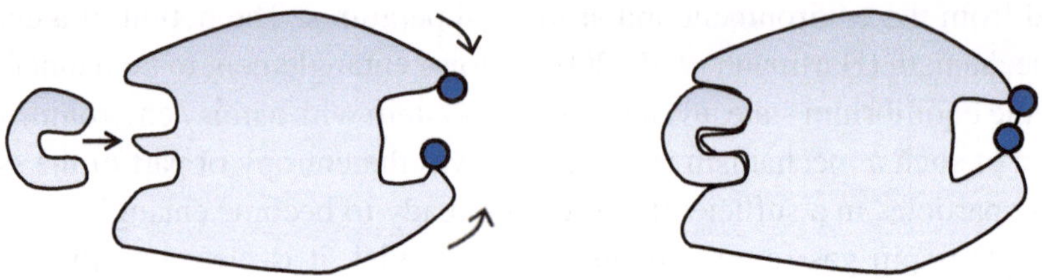

Figure 13.5 Time-dependent quantum interaction generated by molecular motion. A conceivable process is the following. A chemical process that occurs at one binding site of a protein leads to a configuration change of another site of the same protein. This conformation change induces an interaction between two atoms (or more generally, two selected degrees of freedom) that would otherwise be in a shielded site of the molecule. The two atoms depicted in blue are not interacting in the 'rest' state of the protein, where they are at thermal equilibrium and non-entangled. When the conformation change occurs they are brought together and start interacting. The interaction takes the atoms out of their equilibrium state and may entangle them. The entanglement survives until de-coherence kills it and the atoms reach a new thermal equilibrium state. Then the protein reverts to the rest conformation and the atoms revert to the rest mode and then the process starts again. The point here is that, although both thermal equilibrium states may be non-entangled, the state of the atoms during the transition time (which may be relatively long) may well be entangled. Please note that, to complete the cycle, the protein needs to be supplied with free energy; if we stop driving the system, i.e. if we stop the supply of reactant chemicals then the atoms simply reach thermal equilibrium and remain there, and entanglement is lost.

refined description may be required, for example when we are interested in how the conformational change influences the quantum state of a specific *subsystem* of the bulk molecule. This subsystem could be a single electron spin, whose state we want to follow, or a moveable charge. In such a situation, a semi-quantal picture would be more appropriate, where we have a certain number of quantum degrees of freedom which are attached to a 'classical backbone structure' associated with the conformation of the bulk molecule.[2]

At such a level of description, we effectively obtain time-dependent quantum interactions between selected degrees of freedom, whose strengths are varying while the molecule changes its shape (see Figure 13.5). The time dependence of these interactions is imposed by the classical dynamics of the conformational motion, as long as we assume that the quantum degrees of freedom do not couple to the shape of the molecule. The situation is reminiscent of the spin gas described in the previous section, except that here, the system dynamics is not governed by random collisions, but prescribed by the conformational motion.

Molecular machines running through a sequence of conformational changes will, in this picture, implement an intricate time-dependent Hamiltonian which

[2] This picture is somewhat similar, in spirit, to a hybrid quantum and molecular mechanics approach introduced by Warshel (Warshel, 2002) to study enzymatic reactions.

corresponds, in quantum information, with a 'sequence of quantum gates'. It is an intriguing question as to whether molecular machines of this type could, in principle, perform some kind of quantum computation. At first sight, this may look rather unlikely. For, if we want to take this question seriously, we have to consider the important effect of environmental noise which, in any realistic scenario, will accompany the quantum interactions we have just described. It should be pointed out that, for example, in the context of protein folding, people typically pay no attention to such subtle quantum interactions and describe the process classically. Since everything happens at room temperature and since the noise arising from interactions of the protein with its environment is usually substantial, it seems reasonable to assume that any quantum effect(s) will be completely washed out.

However, as we have demonstrated in Cai *et al.* (2010a), this intuitive argument is not always correct. To show this, we have analysed the following simple example of a 'two-spin molecule', where the selected quantum degrees of freedom (indicated in blue) in Figure 13.5 are effective two-level systems, such as spins, interacting with each other by some quantum Hamiltonian, while they are embedded in a thermal and decoherent environment. The interaction between the spins and with the environment will, within the semi-classical picture that we have introduced, depend on the conformational state of the embedding molecule.

Conformational transitions of the molecule introduce an effective time dependence into the system Hamiltonian, as well as into the interaction with the environment. As an example for the system Hamiltonian we have considered the following expression:

$$H_M(t) = J(t)\sigma_x^{(1)}\sigma_x^{(2)} + B(t)\big(\sigma_z^{(1)} + \sigma_z^{(2)}\big), \tag{13.25}$$

where $\sigma_x^{(\alpha)}$ and $\sigma_z^{(\alpha)}$ are Pauli operators of the αth two-level system, $J(t)$ is the strength of their interaction and $\omega_0(t) = 2B(t)$ the local level splitting, as determined by the molecular configuration at time t.

To model the effect of the environment, we have used a master equation of the form,

$$\frac{\partial}{\partial t}\rho(t) = -i[H_M(t), \rho] + \mathcal{D}(t)\rho(t) \equiv \mathcal{L}(t)\rho(t), \tag{13.26}$$

where $\rho(t)$ denotes the density operator of the system (two-spin molecule), $H_M(t)$ their coherent interaction Hamiltonian, and $\mathcal{D}(t)$ summarizes the dissipative effect of the system–environment coupling (Cai *et al.*, 2010a). The purpose of this master equation is *not* to give a concise description of a realistic biological environment. Instead, it provides a framework to analyse a *worst-case scenario* where, for any time t, the corresponding Liouville operator $\mathcal{L}(t)$, is the generator of an *entanglement breaking channel* (Horodecki *et al.*, 2003). In other words, we consider a

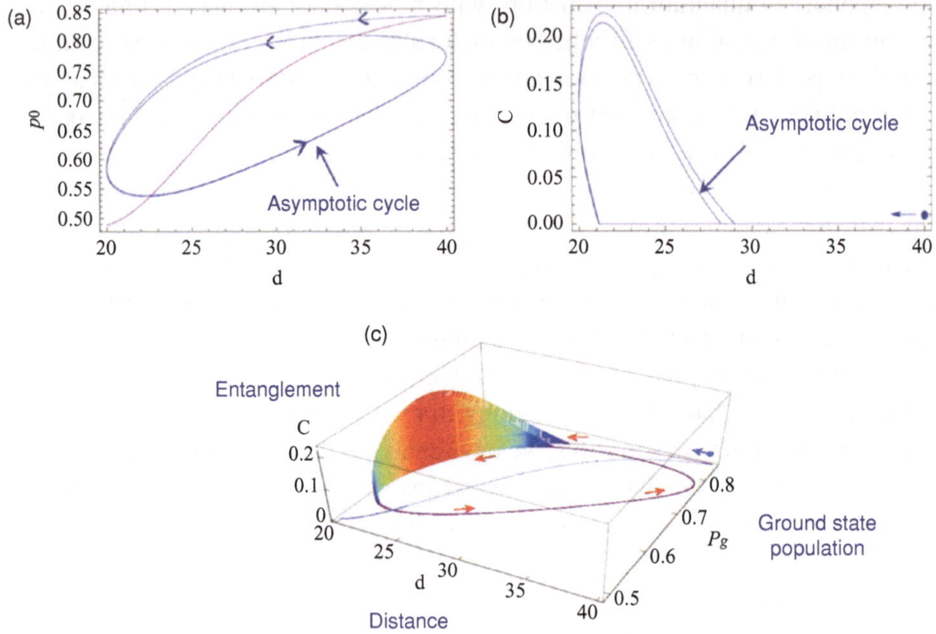

Figure 13.6 (a) Ground state population p_g and (b) entanglement C versus the molecular configuration, characterized by distance d between the spins. (c) Three-dimensional representation. For explanation see text. The parameters are the same as in Figure 3 of Cai *et al.* (2010a).

situation where for any *static* configuration, the steady state is completely separable, i.e. not entangled. An example of such a situation is given by an environment consisting of bosonic heat baths, where each spin is coupled to an individual thermal bath of harmonic oscillators (Breuer and Petruccione, 2002; Gilmore and McKenzie, 2005). For a static scenario, it is well known that, above a certain critical temperature, no initial entanglement can survive in such an environment.

In Cai *et al.* (2010a) we could show that, under such a well-established *entanglement breaking* situation for the static case, entanglement can still be *generated* if the quantum degrees of freedom start *moving* relative to each other, introducing a time dependence in the Hamiltonian. In the simplest case, we considered an oscillatory motion which leads to a time-dependent dipole coupling of the form $J(t) = J_0/|x_1(t) - x_2(t)|^3$ and time-dependent local fields $B(t) = B_\alpha(t) = B_0 - B_1 e^{-x_\alpha^2(t)/\sigma}$, with constants J_0, B_0, B_1, σ and oscillatory functions for the spin positions $x_\alpha(t) = x_\alpha(0) + (-1)^\alpha a(\cos \frac{2\pi t}{\tau} - 1)$, $\alpha = 1, 2$ with period τ and amplitude a (Cai *et al.*, 2010a).

Figure 13.6 shows the entanglement C and the ground-state population p_g, versus the molecular configuration, characterized by the distance $d = x_1 - x_2$ between the spins. We used the concurrence $C(\rho)$ as a measure of entanglement of two qubits in state ρ (Wootters, 1998). This vanishes for a non-entangled (separable) state and reaches the value unity for a maximally entangled state, that is, $0 \leq C(\rho) \leq 1$. The temperature, which enters into the dissipator $\mathcal{D}(t)$ of the master equation, was chosen to be so high that no entanglement could survive in any static configuration. That is, $C(\rho_{th}) = 0$ for all static thermal states defined by $\mathcal{L}(t)\rho_{th} = 0$ (here the variable t effectively parameterizes the molecular configuration $d(t)$).

Figure 13.6a shows the ground state population p_g as a function of the molecular distance d, illustrating the non-equilibrium features of the process. The (purple) curve near the diagonal corresponds to the thermal state of the static configuration associated with distance d. The thermal population of the ground state changes from about $p_g \sim 85\%$, when the spins are spatially separated ($d = 40$) to about $p_g \sim 50\%$ when they are close ($d = 20$). This is because the energy separation of the two lowest-lying eigenstates of the Hamiltonian $H_M(t) = H_M(d(t))$ depends on the molecular distance and becomes smaller when the spins are close. In the limit when the molecule changes its shape very slowly (meaning that the oscillation period is much larger than the thermalization time of the baths), then the system will remain in its thermal state $\rho_{th}(d(t))$, with the ground state population decreasing as described. Under these conditions the state always remains separable (non-entangled), i.e. $C(\rho_{th}(d)) = 0$ for all values of d, which corresponds to the base line in Figure 13.6b.

For faster conformational changes, when the oscillation period is comparable with the thermalization time, the system will be driven out of the thermal state. That is, the time it spends in each configuration is too short to allow its state to thermalize. This is indicated by the blue cycling curve in Figure 13.6a. Starting from an initial value of $p_g \sim 85\%$, the system follows the blue curve and soon, after about one oscillation period, enters into the *asymptotic cycle*, as is indicated by the arrows. One can see that when the spins approach each other, the population at each distance is higher than in the corresponding thermal equilibrium; while when the spins separate, it is smaller. In this sense, the system is, through the motion of the spins, always dragged out of its thermal equilibrium. If the motion were stopped at any distance d, then the population would relax to its thermal equilibrium value (from above or below, on a vertical line). In Figure 13.6b the corresponding blue curve shows the entanglement between the spins as they go through the cyclic motion. One can see that entanglement builds up when the spins come closer (in the first cycle at $d = 29$ and then, in all subsequent cycles, at $d = 28$) and disappears again when they start separating (at $d = 21$). Thus, in spite of the entanglement-breaking property of the 'hot & noisy' environment (for every static configuration), entanglement is recurrently generated on the asymptotic cycle as long as the molecule keeps moving. Again, if the motion were to stop at any configuration, the blue curve would collapse to the purple line at $C = 0$ and the entanglement would disappear. For convenience, the situation is summarized in a three-dimensional plot in Figure 13.6c, where the region of recurrent entanglement can clearly be seen.

The effect, as we observe in Figure 13.6, can be explained by a subtle interplay between a sort of adiabatic dragging of the population in the ground state when the spins approach each other, and a re-thermalization when the spins are close to or far from each other, effectively resetting the system. Remarkably, the recurrent generation of entanglement only happens when the timescale of the molecular motion is comparable with the timescale of thermalization. In contrast, any entanglement disappears permanently, after some transient phase, when the molecule moves either too slowly (system remains at every time in the non-entangled thermal state) or too fast (environment cannot reset the system into a low-entropy separable state). For further details, we refer the interested reader to the original literature (Cai *et al.*, 2010a).

We have now discussed the mechanism in its simplest form, assuming, for example, periodic motion and a specific form of an environment (described as Markovian bosonic heat bath). However, the effect is rather robust and many of the simplifying assumptions can be relaxed (Cai *et al.*, 2010a). In Guerreschi *et al.* (2012), we have analysed the same model for stochastic motion and different motion profiles for $d(t)$. We have also taken into account additional dephasing processes that are particularly destructive for the entanglement, and we have extended the analysis to non-Markovian models for the environment. In all of these cases, the feature of recurrent entanglement generation persists, as long as the system has the opportunity to undergo a certain resetting phase in its noisy evolution.

The take-home message of this study is twofold. Firstly, it shows that even high levels of noise may not necessarily 'wash out' the quantum features of an underlying Hamiltonian interaction. The common belief that quantum interactions that are accompanied by sufficiently high levels of noise can always be replaced by a classical stochastic process (i.e. in quantum information language, a process that generates a separable completely positive map), is *wrong* in the present context. Secondly, this provides an explicit model of a molecular machine that can act as an 'entanglement generator' in a noisy environment. Finally, it also illustrates the notion of 'live entanglement' which we discussed in Section 13.4, since the occurrence of entanglement is entirely dependent on the delivery of free energy (food) to the system, which drives the conformational motion.

It will be interesting to see whether such an entanglement generator exists in a concrete bio-molecular context, and whether nature could use more complex molecular machines as some kind of (noisy) quantum information processors. This example shows that this possibility can no longer be discounted out of hand.

At the end of this section, we would like to mention that a similar mechanism, also based on conformational motion, can lead to intra-molecular refrigeration (Briegel and Popescu, 2013). This is a process where parts of a molecule can be cooled below the environmental temperature. In the context of enzymatic reactions, this could in principle lead to increased efficiency of catalysis at high temperatures. It could also be exploited by molecular machines to cool some of their parts and thus reduce the deteriorating effect of thermal noise. While the mechanisms underlying entanglement generation and intramolecular refrigeration are similar, the latter process is far more general and does not require any quantum coherence. We shall not treat it in the present discussion, which focuses on entanglement. The interested reader is referred to (Briegel and Popescu, 2013).

13.5.4 Concrete examples of biological systems

In previous sections, we have discussed what type of entanglement and coherent processes we are looking for. We have shown how entanglement can be generated and sustained in 'warm, wet and noisy' environments, despite the deteriorating effects of decoherence, as long as there is a source of free energy that is used to keep the system away from thermal equilibrium.

The arguments that we presented were abstract and general, discussing the very possibility of entanglement generation in such hostile environments. Recently, several concrete biological systems in which non-trivial quantum effects seem to play a role have been discussed in the literature. The two most prominent examples upto today are (a) energy transport in photosynthesis and (b) magneto-reception in bird navigation based on the radical pair mechanism, both of which are discussed extensively in this book. What role, if any, does entanglement play in these systems?

Photosynthesis

As described in previous chapters, there is by now very strong experimental evidence (Engel *et al.*, 2007; Collini *et al.*, 2010; Panitchayangkoon *et al.*, 2010) that there are non-trivial quantum effects during the energy transport in photosynthesis (for a recent review, including further references see for example, Ishizaki and Fleming (2012)). The most studied system is the so called Fenna–Matthews–Olson (FMO) complex (Fenna and Matthews, 1975), which is a molecular structure present in green sulphur bacteria. The role of the FMO complex is essentially that of a cable which connects the antenna that absorbs photons to the reaction centre, where the energy is used to initiate the first part of the electro-chemical process of photosynthesis. Experiments made on the FMO complex have shown strong evidence of quantum coherence.

While the existence of quantum coherence in the FMO complex is by now almost certain, an even more intriguing possibility has been raised, namely that entanglement is also present and that perhaps this is the crucial element behind the surprisingly high efficiency of energy transport through the FMO complex. Raising this question was a dramatic development, and has generated an intensive area of research (Bradler *et al.*, 2010; Caruso *et al.*, 2010; Fassioli and Olaya-Castro, 2010; Hossein-Nejad and Scholes, 2010; Ishizaki and Fleming, 2010; Mukamel, 2010; Sarovar *et al.*, 2010; Scholak *et al.*, 2011; Whaley *et al.*, 2011; Zhu *et al.*, 2012).

Let us start by describing the FMO complex. The FMO complex (Fenna and Matthews, 1975) consists of three identical substructures (monomers), each of which connects the antenna with the reaction centre. The main constituents of each substructure are seven chlorophyll molecules, some of them next to the antenna, others next to the reaction centre, and yet others placed somewhere in between, in a specific pattern. The role of these chlorophyll molecules is to pass the excitation from the antenna to the reaction centre. What we are interested in is the state of excitation of these chlorophyll molecules.

To see where entanglement may come in, let us make two simplifying assumptions. Firstly, since we have experimental evidence of quantum coherence in this system, let us assume that the state of the excitations is pure. Secondly, since such bacteria live under conditions of extremely low levels of light, we shall assume that there is never more than a single excitation propagating in the FMO complex at any time.

With the above two assumptions the state of the chlorophyll molecules can be written as a superposition of the form:

$$|\Psi\rangle = c_1|1\rangle_1|0\rangle_2|0\rangle_3|0\rangle_4|0\rangle_5|0\rangle_6|0\rangle_7 + c_2|0\rangle_1|1\rangle_2|0\rangle_3|0\rangle_4|0\rangle_5|0\rangle_6|0\rangle_7 + \cdots$$
$$+ c_7|0\rangle_1|0\rangle_2|0\rangle_3|0\rangle_4|0\rangle_5|0\rangle_6|1\rangle_7, \qquad (13.27)$$

where $|0\rangle_i$ and $|1\rangle_i$ represent chlorophyll molecule i in the ground and excited state, respectively, and the coefficients c_i are time dependent and encode how the excitation propagates through the system.

The state (13.27) is, for most values of the coefficients c_i, indeed an entangled state (except, for example, for the case when there is only a single non-zero coefficient). It is entangled in a similar way to the states of two systems described in Section 13.2, capable, in principle, of exhibiting non-local correlations between the results of measurements on different chlorophyll molecules.

Both of the above simplifying assumptions have to be treated with care. Obviously, there is also noise in the system, so the state is definitely not pure. How mixed it is, is not yet clear. Secondly, the assumption that there is precisely a single excitation in the system has also been disputed (Tiersch *et al.*, 2012).

More importantly, even if entanglement does exist in the FMO complex, its *role* is not yet clear. Does it have an essential role – does it ensure that the energy transport through the complex is so efficient – or is it just a secondary and irrelevant aspect? As we have seen in the introduction to this chapter, one needs to implement rather sophisticated protocols in order to make use of the non-local properties of entangled states – just the very existence of entanglement is not enough.

All of the above questions on entanglement in photosynthesis are very important but still open. For a recent review of the current state of the art, see, for example, Whaley *et al.* (2011).

Finally, let us note that, as far as our classification of quantum biological processes goes, this is of a 'dead' quantum coherence type – functioning metabolism is not needed, and the FMO complex can operate while initially at thermal equilibrium, very much as a piece of optical fibre.

Avian magnetic compass

The second example concerns the avian magnetic compass based on the radical pair mechanism, which is one of the main hypotheses to explain the remarkable ability of certain birds to sense the Earth's magnetic field during migration (Wiltschko and Wiltschko, 1972, 1996; Mouritsen, 2001; Johnsen and Lohmann, 2008). For a detailed exposition of this interesting proposal, we refer the reader to Chapter 10 of this book and to the original literature cited therein. Here we shall just concentrate on the role of entanglement in the avian compass model.

The basic mechanism underlying the avian inclination compass, as was proposed by Schulten and co-workers (Schulten *et al.*, 1978; Ritz *et al.*, 2000b), is the radical pair mechanism (Steiner and Ulrich, 1989; Rodgers and Hore, 2009). It is a spin-dependent chemical process, in which a pair of correlated electron spins is created by a photo-induced electronic excitation and transfer process in a pair of molecules. As a result, the two electrons are separated, for some time, on two different molecules, forming a radical pair, before they can recombine. The spin state of the electrons right after the (fast) photo-excitation and transfer is usually assumed to be a pure quantum state (singlet or triplet state). This quantum state is changed by the magnetic interactions of the two electron spins with surrounding nuclear spins on the molecules and with the Earth's magnetic field. The resulting change of the quantum spin state influences the probabilities for recombination through different reaction channels, presumably giving rise to two different chemical processes on the retina of the bird (Ritz *et al.*, 2000b).

There is by now a large body of experimental and theoretical work investigating the role of the radical pair mechanism for avian magneto-reception (for a recent review see, for example, Rodgers and Hore (2009) and Chapter 10 of this book).

This remarkable proposal seems to be a 'quantum effect in biology' par excellence. It certainly goes beyond the regime of the 'quantum structure of basic constituents' as discussed in Section 13.4. It has well-defined quantum degrees of freedom (electron and nuclear spins on the radical pair molecules) with a clearly mapped-out biological function, and it is a quantum non-equilibrium effect. Indeed, the functioning of the magnetic compass depends crucially on the fact that, during the spatial separation of the two electrons, i.e. while they reside on the different radical molecules, the system has no time to thermalize. Since the energy spacing between the singlet and the different triple states is much smaller than the thermal energy, thermalization would completely mix the different spin states over the relevant range of the external magnetic field, thereby making the singlet fraction insensitive to the external field and thus destroying the compass function. Fortunately, the time for thermalization of the electron spins seems to be much longer than the entire lifetime of the radical pair (time between electron separation and recombination). On the timescale of the separation and recombination process, the dominant part of the environment seen by the effective two-spin electron pair – which is our system of interest – consists of the nuclear spins residing on the radial pair molecules. It is the hyperfine interaction between the electron spins and their surrounding nuclear spins, in comparison with the interaction of the electron spins with the external magnetic field, which is responsible for the inter-conversion and mixing of the singlet with the triplet states, thereby controlling their recombination through different channels.

So what is then the role of entanglement in the avian compass? To appreciate the motivation of this questions, one must understand the difference between entanglement and mere spin coherence. A state can have an appreciable amount of two-particle spin coherence, as quantified by the singlet fraction (or singlet fidelity), while having zero entanglement! Since the presence of entanglement is a necessary resource for many of the applications in quantum information science, such a situation means that an interpretation of the avian compass as some kind of quantum information processor becomes unlikely. Given the fruitful connection that has been established between quantum physics and information theory on the one side, and the notion of information processing in modern biology, on the other, it is natural if not compelling to ask whether nature would, on the level of molecular biology, make use of some kind of quantum information processing. For this to be the case, the presence of entanglement between the functionally relevant degrees of freedom seems to be mandatory.

The first question is then: Are the electron spins entangled after photo-induced excitation and their separation onto the different radicals? (That is, are they truly in a pure singlet, state or should they rather be described as a mixed state, as would result, for example, from some de-phasing process?) If the initial state is really the singlet state, does the entanglement persist over an appreciable range of time? Clearly, the hyperfine interaction with the nuclear spins leads to de-coherence on the state space of the electron spins, resulting sooner or later in mere classical correlations. Finally, does the presence of entanglement make any difference, i.e. is it a relevant signature to look at? Are the notions of non-locality and quantum correlations as described in the introduction, relevant or can the compass function and sensitivity also be explained by mere classical correlations?

Some of these questions have recently been addressed in Cai *et al.* (2010b) (see also (Gauger *et al.*, 2011; Hogben *et al.*, 2012; Tiersch and Briegel, 2012)). The answers to these questions depend on the details of the specific radical pair reaction under consideration, but in particular on the lifetime of the radical pairs.

For example, for the well-studied radical pair reaction involving pyrene (Py) and dimethylaniline (DMA) molecules in solution, most of the questions raised above can be answered positively. Here, the time during which entanglement persists between the electron spins (for an initial singlet state) is of the same order as the lifetime of the radical pair reaction itself (a few nanoseconds). Furthermore, the sensitivity of the reaction against variations of the external magnetic field is significantly increased by the presence of entanglement (Cai *et al.*, 2010b), as compared with the sensitivity reachable by any separable state, showing only classical spin correlations. Unfortunately, Py–DMA plays no role for the avian compass. Furthermore, its hyperfine interactions are isotropic and cannot provide directional information. It could be used as a magnetometer but not as a compass.

As a second example, a similar analysis was performed for a radical pair ($FADH^{\bullet}$–$O_2^{\bullet -}$ in cryptochrome) which is considered as a molecular candidate for the avian compass. Here the results are different. While the lifetime of this radical pair reaction seems to be of the order of microseconds, the entanglement between the electron spins still decays on a short timescale of a few nanoseconds. This means that the entanglement between the electron spins of the initial singlet state exists only for a tiny fraction of the entire reaction time, thus one would not expect it to play an important role in directional sensitivity of the compass. Consistent with this observation is the finding that a compass with a similar sensitivity – based on the same molecular candidate – can be achieved with a large fraction of initial states that are separable, i.e. not entangled at all (Cai *et al.*, 2010b).

In summary, the intriguing question of whether birds exploit entanglement for navigation, with the radical pair mechanism representing some kind of quantum information processing, remains open at present. The answer will ultimately depend on the specific molecular realization of their chemical compass, which still needs to be identified. Notwithstanding this reservation, the radical pair mechanism, with its supposed role for magneto-reception in animals, clearly belongs to the few outstanding examples that highlight the potential scope and relevance of non-trivial quantum effects in biology.

13.6 Conclusions

Biological systems are of extraordinary complexity and diversity. As such, at the moment we do not know where to start searching for entanglement, be it an experimental search or a theoretical one. Furthermore, the specific toy models presented here are almost certainly of very little direct relevance to biology. However, our goal here is far more limited – it is to argue that the presence of controlled entanglement with biological functionality cannot be discounted automatically, without a careful study. Indeed, although our specific toy models may well have very little direct relevance, we are confident that the processes we describe (entanglement pumping, resetting, etc.) are to be found in one way or another; the same applies to the idea of molecular cooling. Ultimately, the power of biological evolution, coupled with the fact that biological organisms are open, driven systems, may open the door to many unexpected quantum phenomena. Similarly, they also open the door to highly non-trivial thermodynamic phenomena.

14

Design and applications of bio-inspired quantum materials

MOHAN SAROVAR, DÖRTHE M. EISELE AND K. BIRGITTA WHALEY

In this chapter, we explore the opportunities that the dynamical quantum effects recently revealed as components of key functions of plants and higher organisms and described in previous chapters, might offer for the design of new nano-scale materials possessing quantum-enhanced functionality. We discuss the potential applications of such biomimetic materials with engineered quantum properties, and present a review of progress thus far on two prototypical systems: biomimetic light-harvesting materials and biomimetic magnetic sensors.

14.1 Potential applications of bio-inspired quantum materials

It is well appreciated that quantum dynamics can lead to enhanced performance in tasks such as metrology (Giovannetti *et al.*, 2011), computing (Nielsen and Chuang, 2001) and communication (Gisin and Thew, 2007). However, such enhancements have yet to be realized for artificial systems in the biological domain. As discussed in earlier chapters of this book, it has been demonstrated or hypothesized that quantum processes are critical to the accurate description of the functional dynamics of several biological systems. What opportunities do these observations present for the motivation and development of biomimetic materials? The possibility of constructing artificial materials with the ability to mimic natural systems leads to a diverse range of potential applications. A key question is thus whether we can use nature's ingenuity as inspiration and incorporate quantum effects into synthetic systems to provide quantum-enhanced function? Such explorations also hold out the tandem promise of achieving greater understanding of the role of quantum mechanics in biological function, since in contrast with the traditional top-down approach to investigating natural systems, this perspective requires the development

Quantum Effects in Biology, ed. Masoud Mohseni, Yasser Omar, Gregory S. Engel and Martin B. Plenio. Published by Cambridge University Press.

of a bottom-up approach to synthesis of unnatural systems possessing capability to mimic all or part of some biological role.

Several intriguing possible directions for such applications arise from the developments in understanding of how quantum effects affect the dynamics of biological phenomena described in earlier chapters. These include several sensing applications, e.g. of single molecules, of light or of weak magnetic fields, as well as the synthesis of robust, 'fault forgiving' devices for energy transfer, and smart pixels controlled by dendrimer states for imaging (Tully and Fréchet, 2001). In what follows, we focus on the opportunities presented by artificial light-harvesting systems and radical pair materials for sensing of light and magnetic fields, respectively. We describe here the motivations for these two directions, as well as candidate architectures and challenges to be overcome in order to be able to realize such applications.

14.1.1 Opportunities for novel energy and sensing applications with artificial light-harvesting systems

As we have seen in earlier chapters of this book, natural light-harvesting systems are both highly complex in structure and highly specialized in function. Taken together with the recent observations of quantum coherences in the dynamics of electronic energy transport through these biological units, these features raise several intriguing questions. On the one hand, it prompts us to ask whether the coherent quantum dynamics associated with energy transport in these systems are the result of a finely tuned evolutionary process, that has maintained and even refined excitonic coherence in the face of mounting structural complexity, in order to enhance light-harvesting? Or are these manifestations of coherent quantum dynamics merely a by-product of structural evolution that is driven by quite different adaptive pressures, or even just by response to genetic drift? Looking backwards for answers to the biological origin of coherence in natural systems constitutes a major question that will require a significant interdisciplinary effort between biology, physics and chemistry to make progress. Quite a different set of questions is generated if we look forwards from the natural light-harvesting systems prevalent in biology today and ask, "what can we do with artificial light-harvesting systems that are designed to emulate one or all functions of natural systems?"

We can envisage artificial light-harvesting systems that range from complete biomimetic structures, to systems that either emulate just one feature of the natural systems or translate some natural function to a different, unnatural regime. Given the possibility that natural structures have been optimized by aeons of evolutionary development, development of functionally specialized artificial systems appears more realistic and also more straightforward. The growth of supramolecular chemistry and self-assembly methods over the last twenty years allows us to consider a wide variety of structures as viable platforms for such artificial light-harvesting systems. A key question is then, what are the revolutionary applications that become possible with such biomimetic systems?

One immediate application is robust, high efficiency solar energy conversion. Such a goal, of achieving a complete biomimetic of natural photosynthetic systems, is clearly a major and high profile application for our society today, given the need to find alternative energy sources to fossil fuels. Significant research efforts are going into solar cells and a wide variety of solid-state photovoltaic (Li *et al.*, 2006) and thin film, solid–liquid photoelectrochemical cells (Grätzel, 2001) are under development. Artificial light-harvesting arrays consisting of organized assemblies of chromophores and proteins constitute an alternative to conventional solar cell technologies (Gust *et al.*, 2009). Within the general framework of tailored environments for light-harvesting achieved by molecular self-assembly, the role of quantum coherence in facilitating energy transport through complex pigment–protein structures can provide an additional 'knob' with which to tune the functionality of artificial light-harvesting complexes. This capability can be explored by systematic, 'bottom-up' studies of pigment–protein complexes with controllable inter-pigment distances and relative orientation, which determine the coupling between chromophores and are hence a primary factor in the extent of quantum coherence. However, when used as an energy source, such artificial systems also require the incorporation of catalysts that facilitate the charge separation and redox chemical reaction steps of photosynthesis (Gust *et al.*, 2001). Whether this is realized as an integral part of the self-assembled structure, or whether the light-harvesting array is integrated within a solid-state device, the scale-up from light-harvesting to a fully functional photosynthetic unit that produces chemical energy from sunlight is extremely challenging.

More diverse opportunities are presented when attention is focused only on mimicking the light-harvesting functionality of a pigment–protein complex, without also incorporating the additional requirements of a complete photosynthetic biomimetic system. One promising direction is to build on the exquisite ability of some light-harvesting antenna systems to collect light and function under very low intensity conditions. For example, green sulphur bacteria are able to harvest light deep in the Black Sea under conditions corresponding to an average of one photon incident every eight hours (Manske *et al.*, 2005). This suggests that an artificial light-harvesting apparatus might provide an ultra high performance photon sensor. Natural systems are effective at harvesting light in the optical and near infrared. While advanced technology for low intensity light detection already exists in these regimes, a biomimetic light-harvesting photon detector would offer several unique capabilities. These potentially include customizability and versatility, low weight, low cost, the ability to operate at room temperatures with high quantum efficiency, and the ability to respond to a broad band of photon wavelengths. Such artificial structures can in principle also be tailored for sensitivity deeper into the infrared. Indeed, such a biomimetic device would be of greatest interest if it could operate over a broad range of infrared wavelengths, where single photon detection tools are currently extremely restricted.

Although there have been previous efforts to integrate natural light-harvesting structures into solid-state environments for photodetection, there have so far been no systematic efforts to mimic light-harvesting structures and take advantage of their quantum coherent properties in order to create sensitive light sensors. The use of biomimetic synthetic structures possesses additional advantages over the simple integration of natural photosynthetic complexes. For example, the constraints of device integration severely restrict the list of possible candidate natural light-harvesting arrays, yet synthetic supramolecular arrays present a vastly larger landscape of possibilities. Furthermore, given our rapidly increasing understanding of the relation between excitonic quantum coherence and structural features of light-harvesting arrays, we can now begin to tune the properties of synthetic arrays as desired to optimize the performance of light sensing. The protein scaffolds of viruses provide one attractive option for synthetic design of such artificial systems, both on account of their highly ordered structures and also because of the ability to generate a range of well-defined structures by self-assembly under unnatural conditions. Such systems also possess the potential for attachment to electroactive surfaces via electron transfer functional groups, which would enable the energy collected by the light-harvesting system to initiate electron transfer, thereby allowing electrical detection of the incident photon. Progress in constructing such protein-scaffolded light-harvesting antennas will be reviewed in the next section.

The end goal of this research is to produce robust photovoltaic or light sensing technologies with control over the nanoscale structure. Manipulating this structure allows tuning of light capture, energy transfer and free-carrier generation processes; engineering of these processes is expected to allow improvements in some combination of: quantum efficiency, utilized solar spectrum, sensitivity to specific wavelengths, durability or production costs.

14.1.2 Magnetic sensing with radical pair materials

During great migrations, birds navigate by detecting the inclination of the Earth's magnetic field, which varies predictably with latitude (Rodgers and Hore, 2009). As discussed in Chapter 10 of this book, one hypothesis for exactly how birds sense the Earth's magnetic field is that they exploit quantum-mechanical evolution of photo-excited radical pairs. Although a definitive experimental confirmation of this radical pair mechanism in bird navigation is still outstanding, this hypothesis has inspired researchers to consider artificial devices for magnetometry based on radical pair reactions (see also (Weaver *et al.*, 2000)).

Magnetometry is a well-developed field, and there exist many commercial magnetic field sensing devices based on diverse physical phenomena. Sensing magnetic fields is in fact a critical component of many everyday technologies, including aeroplanes, navigation tools and medical diagnostic tools. See Edelstein (2007) for a review of modern magnetometry methods and technologies. The most sensitive detector of a DC, or low frequency, magnetic field, is the SQUID magnetometer which can detect fields of order 10^{-14} tesla (T). Magnetometry is also a field where quantum-mechanical effects have been exploited to increase sensitivity – e.g. the Josephson effect underlies SQUID magnetometers, while nanoscale magnetometers reliant on manipulating spin coherence have recently been constructed using colour centres in diamond (Balasubramanian *et al.*, 2008; Le Sage *et al.*, 2013). A useful distinction to bear in mind is that there are two types of magnetic field sensing tasks: one, the sensing of the overall magnitude of a magnetic field, and two, the sensing of the vector components of a magnetic field. The latter task, that of sensing a field direction, is typically more difficult.

What is the motivation for building a biomimetic magnetic field sensor – a *chemical compass* – based on the radical pair mechanism?[1] Given the remarkable sensitivities of modern magnetometers (Edelstein, 2007), it would be quite ambitious to suppose that a biomimetic magnetic field sensor could match these standards. However, there are factors other than sensitivity that dictate the utility of a magnetic sensing device, including: cost of production, cost of maintenance, operating temperature, frequency response, stability, reliability, size and power requirements. Although the development of a sensor based on the radical pair mechanism is still nascent, the hope is that such a sensor will be superior in some of the above measures of utility. Specifically, it is very plausible that a radical pair based magnetic field sensor constructed from organic components will have lower production costs and lower power requirements, and have higher operating temperatures than many of today's commercial magnetometers.

14.2 Progress in designing biomimetic quantum materials

Having provided an overview of the motivation and challenges in the field of biomimetic quantum-enhanced material design, we now present a review of progress made thus far on the design and synthesis of such materials.

[1] We use the term 'biomimetic' somewhat loosely here, since to date it has not been proven that the radical pair mechanism underlies bird navigation.

14.2.1 Structured chromophoric assemblies for light-harvesting

Every one of Nature's light-harvesting complexes (LHCs) is at its core an assembly of light absorbing pigments. The wondrous aspect is how such assemblies are accurately and reliably constructed from basic building blocks such as molecules from the chlorophyll, carotenoid or porphyrin families, with protein structures providing scaffolding when necessary. Particularly remarkable are the photo-stability of natural light-harvesting antennas and the mechanisms involved in *photo-protection*, the range of defence mechanisms by which LHCs regulate energy absorption and transfer to prevent oxidation damage, for example, see Murchie and Niyogi (2011). In fact, current organic photovoltaics are sorely lacking such photo-stability and regulatory processes, and this is one of the primary obstacles to their commercial success (Norrman *et al.*, 2006).

Over the past two decades, much effort has been invested in producing artificial light-harvesting units that mimic the structure and properties of natural LHCs. The aspect that distinguishes these efforts from organic photovoltaic devices based on polymers or quantum dots is the driving force of their biomimicry, as well as their attempts to use the same, or similar, building blocks seen in Nature. This sometimes results in directly integrating entire sections of natural LHCs with an inorganic device framework. There are at least three specific properties that such biomimetic artificial light-harvesting or light-sensing devices strive to replicate from the model of natural light-harvesting complexes:

(1) A near unit efficiency of energy transfer from the region of light absorption to the region of charge separation, where the exciton that results from light absorption is converted into a charge separated electron and hole.
(2) A near unit efficiency of charge separation that occurs at the reaction centre in natural light-harvesting complexes.
(3) The above mentioned photo-stability and photoprotection mechanisms of natural light-harvesting complexes.

To date, no biomimetic approach has been successful in replicating all of the above properties in a single device. This is perhaps not surprising, since replicating the performance of natural LHCs is a formidable task. The initial steps of light capture and conversion involve fragile excited states and charge-separated states that are vulnerable to numerous reactions that dissipate the captured energy as wasted heat. Precise control over nanoscale structure is needed to regulate and inhibit these dissipation channels, and this is one of the great challenges for the fields of synthetic and supramolecular chemistry. Ideally, one would like to construct an artificial light-harvesting complex from the molecular level up. However, present – covalent or supramolecular – synthesis techniques, advanced as they are, are not

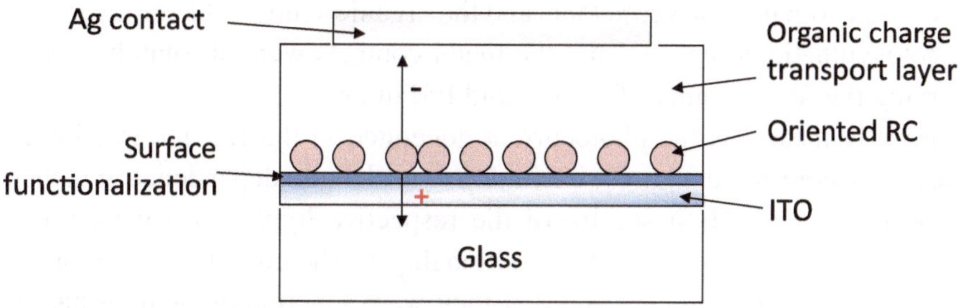

Figure 14.1 A cross-section of the device integration of reaction centres from (Das *et al.*, 2004). Figure provided courtesy of Professor Marc Baldo, M.I.T.

quite at the stage where one can both tailor nanoscale structure and at the same time produce large-scale molecular arrays.

In this section we will review some of the demonstrated techniques, as well as some promising proposals, for constructing biomimetic light-harvesting antennas for photovoltaic or light-sensing applications.

Direct integration of natural antennas into solid-state environments

The most obvious way to leverage the superior light capture and conversion capabilities of natural LHCs, is to directly 'wire up' such structures to a charge collection device that extracts the electrons produced by charge separation. These electrons then constitute the detection signal for light sensors or the electrical current in a photovoltaic. Of course, this approach has the potential to preserve all the quantum properties of the natural LHC that is integrated, but great care must be taken to ensure that the functioning of the LHC is not adversely effected by integration into an extremely foreign solid-state environment.

Such an integration of a natural light-harvesting complex and a solid-state electron collection device has been accomplished by the Baldo group at M.I.T. (Das *et al.*, 2004) (see also Trammell *et al.* (2004) for a supporting study). This group has integrated two biological LHC components, the reaction centre (RC) from the purple bacterium species *Rhodobacter sphaeroides* and Photosystem I (PSI) from spinach, into solid-state electronic devices. See Figure 14.1 for a cross-section of the constructed device. The challenge of preserving the light-harvesting functionality of the biological complexes within the solid-sate environment was overcome by careful design of the assembly process. The molecular complexes were self-assembled on a gold-coated surface which was functionalized with an affinity resin to immobilize and orient the pigment–proteins. Then the complexes were stabilized using customized peptide surfactants. This self-assembled layer was finally covered with an amorphous organic semiconductor, before deposition of top metal contacts for collection of the generated charges. The authors of the study stress that the use of custom peptide surfactants and the organic semiconductor layer between the photosynthetic complexes and the metal contacts were absolutely crucial for preserving the functionality of the RC and PSI units.

The study measured the photocurrent generated in the RC and PSI integrated devices and confirmed that its excitation wavelength dependence matches the solution phase absorption spectra of the respective light-harvesting complexes. This is a good indication that the functionality of the complexes was preserved during the integration process and that they were functioning as light-harvesting modules of the device. In addition, the authors estimate that the internal quantum efficiency of the device is 12% and predict that it could be increased to 20% by improvements in the integration process. However, it should be noted that there is another limiting factor for such devices: the self-assembled light-harvesting components (RC or PSI) absorb very little of the available light (<1%) because they form a thin monolayer. This limitation leads to low overall power efficiencies and has to be overcome before such devices can be considered true photovoltaics. Related recent experiments on biohybrid device development have focused on the use of the cholorosome of green sulphur bacteria (see below), which offers greater light absorbance. Encouraging results have been obtained using both columnar structures of the chlorosome, formed by electrospray deposition onto dye-coated TiO_2 surfaces (Modesto-Lopez *et al.*, 2010) and chlorosomes that were weakly associated with graphite electrodes (Sridharan *et al.*, 2009).

In addition to the above experiments measuring optoelectronic energy transfer, several studies have focused on the controlled placement and immobilization of biological light-harvesting complexes onto solid surfaces. These studies achieve the immobilization, control over orientation and patterning of LHCs by complex surface lithography and chemical functionalization techniques, which show potential for tailoring nanoscale structure of biological LHCs adsorbed onto surfaces. Similarly to Das *et al.* (2004), the most commonly used technique is to form self-assembled monolayers on gold or glass substrates, and then to selectively chemically modifiy (e.g. CH_3-terminate) parts of these monolayers so that they resist or attract adsorption by LHC proteins. The resolution at which this chemical modification can be done determines the length scales at which the patterning of LHCs can be achieved. This technique was utilized by Ciobanu *et al.* to immobilize PSI complexes (Ciobanu *et al.*, 2005), by Reynolds *et al.* (Reynolds *et al.*, 2007) to immobilize LH2 complexes from purple bacteria and by Escalante *et al.* (Escalante *et al.*, 2008) to immobilize LH1 and LH2 complexes from purple bacteria. Microscopy was used in these studies to confirm the patterned adsorption of light-harvesting complexes onto a surface, and Reynolds *et al.* and Escalante *et al.* also used fluorescence spectroscopy to confirm that the optical properties of the LHCs were not significantly modified (compared with solution phase spectra) by

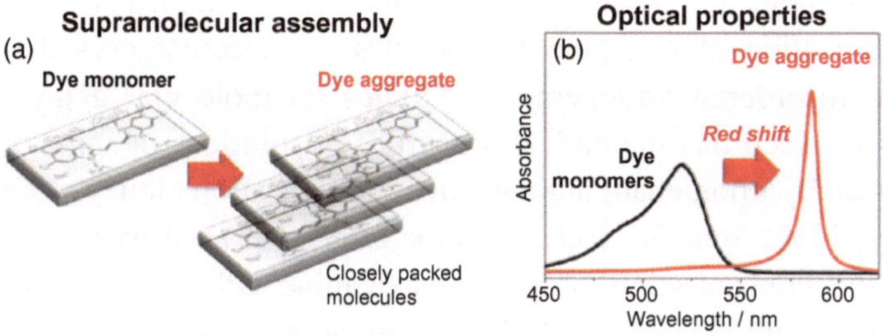

Figure 14.2 Closely packed dye molecules that form a supramolecular assembly, i.e. a molecular aggregate. (a) Schematic of a molecular aggregate self-assembled from molecular sub-units connected through weak non-covalent linkages to create fragile and reversible monomer-to-assembly transitions. (b) Absorption spectra of dye molecules (cyanine dye derivative TDBC) in solution (black line) and after self-assembly into molecular aggregation (red line): the spectral red-shift upon aggregation is characteristic for J-aggregates. Figure adapted from (Eisele, 2010).

adsorption onto the surface. From this retention of optical properties they conclude that the structural integrity of the LHCs is preserved through the adsorption and patterning processes. However, it should be noted that in contrast to the Das *et al.* study, discussed above, these adsorption studies did not have a scheme to collect photocurrents and therefore demonstrate photovoltaic functionality.

The experiments detailed in this section, especially that in (Das *et al.*, 2004), are promising demonstrations of an approach to construct devices that can leverage the excellent light capture and exciton transport performance of biological light-harvesting complexes. However, in the context of this book, one must ask whether the quantum properties of the light-harvesting components (RC, PSI, LH1 and LH2) were preserved during the integration process, and are still present in the final devices. The matching of solution phase absorption spectra by device photocurrents in Das *et al.* (2004), and a similar matching of fluorescence spectra of surface adsorbed samples to solution spectra in Reynolds *et al.* (2007) and Escalante *et al.* (2008) are encouraging evidence that the structure and hence excited state dynamics of the photosynthetic pigments were largely preserved. However, it is impossible to conclude at this stage whether the excitonic properties and energy transfer properties were unaltered (or enhanced). Further experiments examining energy transfer in integrated photosynthetic complexes will be necessary to determine this.

Molecular aggregates

The term molecular aggregate is commonly reserved for densely packed molecular structures (Figure 14.2a) which result from non-covalent intermolecular interactions such as pi–pi stacking, hydrogen bonding and metal ion coordination:

so-called supramolecular assemblies. Such interactions can lead to both extremely regular organization of constituent molecules – molecular crystals – and also less regular, disordered arrangements. Studies on molecular aggregates in the early 1930s revealed that the optical properties of solutions of some organic dyes change drastically upon changing their dye concentration (Jelley, 1936; Scheibe, 1936; Jelley, 1937; Scheibe, 1937; Scheibe *et al.*, 1937); it was found that upon increasing the concentration of pseudoisocyanine (PIC) dye in solution, the absorption spectrum changed such that the broad absorption band at wavelength ~525 nm disappeared and was replaced by a narrow absorption band at 570 nm. Because this effect was first discovered independently by Jelley and Scheibe, such a new, narrowed and red-shifted absorption band (Figure 14.2b) is now generally known as the J (Jelley) band or the S (Scheibe) band. This new absorption band, the J- or S-band, is associated with extended electronic excitations of aggregated PIC molecules (Franck and Teller, 1938), much like the well-known *Frenkel* excitons of molecular bulk crystals (Frenkel, 1931). Molecular aggregates with such new, narrowed and red-shifted absorption bands are therefore referred to as J-aggregates (or, less commonly, as S-aggregates).

In the past, such J-aggregates have found significant applications in the field of photography where they are used to increase spectral sensitivity of films. More recently, molecular aggregation has been recognized as a useful framework for describing, and imitating, many natural photosynthetic light-harvesting complexes (Scholes, 2002). The techniques and language used for describing the molecular aggregates found in these natural light-harvesting systems is exactly the same as those applied to molecular J-aggregates, which has caused a rapid merging of these two fields. The increasing interest in J-aggregates is mainly due to their unique collective optical properties, whereby the optical response of the aggregated dye molecules is fundamentally different from the addition of the optical response of the individual dye molecules. This behaviour results from the collective Frenkel excitons that describe the excited states of the aggregates.

Perhaps Nature's most efficient light-harvesting antenna system is the *chlorosome* antenna of green sulphur bacteria, one of only few phototropic organisms that can survive by photosynthesis in the extremely light-sparse environment of deep sea (Frigaard *et al.*, 2004; Balaban *et al.*, 2005; Bryant *et al.*, 2007). Typically these bacteria live in conditions with very little light and hence their light capture apparatus has to be extremely effective at absorbing the available light and transporting resulting photo-excitations to reactions centres. The chlorosome antennas are almost 100% efficient at these tasks. In contrast with many other antenna structures, especially those in higher plants, the chlorosome consists solely of pigment molecules (bacteriochlorophyll *c, d* and *e*), with no protein scaffolding. The unique ability of the cholorosome to harvest and transport energy efficiently

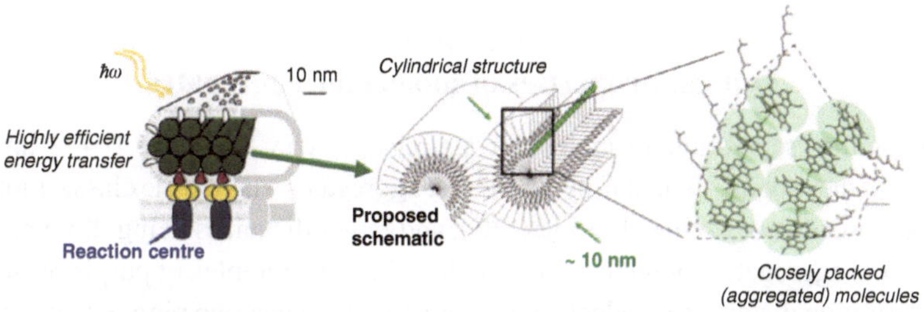

Figure 14.3 Illustration of the proposed geometry and structure of light-harvesting apparatus of green sulphur bacteria. Left panel shows a schematic view of a model of a chlorosome (green) sitting on top of the FMO-complex (yellow) which attaches the chlorosomes to the cytoplasmic membrane containing the reaction centres (blue). This model follows the early electron microscopy picture of (Staehelin *et al.*, 1978, 1980). Schematic view adapted from (Balaban *et al.*, 2005). Middle panel shows a rendition of the proposed cylindrical structure of chlorosomes. Schematic view adapted from (van Rossum *et al.*, 2001). Right panel shows the proposed packing of bacteriochlorophyll molecules within a chlorosome. Schematic view adapted from (van Rossum *et al.*, 2001).

derives from the closely packed aggregates of pi-conjugated molecules that constitute the supramolecular assembly (Balaban *et al.*, 2005). Although the chlorosome's morphology, as well as details of the molecular packing at the atomic level, are still under debate (Staehelin *et al.*, 1980; Steensgaard *et al.*, 2000; Holzwarth, 2004; Oostergetel *et al.*, 2007; Linnanto and Korppi-Tommola, 2008), recent studies propose that these highly efficient antennas potentially consist of cylindrically-shaped aggregates (Figure 14.3) with a diameter of the order of 10 nm and lengths that may extend to several hundred nanometres, containing several hundred thousand bacteriochlorophyll molecules. (Balaban *et al.*, 2005).

Partly motivated by the desirable performance of Nature's light-harvesting molecular aggregates, several researchers have attempted to synthesize customized molecular aggregates for light-harvesting and solar energy technologies. The first challenge with this approach for producing artificial light-harvesting antennas is that it is essential to have a theoretical understanding of how aggregate structure influences the optical and energy transfer properties crucial to light-harvesting. It is clear from the close proximity of pigments in typical molecular aggregates, that quantum properties will be relevant to both optical properties and energy transfer. At separation distances $\lesssim 4$ nm it is important to describe the electronic coupling between pigments in detail and fully quantum mechanically. The distribution and orientation of exciton dipoles, which can only be understood when the coupled system is modelled quantum mechanically, dictate optical properties such as absorption (see Box 14.1). Furthermore, the formation, extent of delocalization and transport of excitons governs how efficient the aggregate is at transferring

Box 14.1
Optical properties of molecular aggregates

Inter-molecular separation in molecular aggregates is in the range 0.4–2 nm, with smaller distances being typical in synthetic dye aggregates, such as the classic molecular aggregate pseudoisocyanine chloride (PIC), and larger distances being more prevalent in molecular aggregates found in nature, such as the LH2 complex of purple bacteria. At these separation distances the electronic interaction between two pigments is primarily Coulombic (i.e. exchange and other short-range interactions are negligible) (Scholes, 2003). Furthermore, it is common to use a point–dipole approximation for the Coulomb interaction. Within this approximation, the redistribution of charge that results from molecular excitation is represented by a single transition dipole moment, $\vec{\mu}$, and interaction between molecules is modelled by a dipole–dipole interaction between these transition dipoles. This approximation to a full Coulomb integral is appropriate if the intermolecular separation is larger than the physical extent of both molecules. If this condition does not hold then higher-order expansions of the Coulomb integral – multipole expansions – should be employed.

Within the dipole approximation, the interaction between two neighbouring pigments is

$$V(r, \mu_1, \mu_2) = \frac{1}{4\pi\epsilon_0}\left(\frac{\vec{\mu_1}\cdot\vec{\mu_2} - 3(\vec{\mu_1}\cdot\hat{r})(\vec{\mu_2}\cdot\hat{r})}{r^3}\right)$$
$$= \frac{|\mu_1||\mu_2|}{4\pi\epsilon_0 r^3}(-2\cos\theta_1\cos\theta_2 + \sin\theta_1\sin\theta_2\cos\phi),$$

where ϵ_0 is the permittivity of free space, $\vec{r}$ is the vector between the two molecules and $\hat{r}$ is its normalized version (i.e. $\vec{r} = r\hat{r}$). The angles used in the second form of the coupling are the angles between the dipoles and the vector separating them (see figure below).

Consider a transition from the singlet electronic ground state of a molecule, $|S_0\rangle$, to a singlet excited state, $|S_1\rangle$, separated in energy by ε. Once excited, the two molecules will interact and the Hamiltonian describing this interaction (under the Heitler–London approximation which is appropriate when $\varepsilon \gg V$) is

$$\mathcal{H} = \varepsilon_1|S_1^{(1)}\rangle\langle S_1^{(1)}| + \varepsilon_2|S_1^{(2)}\rangle\langle S_1^{(2)}| + V\left(|S_1^{(1)}\rangle\langle S_0^{(2)}| + |S_0^{(1)}\rangle\langle S_1^{(2)}|\right),$$

where the superscript indexes molecule number, and we have allowed the excited states of the two molecules to have different energies (due to local environment shifts or if the molecules are two different species). We have set the energy of the ground state of both molecules to zero for convenience and we are not describing any excited states with greater energy, because it is only the lowest excited states that dictate the linear optical properties which we are interested in. Diagonalizing this Hamiltonian yields new excited states ($|E_1\rangle, |E_2\rangle$) that have some contribution from $|S_1^{(1)}\rangle$ and $|S_1^{(2)}\rangle$: $|E_{1/2}\rangle = \alpha_{1/2}|S_1^{(1)}\rangle + \beta_{1/2}|S_1^{(2)}\rangle$.

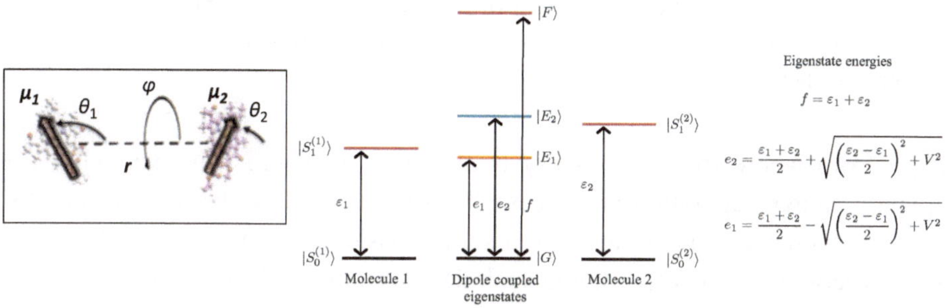

Since the wavelength of light absorbed by such molecules is much larger than the inter-molecule separation, r, the light interacts with the combined dipole operator $\vec{D} = \vec{\mu_1} + \vec{\mu_2}$. Then the light absorbed will be concentrated around the frequencies $e_1/\hbar$ and $e_2/\hbar$, and the magnitude of the absorption profile at these frequencies is proportional to $M_1 \equiv |\langle E_1|D|G\rangle|^2$ and $M_2 \equiv |\langle E_2|D|G\rangle|^2$, respectively, where $|G\rangle = |S_0^{(1)}\rangle \otimes |S_0^{(2)}\rangle$ is the ground state. It is interesting to look at how these features of the absorption profile change with the structural details of aggregation. Obviously, as the distance between the chromophores increases, their coupling decreases ($V \to 0$) and hence the absorption profile approaches the monomer absorption profile. The dependence on the angles defining the aggregate structure is less trivial, and in the following figures we show how the quantities e_1, e_2, M_1, M_2 change with the structural angles. For this example, we assume that the magnitude of both monomer dipoles is the same ($|\mu_1| = |\mu_2| = |\mu| = 10$ debye), $\varepsilon_1 = 12\,000\ \text{cm}^{-1}$, $\varepsilon_2 = 12\,200\ \text{cm}^{-1}$, and $r = 15$ Å. Finally, for simplicity we will fix $\theta_1 = \pi/4$ and look at the dependence on the other two free angles.

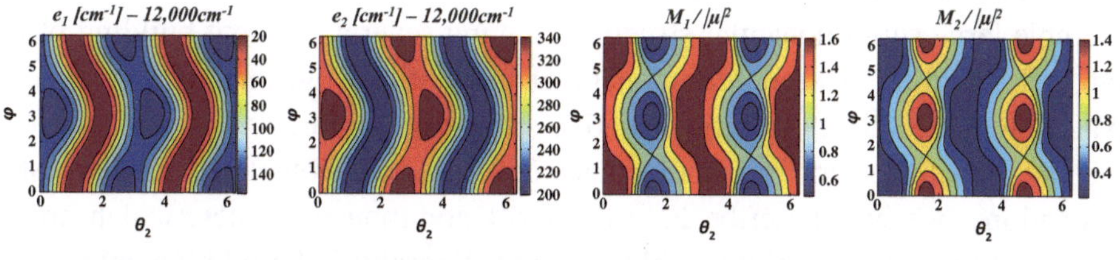

There are several things to note from these surfaces. Firstly, the transition energies of the two excitons (eigenstates of $\mathcal{H}$) can be greater, or less, than the original monomer transition energies depending on the dipole angles. Also, from the plots of $M_{1/2}$ we see that the magnitude of the combined dipole is distributed across the two excitonic states. Depending on the values of the angles, either the lower eigenstate or the upper eigenstate can have the dominant dipole strength. This is an example of how the underlying structure strongly influences the optical properties of the aggregate. This redistribution of dipole strength (or oscillator strength) is a key feature of molecular aggregation, and can lead to striking features in linear spectra. For example, in the symmetric case where $\varepsilon_1 = \varepsilon_2$, and the dipoles are lined up head-to-tail (i.e. $\phi = \theta_1 = \theta_2 = 0$), *all* the dipole strength will be concentrated on the low energy exciton, and $M_2 = 0$. Large aggregates with this structure are known as *J*-aggregates (Würthner *et al.*, 2011), and this concentration of oscillator strength leads to a dominant narrow peak in their absorption spectrum.

energy from where the light energy is absorbed to where it is eventually used (typically charge separation interfaces). These effects can only be described by a quantum-mechanical model and not a semi-classical one such as Förster energy transfer theory (see Chapter 3); and such descriptions can be complex.

The second challenge with the molecular aggregate approach is that one must gain enough experimental control over the aggregation process so that the final product can be manipulated to have desirable light capture and energy transfer properties. Typically such aggregates are produced by a solution-based self-assembly process, and the final morphology of an aggregate is dependent on many factors including the species and concentration of the constituent molecules, temperature and solvent properties such as pH. Hence there are many control 'knobs' in the assembly process and part of the relatively new field of supramolecular chemistry (Steed and Atwood, 2009) is concerned with understanding how each of these influences structure and morphology.

In spite of these challenges, many attempts have been made to produce synthetic molecular aggregates for light-harvesting, using a broad range of constituent pigment molecules. The extraordinary chlorosome antenna system has inspired much work on cylindrical supramolecular systems that self-assemble from various artificial pi-conjugated molecules that can interact with light (Vlaming *et al.*, 2009; Shao *et al.*, 2011; Aida *et al.*, 2012). Also notable are ambitious attempts to self-assemble large supramolecular structures having several covalent building blocks that each function as some component of an artificial light-harvesting complex (Wasielewski, 2009). As a representative survey of the large variety of pigments that have been used to form molecular aggregates for potential light-harvesting applications, we will describe here the most common examples, which are the families of (a) cyanine, (b) zinc chlorin, and (c) porphyrin based pigments.

(a) *Cyanine dyes and amphiphilic derivatives.* Even before their discovery by Jelley and Scheibe in the early 1930s, the cyanine dye derivatives had long been recognized as important spectral sensitizers for photographic imaging (Kobayashi, 1996) and non-linear optics applications, on account of (1) their extraordinarily sharp absorption bands arising from the large, highly delocalized pi-conjugated systems, and (2) their unique aggregation properties. They are also well established as model systems for studies of photo-activated electron transport processes using monolayer assemblies at solid state and liquid interfaces (Scholes *et al.*, 2011; Aida *et al.*, 2012). When cyanine dyes are

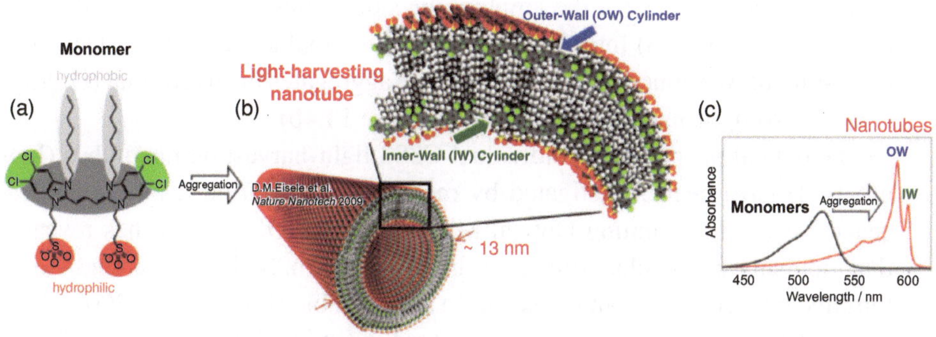

Figure 14.4 Artificial light-harvesting nanotube self-assembled from amphiphilic cyanine dye molecules (a) Amphiphilic 3,3'-bis(2-sulphopropyl)-5,5',6,6'-tetrachloro-1,1'-dioctylbenzimidacarbocyanine (C8S3) monomer (De Rossi *et al.*, 1995). (b) Schematic of the self-assembled light-harvesting nanotube (for clarity using only one molecule per unit cell): double-walled structure with the hydrophilic sulfonate groups (red) on the exterior, the hydrophobic alkyl chains (light grey) in the interior of the bilayer and the cyanine dye chromophore (dark grey). The nanotube has a diameter ~ 13 nm. Distance between the two concentric cylinders is ~ 4 nm. Figure adapted from (Eisele *et al.*, 2009). (c) Absorption spectra. Black: dye monomers dissolved in methanol (no aggregation). Red: nanotubular dye aggregates prepared in water/methanol; the two main J-bands can be unambiguously assigned to inner-wall and outer-wall cylinder, respectively (Eisele *et al.*, 2012).

dispersed in polar solvents above a critical concentration level (usually low mM), they self-assemble into molecular aggregates. Although the unique properties of cyanine dye J-aggregates, such as their highly efficient energy transfer properties, have been realized for decades (Knoester, 2002), most recently they have received attention for use in artificial LHCs.

In particular, amphiphilic cyanine dye derivatives can be used to generate structures similar to the cylindrical chromosomes of photosynthetic green sulphur bacteria. Consequently, artificial light-harvesting antennas composed of cyanine dye J-aggregates that mimic natural systems have been proposed. Optical and energy transfer properties can vary widely based on the aggregate's morphology and details of the spatial arrangement of molecules within the aggregate; it is well known that both parameters are controlled by the chemical structure and functional groups of the particular cyanine dye. An entire new class of cyanine dye derivatives was synthesized in the early 1990s by Siegfried Dähne and co-workers. They synthesized a library of amphiphilic cyanine dye molecules by adding various hydrophilic and hydrophobic functional groups to a cyanine dye molecule (De Rossi *et al.*, 1995, 1996). One example of these molecules, abbreviated as C8S3, is shown in Figure 14.4a; when dissolved in a polar solvent these molecules self-assemble due to superposition of hydrophobic forces (of the amphiphilic side chains) and the pi–pi stacking (of the chromophores) into double-walled nanotubular J-aggregate structures, consisting of two concentric cylindrical dye aggregates in an artificial (Didraga *et al.*, 2004) light-harvesting nanotube (Figure 14.4b).

The optical properties of individual C8S3 light-harvesting nanotubes (Figure 14.4) have been investigated by means of polarization-resolved fluorescence Near-field Scanning Optical Microscopy (NSOM) which has revealed that their supramolecular structure is highly uniform, both along an individual tube and between different tubes within an ensemble (Eisele *et al.*, 2009). This makes them an excellent model system for fundamental investigations of exciton transport in individual quasi-one-dimensional nanostructures, as well as for light-harvesting and other optoelectronic applications in future solid-state devices. Most importantly, it allows for study of the properties of the individual nanotube by investigating the ensemble in a solution sample. However, in order to control such large supramolecular systems it is essential to understand both how their various parts interact with each other and also whether these interactions result in coherently shared excited states (excitons), or in diffusive energy transport between them. Through selective chemistry it was possible to unambiguously determine the supramolecular origin of the excitonic transitions in the nanotubes (Figure 14.4c). It was found that the two cylinders of the nanotube, while being only $\sim$4 nm apart from each other, have distinct spectral responses and are best described as two separate, at most weakly coupled excitonic systems (Eisele *et al.*, 2012). Understanding such interactions is critical to the control of energy transfer on a molecular scale, a goal for diverse applications ranging from artificial photosynthesis to molecular electronics. Obtaining a detailed understanding of the linear spectrum of the aggregates is also a key step towards developing a fundamental understanding of how the details of the supramolecular structure influence the optical and energy transfer properties that are critical for light-harvesting.

(b) *Zinc chlorin and its derivatives*. ZnChl is very similar to BChl *c* in form, with the Mg metal centre of BChl *c* being replaced by Zn. It has been shown that derivatives of ZnChl self-assemble into nanoscale rods similar to the chlorosome, for example, Tamiaki *et al.* (1996). In fact, some such aggregates have spectroscopic properties that closely match those of natural chlorosomes (Würthner *et al.*, 2011). Furthermore, using time-resolved fluorescence it was determined that excitons are delocalized among at least 10–15 molecules and that exciton transport timescales are of 1–10 ps (Prokhorenko *et al.*, 2002). Thus it would seem that the quantum properties of such aggregates are important, as would be expected from the large density of pigments and their similarity to the chlorosome. More recently, such ZnChl aggregates have been appended with other organic components (in a self-assembling manner not requiring the assistance of structure stabilizing proteins) to enhance their capabilities. Katterle *et al.* formed a co-aggregate of ZnChl and BChl which is covalently linked to a C60 molecule (Katterle *et al.*, 2007). This combined self-assembling structure functions as a complete 'light-harvesting unit' with the ZnChl aggregate functioning as a light collection antenna, the C60 as an artificial reaction centre performing charge separation, and the intermediary BChl functioning simultaneously as an energy trap for the ZnChl antenna and an electron donor for the C60 reaction centre. This combined structure achieves very fast energy transfer and charge separation and an overall quantum efficiency of 90%. Another modification of ZnChl aggregates was performed by Röger *et al.*, who appended additional peripheral pigments to the tubular structure in order to harvest the green and orange portion of natural sunlight (Röger *et al.*, 2006, 2008). This modification was shown to increase harvesting efficiency by 63% (as compared with the bare ZnChl aggregate) by enhancing spectral coverage.

(c) *Porphyrin based dyes.* Porphyrin based dyes are another family from which aggregates are commonly built for light-harvesting functionality. Perhaps the most studied aggregates from the porphyrin family are the self-assembling nanoscale rods formed from the Tetrakis (4-sulfonatophenyl)-porphyrin ($TPPS_4$) derivative (e.g. Schwab *et al.* (2003)). Spectroscopic studies suggest that excitons in the $TPPS_4$ aggregate are initially delocalized over $\sim$10 molecules but then rapidly become more localized to one or two molecules (Gulbinas *et al.*, 2007). The Shelnutt group has also constructed several aggregates of porphyrin derivatives that are unique in that they utilize the supramolecular technique of ionic self-assembly (Faul and Antonietti, 2003) for aggregation. This enables the synthesis of a variety of aggregate morphologies with multiple constituent monomer porphyrins. Structures synthesized including tubular aggregates with a mixture of two porphyrin derivatives (Wang *et al.*, 2004b), and clover-leaf shaped dendritic structures composed of two types of metallic porphyrins (Martin *et al.*, 2010). The latter are particularly intriguing because the authors of that work show that it is possible to change the electronic character of the constituent molecules without affecting the morphology. This allow one to tune electronic properties of the binary solid (from photoconductor to insulator for example) by altering the metal ions of the constituent porphyrins. Inter-molecular distances suggest that quantum effects will be important for describing optical and energy transfer properties in these porphyrin aggregates, however no spectroscopic data is available for them to date.

Finally, it is important to note that at this stage, unlike the development of prototype biohybrid devices with natural light-harvesting systems described above, none of these supramolecular aggregates have been integrated into a device with photovoltaic capability. Such integration and the stability of the molecular aggregates during it will be crucial to the viability of this approach to artificial light harvesting.

Virus-templated chromophore assemblies

As we saw in the last section, a major challenge in utilizing biologically inspired molecular aggregates for light-harvesting is controlling their nanoscale structure and morphology. A variety of non-covalent interactions must be controlled and precisely utilized to synthesize well-structured aggregates and the field of supramolecular chemistry is tackling this challenge. An alternative approach to overcoming this challenge is to turn to another one of Nature's marvellous processes: protein self-assembly. Self-assembly is a ubiquitous process in biology taking place at all length scales, from molecular to organismic (Whitesides *et al.*, 1991). Many of the synthetic molecular aggregates discussed in the last section are constructed using molecular self-assembly, but the self-assembling process occurs over larger lengths scales as well. In particular individual protein subunits often self-assemble into aggregate structures with a wide range of structural motifs (sheets, ribbons, helices). Some of the best understood protein self-assembly processes occur in viruses where *viral capsids*, protein shells that enclose and protect the genetic material of the virus, self-assemble into a variety of shapes. Recently, much work has focused on exploiting these viral capsids in a *templated self-assembly* process where the geometric arrangement of the assembled capsid is used as a rigid scaffold that guides the organization of other molecules. The use of viruses to perform such templated self-assembly has been demonstrated with synthesis of structures for battery technologies (Nam *et al.*, 2006; Royston *et al.*, 2008), for production of template spheres for targeted drug delivery (Wu *et al.*, 2009) and synthesis of pigment–protein complexes with potential to act as artificial light-harvesting systems (Endo *et al.*, 2007; Miller *et al.*, 2007; Nam *et al.*, 2010).

A popular virus scaffold that is used for templating such self-assembly is the tobacco mosaic virus (TMV), see Box 14.2. TMV is particularly appealing as a light-harvesting structure because of the cylindrical and disc-like geometries that it self-assembles into, which closely resemble natural light-harvesting complexes, such as the chlorosome antenna of green sulphur bacteria (see Figure 14.3), and the LH2 complex of purple bacteria. To use this protein for the assembly of chromophore arrays for light-harvesting, the strategy is to introduce a cysteine residue at certain positions in the capsid protein monomer, which facilitates covalent attachment of a wide variety of commercially available chromophores.

Box 14.2
The tobacco mosaic virus

The tobacco mosaic virus (TMV) is a paradigmatic example for studying self-assembly in biology. It was one of the earliest viruses studied and is structurally the best characterized. In 1955 H. Fraenkel-Conrat and R. C. Williams established that TMV self-assembles (Fraenkel-Conrat and Williams, 1955) by breaking apart the virus and then remixing the constituents. The mixture assembled over a period of 24 hours into intact TMV structures. Since then, much research has been conducted to understand the microscopic structure of TMV and exactly how its constituents self-assemble. TMV self-assembles into several structures depending on the conditions of solution, including a helix, two-layer disc or stacked disc (Klug, 1999). This variety of assembled structures, and the large amount of virus that can be synthesized, are both extremely useful attributes for TMV based templated synthesis.

Structurally, the tobacco mosaic virus is composed of a tight packing of protein sub-units or monomer. These monomers self-assemble into tubular or disc like structures that encase the virus RNA (Klug, 1999). Each disc has 17 protein sub-units, and tubular structures can either be composed of stacked discs or a helical arrangement of protein sub-units with $16\frac{1}{3}$ monomers per turn of the helix. The radius of the central ring in the tubular and disc structure is around 20 Å. It is within this central pocket or elsewhere along the protein monomer that chromophores can be covalently attached using cysteine residues. The cysteine residues are not present in the natural TMV organism but can be introduced in recombinant versions.

The figure below (courtesy of Professor Matthew Francis) illustrates an example of TMV-templated assembly.

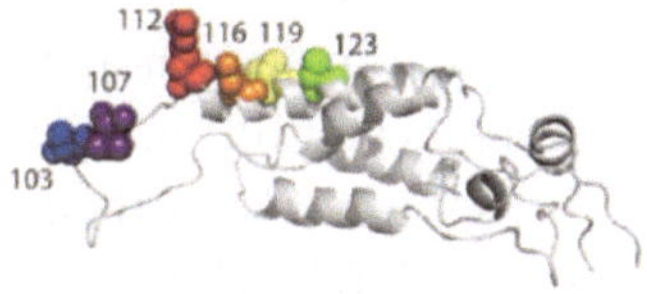

(a) Variety of sites on a TMV monomer at which mutations can be introduced to facilitate covalent attachment of chromophores

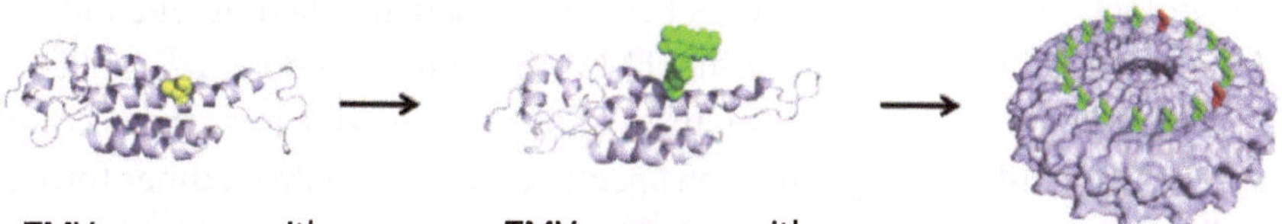

(b) An example of TMV-templated assembly of chromophore arrays

Then as the TMV capsids self-assemble, the attached chromophores gather into a well-defined structure (cylinder, helix, disc) with predictable inter-chromophoric distances. Chromophores of different species can be attached to mimic donor and acceptor species and produce energetic landscapes for funnelling excitation energy. Furthermore, there is a large amount of freedom in which sites of the TMV capsid monomer are modified with the cysteine residue for chromophore attachment. Using such techniques, efficient and robust light-harvesting behaviour has already been demonstrated with broad spectrum absorption and up to 90% efficiency of energy transfer (Endo *et al.*, 2007; Miller *et al.*, 2007, 2010).

Miller *et al.* (2007) attached three species of chromophores to various sites along the TMV capsid protein monomer. Then these monomers were allowed to self-assemble into disc and tubular (rod) structures. The authors estimate that such assemblies contain about 700 chromophores per 100 nm of rod length, and with inter-chromophoric distances of about 20 Å. The attached chromophores were chosen to be two donor-like pigments and one acceptor-like pigment. This allowed the investigation of energy transfer and its efficiency with different ratios of donors and acceptors. Similarly, Endo *et al.* have also incorporated two types of porphyrin-based chromophores into selective sites of TMV protein monomers (Endo *et al.*, 2007). These monomers self-assembled into disc and rod-like structures and energy transfer in both structures was assessed. The authors estimate that the inter-chromophoric distances in their structures are about 20 Å, and posit that the energy transfer occurs primarily through a semi-classical Förster mechanism at these length scales. This conclusion is supported by the fact that the absorption spectra of the assembled structures closely resemble monomeric absorption spectra of the chromophores (i.e. signatures of molecular aggregation, which assist coherent energy transfer, such as peak shifts and side peaks are absent). Similarly, Miller *et al.* also conclude that the energy transfer in their TMV structures is through a Förster mechanism (Miller *et al.*, 2007). However, it should be noted that the attachment sites of the chromophores can be changed such that their density upon self-assembly of the TMV capsids is higher than in the above two experiments, and in such structures it is expected that energy transfer will involve coherent channels. In fact, in earlier experiments Endo *et al.* had attached pyrene molecules to the inner-most attachment site of the TMV monomer (Endo *et al.*, 2006). The resulting self-assembled rod-like structures showed clear signs of strong coupling of the pyrenes, as evidenced by emission spectra consistent with excimer formation (a short-lived association of two or more molecules that is even stronger than aggregation). Therefore, we expect that by varying the attachment sites and types of chromophores attached, a TMV templated structure can interpolate between energy transfer via semi-classical mechanisms and via strongly quantum coherent mechanisms.

Another virus that has been used to template structures with potential for light-harvesting is the M13 bacteriophage. The M13 virus has ~2700 pVIII proteins that run the length of the virus and create a self-assembling cylindrical coat. The cylinder is roughly 880 nm in length and 6.5 nm in diameter. By chemically modifying the pVIII coat proteins Nam *et al.* were able to attach zinc porphyrin pigments to them (Nam *et al.*, 2010). By changing the ratio of zinc porphyrins to pVIII proteins they were able to modify the density of conjugated pigments on the assembled M13 virus. The average inter-pigment separation is estimated to vary between 10–24 Å. In comparison with the TMV system, the density of pigments achievable using the M13 virus is slightly smaller, and the number of pigment attachment sites that have been explored is smaller. After construction of their array of zinc porphyrins on the M13 scaffold, Nam *et al.* performed fluorescence, absorption and transient absorption measurements on the aggregate structure. From these measurements they conclude that there is negligible exciton delocalization in the structures due to the low density of pigments; the energy transfer was determined to take place by the Förster mechanism (Nam *et al.*, 2010). In addition, the authors point out that there is significant fluorescence quenching in the aggregate structures, which is conjectured to be the result of electronic coupling of multiple pigments and the creation of trap sites (although no rigorous model of this mechanism for quenching is provided) (Nam *et al.*, 2010).

The true power of such virus-templated light-harvesting structures may come from their potential to function as artificial light-harvesting antennas with malleable properties. Chromophore types, densities and orientations can in principle be tuned and adjusted. This results in an exquisite amount of control over the nanoscale structural and energetic properties of these organic assemblies. A recent theoretical analysis based on TMV-templated chromophore assemblies and employing multi-objective optimization has identified a critical trade-off that emerges between efficient energy transfer and bandwidth of photons absorbed in cylindrical chromophore assemblies (Sarovar and Whaley, 2013). The trade-off is that one property cannot be maximized without sacrificing the other by modifying the arrangement and energetics of the chromophores on the cylinder. The work shows that having closely spaced chromophores and consequently strong interchromophore coupling is beneficial for negotiating this trade-off and hence achieving high efficiency energy transfer, while maintaining the spectral width of absorption in the presence of energetic and structural disorder. Therefore, strong coupling and moderately delocalized excitons may be a critical tool in maintaining good light-harvesting characteristics in the presence of disorder. Understanding and developing such rational design strategies to negotiate the fundamental trade-offs involved in light-harvesting are essential for developing effective and customized artificial devices that collect or sense electromagnetic radiation under diverse conditions.

It is the subject of ongoing work to explore the large space of virus-templatable structures and identify optimal and robust structures that yield enhanced light-harvesting or sensing performance. The key challenge in this undertaking is understanding the precise structural states of the virus-templated assemblies, including a quantitative understanding of the amount of structural disorder present and how this can be controlled. Although the assembled structures of the TMV protein and similar virus capsids are well understood, the assembled state after attachment of chromophores is less well characterized. Depending on attachment location, linker group used to attach and assembly conditions, the chromophore array can be more or less disordered in structure. Controlling this disorder will be essential to producing replicable and reliable assemblies using this technique.

As with direct molecular aggregates, to date, no virus-templated assembly of chromophores has been integrated into a device capable of utilizing the captured light energy. Schemes exist for integrating TMV-templated assemblies with electron accepting surfaces, however these have not been implemented to date and a key step that needs further investigation is the efficiency with which charge separation can be performed in such systems.

Engineered maquette proteins

A very different approach to protein self-assembly for generation of artificial functional biomaterials is the use of maquettes, protein scaffolds that are designed from scratch to ensure the simplest protein scaffold for a given functional task (Dutton and Moser, 2011). This is a synthetic biology approach that employs iterative and reversible design steps and is deliberately non-biomimetic in its means and goals. To date this emerging approach has been used to construct maquettes consisting of several (three or four) bundles of helices that are specifically engineered to possess particular characteristics for electron transfer (Page *et al.*, 1999), proton coupled electron transfer (Chen *et al.*, 2002), or oxygen transport (Koder *et al.*, 2009). The explicit incorporation of quantum characteristics as a design step in such a rational synthetic approach may provide additional opportunities for generating quantum enhanced functional biomaterials in the future.

14.2.2 Radical pair materials for magnetometry

As discussed in Chapter 10, several theoretical and simulation studies have demonstrated the viability of the radical pair mechanism for detecting weak magnetic fields. Inspired by these results, Maeda *et al.* constructed a proof of principle experiment for demonstrating that photo-activated radical pair reactions in an artificial molecule can be sensitive to the magnitude and direction of weak magnetic fields (Maeda *et al.*, 2008). The molecule used in their experiment was a triad

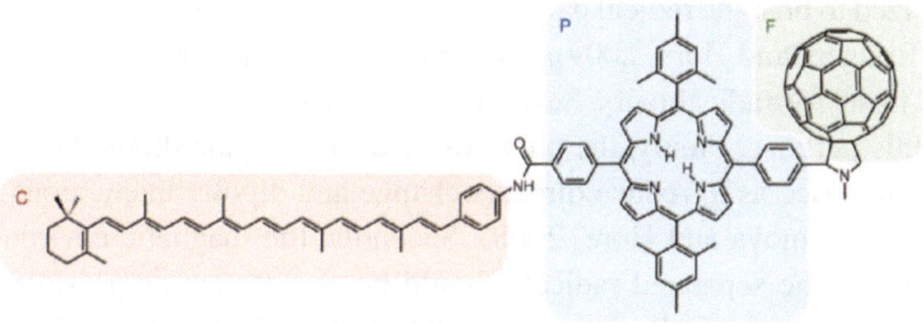

Figure 14.5 The triad molecular complex used by Maeda *et al.* (Maeda *et al.*, 2008) for demonstration of weak magnetic field sensing by radical pair reaction chemistry. Adapted with permission from Macmillan Publishers Ltd: Nature, vol. **453**, p. 387, copyright 2008.

composed of linked carotenoid (C), porphyrin (P), and fullerene (F) groups, see Figure 14.5. Upon green-light irradiation, this triad is capable of supporting a radical pair of electrons separated across almost the whole physical extent of the extended molecule. The magnetic field dependence of the recombination rate of this radical pair state was recorded through transient absorption measurements. The authors immobilized and oriented the CPF triad by freezing it in the nematic phase of a liquid crystal. This orientation allowed the authors to demonstrate that the radical pair recombination rates were substantially altered by varying the magnitude *or* direction of magnetic fields at temperatures around 113 K. This mechanism could detect magnetic field magnitude variations down to ~50 μT, and magnetic field directional variations down to ~3 mT. For comparison, the magnitude of the Earth's magnetic field is roughly 50 μT (with geographic variation in the range 20 μT–65 μT). The authors comment that the molecular triad they chose is by no means an optimum magnetic field detector and therefore they expect even greater sensitivities and higher operating temperatures for optimized structures. It should be noted that there have been other experiments demonstrating the sensitivity of radical pair recombination dynamics to weak magnetic fields – e.g. Rodgers *et al.* (2007). However, the Maeda *et al.* experiment was the first to show that radical pair reactions are not only useful for detecting magnetic field magnitudes, but also magnetic field directionality.

Although the Maeda *et al.* experiment is an encouraging start for constructing a biomimetic chemical compass based on the radical pair mechanism, many questions remain. The chief question is about exactly what type of molecular structure is best suited for hosting the photo-excited radical pair. Maeda *et al.* demonstrated that the carotenoid–porphyrin–fullerene triad was effective, but also commented that it was not optimal. What could be the optimal molecular structure for building a magnetometer based on the radical pair mechanism? The specialized photoreceptor hypothesized to host the radical pair mechanism in birds is the cryptochrome flavoprotein (Rodgers and Hore, 2009), and it is possible that variants of this structure are good hosts for radical pairs. Several factors are desirable in a molecular structure for this purpose. Firstly, the photo-excited radical pair should be sufficiently separated in space as to reduce direct exchange and dipolar interactions (Timmel *et al.*, 1998; Efimova and Hore, 2008). Secondly, the magnetic environments in the vicinity of the separated radicals should be as different as possible, so as to maximize the sensitivity (Rodgers *et al.*, 2007). It has been proposed to introduce magnetic nanostructures into the design of a biomimetic chemical compass in order to maximize this asymmetry of magnetic environments (Cai, 2011). Finally, the radical pair lifetime should be long enough for the weak magnetic field to affect recombination dynamics, but not so long that spin relaxation effects dominate the dynamics (Rodgers and Hore, 2009). The optimal lifetime will in fact depend on many of the other design details, e.g. asymmetry in magnetic environment, as well as the magnitude of the magnetic field to detect. These open questions highlight the need for a systematic study of the design rules for building an efficient radical pair based chemical compass. Such a study will greatly inform the search for molecular structures suitable for hosting the radical pairs that enable magnetic field sensing.

15

Coherent excitons in carbon nanotubes

LEONAS VALKUNAS AND DARIUS ABRAMAVICIUS

15.1 Structure

Carbon nanotubes (CNT) and fullerenes are large molecules constructed entirely of carbons. Single-walled carbon nanotubes (SWNT) can be viewed as a strip cut from an infinite graphene sheet rolled up into a tube (see Figure 15.1). The diameter and helicity of a SWNT are uniquely defined by the roll-up vector $\boldsymbol{C}_k = n\boldsymbol{a}_1 + m\boldsymbol{a}_2$ that connects crystallographically equivalent sites on the graphene lattice, where $\boldsymbol{a}_1$ and $\boldsymbol{a}_2$ are the graphene lattice vectors and n and m are integers. Translation vector $\boldsymbol{T}$ is along the tube axis and, thus, orthogonal to $\boldsymbol{C}_k$. In terms of such definitions, integers n and m characterize the rolling directions, chirality and diameter $d = |\boldsymbol{C}_k|/\pi$ of a particular carbon nanotube, therefore, SWNTs are usually defined by these two numbers as (n, m).

15.2 Electronic properties in 1D systems

Translation symmetry is the main feature of solid states, which permits classification of the wavefunctions of any electronic states. According to the so-called Bloch theorem, the wavefunctions of a periodic system $\psi(\boldsymbol{r})$ are given as products of a periodic function $u_{nk}(\boldsymbol{r})$ and the exponential phase function $\exp(i\boldsymbol{kr})$, $\psi_{nk}(\boldsymbol{r}) = \exp(i\boldsymbol{kr})u_{nk}(\boldsymbol{r})$. The quantum number n is a property of the unit cell. The wavenumber, $\boldsymbol{k}$, is the main quantum number of the periodic systems, which satisfies the translation symmetry (Peierls, 1995). Thus, the state corresponding to any $\boldsymbol{k}$-number is the stationary state with the energy eigenvalue $E(\boldsymbol{k})$.

One-dimensional (1D) systems are the simplest models that possess the translation symmetry. Any symmetry breaking should perturb the energy spectrum $E(\boldsymbol{k})$. Indeed, for a 1D periodic potential with period described by the lattice constant a,

Quantum Effects in Biology, ed. Masoud Mohseni, Yasser Omar, Gregory S. Engel and Martin B. Plenio.
Published by Cambridge University Press.

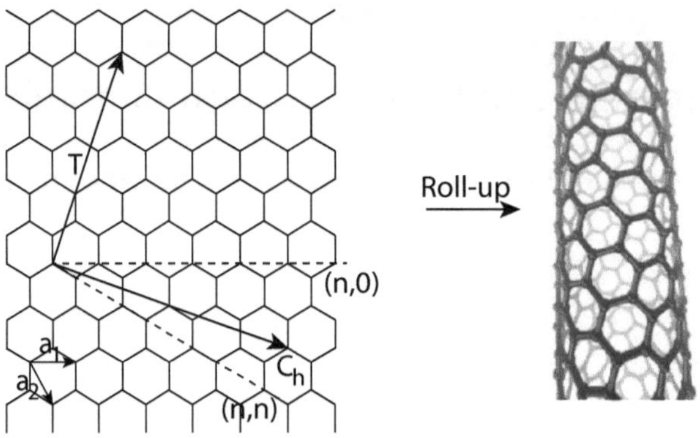

Figure 15.1 Relationship between the graphene sheet (left) characterized by unit vectors $\boldsymbol{a}_1$ and $\boldsymbol{a}_2$. Direction of the roll-up vector $\boldsymbol{C}_h$ and the related translation vector $\boldsymbol{T}$ defines the structural organization (either zigzag or armchair) of a particular SWNT (right).

the basic Brillouin zone is defined in the interval

$$-\frac{\pi}{a} < k < \frac{\pi}{a}. \tag{15.1}$$

In the presence of an additional periodic distortion with period s-times that of the lattice constant a, the reduced basic zone would be defined in the shorter interval of the k-space:

$$-\frac{\pi}{sa} < k < \frac{\pi}{sa}. \tag{15.2}$$

Because of this, the energy band of the unperturbed system splits into sub-bands separated by energy gaps caused by the periodic distortion. This circumstance can substantially change the properties of the 1D system. For instance, if the initial band is partly filled by electrons with total energy coinciding (exactly or nearly) with the edge of the sub-band after distortion of the system, then the lowest sub-bands are occupied and the upper sub-bands are empty. Thus, the mobile particles become immobile because of the distortion, and the system experiences the so-called Peierls instability (Peierls, 1995), which was originally attributed to the instability of the Fermi energy states of the 1D metals. The gain of energy, and hence magnitude of the distortion, are greatest when s is smallest. This means that in the case of one electron per unit cell of the 1D lattice and taking into account the electron spin, the most favourable case relates to the distortion corresponding to $s = 2$, i.e. when every second unit cell is displaced. The Peierls instability leads to the metal–semiconductor transition. The possibility of this transition in conjugated polymers (Heeger *et al.*, 1988) and metallic carbon nanotubes (Sato *et al.*, 1998; Figge *et al.*,

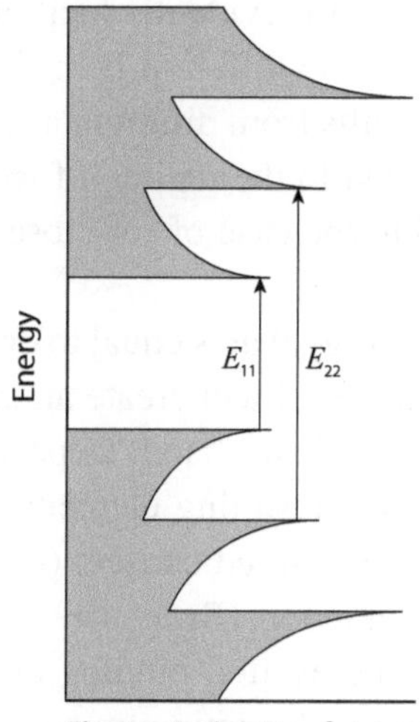

Figure 15.2 Schematic view of electronic density of states of the semiconducting SWNT. The energy gap separates the valance and conduction bands. Arrows indicate optical transitions between the corresponding van Hove singularities (indicated as E_{11} and E_{22}), which should be present in (quasi)-1D systems according to the tight-binding approximation.

2001) has been considered. In contrast with polymers the distortion-induced energy gap in SWNTs is small, smaller than the thermal fluctuations.

Two viewpoints are used to describe the electron energy spectrum of semiconducting nanotubes. The first, which originates from calculations in a so-called tight binding approximation, attributes the electronic spectral bands to the transitions between the van Hove singularities of the valence and conduction bands of (quasi)1D systems, as shown in Figure 15.2 (Dresselhaus *et al.*, 1996; Sato *et al.*, 1998). The other approach takes multi-particle correlation effects into account, such as, for instance, electron–hole Coulomb coupling, resulting in the exciton origin of the spectra (Spataru *et al.*, 2004a, b).

Both quasiparticles, electrons/holes or excitons can move through the system collectively, with the lattice distortion arranging the so-called solitons and/or polarons. The same type of excitation is also invoked by explaining the spectral properties and excitation dynamics in conjugated polymers (Su *et al.*, 1980; Heeger *et al.*, 1988). For semiconducting SWNTs, Peierls instability causing dimerization through the entire nanotube in the ground state is also predicted on the basis of semiempirical quantum chemical techniques (Tretiak *et al.*, 2007). However, the local effect caused by vibrational relaxation in the excited states produces a competitive mechanism of the exciton self-trapping and, thus, results in a vanishing of the Peierls instability. The self-trapping effect is evidently facilitated by the inhomogeneity of the environment of the SWNT. The different behaviour of the excitation in CNT and in conjugated polymers results from different strengths in the exciton–phonon coupling: the SWNTs are related to the regime of weak coupling, while the conjugated polymers correspond to the case of (or close to) strong coupling (Tretiak *et al.*, 2007).

The absorbed photon with energy that is equal to or larger than the characteristic bandgap of a semiconductor material will create an electron (e) in the conduction band and leave a hole (h) in the valence band. Depending on the magnitude of the Coulombic $e - h$ interaction, the resulting elementary excitation can either be a neutral exciton, or a pair of free charged carriers (e and h) with weak correlation caused by the Coulombic interaction (Brus, 1991). A well-defined measure of the Coulombic interaction is the exciton binding energy (E_b), quantified by the energy difference between the $e - h$ continuum and the corresponding exciton. If the exciton binding energy is larger than the thermal energy ($E_b \gg k_B T$), then the bound excitons are stable. On the other hand, if E_b is less than or comparable to $k_B T$, the excitons are not stable and dissociate promptly into charged carriers.

Effects caused by geometrical restrictions have been widely considered for the purpose of understanding the possible influence on the exciton binding energy and redistribution of oscillator strengths among various excitonic transitions (Brus, 1991; He, 1991; Ogawa and Takagahara, 1991; Alivisatos, 1996; Zhang and Mascarenhas, 1999). Considering the Coulomb $e - h$ interaction in a system of d-dimensions it was shown that the exciton binding energy and, correspondingly, the exciton radius in the lowest exciton state was equal to (He, 1991; Ogawa and Takagahara, 1991)

$$E_b = \left(\frac{2}{d-1}\right)^2 E_e, \tag{15.3}$$

$$r_{ex} = \left(\frac{d-1}{2}\right)^2 a_e, \tag{15.4}$$

where $E_e = \frac{\mu}{m_e \epsilon} Ry$ is the effective Rydberg constant and $a_e = \frac{m_e \epsilon}{\mu} a_B$ is the effective Bohr radius. Here m_e and μ are the electron mass and the reduced mass for the $e - h$ relative motion, respectively, ϵ is the dielectric constant and Ry and a_B are the Rydberg energy and the Bohr radius. For typical values of $\frac{m_e \epsilon}{\mu}$ the value of the effective Rydberg constant E_e is of the order of tens and hundreds of meV. According to Equations (15.3) and (15.4), in the case of $d = 3$, the exciton binding energy equals E_e and $r_{ex} = a_B$; in the case of $d = 2$, the exciton binding energy increases four times and the exciton radius decreases four times, correspondingly, while for $d = 1$ $E_b = \infty$ and $r_{ex} = 0$. Divergence of the exciton binding energy and the δ-type wavefunction in the case of $d = 1$ were already noted by Elliott and Loudon at the very beginning of exciton spectra studies (Elliott and Loudon, 1959, 1960). This could be understood as follows: in the cases $d = 2$ and $d = 3$, a charged particle can freely move around the origin of the Coulomb potential, while in the case of $d = 1$ it should move through the origin because of the spatial restriction.

The Coulomb attraction also exerts an influence on characteristics of the band–band transition (generation of unbound $e - h$ pairs) due to the so-called Sommerfeld factor, which is determined as the ratio of the oscillator strengths for the transition, calculated by taking into account the Coulomb correlation to the oscillator strength, calculated by neglecting this correlation (Ogawa and Takagahara, 1991). In the case of $d = 2$ and $d = 3$ this factor is always larger than unity, indicating that the Coulomb correlation enhances the corresponding optical transitions. However, in the case of $d = 1$, this factor is less than unity for the optically allowed transitions, and the transition strength is essentially concentrated in the transition to the lowest exciton state (Ogawa and Takagahara, 1991).

Physical properties of SWNTs depend on their diameter, chiral angle and aggregation state (Dresselhaus *et al.*, 1996; Sato *et al.*, 1998). The electronic band structure of the SWNT can display either metallic or semiconducting properties, as confirmed by calculations of the electronic band structure. As follows from calculations, based on a one-electron tight binding (Hückel-type approach) approximation, the SWNT is metallic if $(n - m)/3$ is an integer, otherwise it is a semiconductor. Because of the spatial quantization along a cross-section of the nanotube, the electron energy bands split into sub-bands (Figure 15.2) resembling a one-dimensional (1D) character of these materials.

15.3 Exciton–exciton interactions

Experimental studies of the exciton dynamics in SWNT employing ultra fast pump–probe and time-resolved fluorescence techniques reveal very fast exciton decay kinetics on the timescale of a few hundreds of femtoseconds (Ma *et al.*, 2008). As follows from the analysis of the decay behaviour of the kinetics and the dependence of the corresponding amplitude on excitation intensity, the occurrence of a predominant exciton–exciton annihilation process in semiconducting SWNTs was concluded (Ma *et al.*, 2004, 2005; Huang and Krauss, 2006; Valkunas *et al.*, 2006; Ma *et al.*, 2008; Abramavicius *et al.*, 2009). The remarkable intensity dependence of the photon-echo peakshift sets SWNTs apart from almost all other systems studied by this technique to date. A decrease in the photon-echo peakshift with increasing excitation intensity was reported for GaAs quantum wells as well (Carter *et al.*, 2007). The strong dependence of the initial peak shift value, the absence of dependence on population time (Graham *et al.*, 2008), and the marked deviation from the typically observed cubic dependence on excitation intensity found from the semiconducting SWNTs, all imply the existence of a non-linear dephasing mechanism dependent on the interaction of multiple excitons.

Theoretical description of exciton–exciton annihilation caused by correlations of multiple excitons in semiconductors is a challenging many-body problem (Axt and Mukamel, 1988; Huag and Koch, 2004). At the third order in the excitation field, description of a two-exciton scattering process must account for not only the Coulomb interactions between the electrons and holes but also for their Pauli exclusion. The simplest tight-binding model Hamiltonian of electrons and holes in a semiconductor (no phonons) in real space can be given by (Axt and Mukamel, 1988; Chernyak *et al.*, 1998)

$$\begin{aligned}\hat{H}_{eh} &= \sum_i \left(\varepsilon_i^{(c)}\hat{c}_i^\dagger\hat{c}_i + \varepsilon_i^{(d)}\hat{d}_i^\dagger\hat{d}_i\right)\\ &+ \sum_{ij}^{i\neq j}\left(V_{ij}^{(c)}\hat{c}_i^\dagger\hat{c}_j + V_{ij}^{(d)}\hat{d}_i^\dagger\hat{d}_j\right)\\ &+ \sum_{ij}\left(\frac{1}{2}W_{ij}^{(cc)}\hat{c}_i^\dagger\hat{c}_j^\dagger\hat{c}_j\hat{c}_i + \frac{1}{2}W_{ij}^{(dd)}\hat{d}_i^\dagger\hat{d}_j^\dagger\hat{d}_j\hat{d}_i - W_{ij}^{(cd)}\hat{c}_i^\dagger\hat{d}_j^\dagger\hat{d}_j\hat{c}_i\right). \qquad (15.5)\end{aligned}$$

Here, $\hat{c}_i^\dagger$ is an electron creation operator at site i, $\hat{d}_i^\dagger$ is a hole creation operator; the operators without daggers are the conjugate annihilation operators. The electron on-site energies are $\varepsilon_i^{(c)}$, while those of holes are $\varepsilon_i^{(d)}$. Electron and hole hopping parameters are $V_{ij}^{(c)}$ and $V_{ij}^{(d)}$, respectively. These parameters characterize non-interacting electrons and holes. The rest of the parameters are the Coulomb monopole–monopole interaction energies: $W_{ij}^{(cc)}$ between two electrons at sites i and j, $W_{ij}^{(dd)}$ between two holes, and $W_{ij}^{(cd)}$ between an electron and a hole. This Hamiltonian neglects exchange and four-point Coulomb integrals, however it captures the main exciton properties. The electrons and holes are fermions with non-zero anticommutators:

$$\begin{aligned}\hat{c}_i\hat{c}_j^\dagger + \hat{c}_j^\dagger\hat{c}_i &= \delta_{ij},\\ \hat{d}_i\hat{d}_j^\dagger + \hat{d}_j^\dagger\hat{d}_i &= \delta_{ij}. \qquad (15.6)\end{aligned}$$

The Hamiltonian of Equation (15.5) has been used to derive equations of motion for the electrons and holes in semiconductors, where each atom is taken as a site. Hierarchy of equations of motion for the electron and hole variables is obtained using the Heisenberg equation of motion. The hierarchy is truncated exactly at four-particle correlation functions when considering the third-order non-linear optical signals. In this case only two electron–hole pairs (bi-excitons) need to be considered. Such an approach has been successfully used for semiconductors with Hamiltonian parameters obtained from ab initio simulations (Axt and Mukamel, 1988; Khitrova *et al.*, 1999; Oszwaldowski *et al.*, 2005). Non-linearities of the final equations of motion originate from two sources: the Coulomb interaction and the Pauli exclusion coming from the fermion commutation relations.

This truncation scheme is valid for calculating the third-order non-linear excitation properties since the observed phenomena then include up to bi-exciton resonances. Higher-order non-linearities require including higher-order electron–electron correlations induced by the Pauli exclusion. This would result in a very complicated hierarchy of equations which may not be necessary if the experimental set-up cannot record such high-order exciton resonances (Abramavicius *et al.*, 2009). Instead of using the fermion particle commutation relations for deriving equations of motion, the Pauli exclusion requirements in real space may be mapped into a non-linear interaction potential of bosons as basic particles. While the boson commutation relations allow unphysical states, where two electrons (or two holes) as bosons may occupy the same state, such states may be excluded later by forcing their energies out of the physical detection window, e.g. to infinity. The anharmonic potential containing such divergencies is then solely responsible for non-linear signals. In the basis of delocalized excitons, we can treat the exciton bands as an anharmonic ladder of single-exciton, two-exciton, etc. delocalized exciton bands.

The starting point of such phenomenological modelling is the Hamiltonian of single-type bosons (resonant excitons, relevant to a specific problem), coupled to a phonon bath and an optical field $E(t)$:

$$\hat{H} = \sum_{mn}^{m \neq n} (\delta_{mn} h_{mm} + h'_{mn}(Q))(\hat{b}^\dagger)^m (\hat{b})^n - \hat{P} \cdot E(t) + \hat{H}_B(Q), \tag{15.7}$$

where $\hat{b}^\dagger$ is the exciton creation operator, and $\hat{b}$ is the conjugate annihilation operator. For bosons we have $[\hat{b}, \hat{b}^\dagger] = 1$. h_{00} is the ground state energy which can be selected arbitrarily and we set $h_{00} = 0$. h_{11} is the energy of a single exciton, h_{22} is the bi-exciton energy shift from a harmonic ladder, etc. $h'_{mn}(Q)$ describes the fluctuations induced by phonon coordinates Q, characterized by the bath Hamiltonian $\hat{H}_B(Q)$. These fluctuations induce pure dephasings via h'_{mm} and energy relaxation via h'_{mn}, $m \neq n$ between states m and n. High orders of h_{mm} determine non-linear exciton properties as they appear in equations for optical signals at mth order in the optical field. This demonstrates the level of complexity at strong laser intensities.

This model is sufficient for resonant optical signals, where the laser is tuned to a single system resonance. The interaction of the system with the optical field is represented by a polarization operator,

$$\hat{P} = \mu(\hat{b}^\dagger + \hat{b}), \tag{15.8}$$

where μ is the transition dipole.

The response of this system to the optical field can be described by using the perturbation theory with respect to the optical field. The Heisenberg equation is then used to determine the time evolution of the polarization operator, $i\hbar\dot{\hat{P}} = [\hat{P}, \hat{H}]$. Using the above definition of the polarization operator and bosonic commutation relations we obtain an equation for $\hat{b}$ (we neglect bath-related terms at the moment):

$$\dot{\hat{b}} = -\frac{i}{\hbar}\sum_m m h_{mm}(\hat{b}^\dagger)^{m-1}(\hat{b})^m + \frac{i}{\hbar}\mu \cdot E(t). \tag{15.9}$$

The induced polarization expectation value is then $P = \mu\langle\hat{b}\rangle + c.c.$ However, system non-linearities lead to an infinite hierarchy of coupled differential equations. These can be closed at a certain order in the incoming field, since at mth order only m particle correlations are important. So, for instance, at the third order in the field, only $m = 2$ and terms $\langle\hat{b}^\dagger\hat{b}\hat{b}\rangle$ are necessary to be considered and all higher-order terms can be safely neglected.

When exciton multiple quanta resonant frequencies are not directly observable, the mean field approximation (MFA) is convenient: we factorize all correlation terms as follows,

$$\langle(\hat{b}^\dagger)^m(\hat{b})^n\rangle = \langle(\hat{b}^\dagger)^m\rangle\langle(\hat{b})^n\rangle \equiv (b^*)^m(b)^n, \tag{15.10}$$

where we denote $b \equiv \langle\hat{b}\rangle$. In the MFA we then get a single equation of motion,

$$\dot{b} = -\frac{i}{\hbar}b(h_{11} + 2h_{22}n + 3h_{33}n^2 + \cdots) + \frac{i}{\hbar}\mu \cdot E(t). \tag{15.11}$$

Note that here, $n \equiv |b|^2 = \langle\hat{b}^\dagger\hat{b}\rangle = \langle\hat{n}\rangle$ is the number of quanta, so the higher non-linear terms become necessary when a larger number of particles is present. Moreover, we may interpret the brackets as the series expansion of a non-linear potential carrying information on how the system energy behaves with the number of particles. If we consider highly excited states, a continuous function may be used instead. At the third order in the field, by including the Markovian relaxation the following equation of motion is obtained (Axt and Mukamel, 1988; Chernyak *et al.*, 1998):

$$\dot{b} = -i\left(\varepsilon - i\frac{\gamma}{2}\right)b - i\left(\Delta - i\frac{3}{2}\bar{\gamma}\right)b|b|^2 + i\mu \cdot E(t), \tag{15.12}$$

where $\Delta = 2h_{22}$ is the so-called biexciton binding energy, γ is the linear exciton decay rate, and $\bar{\gamma}$ stands for the non-linear decay of excitons. These rates can be derived using second-order perturbation theory in the Markovian approximation from the off-diagonal fluctuation parameters according to the Fermi golden rule:

$$\gamma = \frac{2\pi}{\hbar}|h'_{10}|^2\rho(\varepsilon), \tag{15.13}$$

$$\bar{\gamma} = \frac{2\pi}{\hbar}|h'_{21}|^2\rho(\varepsilon + \Delta), \tag{15.14}$$

where $\rho(\omega)$ represents the bath spectral density. The term proportional to $\bar{\gamma}$ is related to $\hat{b}^\dagger\hat{b}\hat{b}$ and thus describes the non-linear decay of two excitons into a single exciton, i.e. it reflects, e.g. exciton–exciton annihilation.

15.4 Non-linear optical response of excitons

We consider a four-wave mixing non-linear signal performed using three excitation optical laser fields. The excitation field then consists of three Gaussian pulses,

$$E(t) = \sum_j \mathcal{E}_j(t - \tau)\exp(i\boldsymbol{k}_j\boldsymbol{r} - i\omega_j(t - \tau_j)) + c.c., \tag{15.15}$$

where $\boldsymbol{k}$ and τ_j are the wavevector and the central time of pulse j with a Gaussian envelope $\mathcal{E}_j(t - \tau)$. For the case of short-pulse excitations it is useful to consider chronologically ordered pulses with $\boldsymbol{k}_1$ incoming first, and the second pulse $\boldsymbol{k}_2$ arrives delayed by time τ (the delay between pulses 1 and 2). The third pulse $\boldsymbol{k}_3$ comes after the second pulse delayed by T (the delay between 2 and 3). This set of pulses generates a four-wave-mixing signal characterized by the induced non-linear polarization, $P(t)$. Time t is measured starting with the third pulse. The polarization is given by the expectation value of the polarization operator, defined by Equation (15.8) and its time evolution may be calculated from the exciton equation of motion for a certain configuration of pulse delays.

Because of non-linear system properties the incoming optical pulses are mixed and the signal is generated in all possible directions $u\boldsymbol{k}_1 + v\boldsymbol{k}_2 + w\boldsymbol{k}_3$, where u, v, w are integer numbers determining the signal configuration. If we consider the perturbation expansion in the field, we need at least $|u| + |v| + |w|$ order perturbation theory in the field to generate such a signal. However, higher orders may contribute to this signal, which causes dependence on the excitation intensity. The commonly studied photon-echo signal is detected in the $-\boldsymbol{k}_1 + \boldsymbol{k}_2 + \boldsymbol{k}_3$ phase matching signal direction (Mukamel, 1995), which requires at least the third order in the field.

The numerical non-perturbative propagation of Equation (15.12) can be performed to calculate the induced polarization at a given excitation configuration, defined by the delay times (τ, T, t). However, the numerically propagated polarization now mixes all phase matching signals. The specific signal, defined by (u, v, w), can be extracted using the phase cycling procedure (Tian *et al.*, 2003), which is equivalent to the spatial Fourier transformation. In general, the phase-cycling procedure to extract Nth order contributions is as follows. A 2π interval of a jth pulse phase $\Phi_j = \boldsymbol{k}_j \boldsymbol{r}$ is divided into N points, $\Phi_{js} = 2\pi s/N$, where $s = 0, 1, \ldots, N-1$. The time evolution of the variable b is calculated for a certain pulse delay configuration (τ, T, t) according to Equation (15.12) as a function of phases of the three pulses, $b(\{\Phi_{js_j}\}, \tau, T, t)$. The $u\boldsymbol{k}_1 + v\boldsymbol{k}_2 + w\boldsymbol{k}_3$ phase matching contribution to b may then be given as a discrete Fourier transformation,

$$\bar{b}_{u,v,w}(\tau, T, t) = \sum_{s_1 s_2 s_3} b(\Phi_{1s_1}, \Phi_{2s_2}, \Phi_{3s_3}, \tau, T, t) \times \exp(-iu\Phi_{1s_1} - iv\Phi_{2s_2} - iw\Phi_{3s_3}). \tag{15.16}$$

For instance, the photon-echo signal is given by $(u, v, w) = (-1, 1, 1)$ and using Equation (15.16) it can be calculated at any order in the incoming field as $P_k(\tau, T, t) \equiv \mu \bar{b}_{-1,1,1}(\tau, T, t)$.

The parameters γ and $\bar{\gamma}$ entering the exciton equation of motion characterize the exciton decay of a single nanotube, and thus induce homogeneous spectral broadenings. Considering an ensemble of nanotubes, we have to account for structural fluctuations on various timescales. The fast fluctuations are responsible for motional narrowing (this would appear as additional homogeneous spectral linewidth) and the static fluctuations (in the limit of ergodicity) induce inhomogeneous broadening. Considering the exciton model Hamiltonian, these fluctuations may be mapped onto *diagonal* fluctuation parameters h'_{11}. Using the time correlation functions of these fluctuations, their action is conveniently expressed through the lineshape functions $g(t)$ by cumulant expansion techniques (Mukamel, 1995). In this model the statistical properties of the fluctuations are characterized by the transition frequency fluctuation correlation function,

$$\langle \tilde{\varepsilon}(t)\tilde{\varepsilon}(0)\rangle = 2\Gamma\delta(t) + \sigma_e^2 \exp(-t/\tau_e), \tag{15.17}$$

where $t > 0$ and τ_e is the correlation time of slow (not necessarily absolutely static; this is supported by experiments of photon-echo) fluctuations. Here Γ and σ_e are the strengths of fast (Markovian) and slow fluctuations, respectively. The lineshape function for this type of fluctuations can be calculated analytically and is

$$g(t) = \Gamma t + (\sigma_e \tau_e)^2[\exp(-t/\tau_e) + t/\tau_e - 1]. \tag{15.18}$$

When $\tau_e > \sigma_e^{-1}$, this model leads to an absorption full linewidth of $\sigma_e\sqrt{8\ln(2)}$. The third-order induced polarization at the photon-echo phase-matching direction for the real lineshape function may be written in the form (Mukamel, 1995),

$$P_s(\tau, T, t) = P_h(\tau, T, t)\exp[-g(\tau) - g(t) + g(T) - g(t+T) - g(T+\tau) + g(t+T+\tau)] \tag{15.19}$$

where $P_h(\tau, T, t)$ and $P_s(\tau, T, t)$ are the homogeneous and inhomogeneous signals, respectively. This expression thus allows us to estimate the statistical properties of the ensemble of nanotubes, which corresponds to experimental conditions of nanotube solutions.

An example of the non-linear optical signal is three-pulse-photon-echo peakshift (3PEPS) spectroscopy, which is a time-integrated intensity signal at a certain pulse delay configuration. The photon-echo intensity is collected as a function,

$$I_s(T, \tau) = \int dt |P_s(\tau, T, t)|^2. \tag{15.20}$$

In simulations this signal needs to be calculated for a grid of T and τ values, and the peakshift $\tau^*(T)$ is extracted as the value of τ, where $I_s(T, \tau)$ has its maximum amplitude, as a function of T. The same formalism can be applied to simulate more complicated non-linear signals at high excitation intensities, such as pump–probe or two-dimensional photon-echo.

15.5 Simulations of intensity-dependent 3PEPS

The simple non-linear oscillator model for carbon nanotubes described above can be used to model exciton annihilation experiments (Abramavicius *et al.*, 2009). In the 3PEPS experiment the central laser frequency is tuned to the main transition frequency of the specific exciton, $\omega_0 = \varepsilon$. The central exciton frequency has been chosen at the absorption maximum $\varepsilon = 9800\ \text{cm}^{-1}$. The bi-exciton resonance shift Δ for estimation of the annihilation effect can be neglected since the bi-exciton binding effect in a 3PEPS signal is a secondary unresolvable effect compared with the exciton annihilation at high intensities. The dephasing parameter γ and the annihilation parameter $\bar{\gamma}$ are fitted to reflect the experimentally observed exciton decay pattern: $\gamma/\varepsilon = 5.4 \times 10^{-5}$ (10 ps timescale) and $\bar{\gamma}/\varepsilon = 6.75 \times 10^{-4}$ (800 fs timescale) (Wang *et al.*, 2004a), respectively. The inhomogeneous linewidth is fitted to the experimental absorption $\sigma_e/\varepsilon = 0.02$ and the diagonal fluctuation parameters τ_c and Γ tuned to best reproduce the experimental 3PEPS data: $\tau_c\varepsilon = 4000$, $\Gamma/\varepsilon = 4.5 \times 10^{-3}$.

The annihilation rate $\bar{\gamma}$ controls the system susceptibility at different excitation intensities via the exciton equation of motion, Equation (15.12). When the ratio

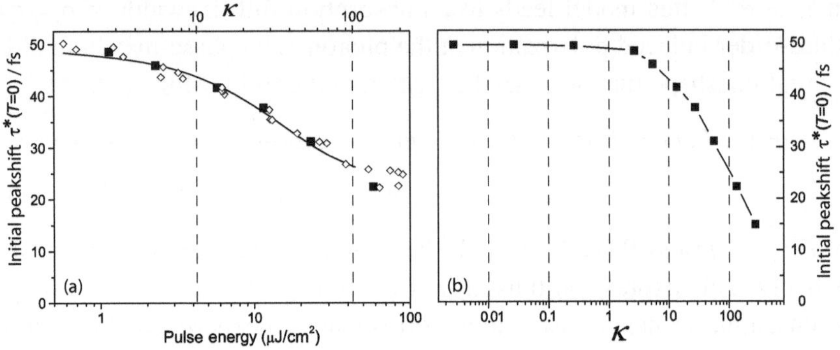

Figure 15.3 The simulated initial peakshift value $\tau^*(T = 0)$ as a function of κ (black squares). (a) Comparison with experiment: κ is depicted on the top scale of the plot, experimental laser intensity is on the bottom scale; the experimental data (open diamonds) are from the SWNT-PVP film (Abramavicius *et al.*, 2009) and the solid line is the exponential fit. (b) The simulated peak shift in the broad range of excitation intensities κ: $\kappa = 1$ separates the two intensity regimes.

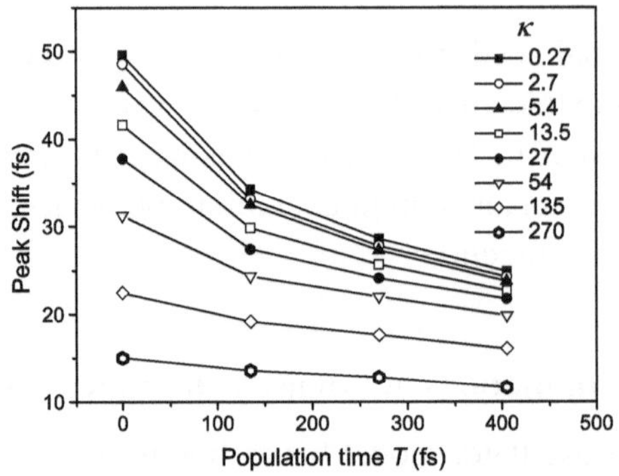

Figure 15.4 Simulated 3PEPS decays of the SWNT consistent with data points of Figure 15.3a. The lines connecting the symbols are for visual purposes.

$\bar{\gamma}|b|^2/\gamma < 1$, the linear decay dominates and the annihilation can be neglected. This ratio depends on the exciton population, given by $|b|^2$. Thus, at high field intensities when the exciton populations are large, the inequality is reversed and the annihilation becomes a dominating effect. This ratio can be approximately estimated by the linear response theory, which induces a first-order exciton response, $b^{(1)}$, thus, giving the maximum at $|b^{(1)}| = \mu \int d\tau \mathcal{E}_1(\tau)$. By approximating the polarization decay as σ_e^{-1} (σ_e is usually much larger than the homogeneous decay rate), we introduce the intensity parameter $\kappa = \frac{\bar{\gamma}}{\sigma_e}\mu^2(\int d\tau \mathcal{E}_1(\tau))^2$. The regime where exciton annihilation becomes important is uniquely defined by the excitation intensity and the ratio of the annihilation rate to the full linewidth.

In Figures 15.3 and 15.4 we present the calculated dependence of the peakshift on the pulse intensity κ (black squares). Figure 15.3 clearly shows transition behaviour from an intensity-independent regime $\kappa < 1$ into a regime where the peakshift rapidly decreases when $\kappa > 1$. The transition around $\kappa = 1$ is the point where exciton population decay rate becomes comparable with the total absorption linewidth. The range of κ values from 1.2 to 230 shows very good agreement with the experiment. Note that this comparison allows us to estimate the number of excitons participating in the signal. The calculated 3PEPS decays at various excitation intensities are shown in Figure 15.4. These decays describe the trend of the observed intensity dependence rather well (Abramavicius *et al.*, 2009).

15.6 Discussion and conclusions

The presented experiment and simulation data demonstrate a striking effect of photon-echo peakshift dependence on laser intensity. According to the non-linear spectroscopy theory, the third-order induced polarization amplitude scales cubically with the incoming laser electric field amplitude and other system parameters are constant (Chapter 4). Thus, at the third order in the field, the 3PEPS signal does not depend on the laser intensity. The experimental observation of highly non-trivial intensity dependence of 3PEPS implies that the response of SWNT to three laser pulses is not limited to the third order. Instead, the non-polynomial dependence of the induced polarization amplitude on the laser intensity makes it impossible to define a perturbative order of the signal. We, thus, used a general scheme based on phase cycling, which enabled us to extract a phase matching signal at a broad range of intensities.

Our model captures the main feature of the experimental findings – the correct trend of intensity dependence of the peakshift, and also offers insights into the physical mechanism underlying this unusual phenomenon. The phenomenological interacting-boson model built directly for delocalized (coherent) excitons is thus advantageous for explaining the underlying physical phenomena. More sophisticated theoretical models are available to describe dephasing and energy relaxation (Chapter 2, for an extended description of the relaxation theory, the reader is referred to Valkunas *et al.* (2013)), however a direct phenomenological approach of ours allows us to directly expose the origin of the non-linear effect. Qualitative agreement of the simulations and experiment demonstrates that at high excitation intensities the system behaves as a simple non-linearly damped oscillator.

An important question that we may address is how many excitations can be pumped into the SWNT before annihilation turns in? From the κ parameter definition we have $n \equiv |b|^2 = \kappa\sigma_e/\bar{\gamma}$. According to our parameter set, which fits the experiments, $\kappa = 1$ corresponds to ~30 excitations in a single nanotube. At these numbers the annihilation starts to affect the 3PEPS signal. Figure 15.3a corresponds to conditions of ~50–300 excitations per SWNT. Thus, a small number of excitons barely 'see' each other and only at ∼50 excitations per nanotube we observe the 'crowding' phenomenon. This implies that at excitons in SWNT weakly interact and the material is an efficient excitation reservoir.

The 3PEPS signal measures dephasing dynamics of the material. More advanced two-dimensional (2D) photon-echo (PE) spectroscopy has the potential to detect quantum coherences, their dephasing and energy relaxation effects (Mukamel, 2000; Valkunas *et al.*, 2013). Based on 2D PE experiments (Engel *et al.*, 2007; Graham *et al.*, 2012) (see also Chapter 6) it has been recently proposed that quantum transport may be driving efficient energy transfer in biological photosynthetic pigment–protein complexes (PPC) (Chapter 7). Quantum transport introduces a specific oscillating character into non-equilibrium state populations via quantum coherences, which maps into diagonal peak oscillations of the 2D PE plots (Abramavicius and Mukamel, 2010).

A concept of excitonic entanglement in the single-exciton manifold has been introduced to quantify coherence in PPC (Chapter 13) (Sarovar *et al.*, 2010). The meaning is essentially equivalent to exciton delocalization, however, it has the additional dynamical aspect, which is related to loss of coherence due to coupling to environment. Note that in the interacting quantum system, an eigenstate wavefunction is always a superposition of some orthogonal basis wavefunctions; such wavefunction entanglement can be removed by proper unitary transformation. This dynamic entanglement in the single-exciton manifold is thus a novel concept. The origins of quantum entanglement are reviewed in Graham *et al.* (2012). The entanglement is defined for non-interacting quantum systems (Einstein *et al.*, 1935): if such systems had some contact before or were specifically prepared, the total wavefunction of the combined system may be impossible to write as a product of wavefunctions belonging solely to these subsystems. Such a state of the combined system is called the entangled state.

As two excitons effectively do not interact in SWNT ($\kappa \ll 1$), it may be possible to observe certain quantum phenomena of excitons in these materials at low excitation levels. The specifically tuned optical field may create a pair of entangled excitons which then propagate quasi-independently, when the field is switched off. That, in principle, may be used for quantum information processing as follows. The detectors may be placed at the ends of the SWNT and detection of one excitation at one end would automatically tell the state of the other exciton if they were entangled, and the total wavefunction are known. This hypothetical data processing unit could also be designed in photosynthetic systems using at least the double-exciton level, which may be achieved by the non-linear optical spectroscopy (Mukamel, 2010). However, the double-excitons in photosynthetic systems lose coherence very quickly due to strong interaction with protein fluctuations. In that sense a SWNT has 'cleaner' quantum states and may be utilized more easily.

Recent experiments have also demonstrated the important role of vibrations which, coupled with electronic degrees of freedom, may be involved in coherence and energy transport in molecular systems (Nemeth *et al.*, 2010; Christensson *et al.*, 2011). Both electronic coherences and vibrational motion result in amplitude beats in the 2D PE spectra, however, possibilities for disentangling different origins of spectral beats have been studied (Tiwari *et al.*, 2013; Butkus *et al.*, 2012). Nanotube structure vibration may be an important contribution in coherent and dissipation dynamics of excitons in carbon nanotubes (Galland *et al.*, 2008; Graham *et al.*, 2012).

15.7 Acknowledgement

L.V. acknowledges the support of EP7 (Marie Curie) project PITN-GA-2012-316633 (POCAONTAS). D.A. acknowledges the support of the European Social Fund under the Global Grant Measure (No. VP1-3.1-ŠMM-07-K-01-020).

References

Abramavicius, D., and Mukamel, S. 2010. Quantum oscillatory exciton migration in photosynthetic reaction centers. *Journal of Chemical Physics*, **133**, 064510.

Abramavicius, D., Ma, Y.-Z., Graham, M. W., Valkunas, L., and Fleming, G. R. 2009. Dephasing in semiconducting single-walled carbon nanotubes induced by exciton-exciton annihilation. *Physical Review B*, **79**, 195445.

Adams, D. M., Brus, L., Chidsey, C. E. D., Creager, S., Creutz, C., Kagan, C. R., Kamat, P. V., Lieberman, M., Lindsay, S., Marcus, R. A., Metzger, R. M., Michel-Beyerle, M. E., Miller, J. R., Newton, M. D., Rolison, D. R., Sankey, O., Schanze, K. S., Yardley, J., and Zhu, X. 2003. Charge transfer on the nanoscale: current status. *Journal of Physical Chemistry B*, **107**, 6668–6697.

Adolphs, J., and Renger, T. 2006. How proteins trigger excitation energy transfer in the FMO complex of green sulfur bacteria. *Biophysical Journal*, **91**, 2778–2797.

Agranovich, V. M., and Galanin, M. D. 1982. *Modern Problems in Condensed Matter Sciences: Vol.3, Electronic Excitation Energy Transfer in Condensed Matter*. Amsterdam: North-Holland.

Agranovich, V. M., and Hochstrasser, E. R. M. 1982. *Modern Problems in Condensed Matter Sciences: Vol.4, Spectroscopy and excitation dynamics of condensed molecular systems*. Amsterdam: North-Holland.

Ahmad, M., Galland, P., Ritz, T., Wiltschko, R., and Wiltschko, W. 2007. Magnetic intensity affects cryptochrome-dependent responses in *Arabidopsis thaliana*. *Planta*, **225**, 615–624.

Aida, T., Meijer, E. W., and Stupp, S. I. 2012. Functional supramolecular polymers. *Science*, **335**(6070), 813–817.

Alberts, B., Johnson, A., Lewis, J., Raff, M., Roberts, K., and Walter, P. 2008. *Molecular Biology of the Cell*. Fifth edn. New York: Garland Science, Taylor & Francis.

Alden, R. G., Parson, W. W., Chu, Z. T., and Warshel, A. 1996. Orientation of the OH dipole of tyrosine (M)210 and its effect on electrostatic energies in photosynthetic bacterial reaction centers. *Journal of Physical Chemistry*, **100**(41), 16761–16770.

Alicki, R., and Lendi, K. 1987. *Quantum Dynamical Semigroups and Applications*. Berlin: Springer-Verlag.

Alivisatos, A. P. 1996. Semiconductor clusters, nanocrystals, and quantum dots. *Science*, **271**, 933–937.

Allemann, R. K., and Scrutton, N. S. 2009. *Quantum Tunnelling in Enzyme-Catalysed Reactions*. Cambridge: Royal Society of Chemistry Publishing.

Andresen, E. R., and Hamm, P. 2009. Site-specific difference 2D-IR spectroscopy of bacteriorhodopsin. *Journal of Physical Chemistry B*, **113**, 6520–6527.

Andrews, D. L., and Demidov, A. A. (eds). 1999. *Resonance Energy Transfer*. Chichester: John Wiley & Sons.

Andrews, D. L., Crutchet, C., and Scholes, G. D. 2011. Resonance energy transfer: Beyond the limits. *Lasers and Photonics Reviews*, **5**, 114.

Aratani, N., Cho, H. S., Ahn, T. K., Cho, S., Kim, D., Sumi, H., and Osuka, A. 2003. Efficient excitation energy transfer in long meso-meso linked Zn(II) porphyrin arrays bearing a 5,15-bisphenylethynylated Zn(II) porphyrin acceptor. *Journal of the American Chemical Society*, **125**, 9668.

Arlt, T., Schmidt, S., Kaiser, W., Lauterwasser, C., Meyer, M., Scheer, H., and Zinth, W. 1993. The accessory bacteriochlorophyll – a real electron carrier in primary photosynthesis. *Proceedings of the National Academy of Sciences, USA*, **90**(24), 11757–11761.

Arndt, M., Nairz, O., Vos-Andreae, J., Keller, C., van der Zouw, G., and Zeilinger, A. 1999. Wave-particle duality of C-60 molecules. *Nature*, **401**, 680–682.

Arnett, D. C., Moser, C. C., Dutton, P. L., and Scherer, N. F. 1999. The first events in photosynthesis: electronic coupling and energy transfer dynamics in the photosynthetic reaction center from *Rhodobacter sphaeroides*. *Journal of Physical Chemistry B*, **103**(11), 2014–2032.

Atkins, P., and de Paula, J. 2002. *Atkin's Physical Chemistry*. New York: Oxford University Press.

Atkins, P., and de Paula, J. 2009. *Physical Chemistry*. Ninth edn. Oxford: Oxford University Press.

Atkins, P. W., and Friedman, R. S. 1999. *Molecular Quantum Mechanics*. Oxford: Oxford University Press.

Axt, V. M., and Mukamel, S. 1988. Nonlinear optics of semiconductor and molecular nanostructures; a common perspective. *Reviews of Modern Physics*, **70**, 145–174.

Bagryansky, V. A., Usov, O. M., Borovkov, V. I., Kobzeva, T. V., and Molin, Yu. N. 2000. Quantum beats in recombination of spin-correlated radical ion pairs with equivalent protons. *Chemical Physics Letters*, **255**, 237–245.

Bagryansky, V. A., Borovkov, V. I., and Molin, Y. N. 2007. Quantum beats in radical pairs. *Russian Chemical Reviews*, **76**, 493–506.

Balaban, T. S., Tamiaki, H., and Holzwarth, A. R. 2005. Chlorins programmed for self-assembly. *Topics in Current Chemistry*, **258**, 1.

Balabin, I. A., and Onuchic, J. N. 2000. Dynamically controlled protein tunneling paths in photosynthetic reaction centers. *Science*, **290**, 114–117.

Balabin, I. A., Beratan, D. N., and Skourtis, S. S. 2008. The persistence of structure over fluctuations in biological electron-transfer reactions. *Physical Review Letters*, **101**, 158102.

Balasubramanian, G., Chan, I. Y., Kolesov, R., Al-Hmoud, M., Tisler, J., Shin, C., Kim, C., Wojcik, A., Hemmer, P. R., Krueger, A., Hanke, T., Leitenstorfer, A., Bratschitsch, R., Jelezko, F., and Wrachtrup, J. 2008. Nanoscale imaging magnetometry with diamond spins under ambient conditions. *Nature*, **455**(7213), 648–651.

Balzani, V., Piotrowiak, P., Rodgers, M. A. J., Mattay, J., and Astruc, D. (eds). 2001. *Electron Transfer in Chemistry Vols. I-V*. Weinheim: Wiley-VCH.

Batchelor, S. N., Kay, C. W. M., McLauchlan, K. A., and Shkrob, I. A. 1993. Time-resolved and modulation methods in the study of the effects of magnetic fields on the yields of free-radical reactions. *Journal of Physical Chemistry*, **97**, 13250–13258.

Beason, R. C., and Semm, P. 1996. Does the avian ophthalmic nerve carry magnetic navigational information? *Journal of Experimental Biology*, **199**, 1241–1244.

Beekman, L. M. P., van Stokkum, I. H. M., Monshouwer, R., Rijnders, A. J., McGlynn, P., Visschers, R. W., Jones, M. R., and van Grondelle, R. 1996. Primary electron transfer in membrane-bound reaction centers with mutations at the M210 position. *Journal of Physical Chemistry*, **100**(17), 7256–7268.

Béjà, O., Aravind, L., Koonin, E. V., Suzuki, M. T., Hadd, A., Nguyen, L. P., Jovanovich, S. B., Gates, C. M., Feldman, R. A., Spudich, J. L., Spudich, E. N., and DeLong, E. F. 2000. Bacterial rhodopsin: evidence for a new type of phototrophy in the sea. *Science*, **289**, 1902–1906.

Beljonne, D., Hennebicq, E., Daniel, C., Herz, L. M., Silva, C., Scholes, G. D., Hoeben, F. J. M., Jonkheijm, P., Schenning, A. P. H. J., Meskers, S. C. J., Phillips, R. T., Friend, R. H., and Meijer, E. W. 2005. Excitation migration along oligophenylenevinylene-based chiral stacks: delocalization effects on transport dynamics. *Journal of Physical Chemistry B*, **109**, 10594.

Ben-Shem, A., Frolow, F., and Nelson, N. 2003. Crystal structure of plant photosystem I. *Nature*, **426**(6967), 630–635.

Bendall, D. S. (ed). 1996. *Protein Electron Transfer*. Oxford, UK: BIOS Scientific Publishers.

Bennett, C. H., and DiVincenzo, D. P. 2000. Quantum Information and Computation. *Nature*, **404**, 247–255.

Beratan, D. N., Betts, J. N., and Onuchic, J. N. 1991. Protein electron transfer rates set by bridging secondary and tertiary structure. *Science*, **252**, 1285–1288.

Beratan, D. N., Skourtis, S. S., Balabin, I. A., Balaeff, A., Keinan, S., Venkatramani, R., and D., Xiao. 2009. Steering electrons on moving pathways. *Accounts of Chemical Research*, **42**, 1669–1678.

Berg, J. M., Tymoczko, J. L., and Stryer, L. 2002. *Biochemistry*. 5th edn. New York: W. J. Freeman.

Berneche, S., and Roux, B. 2001. Energetics of ion conduction through the K^+ channel. *Nature*, **414**, 73–77.

Bezanill, F., and Armstrong, C. M. 1972. Negative conductance caused by entry of sodium and cesium ions into potassium channels of squid axons. *Journal of General Physiology*, **60**, 588–608.

Birge, R. R. 1990. Nature of the primary photochemical event in rhodopsin and bacteriorhodopsin. *Biochimica et Biophysica Acta*, **1016**, 293–323.

Birks, E. J. B. 1976. *Excited States of Biological Molecules*. London: John Wiley & Sons.

Biskup, T., Schleicher, E., Okafuji, A., Link, G., Hitomi, K., Getzoff, E. D., and Weber, S. 2009. Direct observation of a photoinduced radical pair in a cryptochrome blue-light photoreceptor. *Angewandte Chemie – International Edition in English*, **48**, 404–407.

Bittl, R., and Kothe, G. 1991. Transient EPR of radical pairs in photosynthetic reaction centers – prediction of quantum beats. *Chemical Physics Letters*, **177**, 547–553.

Bittner, E. R., Madalan, A., Czader, A., and Roman, G. 2012. Quantum origins of molecular recognition and olfaction in Drosophila. *The Journal of Chemical Physics*, **137**(22), 22A551.

Blankenship, R. E. 2002. *Molecular Mechanisms of Photosynthesis*. Oxford, UK: Blackwell Science.

Blankenship, R. E., M., Tiede D., Barber, J., Brudvig, G. W., Fleming, G., Ghirardi, M., Gunner, M. R., Junge, W., Kramer, D. M., Melis, A., Moore, T. A., Moser, C. C., Nocera, D. G., Nozik, A. J., Ort, D. R., Parson, W. W., Prince, R. C., and Sayre, R. T. 2011. Comparing photosynthetic and photovoltaic efficiencies and recognizing the potential for improvement. *Science*, **13**, 805–809.

Bonačić-Koutecky, V., Kohler, J., and Michl, J. 1984. Prediction of structural and environmental effect on the S1-S0 energy gap and jump probability in double-bond cis-trans photoisomerization. A general rule. *Chemical Physics Letters*, **104**, 440–443.

Bopp, M. A., Sytnik, A., Howard, T. D., Cogdell, R. J., and Hochstrasser, R. M. 1999. The dynamics of structural deformations of immobilized single light-harvesting complexes. *Proceedings of the National Academy of Sciences, USA*, **96**(20), 11271–11276.

Bouly, J.-P., Schleicher, E., Dionisio-Sese, M., Vandenbussche, F., Van Der Straeten, D., Bakrim, N., Meier, S., Batschauer, A., Galland, P., Bittl, R., and Ahmad, M. 2007. Cryptochrome blue light photoreceptors are activated through interconversion of flavin redox states. *Journal of Biological Chemistry*, **282**, 9383–9391.

Bouwmeester, D., Ekert, A., and Zeilinger, A. (eds). 2000. *The Physics of Quantum Information*. Berlin: Springer Verlag.

Boxer, S. G. 1996. Stark spectroscopy of photosynthetic systems. Pages 177–189 in: Hoff, J. Amesz, and J., A. (eds), *Biophysical Techniques in Photosynthesis*. Kluwer Academic Publishers.

Bradforth, S. E., Jimenez, R., van Mourik, F., van Grondelle, R., and Fleming, G. R. 1995. Excitation transfer in the core light-harvesting complex (LH-1) of *Rhodobacter sphaeroides* – an ultrafast fluorescence depolarization and annihilation study. *Journal of Physical Chemistry*, **99**(43), 16179–16191.

Bradler, K., Wilde, M. M., Vinjanampathy, S., and Uskov, D. B. 2010. Identifying the quantum correlations in light-harvesting complexes. *Physical Review A*, **82**, 062310.

Brautigam, C. A., Smith, B. S., Ma, Z., Palnitkar, M., Tomchick, D. R., Machius, M., and Deisenhofer, J. 2004. Structure of the photolyase-like domain of cryptochrome 1 from *Arabidopsis thaliana*. *Proceedings of the National Academy of Sciences, USA*, **101**, 12142–12147.

Brecht, M., Radics, V., Nieder, J. B., and Bittl, R. 2009. Protein dynamics-induced variation of excitation energy transfer pathways. *Proceedings of the National Academy of Sciences, USA*, **106**, 11857–11861.

Brettel, K., and Byrdin, M. 2010. Reaction mechanisms of DNA photolyase. *Current Opinion in Structural Biology*, **20**, 693–701.

Breuer, H. P., and Petruccione, F. 2002. *The Theory of Open Quantum Systems*. New York: Oxford Univerity Press.

Breuer, H.-P., Laine, E. M., and Piilo, J. 2009. Measure for the degree of non-Markovian behavior of quantum processes in open systems. *Physical Review Letters*, **103**, 210401.

Briegel, H. J., and Popescu, S. 2013. Intra-molecular refrigeration in enzymes. *Proceedings of the Royal Society* A, **469**, 20110290.

Brinks, D., Stefani, F. D., Kulzer, F., Hildner, R., Taminiau, T. H., Avlasevich, Y., Müllen, K., and van Hulst, N. F. 2010. Visualizing and controlling vibrational wave packets of single molecules. *Nature*, **465**, 905.

Brixner, T., Mančal, T., Stiopkin, I. V., and Fleming, G. R. 2004. Phase-stabilized two-dimensional electronic spectroscopy. *Journal of Chemical Physics*, **121**(9), 4221.

Brixner, T., Stenger, J., Vaswani, H. M., Cho, M., Blankenship, R. E., and Fleming, G. R. 2005. Two-dimensional spectroscopy of electronic couplings in photosynthesis. *Nature*, **434**, 625–628.

Brocklehurst, B. 1976. Spin correlation in geminate recombination of radical ions in hydrocarbons. 1. Theory of magnetic-field effect. *Journal of the Chemical Society, Faraday Transactions II*, **72**, 1869–1884.

Brocklehurst, B. 2002. Magnetic fields and radical reactions: recent developments and their role in Nature. *Chemical Society Reviews*, **31**, 301–311.

Brookes, J. C. 2008. A microscopic model of signal transduction mechanisms: olfaction. Ph.D. thesis, University College London.

Brookes, J. C. 2010. Science in perception: what can our sense of smell tell us about ourselves and the world around us? *Philosophical Transactions of the Royal Society of London A. (Mathematical, Physical and Engineering Sciences)*, **368**, 3491–3502.

Brookes, J. C., Hartoutsiou, F., Horsfield, A. P., and Stoneham, A. M. 2007. Could humans recognize odor by phonon assisted tunneling? *Physical Review Letters*, **98**, 038101.

Browne, W. R., and Feringa, B. L. 2006. Making molecular machines work. *Nature Nanotechnology*, **1**, 25–35.

Brus, L. 1991. Quantum crystallites and nonlinear optics. *Applied Physics A*, **53**, 465–474.

Bryant, D. A., Costas, A. M. G., Maresca, J. A., Chew, A. G. M., Klatt, C. G., Bateson, M. M., Tallon, L. J., Hostetler, J., Nelson, W. C., Heidelberg, J. F., and Ward, D. M. 2007. Candidatus Chloracidobacterium thermophilum: an aerobic phototrophic acidobacterium. *Science*, **317**(5837), 523–526.

Butkus, V., Zigmantas, D., Valkunas, L., and Abramavicius, D. 2012. Vibrational vs. electronic coherences in 2D spectrum of molecular systems. *Chemical Physics Letters*, **545**, 40–43.

Byler, D. M., and Susi, H. 1986. Examination of the secondary structure of protiens by deconvolved FTIR spectra. *Biopolymers*, **25**, 469–487.

Cai, J.-M. 2011. Quantum Probe and Design for a Chemical Compass with Magnetic Nanostructures. *Physical Review Letters*, **106**(10).

Cai, J.-M., Popescu, S., and Briegel, H. J. 2010a. Dynamical entanglement in oscillating molecules and potential biological implications. *Physical Review E*, **82**, 021921.

Cai, J.-M., Guerreschi, J. J., and Briegel, H. J. 2010b. Quantum control and entanglement in a chemical compass. *Physical Review Letters*, **104**, 220502.

Caldeira, A. O., and Leggett, A. J. 1983. Path integral approach to quantum Brownianmotion. *Physica A*, **121**, 587.

Calderbank, A. R., and Shor, P. W. 1996. Good quantum error-correcting codes exist. *Physical Review A*, **54**, 1098.

Calsamiglia, J., Hartmann, L., Dür, W., and Briegel, H. J. 2005. Spin gases: quantum entanglement driven by classical kinematics. *Physical Review Letters*, **95**, 180502.

Camara-Artigas, A., Blankenship, R. E., and Allen, J. P. 2003. The structure of the FMO protein from Chlorobium tepidum at 2.2 angstrom resolution. *Photosynthesis Research*, **75**, 49.

Canfield, J. M., Belford, R. L., Debrunner, P. G., and Schulten, K. 1994. A perturbation theory treatment of oscillating magnetic fields in the radical pair mechanism. *Chemical Physics*, **182**, 1–18.

Canfield, J. M., Belford, R. L., Debrunner, P. G., and Schulten, K. 1995. A perturbation treatment of oscillating magnetic fields in the radical pair mechanism using the Liouville equation. *Chemical Physics*, **195**, 59–69.

Canfield, J. M., Belford, R. L., Debrunner, P. G., and Schulten, K. 1995a. A perturbation theory treatment of oscillating magnetic fields in the radical pair mechanism – erratum. *Chemical Physics*, **191**, 347.

Canters, G. W., and Vijgenboom, E. (eds). 1997. *Biological Electron Transfer Chains: Genetics, Composition and Mode of Operation*. Dordrecht: Kluwer.

Cao, J. 1997. A phase-space study of Bloch-Redfield theory. *Journal of Chemical Physics*, **107**, 8.

Cao, J., and Silbey, R. J. 2009. Optimal trapping of excitation energy transfer. *Journal of Physical Chemistry A*, **113**, 13826.

Caram, J. R., and Engel, G. S. 2011. Extracting dynamics of excitonic coherences in congested spectra of photosynthetic light harvesting antenna complexes. *Faraday Discussions*, **153**, 93.

Caram, J. R., Lewis, N. H. C., Fidler, A. F., and Engel, G. S. 2012. Signatures of correlated excitonic dynamics in two dimensional spectroscopy of the Fenna-Matthews-Olson photosynthetic complex. *Journal of Chemical Physics*, **136**, 104505.

Carter, S. G., Chen, Z., and Cundiff, S. T. 2007. Echo peak-shift spectroscopy of non-Markovian exciton dynamics in quantum wells. *Physical Review B*, **76**, 121303.

Caruso, F., Chin, A. W., Datta, A., Huelga, S. F., and Plenio, M. B. 2009. Highly efficient energy excitation transfer in light-harvesting complexes: the fundamental role of noise-assisted transport. *Journal of Chemical Physics*, **131**, 105106.

Caruso, F., Chin, A., Datta, A., Huelga, S., and Plenio, M. B. 2010. Entanglement and entangling power of the dynamics in light-harvesting complexes. *Physical Review A*, **81**, 062346.

Castiglione, P. 2000. Diffusion coefficients as function of Kubo number in random fields. *J. Phys. A: Math. Gen.*, **33**, 197.

Catlow, C. R. A., and Stoneham, A. M. 1983. Ionicity in solids. *Journal of Physical Chemistry*, **16**, 4321–4338.

Chachisvilis, M., Pullerits, T., Jones, M. R., Hunter, C. N., and Sundstrom, V. 1994. Vibrational dynamics in the light-harvesting complexes of the photosynthetic bacterium *Rhodobacter sphaeroides*. *Chemical Physics Letters*, **224**(3–4), 345–351.

Chen, C.-H., Liu, K.-Y., Sudhakar, S., Lim, T.-S., Fann, W., Hsu, C.-P., and Luh, T.-Y. 2005. Efficient light harvesting and energy transfer in organic-inorganic hybrid multichromophoric materials. *Journal of Physical Chemistry B*, **109**, 17887.

Chen, X., Discher, B. M., Pilloud, D. L., Gibney, B. R., Moser, C. C., and Dutton, P. L. 2002. De novo design of a cytochrome b maquette for electron transfer and coupled reactions on electrodes. *The Journal of Physical Chemistry B*, **106**(3), 617–624.

Cheng, Y. C., and Fleming, G. R. 2008. Coherence quantum beats in two-dimensional electronic spectroscopy. *Journal of Physical Chemistry A*, **112**, 4254–4260.

Cheng, Y.-C., and Fleming, G. R. 2009. Dynamics of light harvesting in photosynthesis. *Annual Review of Physical Chemistry*, **60**, 241.

Cheng, Y. C., and Silbey, R. J. 2006. Coherence in the B800 ring of purple bacteria LH2. *Physical Review Letters*, **96**, 028103.

Cheng, Y.-C., and Silbey, R. J. 2008. A unified theory for charge-carrier transport in organic crystals. *Journal of Chemical Physics*, **128**, 114713.

Cheng, Y.-C., Engel, G. S., and Fleming, G. R. 2007. Elucidation of population and coherence dynamics using cross-peaks in two-dimensional electronic spectroscopy. *Chemical Physics*, **341**, 285.

Chernyak, V., Zhang, W. M., and Mukamel, S. 1998. Multidimensional femtosecond spectroscopies of molecular aggregates and semiconductor nanostructures: the nonlinear exciton equations. *Journal of Chemical Physics*, **109**, 9587.

Chin, A. W., Rivas, A., Huelga, S. F., and Plenio, M. B. 2010a. Exact mapping between system-reservoir quantum models and semi-infinite discrete chains using orthogonal polynomials. *Journal of Mathematical Physics*, **510**, 092109.

Chin, A. W., Datta, A., Caruso, F., Huelga, S. F., and Plenio, M. B. 2010b. Noise-assisted energy transfer in quantum networks and light-harvesting complexes. *New Journal of Physics*, **12**, 065002.

Chin, A. W., Huelga, S. F., and Plenio, M. B. 2011. Chain representations of open quantum systems and their numerical simulation with time-adapative density matrix renormalisation group methods. Pages 115–144 in: Würfel, U., Thorwart, M., Weber, E. R., and Jagadish, Ch. (eds), *Semiconductors and Semimetals*, vol. 85. Burlington: Academic Press.

Chin, A. W., Huelga, S. F., and Plenio, M. B. 2012. Coherence and decoherence in biological system: principles of noise assisted transport and the origin of long-lived coherences. *Philosophical Transactions of the Royal Society A*, **370**, 3638–3657.

Chin, A. W., Prior, J., Rosenbach, R., Huelga, S. F., and Plenio, M. B. 2013. The role of non-equilibrium vibrational structures in electronic coherence and recoherence in pigment-protein complexes. *Nature Physics*, **9**, 113–118.

Cho, M. 2009. *Two-Dimensional Optical Spectroscopy*. Boca Raton: CRC Press.

Cho, M., Scherer, N. F., Fleming, G. R., and Mukamel, S. 1992. Photon echoes and related four-wave-mixing spectroscopies using phase-locked pulses. *Journal of Chemical Physics*, **96**, 5618.

Cho, M., Yu, J.-Y., Joo, T., Nagasawa, Y., Passino, S. A., and Fleming, G. R. 1996. The integrated photon echo and solvation dynamics. *Journal of Physical Chemistry*, **100**, 11944.

Cho, M., Vaswani, H. M., Brixner, T., Stenger, J., and Fleming, G. R. 2005. Exciton analysis in 2D electronic spectroscopy. *Journal of Physical Chemistry B*, **109**, 10542–10556.

Choi, M.-D. 1975. Completely positive linear maps on complex matrices. *Linear Algebra and its Applications*, **10**, 285–290.

Christensson, N., Milota, F., Hauer, J., Sperling, J., Bixner, O., Nemeth, A., and Kauffmann, H. F. 2011. High frequency vibrational modulations in two-dimensional electronic spectra and their resemblance to electronic coherence signatures. *Journal of Physical Chemistry B*, **115**, 5383–5391.

Christensson, N., Kauffmann, H. F., Pullerits, T., and Mančal, T. 2012. Origin of long-lived coherences in light-harvesting complexes. *Journal of Physical Chemistry B*, **116**, 7449.

Cintolesi, F., Ritz, T., Kay, C. W. M., Timmel, C. R., and Hore, P. J. 2003. Anisotropic recombination of an immobilized photoinduced radical pair in a 50-μT magnetic field: a model avian photomagnetoreceptor. *Chemical Physics*, **294**, 707–718.

Ciobanu, M., Kincaid, H. A., Jennings, G. K., and Cliffel, D. E. 2005. Photosystem I Patterning Imaged by Scanning Electrochemical Microscopy. *Langmuir*, **21**(2), 692–698.

Codgell, R. J., Southall, J., Gardiner, A. T., Law, C. J., Gall, A., Roszak, A. W., and Isaacs, N. W. 2006a. How purple photosynthetic bacteria harvest solar energy. *Comptes Rendus Chimie*, **9**(2), 201–206.

Cogdell, R. J., Gali, A., and Köhler, J. 2006b. The architecture and function of the light-harvesting apparatus of purple bacteria: from single molecules to in vivo membranes. *Quarterly Reviews of Biophysics*, **39**, 227.

Collini, E., Wong, C. Y., Wilk, K. E., Curmi, P. M. G., Brumer, P., and Scholes, G. D. 2010. Coherently wired light-harvesting in photosynthetic marine algae at ambient temperature. *Nature*, **463**, 644–648.

Cowan, M. L., Ogilvie, J. P., and Miller, R. J. D. 2004. Two-dimensional spectroscopy using diffractive optics based phased-locked photon echoes. *Chemical Physics Letters*, **386**, 184.

Craig, D. P., and Thirunamachandran, T. 1984. *Molecular Quantum Electrodynamics: An Introduction to Radiation-Molecule Interactions*. London: Academic Press.

Daizadeh, I., Medvedev, E. S., and Stuchebrukhov, A. A. 1997. Effect of protein dynamics on biological electron transfer. *Proceedings of the National Academy of Sciences, USA*, **94**, 3703–3708.

Daley, A. J., Kollath, C., Schollwöck, U., and Vidal, G. 2004. Time-dependent density-matrix renormalization-group using adaptive effective Hilbert spaces. *Journal of Statistical Mechanics: Theory and Experiment*, P04005.

Damjanović, A., Ritz, T., and Schulten, K. 1999. Energy transfer between carotenoids and bacteriochlorophylls in light-harvesting complex II of purple bacteria. *Physical Review E*, **59**, 3293–3311.

Damjanović, A., Kosztin, I., Kleinekathofer, U., and Schulten, K. 2002. Excitons in a photosynthetic light-harvesting system: a combined molecular dynamics, quantum chemistry, and polaron model study. *Physical Review E*, **65**(Mar), 031919.

Das, R., Kiley, P. J., Segal, M., Norville, J., Yu, A. A., Wang, L., Trammell, S. A., Reddick, L. E., Kumar, R., Stellacci, F., Lebedev, N., Schnur, J., Bruce, B. D., Zhang, S., and Baldo, M. 2004. Integration of photosynthetic protein molecular complexes in solid-state electronic devices. *Nano Letters*, **4**, 1079.

Davies, P. 2004. Does quantum mechanics play a non-trivial role in life? *Biosystems*, **78**, 69–79.

Davydov, A. S. 1982. *Biology and Quantum Mechanics*. Oxford: Pergamon Press.

Dawlaty, J. M., Ishizaki, A., De, A. K., and Fleming, G. R. 2012. Microscopic quantum coherence in a photosynthetic-light-harvesting antenna. *Philosophical Transactions of the Royal Society of London A. (Mathematical, Physical and Engineering Sciences)*, **370**, 3672–3691.

De Rossi, U., Moll, J., Spieles, M., Bach, G., Dähne, S., Kriwanek, J., and Lisk, M. 1995. Control of the J-aggregation phenomenon by variation of the N-alkyl-substituents. *Journal für Praktische Chemie/Chemiker-Zeitung*, **337**(1), 203–208.

De Rossi, U., Dähne, S., Meskers, S. C. J., and Dekkers, H. P. J. M. 1996. Spontane Bildung von optischer Aktivität in J-Aggregaten mit Davydov-Aufspaltung. *Angewandte Chemie*, **108**(7), 827–830.

Decatur, S. M. 2006. Elucidation of residue-level structure and dynamics of polypeptides via isotopp-edited infrared spectroscopy. *Accounts of Chemical Research*, **39**, 169–175.

Decoursey, T. E. 2003. Voltage-gated proton channels and other proton transfer pathways. *Physiology Reviews*, **83**, 475–579.

Deisenhofer, J., Epp, O., Miki, K., Huber, R., and Michel, H. 1984. X-Ray structure-analysis of a membrane-protein complex – electron-density map at 3 Åresolution and a model of the chromophores of the photosynthetic reaction center from *Rhodopseudomonas-viridis*. *Journal of Molecular Biology*, **180**(2), 385–398.

Deisenhofer, J., Epp, O., Mikki, K., Huber, R., and Michel, H. 1985. Structure of the protein subunits in the photosynthetic reaction centre of *Rhodopseudomonas viridis* at 3 Å resolution. *Nature*, **318**, 618–624.

Deisenhofer, J., Epp, O., Sinning, I., and Michel, H. 1995. Crystallographic refinement at 2.3 Åresolution and refined model of the photosynthetic reaction centre from *rhodopseudomonas viridis*. *Journal of Molecular Biology*, **246**(3), 429–457.

Dekker, J. P., and van Grondelle, R. 2000. Primary charge separation in photosystem II. *Photosynthesis Research*, **63**(3), 195–208.

del Rey, M., Chin, A. W., Huelga, S. F., and Plenio, M. B. 2013. The phonon-antennae for efficient transport. *Journal of Physical Chemistry Letters*, **4**, 903–907.

Derebe, M. G., Sauer, D. B., Zeng, W., Alam, A., Shi, N., and Jiang, Y. 2011. Tuning the ion selectivity of tetrameric cation channels by changing the number of ion binding sites. *Proceedings of the National Academy of Sciences of the United States of America*, **108**(2), 598–602.

Desouter-Lecomte, M., and Lorquet, J. C. 1979. Nonadiabatic interactions in unimolecular decay. IV. Transition probability as a function of the Massey parameter. *Journal of Chemical Physics*, **71**, 4391(13 pages).

Devault, D., and Chance, B. 1966. Studies of photosynthesis using a pulsed laser. I. Temperature dependence of cytochrome oxidation rate in chromatium. Evidence for tunneling. *Biophysical Journal*, **6**(6), 825–847.

Devault, D., Parkes, J. H., and Chance, B. 1967. Electron tunnelling in cytochromes. *Nature*, **215**(5101), 642–644.

Dexter, D. L. 1953. A theory of sensitized luminescence in solids. *Journal of Chemical Physics*, **21**, 836.

Didraga, C., Pugzlys, A., Hania, P. R., von Berlepsch, H., Duppen, K., and Knoester, J. 2004. Structure, spectroscopy, and microscopic model of tubular carbocyanine dye aggregates. *The Journal of Physical Chemistry B*, **108**(39), 14976–14985.

Diner, B. A., Schlodder, E., Nixon, P. J., Coleman, W. J., Rappaport, F., Lavergne, J., Vermaas, W. F. J., and Chisholm, D. A. 2001. Site-directed mutations at D1-His198 and D2-His197 of photosystem II in Synechocystis PCC 6803: sites of primary charge separation and cation triplet stabilization. *Biochemistry*, **40**, 9265–9281.

Dirac, P. A. M. 1927. The quantum theory of emission and absorption of radiation. *Proceedings of the Royal Society of London A. (Mathematical, Physical and Engineering Sciences)*, **114**, 243.

Dominy, J. M., Shabani, A., and Lidar, D. A. 2013. A general framework for complete positivity. arXiv: 1312.0908.

Doyle, D. A., Cabral, J. M., Pfuetzner, R. A., Kuo, A. L., Gulbis, J. M., Cohen, S. L., Chait, B. T., and R., MacKinnon. 1998. The structure of the potassium channel: molecular basis of K^+ conduction and selectivity. *Science*, **280**, 69–77.

Dresselhaus, M. S., Dresselhaus, G., and Eklund, P. C. 1996. *Science of Fullerenes and Carbon Nanotubes*. San Diego: Academic Press.

Dreyer, J., Moran, A. M., and Mukamel, S. 2003. Tensor components in three pulse vibrational echoes of rigid dipeptide. *Bulletin of the Korean Chemical Society*, **24**, 1091.

Dunlap, D. H., and Krenke, V. M. 1986. Dynamic localization of a charged-particle moving under the influence of an electric field. *Physical Review B*, **34**, 3625–3633.

Durrant, J. R., Klug, D. R., Kwa, S. L. S., van Grondelle, R., Porter, G., and Dekker, J. P. 1995. A multimer model for P680, the primary electron-donor of photosystem-II. *Proceedings of the National Academy of Sciences, USA*, **92**(11), 4798–4802.

Dutton, P. L., and Moser, C. C. 2011. Engineering enzymes. *Faraday Discussions*, **148**, 443–448.

Dyson, G. M. 1938. The scientific basis of odour. *Journal of the Society of Chemical Industry*, **57**, 647–651.

Dzuba, S. A., Bosch, M. K., and Hoff, A. J. 1996. Electron spin echo detection of quantum beats and double-quantum coherence in spin-correlated radical pairs of protonated photosynthetic reaction centers. *Chemical Physics Letters*, **248**, 427–433.

Edelstein, A. 2007. Advances in magnetometry. *Journal of Physics: Condensed Matter*, **19**, 165217.

Efimova, O., and Hore, P. J. 2008. Role of exchange and dipolar interactions in the radical pair model of the avian magnetic compass. *Biophysical Journal*, **94**, 1565–1574.

Egorov, S. A., Everitt, K. F., and Skinner, J. L. 1999. Quantum dynamics and vibrational relaxation. *Journal of Physical Chemistry A*, **103**(47), 9494–9499.

Einstein, A., Podolsky, B., and Rosen, B. 1935. Can quantum-mechanical description of physical reality be considered complete? *Physical Review*, **47**, 777–780.

Eisele, D. M. 2010. *Optical, Structural and Redox Properties of Nanotubular J-aggregates of Amphiphilic Cyanine Dyes*. Aachen: Shaker Verlag.

Eisele, D. M., Knoester, J., Kirstein, S., Rabe, J. P., and Vanden Bout, D. A. 2009. Uniform exciton fluorescence from individual molecular nanotubes immobilized on solid substrates. *Nature Nanotechnology*, **4**(10), 658–663.

Eisele, D. M., Cone, C. W., Bloemsma, E. A., Vlaming, S. M., van der Kwaak, C. G. F., Silbey, R. J., Bawendi, M. G., Knoester, J., Rabe, J. P., and Vanden Bout, D. A. 2012.

Utilizing redox-chemistry to elucidate the nature of exciton transitions in supramolecular dye nanotubes. *Nature Chemistry*, **4**.

El-Naggar, M. Y., Gorby, Y. A., Xia, W., and Nealson, K. H. 2008. The molecular density of states in bacterial nanowires. *Biophysical Journal*, **95**, L10–L12.

El-Naggar, M. Y., Wanger, G., Leung, K. M., Yuzvinsky, T. D., Southam, G., Yang, J., Lau, W. M., Nealson, K. H., and Gorby, Y. A. 2010. Electrical transport along bacterial nanowires from Shewanella oneidensis MR-1. *Proceedings of the National Academy of Sciences, USA*, **107**, 18127–18131.

Elliott, R. J., and Loudon, R. 1959. Theory of fine structure on the absorption edge in semiconductors. *Journal of Physics and Chemistry of Solids*, **8**, 382–388.

Elliott, R. J., and Loudon, R. 1960. Group theory of scattering processes in crystals. *Journal of Physics and Chemistry of Solids*, **15**, 196.

Endo, M., Wang, H., Fujitsuka, M., and Majima, T. 2006. Pyrene-stacked nanostructures constructed in the recombinant tobacco mosaic virus rod scaffold. *Chemistry – A European Journal*, **12**, 3735.

Endo, M., Fujitsuka, M., and Majima, T. 2007. Porphyrin light-harvesting arrays constructed in the recombinant tobacco mosaic scaffold. *Chemistry – A European Journal*, **13**, 8660.

Engel, G. S., Calhoun, T. R., Read, E. L., Ahn, T. K., Mancal, T., Cheng, Y. C., Blankenship, R. E., and Fleming, G. R. 2007. Evidence for wavelike energy transfer through quantum coherence in photosynthetic systems. *Nature*, **446**, 782–786.

Escalante, M., Maury, P., Bruinink, C. M., van der Werf, K., Olsen, J. D., Timney, J. A., Huskens, J., Hunter, C. N., Subramaniam, V., and Otto, C. 2008. Directed assembly of functional light harvesting antenna complexes onto chemically patterned surfaces. *Nanotechnology*, **19**, 025101.

Eveson, R. W., Timmel, C. R., Brocklehurst, B., Hore, P. J., and McLauchlan, K. A. 2000. The effects of weak magnetic fields on radical recombination reactions in micelles. *International Journal of Radiation Biology*, **76**, 1509–1522.

Falkenberg, G., Fleissner, G., Schuchardt, K., Kuehbacher, M., Thalau, P., Mouritsen, H., Heyers, D., Wellenreuther, G., and Fleissner, G. 2010. Avian magnetoreception: elaborate iron mineral containing dendrites in the upper beak seem to be a common feature of birds. *PLoS ONE*, **5**, e9231.

Fan, C., Wang, S., Hong, J. W., Bazan, G. C., Plaxco, K. W., and Heeger, A. J. 2003. Beyond superquenching: hyper-efficient energy transfer from conjugated polymers to gold nanoparticles. *Proceedings of the National Academy of Sciences, USA*, **100**, 6297.

Fang, C., Senes, A., Cristian, L., DeGrado, W. F., and Hochstrasser, R. M. 2006. Amide vibrations are delocalized across the hydrophobic interface of a transmembrane helix dimer. *Proceedings of the National Academy of Sciences of the United States of America*, **103**, 16740–16745.

Fassioli, F., and Olaya-Castro, A. 2010. Distribution of entanglement in light-harvesting complexes and their quantum efficiency. *New Journal of Physics*, **12**, 085006.

Faul, C. F. J., and Antonietti, M. 2003. Ionic self-assembly: facile synthesis of supramolecular materials. *Advanced Materials*, **15**, 673.

Feenders, G., Liedvogel, M., Rivas, M., Zapka, M., Horita, H., Hara, E., Wada, K., Mouritsen, H., and Jarvis, E. D. 2008. Molecular mapping of movement-associated areas in the avian brain: a motor theory for vocal learning origin. *PLoS ONE*, **3**, e1768.

Fenna, R. E., and Matthews, B. W. 1975. Chlorophyll arrangement in a bacteriochlorophyll protein from *Chlorobium limicola*. *Nature*, **258**, 573.

Ferreira, K. N., Iverson, T. M., Maghlaoui, K., Barber, J., and Iwata, S. 2004. Architecture of the photosynthetic oxygen-evolving center. *Nature*, **303**, 1831–1838.

Feynman, R. P., and Vernon, F. L. 1963. The theory of a general quantum system interacting with a linear dissipative system. *Annals of Physics (N. Y.)*, **24**, 118.

Fidler, A. F., Harel, E., and Engel, G. S. 2010. Dissecting hidden couplings using fifth-order three-dimensional electronic spectroscopy. *Journal of Physical Chemistry Letters*, **1**, 2876.

Fidler, A. F., Harel, E., Long, P. D., and Engel, G. S. 2012. Two-Dimensional spectroscopy can distinguish between decoherence and dephasing of zero-quantum coherences. *Journal of Physical Chemistry A*, **116**, 282.

Figge, M. T., Mostovoy, M., and Knoester, J. 2001. Peierls transition with acoustic phonons and solitwistons in carbon nanotubes. *Physical Review Letters*, **86**, 4572.

Fleissner, G., Holtkamp-Rötzler, E., Hanzlik, M., Winklhofer, M., Fleissner, G., Petersen, N., and Wiltschko, W. 2003. Ultrastructural analysis of a putative magnetoreceptor in the beak of homing pigeons. *Journal of Comparative Neurology*, **458**, 350–360.

Fleissner, G., Stahl, B., Thalau, P., Falkenberg, G., and Fleissner, G. 2007. A novel concept of Fe-mineral based magnetoreception: histological and physicochemical data from the upper beak of homing pigeons. *Naturwissenschaften*, **94**, 631–642.

Fleming, G. R., and Cho, M. 1996. Chromophore-solvent dynamics. *Annual Review of Physical Chemistry*, **47**, 109.

Fleming, G. R., Martin, J. L., and Breton, J. 1988. Rates of primary electron-transfer in photosynthetic reaction centers and their mechanistic implications. *Nature*, **333**(6169), 190–192.

Flynn, C. P., and Stoneham, A. M. 1970. Quantum theory of diffusion with application to light interstitials in metals. *Physical Review B*, **1**, 3966–3987.

Förster, T. 1948. Zwischenmolekulare Energiewanderung und Fluoreszenz. *Annals of Physics (Berlin)*, **2**, 55.

Förster, T. 1965. *Modern Quantum Chemistry, Part III*. New York: Academic Press.

Fraenkel-Conrat, H., and Williams, R. C. 1955. Reconstitution of active tobacco mosaic virus from its inactive protein and nucleic acid components. *Proceedings of the National Academy of Sciences, USA*, **41**, 690.

Franck, J., and Teller, E. 1938. Migration and photochemical action of excitation energy in crystals. *The Journal of Chemical Physics*, **6**, 861.

Franco, M. I., Turin, L., Mershin, A., and Skoulakis, E. M. C. 2011. Molecular vibration-sensing component in Drosophila melanogaster olfaction. *Proceedings of the National Academy of Sciences, USA*, **108**, 3797–3802.

Frauenfelder, H., Wolynes, P. G., and Austin, R. H. 1999. Biological physics. *Reviews of Modern Physics*, **71**, S419.

Fredrickson, J. K., Romine, M. F., Beliaev, A. S., Auchtung, J. M., Driscoll, M. E., Gardner, T. S., Nealson, K. H., Osterman, A. L., Pinchuk, G., Reed, J, L., Rodionov, D. A., Rodrigues, J. L. M., Saffarini, D. A., Serres, M. H., Spormann, A. M., Zhulin, I. B., and Tiedje, J. M. 2008. Towards environmental systems biology of Shewanella. *Nature Reviews Microbiology*, **6**, 592–603.

Freiberg, A., Ratsep, M., Timpmann, K., Trinkunas, G., and Woodbury, N. W. 2003. Self-trapped excitons in LH2 antenna complexes between 5 K and ambient temperature. *Journal of Physical Chemistry B*, **107**(41), 11510–11519.

Freire, R., Munro, U., Rogers, L. J., Sagasser, S., Wiltschko, R., and Wiltschko, W. 2008. Different responses in two strains of chickens (*Gallus gallus*) in a magnetic orientation test. *Animal Cognition*, **11**, 547–552.

Frenkel, J. A. 1931. On the transformation of light into heat in solids. II. *Physical Review*, **37**(10), 1276.

Frigaard, N.-U., Li, H., Milks, K. J., and Bryant, D. A. 2004. Nine mutants of *Chlorobium tepidum* each unable to synthesize a different chlorosome protein still assemble functional chlorosomes. *Journal of Bacteriology*, **186**(3), 646–653.

Frutos, L. M., Andruniów, T., Santoro, F., Ferré, N., and Olivucci, M. 2007. Tracking the excited-state time evoluation of the visual pigment with multiconfigurational quantum chemistry. *Proceedings of the National Academy of Science USA*, **104**, 4464–4469.

Fujimoto, K., Hayashi, S., Hasegawa, J., and Nakatsuji, H. 2007. Theoretical studies on the color-tuning mechanism in retinal proteins. *Journal of Chemical Theory and Computation*, **3**, 605–618.

Fujimoto, K., Hasegawa, J., and Nakatsuji, H. 2009. Color tuning mechanism of human red, green, and blue cone pigments: SAC-CI theoretical study. *Bulletin of the Chemical Society of Japan*, **82**, 1140–1148.

Fujimoto, K. J., Asai, K., and Hasegawa, J. 2010. Theoretical study of the opsin shift of deprotonated retinal Schiff base in the M state of bacteriorhodopsin. *Physical Chemistry Chemical Physics*, **12**, 13107–13116.

Gaab, K. M., and Bardeen, C. J. 2004a. The effects of connectivity, coherence, and trapping on energy transfer in simple light-harvesting systems studied using the Haken-Strobl model with diagonal disorder. *The Journal of Chemical Physics*, **121**, 7813.

Gaab, K. M., and Bardeen, C. J. 2004b. Wavelength and temperature dependence of the femtosecond pump-probe anisotropies in the conjugated polymer MEH-PPV: implications for energy-transfer dynamics. *Journal of Physical Chemistry B*, **108**, 4619.

Gadsby, D. C. 2009. Ion channels versus ion pumps: the principal difference, in principle. *Nature Reviews Molecular Cell Biology*, **10**, 344–352.

Gai, F., Hasson, K. C., McDonald, J. C., and Anfinrud, P. A. 1998. Chemical dynamics in proteins: the photoisomerization of retinal in bacteriorhodopsin. *Science*, **279**, 1886–1891.

Galland, C., Högele, A., Türeci, H. E., and Imamoğlu, A. 2008. Non-Markovian decoherence of localized nanotube excitons by acoustic phonons. *Physical Review Letters*, **101**, 067402.

Gammaitoni, L., Hänggi, P., Jung, P., and Marchesoni, F. 1998. Stochastic resonance. *Reviews of Modern Physics*, **70**(1), 223–287.

Gane, S., Georganakis, D., Maniati, K., Vamvakias, M., Ragoussis, N., Skoulakis, E. M. C., and Turin, L. 2013. Molecular vibration-sensing component in human olfaction. *PLoS One*, **8**, e55780.

Ganim, Z., and Tokmakoff, A. 2006. Spectral signatures of heterogeneous protein ensembles revealed by MD simulations of 2D-IR spectra. *Biophysical Journal*, **91**, 2636–2646.

Ganim, Z., Chung, H. S., Smith, A. W., Deflores, L. P., Jones, K. C., and Tokmakoff, A. 2008. Amide I two-dimensional infrared spectroscopy of proteins. *Accounts of Chemical Research*, **41**, 432–441.

Ganim, Z., Jones, K. C., and Tokmakoff, A. 2010. Insulin dimer dissociation and unfolding revealed by amide I two-dimensional infrared spectroscopy. *Physical Chemistry Chemical Physics*, **12**, 3579–3588.

Ganim, Z., Tokmakoff, A., and Vaziri, A. 2011. Vibrational Excitons in Ionophores; Experimental probes for quantum coherence-assisted ion transport and selectivity in ion channels. *New Journal of Physics*, **13**, 113030.

Gauger, E., Rieper, E., Morton, J. J. L., Benjamin, S. C., and Vedral, V. 2011. Sustained quantum coherence and entanglement in the avian compass. *Physical Review Letters*, **106**, 040503.

Gegear, R. J., Casselman, A., Waddell, S., and Reppert, S. M. 2008. Cryptochrome mediates light-dependent magnetosensitivity in *Drosophila*. *Nature*, **454**, 1014–1018.

Gegear, R. J., Foley, L. E., Casselman, A., and Reppert, S. M. 2010. Animal cryptochromes mediate magnetoreception by an unconventional photochemical mechanism. *Nature*, **463**, 804–807.

Gilmore, J., and McKenzie, R. H. 2005. Spin-boson models for quantum decoherence of electronic excitations of biomolecules and quantum dots in a solvent. *Journal of Physics: Condensed Matter*, **17**, 1735.

Gilmore, J., and McKenzie, R. H. 2008. Quantum dynamics of electronic excitations in biomolecular chromophores: role of the protein environment and solvent. *Journal of Chemical Physics*, **112**, 2162.

Ginsberg, N. S., Cheng, Y. C., and Fleming, G. R. 2009. Two-dimensional electronic spectroscopy of molecular aggregates. *Accounts of Chemical Research*, **42**(9), 1352–1363.

Giovani, B., Byrdin, M., Ahmad, M., and Brettel, K. 2003. Light-induced electronic transfer in a cryptochrome blue-light photoreceptor. *Nature Structural Biology*, **10**, 489–490.

Giovannetti, V., Lloyd, S., and Maccone, L. 2011. Advances in quantum metrology. *Nature Photonics*, **5**(4), 222–229.

Gisin, N., and Thew, R. 2007. Quantum communication. *Nature Photonics*, **1**(3), 165–171.

González-Luque, R., Garavelli, G., Bernardi, F., Merchán, M., Robb, M. A., and Olivucci, M. 2000. Computational evidence in favor of a two-state, two-mode model of the retinal chromophore photoisomerization. *Proceedings of the National Academy of Science USA*, **97**, 9379–9384.

Gorby, Y. A., Yanina, S., McLean, J. S., Rosso, K. M., Moyles, D., Dohnalkova, A., Beveridge, T. J., Chang, I. S., Kim, B. H., Kim, K. S. Culley, D. E., Reed, S. B., Romine, M. F., Saffarini, D. A., Hill, E. A., Shi, L., Elias, D. A., Kennedy, D. W., Pinchuk, G., Watanabe, K., Ishi, S., Logan, B., Nealson, K. H., and K., Fredrickson J. 2006. Electrically conductive bacterial nanowires produced by Shewanella oneidensis strain MR-1 and other microorganisms. *Proceedings of the National Academy of Sciences, USA*, **103**, 11358–11363.

Gorini, V., Kossakowski, A., and Sudarshan, E. C. G. 1976. Completely positive semigroups of N-level systems. *Journal of Mathematical Physics*, **17**, 821.

Gouaux, E., and MacKinnon, R. 2005. Principles of selective ion transport in channels and pumps. *Science*, **310**, 1461–1465.

Goychuk, I., and Hänggi, P. 2005. Quantum dynamics in strong fluctuating fields. *Advances in Physics*, **54**, 52.

Graham, M. W., Ma, Y.-Z., and Fleming, G. R. 2008. Femtosecond photon echo spectroscopy of semiconducting single-walled carbon nanotubes. *Nano Letters*, **8**, 3936–3941.

Graham, M. W., Calhoun, T. R., Green, A. A., Hersam, M. C., and Fleming, G. R. 2012. Two-dimensional electronic spectroscopy reveals the dynamics of phonon-mediated excitation pathways in semiconducting single-walled carbon nanotubes. *Nano Letters*, **12**, 813–819.

Gralnick, J. A., and Newman, D. K. 2007. Extracellular respiration. *Molecular Microbiology*, **65**, 1–11.

Grätzel, M. 2001. Photoelectrochemical cells. *Nature*, **414**(6861), 338–344.

Gray, H. B., and Winkler, J. R. 2005. Long-range electron transfer. *Proceedings of the National Academy of Sciences, USA*, **102**, 3534–3539.

Gray, H. B., and Winkler, J. R. 2003. Electron tunneling through proteins. *Quarterly Reviews of Biophysics*, **36**, 341–372.

Gray, H. B., and Winkler, J. R. 2009. Electron flow through proteins. *Chemical Physics Letters*, **483**, 1–9.

Grigoryants, V. M., Tadjikov, B. M., Usov, O. M., and Molin, Y. N. 1995. Phase-shift of quantum oscillations in the recombination luminescence of spin-correlated radical-ion pairs. *Chemical Physics Letters*, **246**, 392–398.

Groblacher, S., Hammerer, K., Vanner, M. R., and Aspelmeyer, M. 2009. Observation of strong coupling between a micromechanical resonator and an optical cavity field. *Nature*, **460**, 724–727.

Groenhof, G., Schäfer, L. V., Boggio-Pasqua, M., and Robb, M. A. 2009. Handbook of Molecular Biophysics. Methods and Applications. Weinheim: Wiley-VCH.

Gronheid, R., Hofkens, J., Köhn, F., Weil, T., Reuther, E., Müllen, K., and Schryver, F. C. De. 2002. Intramolecular Förster energy transfer in a dendritic system at the single molecule level. *Journal of the American Chemical Society*, **124**, 2418.

Groot, M. L., Pawlowicz, N. P., van Wilderen, L. J. G. W., Breton, J., van Stokkum, I. H. M., and van Grondelle, R. 2005. Initial electron donor and acceptor in isolated photosystem II reaction centers identified with femtosecond mid-IR spectroscopy. *Proceedings of the National Academy of Sciences, USA*, **102**(37), 13087–13092.

Grover, M. 1971. Exciton migration in molecular crystals. *Journal of Chemical Physics*, **54**, 4843.

Grynberg, G., Aspect, A., and Fabre, C. 2010. *Introduction to Quantum Optics: From the Semi-classical Approach to Quantized Light*. Cambridge, UK: Cambridge University Press.

Guenzler, H., and Gremlich, H.-U. 2002. *IR Spectroscopy: An Introduction*. Weinheim: Wiley-VCH.

Guerreschi, G. G., Cai, J.-M., Popescu, S., and Briegel, H. J. 2012. Persistent dynamic entanglement from classical motion: how bio-molecular machines can generate non-trivial quantum states. *New Journal of Physics*, **14**, 053043.

Gulbinas, V., Karpicz, R., Augulis, R., and Rotomskis, R. 2007. Exciton relaxation in nanotubular TPPS_4 aggregates in water solution and in polymeric matrix. *Chemical Physics*, **332**, 255.

Gunn, J. R., and Dawson, K. A. 1989. Microscopic model of amphiphilic assembly. *Journal of Chemical Physics*, **91**, 6393–6403.

Guo, S., and Kim, J. 2010. Dissecting the molecular mechanism of drosophila odorant receptors through activity modeling and comparative analysis. *Proteins*, **78**, 381–399.

Gust, D., Moore, T. A., and Moore, A. L. 2001. Mimicking photosynthetic solar energy transduction. *Accounts of Chemical Research*, **34**(1), 40–48.

Gust, D., Moore, T. A., and Moore, A. L. 2009. Solar fuels via artificial photosynthesis. *Accounts of Chemical Research*, **42**(12), 1890–1898.

Gwan, J. F., and Baumgaertner, A. 2007. Cooperative transport in a potassium ion channel. *Journal of Chemical Physics*, **127**, 045103.

Ha, T., Enderle, T., Ogletree, D. F., Chemla, D. S., Selvin, P. R., and Weiss, S. 1996. Probing the interaction between two single molecules: fluorescence resonance energy transfer between a single donor and a single acceptor. *Proceedings of the National Academy of Sciences, USA*, **93**, 6264.

Haken, H., and Strobl, G. 1973. An exactly solvable model for coherent and incoherent exciton motion. *Zeitschrift für Physik*, **262**, 135.

Hall, L. T., Hill, C. D., Cole, J. H., Städler, B., Caruso, F., Mulvaney, P., Wrachtrup, J., and Hollenberg, L. C. L. 2010. Monitoring ion channel function in real time through quantum decoherence. *Proceedings of the National Academy of Sciences of the United States of America*, **107**, 18777–18782.

Hameroff, S. 1998. Quantum computation in brain microtubules? The Penrose-Hameroff 'Orch OR' model of consciousness. *Philosophical Transactions of the Royal Society of London A. (Mathematical, Physical and Engineering Sciences)*, **356**, 1869–1896.

Hameroff, S., Craddock, T. J. A., and Tuszynski, J. A. 2010. "Memory bytes"–molecular match for CaMKII phosphorylation encoding of microtubule lattices. *Journal of Integrative Neuroscience*, **9**, 253–267.

Hameroff, S. R. 2006. The entwined mysteries of anesthesia and consciousness – is there a common underlying mechanism? *Anesthesiology*, **105**, 400–412.

Hamm, P. 2006. Three-dimensional-IR spectroscopy: beyond the two-point frequency fluctuation correlation function. *Journal of Chemical Physics*, **124**, 124506.

Hamm, P., and Zanni, M. 2011. *Concepts and Methods of 2D Infrared Spectroscopy*. Cambridge: Cambridge University Press.

Hamm, P., Lim, M. H., and Hochstrasser, R. M. 1998. Structure of the amide I band of peptides measured by femtosecond nonlinear-infrared spectroscopy. *Journal of Physical Chemistry B*, **102**, 6123–6138.

Hänggi, P, and Jung, P. 1995. Colored noise in dynamical systems. *Advances in Chemical Physics*, **89**, 239–326.

Hänggi, P., Jung, P., Zerbe, C., and Moss, F. 1993. Can colored noise improve stochastic resonance? *Journal of Statistical Physics*, **70**(1), 25–47.

Harris, R. A., and Silbey, R. J. 1985. Dynamics of an asymmetric tunneling system interacting with a heat bath. *Journal of Chemical Physics*, **83**, 1069.

Harris, S.-R., Henbest, K. B., Maeda, K., Pannell, J. R., Timmel, C. R., Hore, P. J., and Okamoto, H. 2009. Effect of magnetic fields on cryptochrome-dependent responses in *Arabidopsis thaliana*. *Journal of the Royal Society Interface*, **6**, 1193–1205.

Hartmann, L., Calsamiglia, J., Dür, W., and Briegel, H. J. 2005. Spin gases as microscopic models for non-Markovian decoherence. *Physical Review A*, **72**, 052107.

Hartmann, L., Dür, W., and Briegel, H. J. 2006. Steady state entanglement in open and noisy quantum systems at high temperature. *Physical Review A*, **74**, 052304.

Hartmann, L., Dür, W., and Briegel, H. J. 2007. Entanglement and its dynamics in open, dissipative systems. *New Journal of Physics*, **9**, 230.

Hau, H. H. and Gralnick, J. A. 2007. Ecology and biotechnology of the Genus Shewanella. *Annual Review of Microbiology*, **61**, 237–258.

Haugland, R. P., Yguerabide, J., and Stryer, L. 1969. Dependence of the kinetics of singlet-singlet energy transfer on spectral overlap. *Proceedings of the National Academy of Sciences, USA*, **63**, 23.

Hayashi, S., Tajkhorshid, E., Pebay-Peyroula, E., Royant, A., Landau, E. M., Navarro, J., and Schulten, K. 2001. Structural determinants of spectral tuning in retinal proteins – bacteriorhodopsin vs. sensory rhodopsin II. *Journal of Physical Chemistry B*, **105**, 10124–10131.

Hayashi, S., Tajkhorshid, E., and Schulten, K. 2003. Molecular dynamics simulation of bacteriorhodopsin photoisomerization using ab initio forces for the excited chromophore. *Biophysical Journal*, **85**, 1440.

Hayashi, S., Tajkhorshid, E., and Schulten, K. 2009. Photochemical reaction dynamics of the primary event of vision studied by means of a hybrid molecular simulation. *Biophysical Journal*, **96**, 403–416.

Hayes, D., and Engel, G. S. 2011. Extracting the excitonic Hamiltonian of the Fenna-Matthews-Olson complex using three-dimensional third-order electronic spectroscopy. *Biophysical Journal*, **100**, 2043.

Hayes, D., Panitchayangkoon, G., Fransted, K. A., Caram, J. R., Wen, J., Freed, K. F., and Engel, G. S. 2010. Dynamics of electronic dephasing in the Fenna-Matthews-Olson complex. *New Journal of Physics*, **12**, 065042.

Hayes, D., Wen, J. Z., Panitchayangkoon, G., Blankenship, R. E., and Engel, G. S. 2011. Robustness of electronic coherence in the Fenna-Matthews-Olson complex to vibronic and structural modifications. *Faraday Discuss*, **150**, 459.

Hayward, S., and Gö, N. 1995. Collective variable description of native protein dynamics. *Annual Review of Physical Chemistry*, **46**, 223–250.

He, X.-F. 1991. Excitons in anisotropic solids: the model of fractional-dimensional space. *Physical Review B*, **43**, 2063–2069.

Heeger, A. J., Kivelson, S., Schrieffer, J. R., and Su, W. P. 1988. Solitons in conducting polymers. *Reviews of Modern Physics*, **60**, 781–850.

Hein, M., Dür, W., Eisert, J., Raussendorf, R., Van den Nest, M., and Briegel, H. J. 2006. Entanglement in graph states and its applications. In: Zoller, P., Casati, G., Shepelyansky, D., and Benenti, G. (eds), *Proceedings International School of Physics Enrico Fermi on quantum computers, algorithms and chaos (Varenna, Italy)*, vol. 162.

Hemelrijk, P. W., Kwa, S. L. S., Vangrondelle, R., and Dekker, J. P. 1992. Spectroscopic properties of LHC-II, the main light-harvesting chlorophyll-a/b protein complex from chloroplast membranes. *Biochimica et Biophysica Acta*, **1098**(2), 159–166.

Henbest, K. B., Kukura, P., Rodgers, C. T., Hore, P. J., and Timmel, C. R. 2004. Radio frequency magnetic field effects on radical recombination reaction: a diagnostic test for the radical pair mechanism. *Journal of the American Chemical Society*, **126**, 8102–8103.

Henbest, K. B., Athanassiades, E., Maeda, K., Kuprov, I., Hore, P. J., and Timmel, C. R. 2006. Photoionization of TMPD in DMSO solution: mechanism and magnetic field effects. *Molecular Physics*, **104**, 1789–1794.

Henbest, K. B., Maeda, K., Hore, P. J., Joshi, M., Bacher, A., Bittl, R., Weber, S., Timmel, C. R., and Schleicher, E. 2008. Magnetic-field effect on the photoactivation reaction of *Escherichia coli* DNA photolyase. *Proceedings of the National Academy of Sciences, USA*, **105**, 14395–14399.

Heyduk, T. 2002. Recent advances in FRET: distance determination in protein-DNA complexes. *Current Opinion in Biotechnology*, **13**, 292.

Heyers, D., Manns, M., Luksch, H., Güntürkün, O., and Mouritsen, H. 2007. A visual pathway links brain structures active during magnetic compass orientation in migratory birds. *PLoS ONE*, **2**, e937.

Hill, E., and Ritz, T. 2010. Can disordered radical pair systems provide a basis for a magnetic compass in animals? *Journal of the Royal Society Interface*, **7**, S265–S271.

Hille, B. 2001. *Ion Channels of Excitable Membranes*. Third edn. Sunderland, Massachusetts: Sinauer.

Hillisch, A., Lorenz, M., and Diekmann, S. 2001. Recent advances in FRET: distance determination in protein-DNA complexes. *Current Opinion in Structural Biology*, **11**, 201.

Hindin, E., Forties, R. A., Loewe, R. S., Ambroise, A., Kirmaier, C., Bocian, D., Lindsey, J. S., Holten, D., and Knox, R. S. 2004. Excited-state energy flow in covalently linked multiporphyrin arrays: the essential contribution of energy transfer between nonadjacent chromophores. *Journal of Physical Chemistry B*, **108**, 12821.

Hoang, N., Schleicher, E., Kacprzak, S., Bouly, J.-P., Picot, M., Wu, W., Berndt, A., Wolf, E., Bittl, R., and Ahmad, M. 2008. Human and *Drosophila* cryptochromes are light activated by flavin photoreduction in living cells. *PLoS Biology*, **6**, 1559–1569.

Hochstrasser, R. M. 2001. Two-dimensional IR-spectroscopy: polarization anisotropy effects. *Chemical Physics*, **266**, 273.

Hochstrasser, R. M. 2007. Two-dimensional spectroscopy at infrared and optical frequencies. *Proceedings of the National Academy of Sciences of the United States of America*, **104**, 14190–14196.

Hoffmann, M., Wanko, M., Strodel, P., König, P. H., Frauenheim, T., Schulten, K., Thiel, W., Tajkhorshid, E., and Elstner, M. 2006. Color tuning in rhodopsins: the mechanism for the spectral shift between bacteriorhodopsin and sensory rhodopsin II. *Journal of the American Chemical Society*, **128**, 10808–10818.

Hogben, H. J., Efimova, O., Wagner-Rundell, N., Timmel, C. R., and Hore, P. J. 2009. Possible involvement of superoxide and dioxygen with cryptochrome in avian magnetoreception: Origin of Zeeman resonances observed by *in vivo* EPR spectroscopy. *Chemical Physics Letters*, **490**, 118–122.

Hogben, H. J., Biskup, T., and Hore, P. J. 2012. Entanglement and sources of magnetic anisotropy in radical pair-based avian magnetoreceptors. *Physical Review Letters*, **109**, 220501.

Holstein, T. 1959a. Studies of polaron motion: part I. The molecular-crystal model. *Annals of Physics*, **8**, 325.

Holstein, T. 1959b. Studies of polaron motion: part II. The "small" polaron. *Annals of Physics*, **8**, 343.

Holt, N. E., Fleming, G. R., and Niyogi, K. K. 2004. Toward an understanding of the mechanism of nonphotochemical quenching in green plants. *Biochemistry*, **43**(26), 8281–8289.

Holthaus, M., and Hone, D. W. 1996. Localization effects in ac-driven tight-binding lattices. *Philosophical Magazine B – Physics of Condensed Matter Statistical Mechanics Electronic Optical and Magnetic Properties*, **74**, 105–137.

Holzapfel, W., Finkele, U., Kaiser, W., Oesterhelt, D., Scheer, H., Stilz, H. U., and Zinth, W. 1989. Observation of a bacteriochlorophyll anion radical during the primary charge separation in a reaction center. *Chemical Physics Letters*, **160**(1), 1–7.

Holzapfel, W., Finkele, U., Kaiser, W., Oesterhelt, D., Scheer, H., Stilz, H. U., and Zinth, W. 1990. Initial electron-transfer in the reaction center from *Rhodobacter sphaeroides*. *Proceedings of the National Academy of Sciences, USA*, **87**(13), 5168–5172.

Holzwarth, A. R. 2004. Light absorption and harvesting. *Molecular to Global Photosynthesis*, 43–115.

Hopfield, J. J. 1974. Electron transfer between biological molecules by thermally activated tunneling. *Proceedings of the National Academy of Sciences, USA*, **71**, 3640–3644.

Horodecki, M., Shor, P. W., and Ruskai, M. B. 2003. General entanglement breaking channels. *Reviews in Mathematical Physics*, **15**, 629.

Horodecki, R., Horodecki, P., Horodecki, M., and Horodecki, K. 2009. Quantum entanglement. *Reviews of Modern Physics*, **81**, 865–942.

Hossein-Nejad, H., and Scholes, G. D. 2010. Energy transfer, entanglement and decoherence in a molecular dimer interacting with a phonon bath. *New Journal of Physics*, **12**, 065045.

Hossein-Nejad, H., Curutchet, C., Kubica, A., and Scholes, G. D. 2011. Delocalization-enhanced long-range energy transfer between cryptophyte algae PE545 antenna proteins. *Journal of Physical Chemistry B*, **115**, 5243.

Hsiao, J.-S., Krueger, B., Wagner, W., Johonson, T. E., Delaney, J. K., Mazuerall, D., Fleming, G. R., Lindsey, J. S., Bocian, D. F., and Donohoe, R. J. 1996. Soluble synthetic multiporphyrin arrays. 2. Photodynamics of energy-transfer processes. *Journal of the American Chemical Society*, **118**, 11181.

Hsu, C.-P. 2009. The electronic couplings in electron transfer and excitation energy transfer. *Accounts of Chemical Research*, **42**, 509–518.

Hu, X., and Schulten, K. 1997. How nature harvests sunlight. *Physics Today*, **50**, 28–34.

Hu, X., and Schulten, K. 1998. A model for the light-harvesting complex I (B875) of *rhodobacter sphaeroides*. *Biophysical Journal*, **75**, 683–694.

Hu, X., Damjanović, A., Ritz, T., and Schulten, K. 1998. Architecture and function of the light harvesting apparatus of purple bacteria. *Proceedings of the National Academy of Sciences, USA*, **95**, 5935–5941.

Hu, X., Ritz, T., Damjanović, A., Autenrieth, F., and Schulten, K. 2002. Photosynthetic apparatus of purple bacteria. *Quarterly Reviews of Biophysics*, **35**(1), 1–62.

Haug, H., and Koch, S. W. 2004. *Quantum Theory of the Optical and Electronic Properties of Semiconductors*. Fourth edn. Singapore: World Scientific.

Huang, K., and Rhys, A. 1950. Theory of light absorption and non-radiative transitions in F-centres. *Proceedings of the Royal Society of London A. (Mathematical, Physical and Engineering Sciences)*, **A204**, 406–423.

Huang, L., and Krauss, T. D. 2006. Quantized bimolecular Auger recombination of excitons in single-walled carbon nanotubes. *Physical Review Letters*, **96**, 057407.

Hudson, B. S., Kohler, B. E., and Schulten, K. 1982. Excited States. vol. 6. New York: Academic Press.

Hughes, A. L. 1999. *Adaptive Evolution of Genes and Genomes*. New York: Oxford University Press.

Hunter, C. N., Daldal, F., Thurnauer, M. C., and Beatty, J. T. (eds). 2009. *The Purple Phototrophic Bacteria*. Netherlands: Springer.

Ihalainen, J. A., Linnanto, J., Myllyperkio, P., van Stokkum, I. H. M., Ucker, B., Scheer, H., and Korppi-Tommola, J. E. I. 2001. Energy transfer in LH2 of Rhodospirillum molischianum, studied by subpicosecond spectroscopy and configuration interaction exciton calculations. *Journal of Physical Chemistry B*, **105**(40), 9849–9856.

Iozzi, M. F., Mennucci, B., Tomasi, J., and Cammi, R. 2004. Excitation energy transfer (EET) between molecules in condensed matter: a novel application of the polarizable continuum model (PCM). *Journal of Chemical Physics*, **120**, 7029.

Ishizaki, A., and Fleming, G. R. 2009a. Theoretical examination of quantum coherence in a photosynthetic system at physiological temperature. *Proceedings of the National Academy of Sciences, USA*, **106**, 17255–17260.

Ishizaki, A., and Fleming, G. R. 2009b. Unified treatment of quantum coherent and incoherent hopping dynamics in electronic energy transfer: reduced hierarchy equation approach. *Journal of Chemical Physics*, **130**, 234111.

Ishizaki, A., and Fleming, G. R. 2010. Quantum superpositions in photosynthetic light harvesting: delocalization and entanglement. *New Journal of Physics*, **12**, 055004.

Ishizaki, A., and Fleming, G. R. 2011. On the interpretation of quantum coherent beats observed in two-dimensional electronic spectra of photosynthetic light harvesting complexes. *Journal of Physical Chemistry B*, **115**, 6227–6233.

Ishizaki, A., and Fleming, G. R. 2012. Quantum coherence in photosynthetic light harvesting. *Annual Review of Condensed Matter Physics*, **3**, 333–361.

Ishizaki, A., and Tanimura, Y. 2005. Quantum dynamics of system strongly coupled to low-temperature colored noise bath: reduced hierarchy equations approach. *Journal of the Physical Society of Japan*, **74**, 3131.

Ishizaki, A., Calhoun, T. R., Schlau-Cohen, G. S., and Fleming, G. R. 2010. Quantum coherence and its interplay with protein environments in photosynthetic electronic energy transfer. *Physical Chemistry – Chemical Physics*, **12**, 7319.

Jackson, B., and Silbey, R. J. 1983. On the calculation of transfer rates between impurity states in solids. *Journal of Chemical Physics*, **78**, 4193.

Jackson, M., and Mantsch, H. H. 1995. The use and misuse of FTIR spectroscopy in the determination of protein-structure. *Critical Reviews in Biochemistry and Molecular Biology*, **30**, 95–120.

Jang, S. 2007. Generalization of the Förster resonance energy transfer theory for quantum mechanical modulation of the donor-acceptor coupling. *Journal of Chemical Physics*, **127**, 174710.

Jang, S. 2009. Theory of coherent resonance energy transfer for coherent initial condition. *Journal of Chemical Physics*, **131**, 164101.

Jang, S. 2011. Theory of multichromophoric coherent resonance energy transfer: a polaronic quantum master equation approach. *Journal of Chemical Physics*, **135**, 034105.

Jang, S., and Cheng, Y.-C. 2013. Resonance energy flow dynamics of coherently delocalized excitons in biological and macromolecular systems: Recent theoretical advances and open issues. *WIREs Computational Molecular Science*, **3**, 84.

Jang, S., and Silbey, R. J. 2003. Single complex line shapes of the B850 band of LH2. *Journal of Chemical Physics*, **118**, 9324.

Jang, S., Cao, J., and Silbey, R. J. 2002a. Fourth order quantum master equation and its Markovian bath limit. *Journal of Chemical Physics*, **116**, 2705.

Jang, S., Jung, Y. J., and Silbey, R. J. 2002b. Nonequilibrium generalization of Förster-Dexter theory for excitation energy transfer. *Chemical Physics*, **275**, 319.

Jang, S., Jung, Y. J., and Silbey, R. J. 2002c. Nonequilibrium generalization of Förster-Dexter theory for excitation energy transfer. *Chemical Physics*, **275**, 319.

Jang, S., Newton, M. D., and Silbey, R. J. 2004. Multichromophoric Förster resonance energy transfer. *Physical Review Letters*, **92**, 218301.

Jang, S., Newton, M. D., and Silbey, R. J. 2007. Multichromophoric Förster resonance energy transfer from B800 to B850 in the light harvesting complex 2: evidence for subtle energetic optimization by purple bacteria. *Journal of Physical Chemistry B*, **111**, 6807.

Jang, S., Chen, Y.-C., Reichman, D. R., and Eaves, J. D. 2008. Theory of coherent resonance energy transfer. *Journal of Chemical Physics*, **129**, 101104.

Janosi, L., Kosztin, I., and Damjanović, A. 2006. Theoretical prediction of spectral and optical properties of bacteriochlorophylls in thermally disordered LH2 antenna complexes. *Journal of Chemical Physics*, **125**(1), 014903.

Jelley, E. E. 1936. Spectral absorption and fluorescence of dyes in the molecular state. *Nature*, **138**, 1009.

Jelley, E. E. 1937. Molecular, nematic and crystal states of I: I-Diethyl–Cyanine Chloride. *Nature*, **139**, 631.

Jiang, X.-P., and Brumer, P. 1991. Creation and dynamics of molecular states prepared with coherent vs partially coherent pulses of light. *Journal of Chemical Physics*, **94**, 5833–5843.

Jimenez, R., Dikshit, S., Bradforth, S. E., and Fleming, G. R. 1996. Electronic excitation transfer in the LH2 complex of rhodobacter sphaeroides. *Journal of Physical Chemistry*, **100**, 6825.

Johnsen, S., and Lohmann, K. J. 2008. Magnetoreception in animals. *Physics Today*, **61**, 29–35.

Jonas, D. M. 2003. Two-dimensional femtosecond spectroscopy. *Annual Review of Physical Chemistry*, **54**, 425.

Jonas, D. M., Lang, M. J., Nagasawa, Y., Joo, T., and Fleming, G. R. 1996. Pump-probe polarization anisotropy study of femtosecond energy transfer within the photosynthetic reaction center of *Rhodobacter sphaeroides* R26. *Journal of Physical Chemistry*, **100**(30), 12660–12673.

Joo, T., Jia, Y., Yu, J.-Y., Lang, M. J., and Fleming, G. R. 1996. Third-order nonlinear time domain probes of solvation dynamics. *Journal of Chemical Physics*, **104**, 6089.

Jortner, J. 1976. Temperature-dependent activation-energy for electron-transfer between biological molecules. *Journal of Chemical Physics*, **64**(12), 4860–4867.

Jortner, J., and Bixon, M. 1999. Electron Transfer: From Isolated Molecules to Biomolecules. *Advances in Chemical Physics*, **106**, 35–202.

Jortner, J., and Ratner, M. A. 1997. *Molecular Electronics*. Oxford: Blackwell Science.

Juzeliūnas, G. and Andrews, D. L. 1994. Quantum electrodynamics of resonant energy transfer in condensed matter. *Physical Review B*, **49**, 8751.

Kao, Y.-T., Tan, C., Song, S.-H., Öztürk, N., Li, J., Wang, L., Sancar, A., and Zhong, D. 2008. Ultrafast dynamics and anionic active states of the flavin cofactor in cryptochrome and photolyase. *Journal of the American Chemical Society*, **130**, 7695–7701.

Katterle, M., Prokhorenko, V. I., Holzwarth, A. R., and Jesorka, A. 2007. An artificial supramolecular photosynthetic unit. *Chemical Physics Letters*, **447**, 284.

Keary, N., Ruploh, T., Voss, J., Thalau, P., Wiltschko, R., Wiltschko, W., and Bischof, H.-J. 2009. Oscillating magnetic field disrupts magnetic orientation in Zebra finches, *Taeniopygia guttata*. *Frontiers in Zoology*, **6**, 25.

Khalil, M., Demirdoven, N., and Tokmakoff, A. 2003. Coherent 2D-IR spectroscopy: molecular structure and dynamics in solution. *Journal of Physical Chemistry A*, **107**, 5258–5279.

Khitrova, G., Gibbs, H. M., Jahnke, F., Kira, M., and Koch, S. W. 1999. Nonlinear optics of normal-mode-coupling semiconductor microcavities. *Reviews of Modern Physics*, **71**, 1591–1639.

Khorana, H. G. J. 1992. Rhodopsin, photoreceptor of the rod cell. An emerging pattern for structure and function. *Journal of Biological Chemistry*, **267**, 1–4.

Kim, D., and Osuka, A. 2004. Directly linked porphyrin arrays with tunable excitonic interactions. *Accounts of Chemical Research*, **37**, 735.

Kimura, A. 2009. General theory of excitation energy transfer in donor-mediator-acceptor systems. *Journal of Chemical Physics*, **130**, 154103.

Kirmaier, C., Weems, D., and Holten, D. 1999. M-side electron transfer in reaction center mutants with a lysine near the nonphotoactive B bacteriochlorophyll. *Biochemistry*, **38**(35), 11516–11530.

Kirschvink, J., Walker, M., and Diebel, C. 2001. Magnetite-based magnetoreception. *Current Opinion in Neurobiology*, **11**, 462–467.

Kirschvink, J. L., and Gould, J. L. 1981. Biogenic magnetite as a basis for magnetic-field detection in animals. *Biosystems*, **13**, 181–201.

Kirtley, J., Scalapino, D. J., and Hansma, P. K. 1976. Theory of vibrational mode intensities in inelastic electron tunneling spectroscopy. *Physical Review B*, **14**, 3177–3184.

Kleiman, V. D., Melinger, J. S., and McMorrow, D. 2001. Ultrafast dynamics of electronic excitations in a light-harvesting phenylacetylene dendrimer. *Journal of Physical Chemistry B*, **105**, 5595.

Klopping, H. L. 1971. Olfactory theories and the odors of small molecules. *Journal of Agricultural and Food Chemistry*, **19**, 999–1004.

Klug, A. 1999. The tobacco mosaic virus particle: structure and assembly. *Philosophical Transactions of the Royal Society of London B. (Biological Sciences)*, **354**, 531.

Knoester, J. 2002. Optical properties of molecular aggregates. Pages 149–186 in: *Proceedings – International School of Physics Enrico Fermi*, vol. 149. IOS Press; Ohmsha; 1999.

Knox, R. S., and van Amerongen, H. 2002. Refractive index dependence of the Förster resonance excitation transfer rate. *Journal of Physical Chemistry B*, **106**, 5289.

Kobayashi, T. 1996. *J-Aggregates*. Singapore: World Scientific Publishing Company.

Kobayashi, T., Saito, T., and Ohtani, H. 2001. Real-time spectroscopy of transition states in bacteriorhodopsin during retinal isomerization. *Nature*, **414**, 531–534.

Kobus, M., Nguyen, P. H., and Stock, G. 2011. Coherent vibrational energy transfer along a peptide helix. *The Journal of chemical physics*, **134**, 124518.

Koch, C., and Hepp, K. 2006. Quantum mechanics in the brain. *Nature*, **440**, 611.

Kochendoerfer, G. G., and Mathies, R. A. 1996. Spontaneous emission study of the femtosecond isomerization dynamics of rhodopsin. *Journal of Physical Chemistry*, **100**, 14526–14532.

Koder, R. L., Anderson, J. L. R., Solomon, L. A., Reddy, K. S., Moser, C. C., and Dutton, P. L. 2009. Design and engineering of an O2 transport protein. *Nature*, **458**(7236), 305–309.

Koepke, J., Hu, X. C., Muenke, C., Schulten, K., and Michel, H. 1996. The crystal structure of the light-harvesting complex II (B800-850) from Rhodospirillum molischianum. *Structure*, **4**(5), 581–597.

Kohler, S., Lehmann, J., and Hänggi, P. 2005. Driven quantum transport on the nanoscale. *Physics Reports*, **406**, 379–443.

Kolber, Z. S., Dover, C. L. Van, Niederman, R. A., and Falkowski, P. G. 2000. Bacterial photosynthesis in surface waters of the open ocean. *Nature*, **407**, 177–179.

Kolli, A., Nazir, A., and Olaya-Castro, A. 2011. Electronic excitation dynamics in multichromophoric systems described via a polaron-representation master equation. *Journal of Chemical Physics*, **135**, 154112.

Kosztin, I., and Schulten, K. 2008. Molecular dynamics methods for bioelectronic systems in photosynthesi. Chap. 22, pages 445–464 of: Aartsma, T. J., and Matysik, J. (eds), *Biophysical Techniques in Photosynthesis*. Advances in Photosynthesis and Respiration, vol. 26. Dordrecht: Springer.

Kothe, G., Weber, S., Bittl, R., Ohmes, E., Thurnauer, M. C., and Norris, J. R. 1991. Transient EPR of light-induced radical pairs in plant photosystem-I – observation of quantum beats. *Chemical Physics Letters*, **186**, 474–480.

Koushik, S. V., Chen, H., Thaler, C., Puhl III, H. L., and Vogel, S. S. 2006. Cerulean, venus, and venus$_{Y67C}$ FRET reference standards. *Biophysical Journal*, **91**, L99.

Koyama, Y., Kubo, K., Komori, M., Yasuda, H., and Mukai, Y. 1991. Effect of protonation on the isomerization properties of n-butylamine Schiff base of isomeric retinal as revealed by direct HPLC analyses: selection of isomerization pathways by retinal proteins. *Photochemistry and Photobiology*, **54**, 433–443.

Krameri, B., and MacKinnont, A. 1993. Localization: theory and experiment. *Reports on Progress in Physics*, **56**, 146.

Kraus, K. 1983. *States, Effects and Operations*. Berlin: Springer-Verlag.

Krueger, B. P., Scholes, G. D., and Fleming, G. R. 1998. Calculation of couplings and energy-transfer pathways between the pigments of LH2 by the ab initio transition density cube method. *Journal of Physical Chemistry B*, **102**, 5378.

Kruger, T. P. J., Novoderezhkin, V. I., Ilioaia, C., and van Grondelle, R. 2010. Fluorescence spectral dynamics of single LHCII trimers. *Biophysical Journal*, **98**(12), 3093–3101.

Kubo, R. 1962. Generalized cumulant expansion method. *Journal of the Physical Society of Japan*, **17**, 1100.

Kubo, R. 1963. Stochastic liouville equations. *Journal of Mathematical Physics*, **4**, 174.

Kubo, R. 1966. The fluctuation-dissipation theorem. *Reports on Progress in Physics*, **29**, 255.

Kubo, R., Toda, M., and Hashitsume, N. 2003. *Statistical Physics II: Nonequilibrium Statistical Mechanics*. Berlin: Springer Verlag.

Kukura, P., McCamant, D. W., Yoon, S., Wandschneider, D. B., and Mathies, R. A. 2005. Structural observation of the primary isomerization in vision with femtosecond-stimulated Raman. *Science*, **310**, 1006–1009.

Kuznetsov, A. M., and J., Ulstrup. 1999. *Electron Transfer in Chemistry and Biology: An Introduction to the Theory*. Chichester, UK: Wiley.

Lambe, J., and Jaklevic, R. C. 1968. Molecular vibration spectra by inelatic electron tunneling. *Physical Review*, **165**(3), 821–832.

Lau, J. C. S., Wagner-Rundell, N., Rodgers, C. T., Green, N. J. B., and Hore, P. J. 2010. Effects of disorder and motion in a radical pair magnetoreceptor. *Journal of the Royal Society Interface*, **7**, S257–S264.

Le Sage, D., Arai, K., Glenn, D. R., DeVience, S. J., Pham, L. M., Rahn-Lee, L., Lukin, M. D., Yacoby, A., Komeili, A., and Walsworth, R. L. 2013. Optical magnetic imaging of living cells. *Nature*, **496**(7446), 486–489.

Lee, H., Cheng, Y.-C., and Fleming, G. R. 2007. Coherence dynamics in photosynthesis: protein protection of excitonic coherence. *Science*, **316**, 1462–1465.

Leegwater, J. A. 1996. Coherent versus incoherent energy transfer and trapping in photosynthetic antenna complexes. *The Journal of Physical Chemistry*, **100**, 14403.

Leggett, A. J., Chakravarty, S., Dorsey, A. T., Fisher, M. P. A., Garg, A., and Zwerger, W. 1987. Dynamics of the dissipative two-state system. *Reviews of Modern Physics*, **59**, 1.

Li, B., Wang, L., Kang, B., Wang, P., and Qiu, Y. 2006. Review of recent progress in solid-state dye-sensitized solar cells. *Solar Energy Materials and Solar Cells*, **90**(5), 549–573.

Li, Y. F., Zhou, W., Blankenship, R. E., and Allen, J. P. 1997. Crystal structure of the bacteriochlorophyll a protein from Chlorobium tepidum. *Journal of Molecular Biology*, **271**, 456.

Liedvogel, M., Maeda, K., Henbest, K., Schleicher, E., Simon, T., Timmel, C. R., Hore, P. J., and Mouritsen, H. 2007a. Chemical magnetoreception: bird cryptochrome 1a is excited by blue light and forms long-lived radical-pairs. *PLoS ONE*, **2**, e1106.

Liedvogel, M., Feenders, G., Wada, K., Troje, N. F., Jarvis, E. D., and Mouritsen, H. 2007b. Lateralised activation of Cluster N in the brains of migratory songbirds. *European Journal of Neuroscience*, **25**, 1166–1173.

Lin, C., and Todo, T. 2005. The cryptochromes. *Genome Biology*, **6**, 220.

Lin, J. P., Balabin, I. A., and Beratan, D. N. 2005. The nature of aqueous tunneling pathways between electron-transfer proteins. *Science*, **301**, 1311–1313.

Lin, X., Murchison, H. A., Nagarajan, V., Parson, W. W., Allen, J. P., and Williams, J. C. 1994. Specific alteration of the oxidation potential of the electron-donor in reaction centers from *Rhodobacter sphaeroides*. *Proceedings of the National Academy of Sciences, USA*, **91**(22), 10265–10269.

Lin, Z., Lawrence, C. M., Xiao, D., Kireev, V. V., Skourtis, S. S., Sessler, J. L., Beratan, D. N., and Rubtsov, I. V. 2009. Modulating unimolecular charge transfer by exciting bridge vibrations. *Journal of the American Chemical Society*, **131**, 18060–18062.

Lindblad, G. 1976. On the generators of quantum dynamical semigroups. *Communications in Mathematical Physics*, **48**, 119.

Linnanto, J. M., and Korppi-Tommola, J. E. I. 2008. Investigation on chlorosomal antenna geometries: tube, lamella and spiral-type self-aggregates. *Photosynthesis Research*, **96**(3), 227–245.

Lipman, E. A., Schuler, B., Bakajin, O., and Eaton, W. A. 2003. Single-molecule measurement of protein folding kinetics. *Science*, **301**, 1233.

Litt, A., Aliasmith, C., Kroon, F. W., Weinstein, S., and Thagard, P. 2006. Is the brain a quantum computer? *Cognitive Science*, **30**, 593–603.

Liu, Z. F., Yan, H. C., Wang, K. B., Kuang, T. Y., Zhang, J. P., Gui, L. L., An, X. M., and Chang, W. R. 2004. Crystal structure of spinach major light-harvesting complex at 2.72 angstrom resolution. *Nature*, **428**(6980), 287–292.

Lo, H.-K., Spiller, T., and Popescu, S. (eds). 1998. *Introduction to Quantum Computation and Information*. Singapore: World Scientific.

Logan, B. E. 2009. Exoelectrogenic bacteria that power microbial fuel cells. *Nature Reviews Microbiology*, **7**, 375–381.

Logunov, S. L., Sang, L., and El-Sayed, M. A. 1996. Excited state dynamics of a protonated retinal Schiff base in solution. *Journal of Physical Chemistry*, **100**, 18586–18591.

Loll, B., Kern, J., Saenger, W., Zouni, A., and Biesiadka, J. 2005. Towards complete cofactor arrangement in the 3.0 Åresolution structure of photosystem II. *Nature*, **438**, 1040–1044.

Longuet-Higgins, H. C. 1962. Quantum mechanics and biology. *Biophysical Journal*, **2**, 207–215.

Lovley, D. R., and Phillips, E. J. P. 1988. Novel mode of microbial energy metabolism: organic carbon oxidation coupled to dissimilatory reduction of iron or manganese. *Applied and Environmental Microbiology*, **54**, 1472–1480.

Löwdin, P.-O. 1951. A note on the quantum-mechanical perturbation theory. *Journal of Chemical Physics*, **19**, 1396.

Lu, Q., Lu, H. P., and Wang, J. 2007. Exploring the mechanism of flexible biomolecular recognition with single molecule dynamics. *Physical Review Letters*, **98**, 128105.

Ma, Y.-Z., Stenger, J., Zimmermann, J., Bachilo, S. M., Smalley, R. E., Weisman, R. B., and Fleming, G. R. 2004. Ultrafast carrier dynamics in single-walled carbon nanotubes probed by femtosecond spectroscopy. *Journal of Physical Chemistry*, **120**, 3368–3373.

Ma, Y.-Z., Valkunas, L., Dexhemer, S. L., Bachilo, S. M., and Fleming, G. R. 2005. Femtosecond spectroscopy of optical excitations in single-walled carbon nanotubes: evidence for exciton-exciton annihilation. *Physical Review Letters*, **94**, 157402.

Ma, Y.-Z., Hertel, T., Vardeny, Z. V., Fleming, G. R., and Valkunas, L. 2008. Carbon Nanotubes: Advanced Topics in the Synthesis, Structure, Properties and Applications. Heidelberg: Springer Verlag.

Maeda, K., Henbest, K. B., Cintolesi, F., Kuprov, I., Rodgers, C. T., Liddell, P. A., Gust, D., Timmel, C. R., and Hore, P. J. 2008. Chemical compass model of avian magnetoreception. *Nature*, **453**, 387–390.

Mahan, G. D. 1990. *Many-Particle Physics*. 2nd edn. New York: Plenum Press.

Makarov, D. E., and Makri, N. 1994. Path-integrals for dissipative systems by tensor multiplications – condensed-phase quantum dynamics for arbitrarily long time. *Journal of Chemical Physics*, **221**, 482–491.

Makri, N. 1999. The linear response approximation and its lowest order corrections: an influence functional approach. *Journal of Physical Chemistry B*, **103**(15), 2823–2829.

Makri, N., and Makarov, D. E. 1995a. Tensor propagator for iterative quantum time evolution of reduced density-matrices. 1. Theory. *Journal of Chemical Physics*, **102**, 4600–4610.

Makri, N., and Makarov, D. E. 1995b. Tensor propagator for iterative quantum time evolution of reduced density-matrices. 2. Numerical methodology. *Journal of Chemical Physics*, **102**, 4611.

Malvankar, N. S., Vargas, M., Nevin, K. P., Franks, A. E.Ching, Leang, C., Kim, B-C., Inoue, K., Mester, T., Covalla, S. F., Johnson, J. P., Rotello, V. M., Tuominen, M. T., and Lovley, D. R. 2011. Tunable metallic-like conductivity in microbial nanowire networks. *Nature Nanotechnology*, **6**, 573–579.

Malvankar, N. S., T., Tuominena M., and Lovley, D. R. 2012. Comment on "On electrical conductivity of microbial nanowires and biofilms" by S. M. Strycharz-Glaven, R. M. Snider, A. Guiseppi-Elie and L. M. Tender, Energy Environ. Sci., 2011, 4, 4366. *Energy & Environmental Science*, **5**, 6247–6249.

Manor, J., Mukherjee, P., Lin, Y. S., Leonov, H., Skinner, J. L., Zanni, M. T., and Arkin, I. T. 2009. Gating mechanism of the influenza A M2 channel revealed by 1D and 2D-IR spectroscopies. *Structure*, **17**, 247–254.

Manske, A. K., Glaeser, J., Kuypers, M. M. M., and Overmann, J. 2005. Physiology and phylogeny of green sulfur bacteria forming a monospecific phototrophic assemblage at a depth of 100 meters in the Black Sea. *Applied and Environmental Microbiology*, **71**(12), 8049–8060.

Mančal, T., and Fleming, G. R. 2004. Probing electronic coupling in excitonically coupled heterodimer complexes by two-color three-pulse photon echoes. *Journal of Chemical Physics*, **121**(21), 10556.

Mančal, T., and Šanda, F. 2012. Quantum master equations for non-linear optical response of molecular systems. *Chemical Physics Letters*, **530**, 140–144.

Mančal, T., Christensson, N., Lukeš, V., Milota, F., Bixner, O., Kauffmann, H. F., and Hauer, J. 2012. System-dependent signatures of electronic and vibrational coherence in electronic two-dimensional spectra. *Journal of Physical Chemistry Letters*, **3**, 1497.

Marcus, R., and Sutin, N. 1985. Electron transfers in chemistry and biology. *Biochimica et Biophysica Acta*, **811**, 265–322.

Marcus, R. A. 1956a. Electrostatic free energy and other properties of states having nonequilibrium polarization II. *Journal of Chemical Physics*, **24**, 979–989.

Marcus, R. A. 1956b. On the energy of oxidation-reduction reactions involving electron transfer I. *Journal of Chemical Physics*, **24**, 966–978.

Marcus, R. A. 1964. Chemical and electrochemical electron-transfer theory. *Annual Review of Physical Chemistry*, **15**, 155–196.

Marcus, R. A. 1993. Electron-transfer reactions in chemistry – theory and experiment. *Reviews of Modern Physics*, **65**(3), 599–610.

Martin, J. L., Breton, J., Hoff, A. J., Migus, A., and Antonetti, A. 1986. Femtosecond spectroscopy of electron-transfer in the reaction center of the photosynthetic bacterium *Rhodopseudomonas-sphaeroides* R-26 – direct electron-transfer from the dimeric bacteriochlorophyll primary donor to the bacteriopheophytin acceptor with a time constant 2.8 +/− 0.2 ps. *Proceedings of the National Academy of Sciences, USA*, **83**(4), 957–961.

Martin, K. E., Wang, Z., Busani, T., Garcia, R. M., Chen, Z., Jiang, Y., Song, Y., Jacobsen, J. L., Vu, T. T., Schore, N. E., Swartzentruber, B. S., Medforth, C. J., and Shelnutt, J. A. 2010. Donor-acceptor biomorphs from the ionic self-assembly of porphyrins. *Journal of the American Chemical Society*, **132**, 8194.

May, V., and Kühn, O. 2000. *Charge and Energy Transfer Dynamics in Molecular Systems*. Berlin: WILEY-VCH.

May, V., and Kühn, O. 2001. *Charge and Energy Transfer Dynamics in Molecular Systems*. Berlin: Wiley-VCH.

May, V., and Kühn, O. 2004. *Charge and Energy Transfer Dynamics in Molecular Systems*. Weinheim: Wiley-VCH.

May, V., and Kühn, O. 2011. *Charge and Energy Transfer Dynamics in Molecular Systems*. Weinheim, Germany: Wiley-VCH.

McConnell, H. M. 1961. Intramolecular charge transfer in aromatic free radicals. *Journal of Chemical Physics*, **35**, 508.

McCutcheon, D. P. S., and Nazir, A. 2011. Consistent treatment of coherent and incoherent energy transfer dynamics using a variational master equation. *Journal of Chemical Physics*, **135**, 114501.

McDermott, G., Prince, S. M., Freer, A. A., Hawthornthwaite-Lawless, A. M., Paplz, M. Z., Cogdell, R. J., and Issacs, N. W. 1995. Crystal structure of an integral membrane light-harvesting complex from photosynthetic bacteria. *Nature*, **374**, 517–521.

McKemmish, L. K., Reimers, J. R., McKenzie, R. H., Mark, A. E., and Hush, N. S. 2009. Penrose-Hameroff orchestrated objective-reduction proposal for human consciousness is not biologically feasible. *Physical Review E*, **80**, 021912.

Medvedev, E. S., and Stuchebrukhov, A. 1997. Inelastic tunneling in long-distance biological electron transfer reactions. *Journal of Chemical Physics*, **107**, 3821–3831.

Meier, C., and Tannor, D. J. 1999. Non-Markovian evolution of the density operator in the presence of strong laser fields. *Journal of Chemical Physics*, **111**, 3365.

Meier, T., Chernyak, V., and Mukamel, S. 1997. Multiple exciton coherence sizes in photosynthetic antenna complexes viewed by pump-probe spectroscopy. *Journal of Physical Chemistry B*, **101**(37), 7332–7342.

Mercer, I. P., Gould, I. R., and Klug, D. R. 1999. A quantum mechanical/molecular mechanical approach to relaxation dynamics: calculation of the optical properties of solvated bacteriochlorophyll-a. *Journal of Physical Chemistry B*, **103**(36), 7720–7727.

Michl, J. 1990. *Electronic Aspects of Organic Photochemistry*. New York: Wiley-Interscience.

Miller, R. A., Presley, A. D., and Francis, M. B. 2007. Self-assembling light-harvesting systems from synthetically modified tobacco mosaic virus coat proteins. *Journal of the American Chemical Society*, **129**, 3104.

Miller, R. A., Stephanopoulos, N., McFarland, J. M., Rosko, A. S., Geissler, P. L., and Francis, M. B. 2010. Impact of assembly state on the defect tolerance of TMV-based light harvesting arrays. *Journal of the American Chemical Society*, **132**, 6068.

Min, W., English, B. P., Luo, G., Cherayil, B. J., Kou, S. C., and Xie, X. S. 2005. Fluctuating enzymes: lessons from single-molecule studies. *Accounts of Chemical Research*, **38**, 923–931.

Misra, B., and Sudarshan, E. C. G. 1977. Zenos paradox in quantum-theory. *Journal of Mathematical Physics*, **18**, 756.

Modesto-Lopez, L. B., Thimsen, E. J., Collins, A. M., Blankenship, R. E., and Biswas, P. 2010. Electrospray-assisted characterization and deposition of chlorosomes to fabricate a biomimetic light-harvesting device. *Energy & Environmental Science*, **3**(2), 216–222.

Mohseni, M., Rebentrost, P., Lloyd, S., and Aspuru-Guzik, A. 2008. Environment-assisted quantum walks in photosynthetic energy transfer. *Journal of Chemical Physics*, **129**, 174106.

Mohseni, M., Shabani, A., Lloyd, S., Omar, Y., and Rabitz, H. 2013. Geometrical effects on energy transfer in disordered open quantum systems. *Journal of Chemical Physics*, **138**, 204309.

Mohseni, M., Shabani, A., Lloyd, S., and Rabitz, H. 2014. Energy-scales convergence for optimal and robust quantum transport in photosynthetic complexes. *Journal of Chemical Physics*, **140**, 035102.

Möller, A., Sagasser, S., Wiltschko, W., and Schierwater, B. 2004. Retinal cryptochrome in a migratory passerine bird: a possible transducer for the avian magnetic compass. *Naturwissenschaften*, **91**, 585–588.

Morais-Cabral, J. H., Zhou, Y. F., and MacKinnon, R. 2001. Energetic optimization of ion conduction rate by the K^+ selectivity filter. *Nature*, **414**, 37–42.

Moser, C. C., and Dutton, P. L. 1992. Engineering protein-structure for electron-transfer function in photosynthetic reaction centers. *Biochimica et Biophysica Acta*, **1101**, 171–176.

Moser, C. C., Keske, J. M., Warncke, K., Farid, R. S., and Dutton, P. L. 1992. Nature of biological electron-transfer. *Nature*, **355**(6363), 796–802.

Mouritsen, H. 2001. Navigation in birds and other animals. *Journal of Image and Vision Computing*, **19**, 713.

Mouritsen, H., and Ritz, T. 2005. Magnetoreception and its use in bird navigation. *Current Opinion in Neurobiology*, **15**, 406–414.

Mouritsen, H., Janssen-Bienhold, U., Liedvogel, M., Feenders, G., Stalleicken, J., Dirks, P., and Weiler, R. 2004. Cryptochromes and neuronal-activity markers colocalize in the retina of migratory birds during magnetic orientation. *Proceedings of the National Academy of Sciences, USA*, **101**, 14294–14299.

Mouritsen, H., Feenders, G., Liedvogel, M., Wada, K., and Jarvis, E. D. 2005. A night vision brain area in migratory songbirds. *Proceedings of the National Academy of Sciences, USA*, **102**, 8339–8344.

Müh, F., Madjet, M. E.-A., Adolphs, J., Abdurahman, A., Rabenstein, B., Ishikita, H., Knapp, E.-W., and Renger, T. 2007. Alpha-helices direct excitation energy flow in the Fenna-Matthews-Olson protein. *Proceedings of the National Academy of Sciences of the United States of America*, **104**, 16862.

Mukai, K., Abe, S., and H. Sumi, J. 1999. Theory of rapid excitation-energy transfer from B800 to optically-forbidden exciton states of B850 in the antenna system LH2 of photosynthetic purple bacteria. *Journal of Physical Chemistry B*, **103**, 6096.

Mukamel, S. 1995. *Principles of Nonlinear Optical Spectroscopy*. New York: Oxford University Press.

Mukamel, S. 1999. *Principles of Nonlinear Optical Spectroscopy*. USA: Oxford University Press.

Mukamel, S. 2000. Multidimensional femtosecond correlation spectroscopies of electronic and vibrational excitations. *Annual Review of Physical Chemistry*, **51**, 691–729.

Mukamel, S. 2010. Communications: signatures of quasiparticle entanglement in multidimensional nonlinear optical spectroscopy of aggregates. *Journal of Chemical Physics*, **132**, 241105.

Mukherjee, P., Kass, I. Arkin, I. T., and Zanni, M. T. 2006. Structural disorder of the CD3 xi transmembrane domain studied with 2D-IR spectroscopy and molecular dynamics simulations. *Journal of Physical Chemistry B*, **110**, 24740–24749.

Murchie, E. H., and Niyogi, K. K. 2011. Manipulation of photoprotection to improve plant photosynthesis. *Journal of Plant Physiology*, **155**, 86.

Muus, L. T., Atkins, P. W., McLauchlan, K. A., and Pedersen, J. B. 1977. *Chemically Induced Magnetic Polarization*. Dordrecht: D. Reidel.

Myers, C. R., and Nealson, K. H. 1988. Bacterial manganese reduction and growth with manganese oxide as the sole electron acceptor. *Science*, **240**, 1319–1321.

Nagarajan, V., Parson, W. W., Gaul, D., and Schenck, C. 1990. Effect of specific mutations of tyrosine-(M)210 on the primary photosynthetic electron-transfer process in *Rhodobacter sphaeroides*. *Proceedings of the National Academy of Sciences, USA*, **87**(20), 7888–7892.

Nagel, Z. D., and Klinman, J. P. 2006. Tunneling and dynamics in enzymatic hydride transfer. *Chemical Reviews*, **106**, 3095–3118.

Nalbach, P., and Thorwart, M. 2010. Multiphonon transitions in the biomolecular energy transfer dynamics. *Journal of Chemical Physics*, **132**, 194111.

Nalbach, P., Ishizaki, A., Fleming, G. R., and Thorwart, M. 2011. Iterative path-integral algorithm versus cumulant time-nonlocal master equation approach for the dissipative biomolecular exciton transport. *New Journal of Physics*, **13**, 063040.

Nam, K. T., Kim, D.-W., Yoo, P. J., Chiang, C.-Y., Meethong, N., Hammond, P. T., Chiang, Y.-M., and Belcher, A. M. 2006. Virus-enabled synthesis and assembly of nanowires for lithium ion battery electrodes. *Science*, **312**, 885.

Nam, Y. S., Shin, T., Park, H., Magyar, A. P., Choi, K., Fantner, G., Nelson, K. A., and Belcher, A. M. 2010. Virus-Templated Assembly of Porphyrins into Light-Harvesting Nanoantennae. *Journal of the American Chemical Society*, **132**(5), 1462–1463.

Nathans, J., Piantanida, T. P., Eddy, R. L., Shows, T. B., and Hogness, D. S. 1986. Molecular genetics of inherited variation in human color vision. *Science*, **232**, 203–210.

Nazir, A. 2009. Correlation-dependent coherent to incoherent transitions in resonant energy transfer dynamics. *Physical Review Letters*, **103**, 146404.

Nealson, K. H., Belz, A., and McKee, B. 2002. Breathing metals as a way of life: geobiology in action. *Antonie Van Leeuwenhoek International Journal of General and Molecular Microbiology*, **81**, 215–222.

Nemeth, A., Milota, F., Mančal, T., Lukeš, V., Kauffmann, H. F., and Sperling, J. 2008. Vibronic modulation of lineshapes in two-dimensional electronic spectra. *Chemical Physics Letters*, **459**(1-6), 94.

Nemeth, A., Milota, F., Mančal, T., Lukeš, V., Hauer, J., Kauffmann, H. F., and Sperling, J. 2010. Vibrational wave packet induced oscillations in two-dimensional electronic spectra. I. Experiments. *Journal of Chemical Physics*, **132**, 184514.

Newton, M. D. 2001. Electron Transfer in Chemistry, Vol 1. New York: Wiley-VCH.

Nguyen, T.-Q., Wu, J., Doan, V., Schwartz, B. J., and Tolbert, S. H. 2000. Control of energy transfer in oriented conjugated polymer-mesoporous silica composites. *Science*, **288**, 652.

Nielsen, M. A., and Chuang, I. L. 2001. *Quantum Computation and Quantum Information*. Cambridge: Cambridge University Press.

Nimigean, C. M., and Allen, T. W. 2011. Origins of ion selectivity in potassium channels from the perspective of channel block. *Journal of General Physiology*, **137**(5), 405–413.

Nishioka, H., and Kakitani, T. 2008. Average electron tunneling route of the electron transfer in protein media. *Journal of Physical Chemistry B*, **112**, 9948–9958.

Nishioka, H., Kimura, A., Yamato, T., Kawatsu, T., and T., Kakitani. 2005. Interference, fluctuation, and alternation of electron tunneling in protein media. 2. Non-Condon theory for the energy gap dependence of electron transfer rate. *Journal of Physical Chemistry B*, **109**, 15621–15635.

Nitzan, A. 2006. *Chemical Dynamics in Condensed Phases*. Oxford: Oxford University Press.

Norrman, K., Larsen, N. B., and Krebs, F. C. 2006. Lifetimes of organic photovoltaics: combining chemical and physical characterisation techniques to study degradation mechanisms. *Solar Energy Materials and Solar Cells*, **90**, 2793.

Noskov, S. Y., and Roux, B. 2006. Ion selectivity in potassium channels. *Biophysical Chemistry*, **124**, 279–291.

Noskov, S. Y., Berneche, S., and Roux, B. 2004. Control of ion selectivity in potassium channels by electrostatic and dynamic properties of carbonyl ligands. *Nature*, **431**, 830–834.

Novoderezhkin, V. I., and Razjivin, A. P. 1994. Exciton states of the antenna and energy trapping by the reaction center. *Photosynthesis Research*, **42**, 9–15.

Novoderezhkin, V. I., and Razjivin, A. P. 1996. The theory of Förster-type migration between clusters of strongly interacting molecules: application to light-harvesting complexes of purple bacteria. *Chemical Physics*, **211**, 203–214.

Novoderezhkin, V. I., and van Grondelle, R. 2002. Exciton-vibrational relaxation and transient absorption dynamics in LH1 of *Rhodopseudomonas viridis*: a Redfield theory approach. *Journal of Physical Chemistry B*, **106**, 6025–6037.

Novoderezhkin, V. I., and van Grondelle, R. 2010. Physical origins and models of energy transfer in photosynthetic light harvesting. *Physical Chemistry – Chemical Physics*, **12**, 7352–7365.

Novoderezhkin, V. I., Monshouwer, R., and van Grondelle, R. 2000. Electronic and vibrational coherence in the core light-harvesting antenna of *Rhodopseudomonas viridis*. *Journal of Physical Chemistry B*, **104**, 12056–12071.

Novoderezhkin, V. I., Yakovlev, A. G., van Grondelle, R., and Shuvalov, V. A. 2004. Coherent nuclear and electronic dynamics in primary charge separation in photosynthetic reaction centers: a Redfield theory approach. *Journal of Physical Chemistry B*, **108**, 7445–7457.

Novoderezhkin, V. I., Palacios, M. A., van Amerongen, H., and van Grondelle, R. 2005. Excitation dynamics in the LHCII complex of higher plants: Modeling based on 2.72 Å crystal structure. *Journal of Physical Chemistry B*, **109**, 10493–10504.

Novoderezhkin, V. I., Andrizhiyevskaya, E. G., Dekker, J. P., and van Grondelle, R. 2005a. Pathways and timescales of primary charge separation in the photosystem II reaction center as revealed by a simultaneous fit of time-resolved fluorescence and transient absorption. *Biophysical Journal*, **89**, 1464–1481.

Novoderezhkin, V. I., Rutkauskas, D., and van Grondelle, R. 2006. Dynamics of the emission spectrum of a single LH2 complex: interplay of slow and fast nuclear motions. *Biophysical Journal*, **90**, 2890–2902.

Novoderezhkin, V. I., Dekker, J. P., and van Grondelle, R. 2007a. Mixing of exciton and charge-transfer states in Photosystem II reaction centers: modeling of Stark spectra with modified redfield theory. *Biophysical Journal*, **93**, 1293–1311.

Novoderezhkin, V. I., Rutkauskas, D., and van Grondelle, R. 2007b. Multistate conformational model of a single LH2 complex: quantitative picture of time-dependent spectral fluctuations. *Chemical Physics*, **341**, 45–56.

Novoderezhkin, V. I., Doust, A. B., Curutchet, C., Scholes, G. D., and van Grondelle, R. 2010. Excitation dynamics in phycoerythrin 545: modeling of steady-state spectra and transient absorption with modified Redfield theory. *Biophysical Journal*, **99**, 344–352.

Novoderezhkin, V. I., Marin, A., and van Grondelle, R. 2011a. Intra- and inter-monomeric transfers in the light harvesting LHCII complex: the Redfield-Förster picture. *Physical Chemistry – Chemical Physics*, **13**, 17093.

Novoderezhkin, V. I., Romero, E., Dekker, J. P., and van Grondelle, R. 2011b. Multiple charge separation pathways in photosystem II: modeling of transient absorption kinetics. *ChemPhysChem*, **12**, 681–688.

O'Day, K. E. 2008. Shedding light on animal cryptochromes. *PLoS Biology*, **6**, 1359–1360.

Odgen, D. 1994. *Microelectrode Techniques – The Plymouth Workshop Handbook*. Cambridge, UK: The Company of Biologists Ltd.

Oettel, M., and Schillinger, E. 1999. *Estrogens and Antiestrogens*. Berlin: Springer-Verlag.

Ogawa, T., and Takagahara, T. 1991. Optical absorption and Sommerfeld factors of one-dimensional semiconductors: an exact treatment of excitonic effects. *Physical Review B*, **44**, 8138.

O'Grady, S. M., and Lee, S. Y. 2003. Chloride and potassium channel function in alveolar epithelial cells. *American Journal of Physiology-Lung Cellular and Molecular Physiology*, **284**, L689–L700.

Olaya-Castro, A., and Scholes, G. D. 2011. Energy transfer from Förster-Dexter theory to quantum coherent light-harvesting. *International Reviews in Physical Chemistry*, **30**, 49.

Olaya-Castro, A., Lee, C. F., Olsen, F. F., and Johnson, N. F. 2008. Efficiency of energy transfer in a light-harvesting system under quantum coherence. *Physical Review B*, **78**, 085115.

Olbrich, C., Strümpfer, J., Schulten, K., and & Kleinekathöfer, U. 2011. Quest for spatially correlated fluctuations in the FMO light-harvesting complex. *Journal of Physical Chemistry*, **115**, 758–764.

Onuchic, J. N., and Da Gama, A. A. S. 1986. Influence of intersite modes on the exchange interaction in electrontransfer at large distances. *Theoretica Chimica Acta*, **69**, 89–100.

Oostergetel, G. T., Reus, M., G. M. Chew, A., Bryant, D. A., Boekema, E. J., and Holzwarth, A. R. 2007. Long-range organization of bacteriochlorophyll in chlorosomes of *Chlorobium tepidum* investigated by cryo-electron microscopy. *FEBS Letters*, **581**(28), 5435–5439.

Oszwaldowski, R., Reichelt, M., Meier, T., Koch, S. W., and Rohlfing, M. 2005. Nonlinear optical response of the Si(111)-(2x1) surface exciton: influence of biexciton many-body correlations. *Physical Review B*, **71**, 00265324.

Page, C. C., Moser, C. C., and Dutton, P. L. 2003. Mechanism for electron transfer within and between proteins. *Current Opinion in Chemical Biology*, **7**, 551–556.

Page, C. C., Moser, C. C., Chen, X., and Dutton, P. L. 1999. Natural engineering principles of electron tunnelling in biological oxidation–reduction. *Nature*, **402**(6757), 47–52.

Palmieri, B., Abramavicius, D., and Mukamel, S. 2009. Lindblad equations for strongly coupled populations and coherences in photosynthetic complexes. *Journal of Chemical Physics*, **130**(20), 204512.

Panitchayangkoon, G., Hayes, D., Fransted, K. A., Caram, J. R., Harel, E., Wen, J., Blankenship, R. E., and Engel, G. S. 2010. Long-lived quantum coherence in photosynthetic complexes at physiological temperature. *Proceedings of the National Academy of Sciences, USA*, **107**, 12766–12770.

Panitchayangkoon, G., Voronine, D. V., Abramavicius, D., Caram, J. R., Lewis, N. H. C., Mukamel, S., and Engel, G. S. 2011. Direct evidence of quantum transport in photosynthetic light-harvesting complexes. *Proceedings of the National Academy of Sciences, USA*, **108**, 20908.

Papiz, M., Prince, S., Howard, T., Cogdell, R., and Isaacs, N. 2003. The structure and thermal motion of the B800-850 LH2 complex from Rps. acidophila at 2.0 $\mathring{A}$ resolution and 100 K: new structural features and functionally relevant motions. *Journal of Molecular Biology*, **326**, 1523–1538.

Pardo, L. A. 2004. Voltage-gated potassium channels in cell proliferation. *Physiology*, **19**, 285–292.

Parson, W. W. 2007. *Modern Optical Spectroscopy*. Berlin: Springer.

Peierls, R. E. 1995. *Quantum Theory of Solids*. London: Oxford University Press.

Pirbadian, S., and El-Naggar, M. Y. 2012. Multistep hopping and extracellular charge transfer in microbial redox chains. *Physical Chemistry – Chemical Physics*, **14**, 13802–13808.

Pisliakov, A. V., Mancal, T., and Fleming, G. R. 2006. Two-dimensional optical three-pulse photon echo spectroscopy II. Signatures of coherent electronic motion and exciton population transfer in dimer two-dimensional spectra. *Journal of Chemical Physics*, **124**, 234505.

Plenio, M. B., and Huelga, S. F. 2002. Entangled light from white noise. *Physical Review Letters*, **88**, 197901.

Plenio, M. B., and Huelga, S. F. 2008. Dephasing-assisted transport: quantum networks and biomolecules. *New Journal of Physics*, **10**, 113019.

Plenio, M. B., and Virmani, S. 2007. An introduction to entanglement measures. *Quantum Information & Computation*, **7**, 1–51.

Polivka, T., and Sundstrom, V. 2004. Ultrafast dynamics of carotenoid excited states – from solution to natural and artificial systems. *Chemical Reviews*, **104**(4), 2021–2071.

Polizzi, N. F., Skourtis, S. S., and Beratan, D. N. 2012. Physical constraints on charge transport through bacterial nanowires. *Faraday Discussions*, **155**, 43–62.

Polli, D., Altoé, P., Weingart, O., Spillane, K. M., Manzoni, C., Brida, D., Tomasello, G., Orlandi, G., Kukura, P., Mathies, R. A., Garavelli, M., and Cerullo, G. 2010. Conical intersection dynamics of the primary photoisomerization event in vision. *Nature*, **467**, 440–443.

Potts, D., and Kunis, S. 2007. *Stability results for scattered data interpolation by trigonometric polynomials.*

Prior, J., Chin, A. W., Huelga, S. F., and Plenio, M. B. 2010. Efficient simulation of strong system-environment interactions. *Physical Review Letters*, **105**, 050404.

Prokhorenko, V. I., Holzwarth, A. R., Müller, M. G., Schaffner, K., Miyatake, T., and Tamiaki, H. 2002. Energy Transfer in Supramolecular Artificial Antennae Units of Synthetic Zinc Chlorins and Co-aggregated Energy Traps. A Time-Resolved Fluorescence Study. *Journal of Physical Chemistry B*, **106**, 5761.

Prokhorenko, V. I., Nagy, A. M., Waschuk, S. A., Brown, L. S., Birge, R. R., and Miller, R. J. D. 2006. Coherent control of retinal isomerization in bacteriorhodopsin. *Science*, **313**, 1257–1261.

Prokof'ev, N. V., and Stamp, P. C. E. 2000. Theory of the spin bath. *Reports on Progress in Physics*, **63**, 669.

Prytkova, T. R., Beratan, D. N., and Skourtis, S. S. 2007. Photoselected electron transfer pathways in DNA photolyase. *Proceedings of the National Academy of Sciences, USA*, **104**, 802–807.

Pullerits, T., Hess, S., Herek, J. L., and Sundström, V. 1997. Temperature dependence of excitation transfer in LH2 of *Rhodobacter sphaeroides*. *Journal of Physical Chemistry B*, **101**, 10560.

Pullman, B., and Pullman, A. 1963. *Quantum Biochemistry*. New York: Wiley-Interscience.

Rackovsky, S., and Silbey, R. J. 1973. Electronic energy transfer in impure solids I. Two molecules embeded in a lattice. *Molecular Physics*, **25**, 61.

Raman, B., and Gutierrez-Osuna, R. 2009. Relating sensor responses of odorants to their organoleptic properties by means of a biologically-inspired model of receptor neuron convergence onto olfactory bulb. Pages 93–108 of: Gutiérrez, Agustín, and Marco, Santiago (eds), *Biologically Inspired Signal Processing for Chemical Sensing*. Berlin: Springer-Verlag.

Rammer, J. 2007. *Quantum Field Theory of Non-equilibrium States*. New York: Cambridge University Press.

Ranasinghe, M., Wang, Y. and Goodson III, T. 2003. Excitation energy transfer in branched dendritic macromolecules at low (4 K) temperatures. *Journal of the American Chemical Society*, **125**, 5258.

Raszewski, G., and Renger, T. 2008. Light harvesting in photosystem II core complexes is limited by the transfer to the trap: can the core complex turn into a photoprotective mode? *Journal of the American Chemical Society*, **130**(13), 4431–4446.

Raszewski, G., Saenger, W., and Renger, T. 2005. Theory of optical spectra of photosystem II reaction centers: location of the triplet state and the identity of the primary electron donor. *Biophysical Journal*, **88**(2), 986–998.

Raszewski, G., Diner, B. A., Schlodder, E., and Renger, T. 2008. Spectroscopic properties of reaction center pigments in photosystem II core complexes: revision of the multimer model. *Biophysical Journal*, **95**(1), 105–119.

Ratner, M. A. 1990. Bridge-assisted electron transfer: effective electronic coupling. *Journal of Physical Chemistry*, **94**, 4877.

Read, E. L., Engel, G. S., Calhoun, T. R., Mančal, T., Ahn, T.-K., Blankenship, R. E., and Fleming, G. R. 2007. Cross-peak-specific two-dimensional electronic spectroscopy. *Proceedings of the National Academy of Sciences, USA*, **104**(36), 14203.

Read, E. L., Schlau-Cohen, G. S., Engel, G. S., Wen, J.-Z., Blankenship, R. E., and Fleming, G. R. 2008. Visualization of excitonic structure in the Fenna-Matthews-Olson photosynthetic complex by polarization-dependent two-dimensional electronic spectroscopy. *Biophysical Journal*, **95**, 847–856.

Read, E. L., Lee, H., and Fleming, G. R. 2009. Photon echo studies of photosynthetic light harvesting. *Photosynthesis Research*, **101**, 233–243.

Rebentrost, P., and Aspuru-Guzik, A. 2011. Exciton-phonon information flow in the energy transfer process of photosynthetic complexes. *Journal of Chemical Physics*, **134**, 101103.

Rebentrost, P., Mohseni, M., Kassal, I., Lloyd, S., and Aspuru-Guzik, A. 2009a. Environment-assisted quantum transport. *New Journal of Physics*, **11**, 033003.

Rebentrost, P., Chakraborty, R., and Aspuru-Guzik, A. 2009b. Non-Markovian quantum jumps in excitonic energy Transfer. *Journal of Physical Chemistry*, **131**, 184102.

Rebentrost, P., Mohseni, M., and Aspuru-Guzik, A. 2009c. Role of quantum coherence and environmental fluctuations in chromophoric energy transport. *Journal of Physical Chemistry B*, **113**, 9942.

Redfield, A. G. 1957. On the theory of relaxation processes. *IBM Journal of Research and Development*, **1**, 19.

Redfield, A. G. 1965. The theory of relaxation processes. *Advances in Magnetic Resonance*, **1**, 1.

Reguera, G., McCarthy, K. D., Mehta, T., Nicoll, J. S., Tuominen, M. T., and Lovley, D. R. 2005. Extracellular electron transfer via microbial nanowires. *Nature*, **435**, 1098–1101.

Ren, L., Martin, C. M., Wise, K. J., Gillespie, N. B., Luecke, H., Lanyi, J. K., Spudich, J. L., and Birge, R. R. 2001. Molecular mechanism of spectral tuning in sensory rhodopsin II. *Biochemistry*, **40**, 13906–13914.

Renger, T. 2009. Theory of excitation energy transfer: from structure to function. *Photosynthesis Research*, **102**, 471.

Renger, T., and Marcus, R. A. 2002. On the relation of protein dynamics and exciton relaxation in pigment–protein complexes: an estimation of the spectral density and a theory for the calculation of optical spectra. *Journal of Chemical Physics*, **116**, 9997.

Renger, T., May, V., and Kühn, O. 2001. Ultrafast excitation energy transfer dynamics in photosynthetic pigment-protein complexes. *Physical Reports*, **343**, 137.

Renger, T., Madjet, M. E., Knorr, A., and Müh, F. 2011. How the molecular structure determines the flow of excitation energy in plant light-harvesting complex II. *Journal of Plant Physiology*, **168**, 1497.

Reynolds, N. P., Janusz, S., Escalante-Marun, M., Timney, J., Ducker, R. E., Olsen, J. D., Otto, C., Subramaniam, V., Leggett, G. J., and Hunter, C. N. 2007. Directed Formation of Micro- and Nanoscale Patterns of Functional Light-Harvesting LH2 Complexes. *Journal of the American Chemical Society*, **129**(47), 14625–14631.

Ritz, T. 2011. Quantum effects in biology: bird navigation. *Procedia Chemistry*, **3**, 262–275.

Ritz, T., Damjanović, A., Schulten, K., Zhang, J.-P., and Koyama, Y. 2000a. Efficient light harvesting through carotenoids. *Photosynthesis Research*, **66**, 125–144.

Ritz, T., Adem, S., and Schulten, K. 2000b. A model for photoreceptor-based magnetoreception in birds. *Biophysical Journal*, **78**, 707–718.

Ritz, T., Park, S., and Schulten, K. 2001. Kinetics of excitation migration and trapping in the photosynthetic unit of purple bacteria. *Journal of Physical Chemistry B*, **105**, 8259.

Ritz, T., Thalau, P., Phillips, J. B., Wiltschko, R., and Wiltschko, W. 2004. Resonance effects indicate a radical-pair mechanism for avian magnetic compass. *Nature*, **429**, 177–180.

Ritz, T., Wiltschko, R., Hore, P. J., Rodgers, C. T., Stapput, K., Thalau, P., Timmel, C. R., and Wiltschko, W. 2009. Magnetic compass of birds is based on a molecule with optimal directional sensitivity. *Biophysical Journal*, **96**, 3451–3457.

Ritz, Th., Damjanovic, A., Schulten, K., Zhang, J.-P., and Koyama, Y. 2000d. Efficient light harvesting through carotenoids. *Photosynthesis Research*, **66**, 125–144.

Rivas, J. C. M., Schwalbe, H., and Lippard, S. J. 2001. Interchain hydrogen-bonding interactions may facilitate translocation of K^+ ions across the potassium channel selectivity filter, as suggested by synthetic modeling chemistry. *Proceedings of the National Academy of Sciences of the United States of America*, **98**, 9478–9483.

Rodgers, C. T. 2009. Magnetic field effects in chemical systems. *Pure and Applied Chemistry*, **81**, 19–43.

Rodgers, C. T., and Hore, P. J. 2009. Chemical magnetoreception in birds: the radical pair mechanism. *Proceedings of the National Academy of Sciences, USA*, **106**, 353–360.

Rodgers, C. T., and Hore, P. J. 2009d. Chemical magnetoreception in birds: the radical pair mechanism. *Proceedings of the National Academy of Sciences, USA*, **106**, 353.

Rodgers, C. T., Henbest, K. B., Kukura, P., Timmel, C. R., and Hore, P. J. 2005. Low-field optically detected EPR spectroscopy of transient photoinduced radical pairs. *The Journal of Physical Chemistry A*, **109**, 5035–5041.

Rodgers, C. T., Norman, S. A., Henbest, K. B., Timmel, C. R., and Hore, P. J. 2007. Determination of radical re-encounter probability distributions from magnetic field effects on reaction yields. *Journal of the American Chemical Society*, **129**, 6746–6755.

Rodgers, C. T., Wedge, C. J., Norman, S. A., Kukura, P., Nelson, K., Baker, N., Maeda, K., Henbest, K. B., Hore, P. J., and Timmel, C. R. 2009. Radiofrequency polarization effects in zero-field electron paramagnetic resonance. *Physical Chemistry Chemical Physics*, **11**, 6569–6572.

Röger, C., Müller, M. G., Lysetska, M., Miloslavina, Y., Holzwarth, A. R., and Würthner, F. 2006. Efficient Energy Transfer from Peripheral Chromophores to the Self-Assembled Zinc Chlorin Rod Antenna: âĂL A Bioinspired Light-Harvesting System to Bridge the "Green Gap". *Journal of the American Chemical Society*, **128**, 6542.

Röger, C., Miloslavina, Y., Brunner, D., Holzwarth, A. R., and Würthner, F. 2008. Self-Assembled Zinc Chlorin Rod Antennae Powered by Peripheral Light-Harvesting Chromophores. *Journal of the American Chemical Society*, **130**, 5929.

Romero, E., van Stokkum, I. H. M., Novoderezhkin, V. I., Dekker, J P., and van Grondelle, R. 2010. Two different charge separation pathways in photosystem II. *Biochemistry*, **49**, 4300–4307.

Romero, E., Diner, B. A., Nixon, P. J., Coleman, W. J., Dekker, J. P., and van Grondelle, R. 2012. Mixed exciton-charge-transfer states in photosystem II: Stark spectroscopy on site-directed mutants. *Biophysical Journal*, **103**, 185–194.

Rösch, N., and Voityuk, A. 2004. Quantum chemical calculation of donor-acceptor coupling for charge transfer in DNA. *Topics in Current Chemistry*, **237**, 37–72.

Roszak, A. W., Howard, T. D., Southall, J., Gardiner, A. T., Law, C. J., Isaacs, N. W., and Cogdell, R. J. 2003. Crystal structure of the RC-LH1 core complex from *Rhodopseudomonas palustris*. *Science*, **302**(5652), 1969–1972.

Roux, B. 2005. Ion conduction and selectivity in K^+ channels. *Annual Review of Biophysics and Biomolecular Structure*, **34**, 153–171.

Royston, E., Ghosh, A., Kofinas, P., Harris, M. T., and Culver, J. N. 2008. Self-assembly of virus-structured high surface area nanomaterials and their application as battery electrodes. *Langmuir*, **24**, 906.

Ruhman, S., Hou, B., Friedman, N., Ottolenghi, M., and Sheves, M. 2002. Following evoluation of bacteriorhodopsin in its reactive excited state via stimulated emission pumping. *Journal of the American Chemical Society*, **124**, 8854–8858.

Russo, V., Curutchet, C., and Mennucci, B. 2007. Towards a molecular scale interpretation of excitation energy transfer in solvated bichromophoric systems. II. The through-bond contribution. *Journal of Physical Chemistry B*, **111**, 853.

Rutkauskas, D., Novoderezhkin, V. I., Cogdell, R. J., and van Grondelle, R. 2004. Fluorescence spectral fluctuations of single LH2 complexes from *Rhodopseudomonas acidophila* strain. *Biochemistry*, **43**(15), 4431–4438.

Rutkauskas, D., Novoderezhkin, V., Gall, A., Oslen, J., Cogdell, R. J., Hunter, C. N., and van Grondelle, R. 2006. Spectral trends in the fluorescence of single bacterial light-harvesting complexes: experiments and modified redfield simulations. *Biophysical Journal*, **90**(7), 2475–2485.

Sacher, M., and Grampp, G. 1997. Magnetic field effects on the luminescence of *p*-phenylenediamine derivatives. *Berichte der Bunsengesellschaft für Physikalische Chemie*, **101**, 971–974.

Saik, V. O., Ostafin, A. E., and Lipsky, S. 1995. Magnetic-field effects on recombination fluorescence in liquid isooctane. *Journal of Chemical Physics*, **103**, 7347–7358.

Saito, K., and Sumi, H. 2009. Unified expression for the rate constant of the bridged electron transfer derived by renormalization. *Journal of Chemical Physics*, **131**, 134101.

Sakata, S. T., Kurokawa, T., Norholm, M. H. H., Takagi, M., Okochi, Y., von Heijne, G., and Okamura, Y. 2010. Functionality of the voltage-gated proton channel truncated in S4. *Proceedings of the National Academy of Sciences of the United States of America*, **107**, 2313–2318.

Sakmann, B., and Neher, E. 1995. *Single Channel Rercording*. New York: Springer-Verlag.

Salem, L. 1979. The sudden polarization effect and its possible role in vision. *Account of Chemical Research*, **119**, 12687–12688.

Salikhov, K. M. 1996. *Magnetic Isotope Effect in Radical Reactions*. Wien: Springer-Verlag.

Sancar, A. 2003. Structure and function of DNA photolyase and cryptochrome blue-light photoreceptors. *Chemical Reviews*, **103**, 2203–2237.

Sarovar, M., and Whaley, K. B. 2013. Design principles and fundamental trade-offs in biomimetic light harvesting. *New Journal of Physics*, **15**, 013030.

Sarovar, M., Ishizaki, A., Fleming, G. R., and Whaley, K. B. 2010. Quantum entanglement in photosynthetic light-harvesting complexes. *Nature Physics*, **6**(6), 462–467.

Sasaki, J., Brown, L. S., Chon, Y.-S., Kandori, H., Maeda, A., Needleman, R., and Lanyi, J. K. 1995. Conversion of bacteriorhodopsin into a chloride ion pump. *Science*, **269**, 73–75.

Sato, R., Dresselhaus, G., and Dresselhaus, M. S. 1998. *Physical Properties of Carbon Nanotubes*. London: Imperial College Press.

Savikhin, S., Buck, D. R., and Struve, W. S. 1997. Oscillating anisotropies in a bacteriochlorophyll protein: evidence for quantum beating between exciton levels. *Chemical Physics*, **223**, 303.

Scharf, B., and Engelhard, M. 1994. Blue halorhodopsin from natronobacterium pharaonis: wavelength regulation by anions. *Biochemistry*, **33**, 6387–6393.

Scheibe, G. 1936. Über die Veränderlichkeit des Absorptionsspektrums einiger Sensibilisierungsfarbstoffe und deren Ursache. *Angewandte Chemie*, **49**, 563.

Scheibe, G. 1937. Über die Veränderlichkeit der Absorptionsspektren in Lösungen und die Nebenvalenzen als ihre Ursache. *Angewandte Chemie*, **50**, 212–219.

Scheibe, G., Kandler, L., and Ecker, H. 1937. Polymerisation und polymere Adsorption als Ursache neuartiger Absorptionsbanden von organischen Farbstoffen. *Naturwissenschaften*, **25**, 75.

Schlau-Cohen, G. S., Calhoun, T. R., Ginsberg, N. S., Read, E. L., Ballottari, M., Bassi, R., van Grondelle, R., and Fleming, G. R. 2009. Pathways of energy flow in LHCII from two-dimensional electronic spectroscopy. *Journal of Physical Chemistry B*, **113**, 15352.

Schlau-Cohen, G. S., Dawlaty, J. M., and Fleming, G. R. 2012. Ultrafast multidimensional spectroscopy: principles and applications to photosynthetic systems. *IEEE Journal of Selected Topics in Quantum Electronics*, **18**(1), 283–295.

Schlosser, M., and Lochbrunner, S. 2006. Exciton migration by ultrafast Förster transfer in highly doped matrixes. *Journal of Physical Chemistry B*, **110**, 6001.

Schmidt am Busch, M., Müh, F., El-Amine Madjet, M., and Renger, T. 2011. The eighth bacteriochlorophyll completes the excitation energy funnel in the FMO protein. *Journal of Physical Chemistry Letters*, **2**, 93–98.

Schoenlein, R. W., Peteanu, L. A., Mathies, R. A., and Shank, C. A. 1991. The first step in vision: femtosecond isomerization of rhodopsin. *Science*, **254**, 412–415.

Scholak, T., de Melo, F., Wellens, T., Mintert, F., and Buchleitner, A. 2011. Efficient and coherent excitation transfer across disordered molecular networks. *Physical Review E*, **83**, 021912.

Scholes, G. D. 2002. Designing light-harvesting antenna systems based on superradiant molecular aggregates. *Chemical Physics*, **275**, 373.

Scholes, G. D. 2003. Long-range resonance energy transfer in molecular systems. *Annual Review of Physical Chemistry*, **54**, 57.

Scholes, G. D., and Fleming, G. R. 2000. On the mechanism of light harvesting in photosynthetic purple bacteria: B800 to B850 energy transfer. *Journal of Physical Chemistry B*, **104**, 1854–1868.

Scholes, G. D., Fleming, G. R., Olaya-Castro, A., and van Grondelle, R. 2011. Lessons from nature about solar light harvesting. *Nature Chemistry*, **3**, 763–774.

Scholes, G. D., Mirkovic, T., Turner, D. B., Fassioli, F., and Buchleitner, A. 2012. Solar light harvesting by energy transfer: from ecology to coherence. *Energy and Environmental Science*, **5**, 9374–9393.

Schrödinger, E. 1944. *What Is Life?* Cambridge: Cambridge University Press.

Schulten, K. 1982. Magnetic field effects in chemistry and biology. Pages 61–83 in: Treusch, J. (ed), *Festkörperprobleme*, vol. 22. Braunschweig: Vieweg.

Schulten, K. 1984. Ensemble averaged spin pair dynamics of doublet and triplet molecules. *Journal of Chemical Physics*, **80**, 3668–3679.

Schulten, K., and Karplus, M. 1972. On the origin of a low-lying forbidden transition in polyenes and related molecules. *Chemical Physics Letters*, **14**, 305–309.

Schulten, K., Staerk, H., Weller, A., Werner, H.-J. and Nickel, B. 1976. Magnetic field dependence of the geminate recombination of radical ion pairs in polar solvents. Zeitschrift für Physikalische Chemie, NF101, 311–390.

Schulten, K., and Tesch, M. 1991. Coupling of protein motion to electron transfer: Molecular dynamics and stochastic quantum mechanics study of photosynthetic reaction centers. *Chemical Physics*, **158**, 421–446.

Schulten, K., and Tavan, P. 1978. A mechanism for the light-driven proton pump of Halobacterium halobium. *Nature*, **272**, 85–86.

Schulten, K., and Windemuth, A. 1986. Model for a physiological magnetic compass. Pages 99–106 in: Maret, G., Boccara, N., and Kiepenheuer, J. (eds), *Biophysical Effects of Steady Magnetic Fields*. Proceedings in Physics, vol. 11. Berlin: Springer.

Schulten, K., Swenberg, C. E., and Weller, A. 1978. A biomagnetic sensory mechanism based on magnetic field modulated coherent electron spin motion. *Zeitschrift für Physikalische Chemie*, **111**, 1–5.

Schulten, K., Dinur, U., and Honig, B. 1980. The spectra of carbonium ions, cyanine dyes, and protonated Schiff base polyenes. *Journal of Chemical Physics*, **73**, 3927–3935.

Schuster, G. B. (ed). 2004. *Topics in Current Chemistry: Long-Range Charge Transfer in DNA I-II*. Vol. 236–237. Berlin: Springer.

Schuster, P. 2009. Free will, information, quantum mechanics, and biology. *Complexity*, **15**, 8–10.

Schwab, A. D., Smith, D. E., Rich, C. S., Young, E. R., Smith, W. F., and de Paula, J. C. 2003. Porphyrin Nanorods. *Journal of Physical Chemistry B*, **107**, 11339.

Scully, M. O., and Zubairy, M. S. 1997. *Quantum Optics*. Cambridge: Cambridge University Press.

Sell, C. S. 2006. On the unpredictability of odor. *Angewandte Chemie – International Edition in English*, **45**, 6254–6261.

Selvin, P. R. 2000. The renaissance of fluorescence resonance energy transfer. *Nature Structural Biology*, **7**, 730.

Sener, M., Olsen, J. D., Hunter, C. N., and Schulten, K. 2007. Atomic level structural and functional model of a bacterial photosynthetic membrane vesicle. *Proceedings of the National Academy of Sciences, USA*, **104**, 15723–15728.

Sener, M., Strümpfer, J., Timney, J. A., Freiberg, A., Hunter, C. N., and Schulten, K. 2010. Photosynthetic vesicle architecture and constraints on efficient energy harvesting. *Biophysical Journal*, **99**(1), 67–75.

Sener, M., Strümpfer, J., Hsin, J., Chandler, D., Scheuring, S., Hunter, C. N., and Schulten, K. 2011. Förster energy transfer theory as reflected in the structures of photosynthetic light-harvesting systems. *ChemPhysChem*, **12**(3), 518–531.

Serin, J. M., Brousmiche, D. W., and Fréchet, J. M. J. 2002. Cascade energy transfer in a conformationally mobile multichromophoric dendrimer. *Chemical Communications*, **22**, 2605.

Shabani, A. 2009. *Open Quantum Systems and Error Correction*. Ph.D. thesis, University of Southern California.

Shabani, A., and Lidar, D. 2009a. Maps for general open quantum systems and a theory of linear quantum error correction. *Physical Review A*, **80**, 012309.

Shabani, A., and Lidar, D. 2009b. Vanishing quantum discord is necessary and sufficient for completely positive maps. *Physical Review Letters*, **102**, 100402.

Shabani, A., Mohseni, M., Rabitz, H., and Lloyd, S. 2012. Efficient estimation of energy transfer efficiency in light-harvesting complexes. *Physical Review E*, **86**, 011915.

Shao, H., Gao, M., Kim, S. H., Jaroniec, C. P., and Parquette, J. R. 2011. Aqueous self-assembly of L-Lysine-based amphiphiles into 1D n-type nanotubes. *Chemistry – A European Journal*, **17**(46), 12882–12885.

Shichida, Y., and Imai, H. 1998. Visual pigment: G-protein-coupled receptor for light signals. *Cellular and Molecular Life Science*, **54**, 1299–1315.

Shih, C., Museth, A. K., Abrahamsson, M., Blanco-Rodriguez, A. M., Di Bilio, A. J., Sudhamsu, J., Crane, B. R., Ronayne, K. L., Towrie, M., Vlček Jr, A., Richards, J. H., Winkler, J. R., and Gray, H. B. 2008. Tryptophan-accelerated electron flow through proteins. *Science*, **320**, 1760–1762.

Shim, S., Rebentrost, P., Valleau, S., and Aspuru-Guzik, A. 2012. Atomistic study of the long-lived quantum coherences in the Fenna-Matthews-Olson complex. *Biophysical Journal*, **102**, 649.

Shimono, K., Iwamoto, M., Sumi, M., and Kamo, N. 2000. Effects of three characteristic amino acid residues of pharaonis phoborhodopsin on the absorption maximum. *Photochemistry and Photobiology*, **72**, 141–5.

Sigala, P. A., Kraut, D. A., Caaveiro, J. M. M., Pybus, B., Ruben, E. A., Ringe, D., A., Petsko. G., and Herschlag, D. 2008. Testing geometrical discrimination within an enzyme active site: constrained hydrogen bonding in the ketosteroid isomerase hydrogen bonding in the ketosteroid isomerase oxyanion hole. *Journal of the American Chemical Society*, **130**, 13696–13708.

Silbey, R. J. 1976. Electronic energy transfer in molecular crystals. *Annual Review of Physical Chemistry*, **27**, 203.

Silbey, R. J., and Harris, R. A. 1984. Variational calculation of the dynamics of a two level system interacting with a bath. *Journal of Chemical Physics*, **80**, 2615.

Skourtis, S. S., and Beratan, D. N. 1999. Theories of structure-function relationships for bridge-mediated electron-transfer reactions. *Advances in Chemical Physics*, **106**, 377–452.

Skourtis, S. S., and Beratan, D. N. 2007. A molecular double slit paradigm. *AIP Conference Proceedings*, **963**, 809–812.

Skourtis, S. S., and Mukamel, S. 1995. Superexchange versus sequential long range electron transfer; density matrix pathways in Liouville space. *Chemical Physics*, **197**, 367–388.

Skourtis, S. S., and Onuchic, J. N. 1993. Effective two-state systems in bridge-mediated electron transfer: a Green function analysis. *Chemical Physics Letters*, **209**, 171–177.

Skourtis, S. S., Beratan, D. N., and Onuchic, J. N. 1993. The two-state reduction for electron and hole transfer in bridge-mediated electron-transfer reactions. *Chemical Physics*, **176**, 501–520.

Skourtis, S. S., Xie, Q., and Archontis, G. J. 2001. Electron transfer through fluctuating bridges: on the validity of the superexchange mechanism and time-dependent tunneling matrix elements. *Journal of Chemical Physics*, **115**, 9444–9462.

Skourtis, S. S., Waldeck, D. H., and Beratan, D. N. 2004. Inelastic electron tunneling erases coupling-pathway interferences. *Journal of Physical Chemistry B*, **108**, 15511–15518.

Skourtis, S. S., Balabin, I. A., Kawatsu, T., and Beratan, D. N. 2005. Protein dynamics and electron transfer: electronic decoherence and non-Condon effects. *Proceedings of the National Academy of Sciences, USA*, **102**, 3552–3557.

Skourtis, S. S., Lin, J., and Beratan, D. N. 2006. Modern Methods for Theoretical Physical Chemistry of Biopolymers. Boston: Elsevier.

Skourtis, S. S., Waldeck, D. H., and Beratan, D. N. 2010. Fluctuations in biological and bioinspired electron-transfer reactions. **61**, 461–485.

Sokolova, O., Kolmakova-Partensky, L., and Grigorieff, N. 2001. Three-dimensional structure of a voltage-gated potassium channel at 2.5 nm resolution. *Structure*, **9**, 215–220.

Solov'yov, I. A., and Greiner, W. 2007. Theoretical analysis of an iron mineral-based magnetoreceptor model in birds. *Biophysical Journal*, **93**, 1493–1509.

Solov'yov, I. A., and Greiner, W. 2009a. Iron-mineral-based magnetoreceptor in birds: polarity or inclination compass? *European Physical Journal D*, **51**, 161–172.

Solov'yov, I. A., and Greiner, W. 2009b. Micromagnetic insight into a magnetoreceptor in birds: on the existence of magnetic field amplifiers in the beak. *Physical Review E*, **80**, 041919.

Solov'yov, I. A., and Schulten, K. 2009. Magnetoreception through cryptochrome may involve superoxide. *Biophysical Journal*, **96**, 4804–4813.

Solov'yov, I. A., and Schulten, K. 2012. Reaction kinetics and mechanism of magnetic field effects in cryptochrome. Journal of Physical Chemistry B, **116**, 1089–1099.

Solov'yov, I. A., Chandler, D., and Schulten, K. 2007. Magnetic field effects in *Arabidopsis thaliana* cryptochrome-1. *Biophysical Journal*, **92**, 2711–2726.

Solov'yov, I. A., Mouritsen, H., and Schulten, K. 2010. Acuity of a cryptochrome and vision-based magnetoreception system in birds. *Biophysical Journal*, **99**, 40–49.

Solov'yov, I. A., Chang, P.-Y., and Schulten, K. 2012. Vibrationally assisted electron transfer mechanism of olfaction: myth or reality? *Physical Chemistry Chemical Physics*, **14**(40), 13861–13871.

Solov'yov, I. A., Domratcheva, T., Shahi, A. R. M., and Schulten, K. (2012). Decrypting cryptochrome: reveraling the molecular identity of the photoactivation reaction. *Journal of the American Chemical Society*, **134**, 18046–18052.

Solov'yov, I. A., Domratcheva, T., and Schulten, K. 2014. Separation of photo-induced radical pair in cryptochrome to a functionally critical distance. Scientific Reports, 4: 3845, doi: 10.1038/srep03845.

Somsen, O. J. G., van Grondelle, R., and van Amerongen, H. 1996. Spectral broadening of interacting pigments: polarized absorption by photosynthetic proteins. *Biophysical Journal*, **71**(4), 1934–1951.

Song, X., and Marcus, R. A. 1993. Quantum correction for electron transfer rates. Comparison of polarizable versus nonpolarizable descriptions of solvent. *The Journal of Chemical Physics*, **99**(10), 7768–7773.

Spataru, C. D., Ismail-Beigi, S., Benedict, L. X., and Louie, S. G. 2004a. Excitonic effects and optical spectra of single-walled carbon nanotubes. *Physical Review Letters*, **92**, 077402.

Spataru, C. D., Ismail-Beigi, S., Benedict, L. X., and Louie, S. G. 2004b. Quasiparticle energies, excitonic effects and optical absorption spectra in small diameter single-walled carbon nanotubes. *Applied Physics A*, **78**, 1129.

Spudich, J. L., and Jung, K.-H. 2005. Hand Book of Photosensory Receptors. Weinheim: Wiley-VCH.

Sridharan, A., Muthuswamy, J., and Pizziconi, V. B. 2009. Optoelectronic energy transfer at novel biohybrid interfaces using light harvesting complexes from *Chloroflexus aurantiacus*. *Langmuir*, **25**(11), 6508–6516.

Staehelin, L. A., Golecki, J. R., and Drews, G. 1980. Supramolecular organization of chlorosomes (Chlorobium vesicles) and of their membrane attachment sites in *Chlorobium limicola*. *Biochimica et Biophysica Acta (BBA)-Bioenergetics*, **589**(1), 30–45.

Staehelin, L. A., Golecki, J. R., Fuller, R. C., and Drews, G. 1978. Visualization of the supramolecular architecture of chlorosomes (*Chlorobium* type vesicles) in freeze-fractured cells of *Chloroflexus aurantiacus*. *Archives of Microbiology*, **119**, 269.

Standfuss, J., van Scheltingal, A. C. T., Lamborghni, M., and Kuhlbrandt, W. 2005. Mechanisms of photoprotection and nonphotochemical quenching in pea light-harvesting complex at 2.5 Åresolution. *EMBO Journal*, **24**, 919.

Stass, D. V., Tadjikov, B. M., and Molin, Y. N. 1995a. Manifestation of quantum coherence upon recombination of radical-ion pairs in weak magnetic fields – systems with equivalent nuclei. *Chemical Physics Letters*, **235**, 511–516.

Stass, D. V., Lukzen, N. N., Tadjikov, B. M., and Molin, Y. N. 1995b. Manifestation of quantum coherence upon recombination of radical-ion pairs in weak magnetic fields – systems with nonequivalent nuclei. *Chemical Physics Letters*, **233**, 444–450.

Steane, A. M. 1998. Quantum Computing. *Reports on Progress in Physics*, **61**, 117–173.

Steane, A. M. 1996. Error correcting codes in quantum theory. *Physical Review Letters*, **77**, 793.

Steed, J. W., and Atwood, J. L. 2009. *Supramolecular Chemistry*. Wiltshire, UK: Wiley.

Steensgaard, D. B., Wackerbarth, H., Hildebrandt, P., and Holzwarth, A. R. 2000. Diastereoselective control of bacteriochlorophyll e aggregation. 3 1-S-BChl e is essential for the formation of chlorosome-like aggregates. *The Journal of Physical Chemistry B*, **104**(44), 10379–10386.

Steffen, M. A., Lao, K., and Boxer, S. G. 1994. Dielectric asymmetry in the photosynthetic reaction center. *Science*, **264**, 810–816.

Steiner, U. E., and Ulrich, T. 1989. Magnetic field effects in chemical kinetics and related phenomena. *Chemical Reviews*, **89**, 51–147.

Stoneham, A. M. 1975. *Theory of Defects in Solids. Electronic Structure of Defects in Insulators and Semiconductors*. Oxford: Oxford University Press.

Streltsov, A. M., Yakovlev, A. G., Shkuropatov, A. Y., and Shuvalov, V. A. 1996. Femtosecond kinetics of electron transfer in the bacteriochlorophyll(M)-modified reaction centers from *Rhodobacter sphaeroides* (R-26). *FEBS Letters*, **383**, 129–32.

Strümpfer, J., and Schulten, K. 2011. The effect of correlated bath fluctuations on exciton transfer. *Journal of Chemical Physics*, **134**(9).

Strümpfer, J., and Schulten, K. 2012. Excited state dynamics in photosynthetic reaction center and light harvesting complex 1. *Journal of Chemical Physics*, **137**, 065101.

Strümpfer, J., Hsin, J., Sener, M., Chandler, D., and Schulten, K. 2011. The light-harvesting apparatus in purple photosynthetic bacteria, introduction to a quantum biological device. Chap. 2, pages 19–48 of: Roux, Benoit (ed), *Molecular Machines*. Singapore: World Scientific Press.

Strümpfer, J., Sener, M., and Schulten, K. 2012. How quantum coherence assists photosynthetic light harvesting. *Journal of Physical Chemistry Letters*, **3**, 536–542.

Strycharz-Glaven, S. M., and Tender, L. M. 2012. Reply to the 'Comment on "On electrical conductivity of microbial nanowires and biofilms" ' by N. S. Malvankar, M. T. Tuominen and D. R. Lovley, Energy Environ. Sci., 2012, 5, DOI: 10.1039/c2ee02613a. *Energy & Environmental Science*, **5**, 6250–6252.

Strycharz-Glaven, S. M., Snider, R. M., Guiseppi-Elie, A., and Tender, L. M. 2011. On the electrical conductivity of microbial nanowires and biofilms. *Energy & Environmental Science*, **4**, 4366–4379.

Stryer, L., and Haugland, R. P. 1967. Energy transfer: a spectroscopic ruler. *Proceedings of the National Academy of Sciences, USA*, **58**, 719.

Stubbe, J., Nocera, D. G., Yee, C. S., and Chang, M. C. Y. 2002. Radical initiation in the class I ribonucleotide reductase: long-range proton-coupled electron transfer? *Chemical Reviews*, **103**, 2167–2201.

Stuchebrukhov, A. A. 2001. Toward ab initio theory of long-distance electron tunneling in proteins: tunneling currents approach. *Advances in Chemical Physics*, **118**, 1–44.

Su, W. P., Schrieffer, J. R., and Heeger, A. J. 1980. Soliton excitations in polyacetylene. *Physical Review B*, **22**, 2099.

Sumi, H. 1999. Theory on rates of excitation-energy transfer between molecular aggregates through distributed transition dipoles with application to the antenna system in bacterial photosynthesis. *Journal of Physical Chemistry B*, **103**, 252–260.

Sumi, H., and Kakitani, T. 2001. Unified theory on rates for electron transfer mediated by a midway molecule, bridging between superexchange and sequential processes. *Journal of Physical Chemistry B*, **105**, 9603–9622.

Sumi, H., and Marcus, R. A. 1986. Dynamical effects in electron transfer reactions. *Journal of Chemical Physics*, **84**, 4894–4914.

Sundstrom, V., van Grondelle, R., Bergstrom, H., Akesson, E., and Gillbro, T. 1986. Excitation energy transport in the bacteriochlorophyll antenna systems of *Rhodospirillum rubrum* and *Rhodobacter sphaeroides*, studied by low-intensity picosecond absorption spectroscopy. *Biochimica et Biophysica Acta*, **851**, 431–446.

Sundstrom, V., Pullerits, T., and van Grondelle, R. 1999. Photosynthetic light-harvesting: reconciling dynamics and structure of purple bacterial LH2 reveals function of photosynthetic unit. *Journal of Physical Chemistry B*, **103**, 2327.

Takagahara, T., Hanamura, E., and Kubo, R. 1977. Stochastic models of intermediate state interaction in second order optical processes – Stationary response. II –. *Journal of the Physical Society of Japan*, **43**, 811.

Tamiaki, H., Amakawa, M., Shimono, Y., Tanikaga, R., Holzwarth, A. R., and Schaffner, K. 1996. Synthetic Zinc and Magnesium Chlorin Aggregates as Models for Supramolecular Antenna Complexes in Chlorosomes of Green Photosynthetic Bacteria. *Photochemistry and Photobiology*, **63**, 92.

Tanimura, Y. 2006. Stochastic Liouville, Langevin, Fokker-Planck, and master equation approaches to quantum dissipative systems. *Journal of the Physical Society of Japan*, **75**, 082001.

Tanimura, Y., and Kubo, R. 1989. Time evolution of a quantum system in contact with a nearly Gaussian-Markoffian noise bath. *Journal of the Physical Society of Japan*, **58**, 101.

Tavan, P., and Schulten, K. 1986. The low-lying excitations in long polyenes: A PPP-MRD-CI study. *Journal of Chemical Physics*, **85**, 6602–6609.

Tavan, P., and Schulten, K. 1987. Electronic excitations in finite and infinite polyenes. *Physical Review B*, **36**, 4337–4358.

Tegmark, M. 2000. Importance of quantum decoherence in brain processes. *Physical Review E*, **61**, 4194–4206.

Teklos, A., and Skourtis, S. S. 2005. Electron transfer through time dependent bridges: differences between Franck-Condon and Born-Oppenheimer breakdown. *Chemical Physics*, **319**, 52–68.

Thalau, P., Ritz, T., Stapput, K., Wiltschko, R., and Wiltschko, W. 2005. Magnetic compass orientation of migratory birds in the presence of a 1.315 MHz oscillating field. *Naturwissenschaften*, **92**, 86–90.

Thalau, P., Ritz, T., Burda, H., Wegner, R. E., and Wiltschko, R. 2006. The magnetic compass mechanisms of birds and rodents are based on different physical principles. *Journal of The Royal Society Interface*, **3**, 583–587.

Thompson, A. L., Gaab, K. M., Xu, J., Bardeen, C. J., and Martínez, T. J. 2004. Variable electronic coupling in phenylacetylene dendrimers: the role of Förster, Dexter, and charge-transfer interactions. *Journal of Physical Chemistry A*, **108**, 671.

Thompson, A. N., Kim, I., Panosian, T. D., Iverson, T. M., Allen, T. W., and Nimigean, C. M. 2009. Mechanism of potassium-channel selectivity revealed by Na^+ and Li^+ binding sites within the KcsA pore. *Nature Structural & Molecular Biology*, **16**(12), 1317–1324.

Thorwart, M., Reimann, P., and Hänggi, P. 2000. Iterative algorithm versus analytic solutions of the parametrically driven dissipative quantum harmonic oscillator. *Physical Review E*, **62**, 5808–5817.

Thorwart, M., Eckel, J., Reina, J. H., Nalbach, P., and Weiss, S. 2009. Enhanced quantum entanglement in the non-Markovian dynamics of biomolecular excitons. *Chemical Physics Letters*, **478**, 234–237.

Tian, P., Keusters, D., Suzaki, Y., and Warren, W. S. 2003. Femtosecond phase-coherent two-dimensional spectroscopy. *Science*, **300**, 155.

Tiersch, M., and Briegel, H. J. 2012. Decoherence in the chemical compass: the role of decoherence for avian magnetoreception. *Philosophical Transactions of the Royal Society A*, **370**, 4517.

Tiersch, M., Popescu, S., and Briegel, H. J. 2012. A critical view on transport and entanglement in models of photosynthesis. *Philosophical Transactions of the Royal Society A*, **370**, 3771.

Timmel, C. R., and Henbest, K. B. 2004. A study of spin chemistry in weak magnetic fields. *Philosophical Transactions of the Royal Society A*, **362**, 2573–2589.

Timmel, C. R., and Hore, P. J. 1996. Oscillating magnetic field effects on the yields of radical pair reactions. *Chemical Physics Letters*, **257**, 401–408.

Timmel, C. R., Till, U., Brocklehurst, B., Mclauchlan, K. A., and Hore, P. J. 1998. Effects of weak magnetic fields on free radical recombination reactions. *Molecular Physics*, **95**, 71–89.

Timmel, C. R., Cintolesi, F., Brocklehurst, B., and Hore, P. J. 2001. Model calculations of magnetic field effects on the recombination reactions of radicals with anisotropic hyperfine interactions. *Chemical Physics Letters*, **334**, 387–395.

Tiwari, V., Peters, W. K., and Jonas, D. M. 2013. Electronic resonance with anticorrelated pigment vibrations drives photosynthetic energy transfer outside the adiabatic framework. *Proceedings of the National Academy of Sciences, USA*, **110**(4), 1203–1208.

Tokmakoff, A. 2000. Two-dimensional line shapes derived from coherent third-order nonlinear spectroscopy. *Journal of Physical Chemistry A*, **104**, 4247.

Tokmakoff, A. 2007. Shining light on the rapidly evolving structure of water. *Science*, **317**, 54–55.

Torres, J., Kukol, A., Goodman, J. M., and Arkin, I. T. 2001. Site-specific examination of secondary structure and orientation determination in membrane proteins: the peptidic C-13==O-18 group as a novel infrared probe. *Biopolymers*, **59**, 396–401.

Trammell, S. A., Wang, L., Zullo, J. M., Shashidhar, R., and Lebedev, N. 2004. Orientated binding of photosynthetic reaction centers on gold using Ni—NTA self-assembled monolayers. *Biosensors & Bioelectronics*, **19**, 1649–1655.

Tretiak, S., Kilina, S., Piryatinski, A., Saxena, A., Martin, R. L., and Bishop, A. R. 2007. Excitons and Peierls distortion in conjugated carbon nanotubes. *Nano Letters*, **7**, 86–92.

Troisi, A., Nitzan, A., and Ratner, M. A. 2003. A rate constant expression for charge transfer through fluctuating bridges. *Journal of Chemical Physics*, **119**, 5782–88.

Troisi, A., Ratner, M. A., and Zimmt, M. B. J. 2004. Dynamic nature of the intramolecular electronic coupling mediated by a solvent molecule: a computational study. *Journal of the American Chemical Society*, **126**, 2215–2224.

Tronrud, D. E., Wen, J., Gay, L., and Blankenship, R. E. 2009. The structural basis for the difference in absorbance spectra for the FMO protein from various green sulfur bacteria. *Photosynthesis Research*, **100**, 79.

Tully, D. C., and Fréchet, J. M. J. 2001. Dendrimers at surfaces and interfaces: chemistry and applications. *Chemical Communications*, 1229–1239.

Tully, J. C. 1990. Molecular dynamics with electronic transitions. *Journal of Chemical Physics*, **93**, 1061–71.

Turin, L. 1996. A spectroscopic mechanism for primary olfactory reception. *The Chemical Senses*, **21/6**, 773–791.

Ulstrup, J. 1979. *Lecture Notes in Chemistry*. Berlin: Springer-Verlag.

Umena, Y., Kawakami, K., Shen, J-R., and Kamiya, N. 2011. Crystal structure of oxygen-evolving photosystem II at a resolution of 1.9 Å. *Nature*, **473**(7345), 55–60.

Vacha, M., Puzová, T., and Kvícalová, M. 2009. Radiofrequency magnetic fields disrupt magnetoreception in American cockroach. *Journal of Experimental Biology*, **212**, 3473–3477.

Valkunas, L., Ma, Y.-Z., and Fleming, G. R. 2006. Exciton-exciton annihilation in single-walled carbon nanotubes. *Physical Review B*, **73**, 115432.

Valkunas, L., Abramavicius, D., and Mancal, T. 2013. *Molecular Excitation Dynamics and Relaxation*. Wiley VCH.

van Amerongen, H., Kwa, S. L. S., van Bolhuis, B. M., and van Grondelle, R. 1994. Polarized fluorescence and absorption of macroscopically aligned light-harvesting complex II. *Biophysical Journal*, **67**, 837–847.

van Amerongen, H., Valkunas, L., and van Grondelle, R. 2000. *Photosynthetic Excitons*. Singapore: World Scientific.

van Brederode, M. E., Jones, M. R., van Mourik, F., van Stokkum, I. H. M., and van Grondelle, R. 1997. A new pathway for transmembrane electron transfer in photosynthetic reaction centers of *Rhodobacter sphaeroides* not involving the excited special pair. *Biochemistry*, **36**(23), 6855–6861.

van Brederode, M. E., van Mourik, F., van Stokkum, I. H. M., Jones, M. R., and van Grondelle, R. 1999. Multiple pathways for ultrafast transduction of light energy in the photosynthetic reaction center of *Rhodobacter sphaeroides*. *Proceedings of the National Academy of Sciences, USA*, **96**, 2054–2059.

van der Laan, H., Schmidt, T., Visschers, R. W., Visscher, K. J., van Grondelle, R., and Volker, S. 1990. Energy transfer in the B800-850 antenna complex of purple bacteria *Rhodobacter sphaeroides* – a study by spectral hole-burning. *Chemical Physics Letters*, **170**, 231–238.

van Grondelle, R. 1985. Excitation energy transfer, trapping and annihilation in photosynthetic systems. *Biochimica et Biophysica Acta*, **811**, 147–195.

van Grondelle, R., and Novoderezhkin, V. I. 2006. Energy transfer in photosynthesis: experimental insights and quantitative models. *Physical Chemistry – Chemical Physics*, **8**(7), 793–807.

van Grondelle, R., and Novoderezhkin, V. I. 2010. Photosynthesis: quantum design for a light trap. *Nature*, **463**(7281), 614–615.

van Grondelle, R., Dekker, J. P., Gillbro, T., and Sundstrom, V. 1994. Energy transfer and trapping in photosynthesis. *Biochimica et Biophysica Acta*, **1187**, 1–65.

van Oijen, A. M., Ketelaars, M., Kohler, J., Aartsma, T. J., and Schmidt, J. 1999. Unraveling the electronic structure of individual photosynthetic pigment-protein complexes. *Science*, **285**(5426), 400–402.

van Rossum, B. J., Steensgaard, D. B., Mulder, F. M., Boender, G.-J., Schaffner, K., Holzwarth, A. R., and de Groot, H. J. M. 2001. A refined model of chlorosomal antennae of *Chlorobium tepidum* from proton chemical shift constraints obtained with High-Field 2-D and 3-D MAS NMR dipolar correlation spectroscopy. *Biochemistry*, **40**, 1587.

van Stokkum, I. H. M., Larsen, D. S., and van Grondelle, R. 2004. Global and target analysis of time-resolved spectra. *Biochimica et Biophysica Acta*, **1657**(2–3), 82–104.

van Vliet, C. M. 2008. *Equilibrium and non-equilibrium statistical mechanics*. Singapore: World Scientific.

Varnavski, O. P., Ostrowski, J., Sukhomlinova, L., Twieg, R. J., Bazan, G. C., and Goodson III, T. 2002. Coherent effects in energy transport in model dendritic structures investigated by ultrafast fluorescence anisotropy spectroscopy. *Journal of the American Chemical Society*, **124**, 1736.

Vaziri, A., and Plenio, M. B. 2010. Quantum coherence in ion channels: resonances, transport and verification. *New Journal of Physics*, **12**, 085001.

Visscher, K. J., Bergstrom, H., Sundstrom, V., Hunter, C. N., and van Grondelle, R. 1989. Temperature dependence of energy transfer from the long wavelength antenna BChl-896 to the reaction center in *Rhodospirillum rubrum, Rhodobacter sphaeroides* (Wt and M21 mutant) from 77 to 177K, studied by picosecond absorption spectroscopy. *Photosynthesis Research*, **22**, 211–217.

Visser, H. M., Somsen, O. J. G., van Mourik, F., Lin, S., van Stokkum, I. H. M., and van Grondelle, R. 1996. Direct observation of subpicosecond equilibration of excitation energy in the light-harvesting antenna of *Rhodospirillum rubrum*. *Biophysical Journal*, **69**, 1083–1099.

Vlaming, S. M., Augulis, R., Stuart, M. C. A., Knoester, J., and Van Loosdrecht, P. H. M. 2009. Exciton spectra and the microscopic structure of self-assembled porphyrin nanotubes. *The Journal of Physical Chemistry B*, **113**(8), 2273–2283.

von Neumann, J., and Wigner, E. P. 1929. Concerning the behavior of eigenvalues in adiabatic processes. *Zeitschrift für Physik*, **30**, 467–470.

Vos, M. H., Rappaport, F., Lambry, J.-Ch., Breton, J., and Martin, J.-L. 1993. Visualization of coherent nuclear motion in a membrane protein by femtosecond spectroscopy. *Nature*, **363**, 320–325.

Vos, M. H., Jones, M. R., Hunter, C. N., Breton, J., Lambry, J.-C., and Martin, J.-L. 1994. Coherent dynamics during the primary electron-transfer reaction in membrane-bound reaction centers of *Rhodobacter sphaeroides*. *Biochemistry*, **33**(22), 6750–6757.

Vulto, S. I. E., de Baat, M. A., Neerken, S., Nowak, F. R., van Amerongen, H., Amesz, J., and Aartsma, T. J. 1999. Excited state dynamics in FMO antenna complexes from photosynthetic green sulfur bacteria: a kinetic model. *Journal of Physical Chemistry B*, **103**(38), 8153.

Wajnberg, E., Acosta-Avalos, D., Cambraia Alves, O., Ferreira de Oliveira, J., Srygley, R. B. and Esquivel, D. M. S. Magnetoreception in eusocial insects: an update. *Journal of the Royal Society Interface*, **7**, S207–S225.

Wald, G. 1933. Vitamin A in the retina. *Nature*, **132**, 316–317.

Walker, G. C., Aakesson, E., Johnson, A. E., Levinger, N. E., and Barbara, P. F. 1992. Interplay of solvent motion and vibrational excitation in electron-transfer kinetics: experiment and theory. *Journal of Physical Chemistry*, **96**, 3728–3736.

Walker, M. M. 2008. A model for encoding of magnetic field intensity by magnetite-based magnetoreceptor cells. *Journal of Theoretical Biology*, **250**, 85–91.

Wang, F., Dukovic, G., Knoesel, E., Brus, L. E., and Heinz, T. F. 2004a. Observation of rapid Auger recombination in optically excited semiconducting carbon nanotubes. *Physical Review B*, **70**, 241403.

Wang, H., Lin, S., Allen, J. P., Williams, J. C., Blankert, S., Laser, C., and Woodbury, N. W. 2007. Protein dynamics control the kinetics of initial electron transfer in photosynthesis. *Science*, **316**(5825), 747–750.

Wang, Q., Schoenlein, R. W., Peteanu, L. A., Mathies, R. A., and Schank, C. V. 1994. Vibrationally coherent photochemistry in the femtosecond primary event of vision. *Science*, **266**, 422–424.

Wang, Z., Medforth, C. J., and Shelnutt, J. A. 2004b. Porphyrin nanotubes by ionic self-assembly. *Journal of the American Chemical Society*, **126**, 15954.

Wang, Z. Y., Tang, J., and Norris, J. R. 1992. The general treatment of dynamic solvent effects in electron-transfer at high-temperature. *Journal of Chemical Physics*, **97**, 7251–7256.

Wanko, M., Hoffmann, M., Strodel, P., Koslowski, A., Thiel, W., Neese, F., Frauenheim, T., and Elstner, M. 2005. Calculating absorption shifts for retinal proteins: computational challenges. *Journal of Physical Chemistry B*, **109**, 3606–3615.

Wannagat, U., Damrath, V., Huch, V., Veith, M., and Harder, U. 1993. Sila-olfactory substances and olfactory substance isosters. Part 12. Odors of homologous organoelement compounds of the fourth main group (C, Si, Ge, Sn). *Journal of Organometallic Chemistry*, **443**, 153–165.

Wannagat, U., Damrath, V., and Harder, U. 1994. Sila perfumes and isoteric perfumes. 13. Isoteric compounds according to the hydride principle of Grimm in the range of linalool scents. *Monatshefte für Chemie*, **125**, 1159–1169.

Warshel, A. 2002. Molecular dynamics simulations of biological reactions. *Accounts of Chemical Research*, **35**, 385.

Warshel, A., Chu, Z. T., and Parson, W. W. 1989. Dispersed polaron simulations of electron transfer in photosynthetic reaction centers. *Science*, **246**, 112–116.

Warshel, A., Chu, Z. T., and Parson, W. W. 2001. Dynamics of biochemical and biophysical reactions: insight from computer simulations. *Quarterly Reviews of Biophysics*, **34**, 563–679.

Wasielewski, M. R. 2009. Self-assembly strategies for integrating light harvesting and charge separation in artifical photosynthetic systems. *Accounts of Chemical Research*, **42**, 1910.

Weaver, J. C., Vaughan, T. E., and Astumian, R. D. 2000. Biological sensing of small field differences by magnetically sensitive chemical reactions. *Nature*, **405**(6787), 707–709.

Wedge, C. J., Rodgers, C. T., Norman, S. A., Baker, N., Maeda, K., Henbest, K. B., Timmel, C. R., and Hore, P. J. 2009. Radiofrequency polarization effects in low-field electron paramagnetic resonance. *Physical Chemistry Chemical Physics*, **11**, 6573–6579.

Weiss, R. M., and Warshel, A. 1979. A new view of the dynamics of singlet cis-trans photoisomerization. *Journal of the American Chemical Society*, **101**, 6131–6133.

Weiss, U. 2008. *Quantum Dissipative Systems*. Singapore: World Scientific.

Werner, H.-J., Schulten, Z., and Schulten, K. 1977. Theory of the magnetic field modulated geminate recombination of radical ion pairs in polar solvents: application to the pyrene N,N-dimethylaniline system. *Journal of Chemical Physics*, **67**, 646–663.

Whaley, B., Sarovar, M., and Ishizaki, A. 2011. Quantum entanglement phenomena in photosynthetic light harvesting complexes. *Procedia Chemistry*, **3**, 152.

Whitesides, G. M., Mathias, J. P., and Seto, C. T. 1991. Molecular self-assembly and nanochemistry: a chemical strategy for the synthesis of nanostructures. *Science*, **254**, 1312.

Wilde, M. M., McCracken, J. M., and Mizel, A. 2010. Could light harvesting complexes exhibit non-classical effects at room temperature? *Philosophical Transactions of the Royal Society of London A. (Mathematical, Physical and Engineering Sciences)*, **466**, 1347–1363.

Wiltschko, R., and Wiltschko, W. 1995a. *Magnetic Orientation in Animals*. Berlin: Springer.

Wiltschko, R., and Wiltschko, W. 2006. Magnetoreception. *BioEssays*, **28**, 157–168.

Wiltschko, R., Stapput, K., Ritz, T., Thalau, P., and Wiltschko, W. 2007. Magnetoreception in birds: different physical processes for two types of directional responses. *HFSP Journal*, **1**, 41–48.

Wiltschko, R., Munro, U., Ford, H., Stapput, K., and Wiltschko, W. 2008. Light-dependent magnetoreception: orientation behaviour of migratory birds under dim red light. *Journal of Experimental Biology*, **211**, 3344–3350.

Wiltschko, R., Stapput, K., Thalau, P., and Wiltschko, W. 2010. Directional orientation of birds by the magnetic field under different light conditions. *Journal of the Royal Society Interface*, **7**, S163–S177.

Wiltschko, W., and Wiltschko, R. 1972. Magnetic compass of european robins. *Science*, **176**, 62.

Wiltschko, W., and Wiltschko, R. 1995b. Migratory orientation of European robins is affected by the wavelength of light as well as by a magnetic pulse. *Journal of Comparative Physiology A*, **177**, 363–369.

Wiltschko, W., and Wiltschko, R. 1996. Magnetic orientation in birds. *Journal of Experimental Biology*, **199**, 29.

Wiltschko, W., Munro, U., Ford, H., and Wiltschko, R. 1993. Magnetic inclination compass: a basis for the migratory orientation of birds in the northern and southern hemisphere. *Cellular and Molecular Life Sciences*, **49**, 167–170.

Wiltschko, W., Munro, U., Ford, H., and Wiltschko, R. 2006. Bird navigation: what type of information does the magnetite-based receptor provide? *Proceedings of the Royal Society of London B. (Biological Sciences)*, **273**, 2815–2820.

Winkler, J. R., and Gray, H. B. 2010. Electron flow through metalloproteins. *Biochimica et Biophysica Acta*, **1797**, 1563–1572.

Winkler, J. R., Nocera, D. G., Yocom, K. M., Bordignon, E., and Gray, H. B. 1982. Electron-transfer kinetics of pentaammineruthenium(III)(histidine-33)-ferricytochrome c. Measurement of the rate of intramolecular electron transfer between redox centers separated by 15 Åin a protein. *Journal of the American Chemical Society*, **104**, 5798–5800.

Wiseman, H. M., and Eisert, J. 2007. Nontrivial quantum effects in biology: a skeptical physicists' view. Singapore: World Scientific.

Woiczikowski, P. B., Steinbrecher, T., Kubar, T., and Elstner, M. J. 2011. Nonadiabatic QM/MM simulations of fast charge transfer in Escherichia coli DNA photolyase. *Journal of Physical Chemistry B*, **115**, 9846–9863.

Wong, C. Y., Alvey, R. M., Turner, D. B., Wilk, K. E., Bryant, D. A., Curmi, P. M. G., Silbey, R. J., and Scholes, G. D. 2012. Electronic coherence lineshapes reveal hidden excitonic correlations in photosynthetic light harvesting. *Nature Chemistry*, **4**, 396–404.

Wong, K. F., Bagchi, B., and Rossky, P. J. 2004. Distance and orientation dependence of excitation transfer rates in conjugated systems: beyond the Förster theory. *Journal of Physical Chemistry A*, **108**, 5752.

Woods, M. P., Groux, R., Chin, A. W., Huelga, S. F., and Plenio, M. B. 2014. *Mappings of open quantum systems onto chain representations and Markovian embeddings. Journal of Mathematical Physics*, **55**, 092101.

Woodward, J. R., Timmel, C. R., McLauchlan, K. A., and Hore, P. J. 2001. Radio frequency magnetic field effects on electron-hole recombination. *Physical Review Letters*, **87**, 077602.

Woodward, R. B., and Hoffmann, R. 1969. The conservation of orbital symmetry. *Angewandte Chemie – International Edition in English*, **8**, 781–932.

Wootters, W. K. 1998. Entanglement of formation of an arbitrary state of two qubits. *Physical Review Letters*, **80**, 2245.

Wrobel, D., Wannagat, U., and Harder, U. 1982. Some methoxysilanes, disiloxanes and digermoxanes with effectiveness of odour sila-substituted perfumes 3. *Monatshefte für Chemie*, **113**, 381–388.

Wu, J., Liu, F., Shen, Y., Cao, J., and Silbey, R. J. 2010. Efficient energy transfer in light-harvesting systems, I: optimal temperature, reorganization energy and spatial-temporal correlations. *New Journal of Physics*, **12**, 105012.

Wu, W., Hsiao, S. C., Carrico, Z. M., and Francis, M. B. 2009. Genome-free viral capsids as multivalent carriers for Taxol delivery. *Angewandte Chemie – International Edition in English*, **48**, 9493.

Würthner, F., Kaiser, T. E., and Saha-Moller, C. R. 2011. J-Aggregates: from serendipitous discovery to supramolecular engineering of functional dye materials. *Angewandte Chemie – International Edition in English*, **50**, 3376.

Xiao, D., Skourtis, S. S., Rubtsov, I. V., and Beratan, D. N. 2009. Turning charge transfer on and off in a molecular interferometer with vibronic pathways. *Nano Letters*, **9**, 1818–1823.

Xie, Q., Archontis, G., and Skourtis, S. S. 1999. Protein electron transfer: a numerical study of tunneling through fluctuating bridges. *Chemical Physics Letters*, **312**, 237–246.

Xu, D., and Schulten, K. 1992. Multi-mode coupling of protein motion to electron transfer in the photosynthetic reaction center: spin-boson theory based on a classical molecular dynamics simulation. Pages 301–312 of: Breton, J., and Vermeglio, A. (eds), *The Photosynthetic Bacterial Reaction Center: II. Structure, Spectroscopy and Dynamics*. NATO ASI Series A: Life Sciences. New York: Plenum Press.

Xu, D., and Schulten, K. 1994. Coupling of protein motion to electron transfer in a photosynthetic reaction center: investigating the low temperature behavior in the framework of the spin-boson model. *Chemical Physics*, **182**, 91–117.

Yakovlev, A. G., Shkuropatov, A. Y., and Shuvalov, V. A. 2000. Nuclear wavepacket motion producing a reversible charge separation in bacterial reaction centers. *FEBS Letters*, **466**, 209–212.

Yakovlev, A. G., Shkuropatov, A. Y., and Shuvalov, V. A. 2002a. Nuclear wave packet motion between P* and P+BA- potential surfaces with a subsequent electron transfer to HA in bacterial reaction centers at 90 K. Electron transfer pathway. *Biochemistry*, **41**(47), 14019–14027.

Yakovlev, A. G., Shkuropatov, A. Ya, and Shuvalov, V. A. 2002b. Nuclear wavepacket motion between P* and P+BA- potential surfaces with subsequent electron transfer to HA in bacterial reaction centers. 1. Room temperature. *Biochemistry*, **41**(8), 2667–2674.

Yan, Y. J., Fried, L. E., and Mukamel, S. 1989. Ultrafast pump-probe spectroscopy: femtosecond dynamics in Liouville space. *Journal of Physical Chemistry*, **93**, 8149.

Yang, M., and Fleming, G. R. 1999. Two-color three-pulse photon echoes as a probe of electronic coupling in molecular complexes. *Journal of Chemical Physics*, **110**, 2983.

Yang, M., and Fleming, G. R. 2002. Influence of phonons on exciton transfer dynamics: comparison of the Redfield, Förster, and modified Redfield equations. *Chemical Physics*, **275**, 355.

Yang, S. I., Lammi, R. K., Seth, J., Riggs, J. A., Arai, T., Kim, D., Bocian, D., Holten, D., and Lindsey, J. 1998. Excited-state energy transfer and ground-state hole/electron hopping in p-phenylene-linked porphyrin dimers. *Journal of Physical Chemistry B*, **102**, 9426.

Yoshii, T., Ahmad, M., and Helfrich-Foerster, C. 2009. Cryptochrome mediates light-dependent magnetosensitivity of *Drosophila*'s circadian clock. *PLoS Biology*, **7**, 813–819.

Zapka, M., Heyers, D., Hein, C. M., Engels, S., Schneider, N.-L., Hans, J., Weiler, S., Dreyer, D., Kishkinev, D., Wild, J. M., and Mouritsen, H. 2009. Visual but not trigeminal mediation of magnetic compass information in a migratory bird. *Nature*, **461**, 1274–1277.

Zener, C. 1932. Non-adiabatic crossing of energy levels. *Proceedings of the Royal Society of London A*, **137**, 696–702.

Zhang, F., Aravanis, A. M., Adamantidis, A., de Lecea, L., and Deisseroth, K. 2007. Circuit-breakers: optical technologies for probing neural signals and systems. *Nature Reviews Neuroscience*, **8**, 205–218.

Zhang, J.-P., Fujii, R., Qian, P., Inaba, T., Mizoguchi, T., and Koyama, Y. 2000. Mechanism of the carotenoid-to-bacteriochlorophyll energy transfer via the S_1 state in the LH2 complexes from purple bacteria. *Journal of Physical Chemistry B*, **104**, 3683–3691.

Zhang, W. M., Meier, T., Chernyak, V., and Mukamel, S. 1998. Exciton-migration and three-pulse femtosecond optical spectroscopies of photosynthetic antenna complexes. *Journal of Chemical Physics*, **108**, 7763–7774.

Zhang, Y., and Mascarenhas, A. 1999. Scaling of exciton binding energy and Virial theorem in semiconductor quantum wells and wires. *Physical Review B*, **59**, 2040.

Zhou, Y. F., Morais-Cabral, J. H., Kaufman, A., and R., MacKinnon. 2001. Chemistry of ion coordination and hydration revealed by a K^+ channel-Fab complex at 2.0 angstrom resolution. *Nature*, **414**, 43–48.

Zhu, J., Kais, S., Aspuru-Guzik, A., Rodriques, S., Brock, B., and Love, P. J. 2012. Multipartite quantum entanglement evolution in photosynthetic complexes. *Journal of Chemical Physics*, **137**, 074112.

Zhuang, W., Hayashi, T., and Mukamel, S. 2009. Coherent multidimensional vibrational spectroscopy of biomolecules: concepts, simulations, and challenges. *Angewandte Chemie – International Edition*, **48**, 3750–3781.

Zigmantas, D., Read, E. L., Mancal, T., Brixner, T., Gardiner, A. T., Cogdell, R. J., and Fleming, G. R. 2006. Two-dimensional electronic spectroscopy of the B800-B820 light-harvesting complex. *Proceedings of the National Academy of Sciences, USA*, **103**, 12672–7.

Zimmermann, J., Oakman, E. L., Thorpe, I. F., Shi, X., Abbyad, P., II, C. L. Brooks, Boxer, S. G., and Romesberg, F. E. 2006. Antibody evolution constrains conformational heterogeneity by tailoring protein dynamics. *Proceedings of the National Academy of Sciences, USA*, **103**, 13722.

Zoller, P., Beth, T., Binosi, D., Blatt, R., Briegel, H., Bruss, D., Calarco, T., Cirac, J. I., Deutsch, D., Eisert, J., Ekert, A., Fabre, C., Gisin, N., Grangiere, P., Grassl, M., Haroche, S., Imamoglu, A., Karlson, A., Kempe, J., Kouwenhoven, L., Kröll, S., Leuchs, G., Lewenstein, M., Loss, D., Lütkenhaus, N., Massar, S., Mooij, J. E., Plenio, M. B., Polzik, E., Popescu, S., Rempe, G., Sergienko, A., Suter, D., Twamley, J., Wendin, G., Werner, R., Winter, A., Wrachtrup, J., and Zeilinger, A. 2005. Quantum information processing and communication. *European Physical Journal D*, **36**, 203.

Zwanzig, R. 2001. *Nonequilibrium Statistical Mechanics*. New York: Oxford University Press.

Index

Arabidopsis thaliana, 225
Drosophila, 225
2D electronic spectroscopy, 147, 150–151
3PEPS, 345

absorption spectra, 137, 139, 141
anharmonic oscillator, 341
artificial light-harvesting systems, 171, 312, 328

bacterial reaction centres, 182–184
bacteriochlorophylls (BChls), 77–79, 125–132, 134–135, 137–138, 140, 174, 180, 191
bacteriorhodopsin, 237–238, 241
bath frequency cut-off, 172
bath spectral density, 25, 31–32, 34, 43, 47–48, 137, 167, 169, 174–175, 343
Bell inequality, 284
bio-inspired quantum materials, 311–334
birds, 219
Bloch theorem, 335
Bloch–Redfield master equation, 27, 32, 34, 79, 81, 183, 185, 194–195
Born–Oppenheimer surface, 202, 213
bosonic bath, 22, 25, 30, 48–50, 302, 304
Brillouin zone, 336
Brownian oscillator model, 37, 47, 103, 118

carbon nanotube, 335
carbon nanotubes, 335–349
carotenoid–porphyrin–fullerene, 229
CD spectrum, 139–140
charge separation, 5, 28, 131, 148, 168, 179–192, 195, 205, 216, 257, 313, 316–317, 324, 327, 332
charge transfer, 131
chemical compass, 218–236, 309, 315, 333–334
circular dichroism, 131
Cluster N, 233
coherence, 144, 158
coherences, 144
completely positive maps, 17, 304
conical intersection, 246–249, 257
Condon approximation, 132
consciousness, 12
cryptochrome, 218, 223
crystal structure, 131, 138, 140
cumulant approximation, 131, 136–137, 141–143

Decoherence, x, xvii, 12, 14, 22, 27, 162–163, 172–173, 175, 220, 293, 305
deep tunnelling, 199
deep-tunnelling ET reactions, 208
delocalization, 145
density matrix renormalization group (DMRG), 47–52
density of states, 132, 134
dipole–dipole coupling, 133
DNA photolyase, 231
Drude–Lorentzian spectral density, 171

Earth's magnetic field, 218
elastic through-bridge tunnelling, 201
elastic tunnelling, 203, 211–214, 269, 271–272
Electron Paramagnetic Resonance, 220
electron transfer, xviii, 6–7, 124–125, 127–131, 140–143, 179–183, 187, 189–190, 194, 197, 198–217, 219, 221, 223–226, 228, 230–231, 268, 270, 275–276, 314, 332
electron transfer rate, 141–143
electron transfer reactions, 223
electron tunnelling, 7, 10, 214, 267, 269, 273, 276
electronic excitations, 131
electronic spectroscopy, 5, 16, 79, 82–120, 147, 150–151, 189
enantiomer, 264–265, 267, 275–276
enantiomers, 265, 267, 275
energy gap time series, 132, 136, 138, 142
energy transfer efficiency, 160, 162, 170–172

entanglement, x, xi, 48–49, 267, 277–310, 348
environment-assisted quantum transport (ENAQT), xi, 10, 159–175
European robins, 233
evolution, x, 11–13
exciton, 133–135, 139, 338
exciton binding energy, 338
exciton–exciton annihilation, 339
exciton–exciton interactions, 339
excitonic coupling, 139
excitonic TDM, 133

Fenna–Matthews–Olson complex, 147–150, 153–156
Fermi golden rule, 343
Fermi's golden rule, 56, 249, 271
Feynman diagrams, 95, 153
flavin adenine dinucleotide, FAD, 224
fluctuations, 131, 134–135, 140–142, 200
fluctuation–dissipation process, 30–31, 34–35, 136
Förster resonance energy transfer (FRET), 32–33, 53–81
Förster's rate expression, 54, 56–59
four-wave mixing, 343
Franck–Condon, 132, 136

Gauss–Markovian fluctuations, 161
generalized Förster resonance energy transfer, 53–81
generalized Förster theories, 65–72
geomagnetic field, 218

Haken–Strobl model, 31, 160, 168
Heisenberg equation, 342
heterodyne detection, 89, 111, 151
hierarchical equation of motion, 31, 35
Hilbert space, 15–18, 20, 31, 90, 100
homogeneous broadening, 75, 103, 194
hopping, 199
Hyperfine interaction, 219

inelastic tunnelling, 212, 271–272
infrared, 267, 270–271, 275
inhomogeneous broadening, 103, 108

J-aggregates, 319–320, 325

Kraus representation, 18, 20–21
Kubo number, 173

LH2, 156
Landau–Zener transitions, 203–204, 250
light-harvesting complex, 131
light-harvesting 2 (LH2) complex of purple bacteria, 77–78, 125–128, 130–132, 134–135, 138–140, 156, 191–192, 194, 196–197, 318–319, 322, 328
light-harvesting complex II (LHCII), 55, 79–81, 192–195
light–matter interaction, 82–84, 89
Lindblad master equation, 17, 27–28, 161–163, 175
Lindblad operator, 161–162
linear absorption, 131
linear response theory, 133
lineshape function, 131, 133–134, 136–138
Liouville pathways, 92
Liouville–von Neumann equation, 15–17, 21–22, 35, 38–40
local correlations, 287
Lorentz absorption lineshape, 94
low field effect, 220

macroscopic polarization, 88
magnetic compass, 218, 222
magnetic field effects, 227
magnetic orientation, 218
magnetometry, 315, 332
magnetoreception, 220
maquette proteins, 332
Markovian bath, 26–27, 52, 160
matrix product states (MPS), 50–51
MD simulation, 131–132, 143
mean field approximation, 342
mixed-state entanglement, 281
mole rats, 236
molecular aggregates, 76, 319–322, 324–325, 328, 332
multichromophoric Förster resonance energy transfer (MC-FRET), 73, 77–81
multi-dimensional electronic spectroscopy, 82–119

nanotube self-assembled, 325
navigation, 218
neurobiology, 233
noise-assisted quantum transport, 164
non-completely positive maps, 17, 19
non-linear optical response, 343
non-local correlations, 277, 281–282, 286, 288, 306
non-Markovian, 26, 33–35, 37, 43, 79, 171–172, 176, 296, 304
non-Markovianity measure, 171
non-rephasing pathways, 97, 104, 151

OD spectrum, 131, 133, 135, 138–141
odourant, 264–276
Ohmic spectral density, 34, 169
olfaction, 264–268, 271, 275
open driven systems, 278, 291–293, 310
open quantum system, v, xi, 11, 14–52, 83, 159–175
optical properties, 131
optical spectra, 131, 136
optical transitions, 100, 125, 131, 269, 337, 339

optimal quantum transport, 169–170
origin of life, xviii, 12

parametric down conversion, 293–295
path integral techniques, 35, 44–47
Pauli master equation, 17, 42
Peierls instability, 336
phase cycling, 344
phonon, 269–271, 275
photochemical, 6, 223, 230–231, 245–247, 252, 257, 260
photo-induced processes, 82, 88
photoisomerization, 237, 239–241, 257–259, 261–262
photon-echo, 150, 343
photon-echo spectroscopy, 155
photosynthesis, 148, 157
photosynthetic charge separation, 179–180, 183
photosynthetic energy conversion, 180
photosynthetic organisms, 131
photosystem I (PSI), 179, 192, 317–319
photosystem II (PSII), 79, 179, 185–188, 190, 194
pigment molecule, 131–133, 139
pigment–protein complexes, 5, 22, 28, 123–140, 197, 313, 328, 348
polaron transformation, 38–44
populations, 145
potential energy crossing, 246
potential energy surfaces, 182, 202–203, 242, 244, 247, 249–250
protein electron transfer, 198, 270
proton tunnelling, 7
pump–probe experiment, 88, 258
pure-dephasing model, 161, 170
purple bacteria, 77, 124–125, 128, 130, 156, 179–180, 190, 318, 322, 328

QC calculations, 131–132, 135
quantum beating, 153
quantum beats, 146, 220
quantum chemistry (QC), 131
quantum coherence, 144
quantum dynamical map, 19
quantum localization, 169, 172, 174
quantum many-body theory, 131
quantum master equations, 17, 21
quantum materials, 311–334
quantum mechanics postulates, 15–16
quantum networks, 159–160, 164
quantum observable, 16
quantum Zeno regime, 163, 169, 172, 174
quasi-adiabatic propagator path integral (QUAPI), 44–45, 47
quenching, 124, 190, 331

radical pair mechanism, 218
radical pair materials, 312, 314, 332
radicals, 223
radiofrequency magnetic field effects, 229, 233
reaction centre, 131, 143, 155
receptor, 264–265, 267–269, 271–273, 275–276
reorganization energy, 10, 29, 31–32, 160, 169, 171–172, 174, 180, 203, 212, 214–215, 269, 272–275
resonant tunnelling, 199
response functions, 84, 86, 104
retinal, 6, 14, 237–263
rhodopsin, 6, 14, 227, 237–241, 245–246, 252, 255–258, 262, 264
rotating wave approximation (RWA), 95, 97
Rydberg constant, 338

scent, 267–268, 276
Schiff base retinal, 252
Schrödinger equation, 14–15, 17, 21, 32, 242, 249–251
second-order perturbative time-convolution master equation, 24, 35–37, 170
second-order perturbative time-convolutionless master equation, 24
self-assembly, 12, 313–314, 319, 324, 327–330, 332
shape, 264
single-walled carbon nanotubes (SWNT), 335–340, 346–348
singlet–triplet interconversion, 219
smell, 264–265, 267, 271, 275
spectral function, 137, 142–143
spectroscopy, 276
spin dynamics, 221
spin Hamiltonian, 221
spin relaxation, 220
spin-selective reaction, 221
Stark spectroscopy, 182, 188–189
static disorder, 138
stochastic resonance, 175
Stokes shift, 31, 154
structured chromophoric assemblies, 316
super-Ohmic spectral density, 42–43

third-order polarization, 88, 95, 97–98
three-pulse-photon-echo peakshift (3PEPS) spectroscopy, 345–348
tight binding model, 337, 340
tobacco mosaic virus (TMV), 328–329
transition density cube method, 62–63
transition dipole interaction, 62, 64, 78
transition dipole moment (TDM), 132
tunnelling, 267, 269–271, 273–274, 276
tunnelling pathways, 7, 205, 207, 209
two colour electronic coherence photon echo spectroscopy, 155
two-dimensional electronic spectroscopy, 150–151

two time anisotropy spectroscopy (TTAS), 156
two-level system, 131

vibration, 266–267, 269–276
vibrational coherences, 117, 190, 194
vibronic mode, 137
vibronic states, 132
virus-templated chromophore assemblies, 328

Wick's theorem, 30, 36

Zeeman interaction, 219